INTRODUCTORY ASTRONOMY

SECOND EDITION

INTRODUCTORY ASTRONOMY

SECOND EDITION

Nicholas A. Pananides
Late of Bakersfield College

Thomas Arny
University of Massachusetts

ADDISON-WESLEY PUBLISHING COMPANY
Reading, Massachusetts • Menlo Park, California
London • Amsterdam • Don Mills, Ontario • Sydney

THIS BOOK IS IN THE
ADDISON-WESLEY SERIES IN PHYSICS

SPONSORING EDITOR: Laura Rich Finney
DESIGNER: Catherine L. Dorin
ILLUSTRATOR: Phil Carver & Friends, Inc., and Kristin Kramer
COVER DESIGN: Beth Anderson

Library of Congress Cataloging in Publication Data

Pananides, Nicholas A
Introductory astronomy.

Includes bibliographical references and index.
1. Astronomy. I. Arny, Thomas T., joint author.
II. Title.
QB45.P27 1979 520 78-55825
ISBN 0-201-05674-7

Printed in the United States of America.
Published simultaneously in Canada.
Library of Congress Catalog Card No. 78-55825.

ISBN 0-201-05674-7

ABCDEFGHIJ-MA-79

To my mother and my wife

T.A.

PREFACE TO THE SECOND EDITION

Since the publication of the first edition of *Introductory Astronomy*, knowledge of the astronomical universe has, like the universe itself, expanded. Our new knowledge has not been restricted to one area, but encompasses all astronomy from the earth to the most distant galaxies. Exploration of the planets by landers, orbiters, and flyby spacecraft has necessitated major revisions in our ideas of the nature and evolution of the terrestrial planets. New techniques have increased the sensitivity of existing earth-based telescopes and allowed us to see and measure much fainter and more distant stars and galaxies. Still more important, we have gained the ability to observe radiation that is unable to penetrate the earth's atmosphere. X-rays, ultraviolet and infrared radiation are all strongly absorbed by our atmosphere. With high-altitude aircraft, balloons, and spacecraft in orbit, entirely new kinds of astronomical objects have been detected. The change in perspective is comparable to what would occur if all one could see was the sun and the moon, and then one discovered the existence of stars.

Many of the new discoveries are so recent that they have not yet been fully explored. It is important to recognize the incompleteness of human knowledge, and in describing many of these new phenomena I have tried to emphasize how fragmentary our understanding is. It should be borne in mind that many of the ideas presented here still engender considerable controversy. The real nature of such phenomena as quasars may be solved only by a later generation.

Four chapters have been particularly affected by recent discoveries: Chapter 8 (on the planets), Chapter 14 (on stellar evolution), Chapter 15 (on galaxies), and Chapter 16 (on cosmology). The chapters on the earth and moon also have been affected, of course, by space age discoveries, but the initial flurry of activity in these areas has subsided slightly (as least from an astronomer's point of view). In revising the text I have drawn extensively from the many fine articles available in such sources as *Scientific American* and *The Annual Review of Astronomy and Astrophysics.* In a field which changes so rapidly, I encourage students and teachers to check these sources for the latest views and information. Articles readable even by beginners can be found in the excellent magazines *Astronomy*, *Science News*, and *Sky and Telescope*.

An introductory astronomy course traditionally covers such a wide range of information and is taken by such a diverse group of students that it is difficult to know how to balance breadth of coverage against depth. This is especially true in discussing the basic ideas from physics (such as light and atoms) that are so important in understanding many of the concepts to be covered in later chapters. My own goal in teaching an introductory course is to try to give a feeling for why things are the way they are: Why is the earth different from Jupiter? Why do stars evolve? etc. This has led me to discuss some topics in greater depth than others. This variation, of course, strongly reflects my own interest.

The basic framework of the book was laid out by the late Nicholas Pananides, and I have in general retained that framework in the revision. The only organizational change I have made is to move the chapter on stellar evolution to a point earlier in the book. I personally feel that peculiar stars and galactic structure can be understood more easily after normal stars and stellar evolution are discussed. Otherwise, my revisions have been mostly updatings. One place where I have changed the emphasis is in the chapter on galaxies. So many kinds of peculiar galaxies have been discovered that I felt the need to greatly enlarge Chapter 15. I was similarly tempted to enlarge Chapter 16 on cosmology, but refrained because I feel that it is perhaps still premature to discuss at length the structure of the universe as a whole.

There are many people whose help I would like to acknowledge in preparing this revision. This assistance has taken the form of changes in wording as well as the correction of misinformation on my part. I would like to thank Drs. Icko Iben, William Kaula, William Livingston, and Joseph Veverka for reading selected chapters. Drs. Donald Taylor and Bruce Fitzpatrick have read the entire manuscript and their comments have been extremely valuable and are deeply appreciated. Needless to say, errors or obscurities which remain are my fault. I will truly welcome comments from teachers and especially students about changes they think would improve the book or any errors which they find.

There are many people at Addison-Wesley who have been not only extremely helpful, but also very patient. Among these I am especially grateful to Emily Arulpragasam, who worked diligently at my chaotic writing style, and Laura Rich Finney, who, as editor, had the patience not to throttle me and made many extremely helpful suggestions.

Amherst, Massachusetts T. A.
February 1979

PREFACE TO THE FIRST EDITION

The past decade has witnessed an unprecedented growth in astronomy which has been brought about by the many new developments in the sciences and by space exploration. Unmanned space vehicles have gone to the vicinity of Venus, Mars, and beyond and have sent back valuable data about the planets and space. On July 20, 1969, Neil Armstrong became the first man to land on another celestial body—the moon—while millions of people on the earth watched in tense expectation. Since the first landing, television has made the average man a member of each flight to the moon. He has traveled with the astronauts through space, walked with them as they explored the strange world of the moon, and shared their hopes, fears, successes, and failures.

Astronomy has taken on a new excitement and relevancy for the average man. It is no longer an esoteric science that is totally divorced from the everyday life of most men. The marked increase in the enrollment of college astronomy courses by nonscience students indicates how deeply astronomy has captured the popular imagination. Very few students in these courses plan to become professional astronomers; most of them are there to gain an insight into the basic concepts of astronomy. It is the responsibility of our colleges, especially the liberal arts and two-year community colleges, to teach astronomy as a vital element in a progressive society and to provide the foundation that will allow the student to intelligently appreciate the events occurring in space.

Introductory Astronomy is an attempt to fill a long-existing gap. Most introductory textbooks have been written for the professional. Overly complex material that is of little value and interest to the nonscience student has been introduced, basic concepts have been inadequately explained, irrelevant material has been introduced, and a scientific and mathematical background on the part of the student has been unduly assumed. These have produced unnecessary confusion and frustration. *Introductory Astronomy* provides a rigorous introduction to astronomy in a language that is easily understood by the nonscience student. The material is not so elementary that it insults his intelligence, nor so difficult that it makes him lose interest in the subject. Nothing has been assumed except the student's interest. Astronomical concepts are discussed in the clearest possible language, and all new terms are defined before they are used. Mathematical concepts are kept at a minimum, and when introduced, are explained in depth. Clarity is the touchstone that has been employed throughout the text. Since the study of science is no easy task for the average nonscience student, *Introductory Astronomy* was written in the belief that the road can be made less difficult and more interesting if the guideposts are clearly and simply presented. Only after the basic principles have been comprehended will the real beauty of astronomy reveal itself.

The text is designed primarily for use in a one-semester or a one-quarter course; however, it may be used as the basic text in a two-semester or two-quarter course by those instructors who desire to supplement the text with their own material, other readings, and observations. This text is the result of the author's experience in teaching a one-semester astronomy course at Bakersfield College for twenty-seven years, and a one-semester navigation course at the University of Michigan for four years.

The text presents the historical development of astronomy, what basic astronomical ideas and concepts were developed, and why they were accepted. The discussions are closely related to the abundant illustrations which should make the text easy to read and understand. This should allow the instructor to devote more time to classroom discussion, and if he desires, to introduce the social and philosophical aspects, thereby enriching the course.

The appendixes provide a practical guide for viewing the sky. Astronomy is not simply a set of concepts and facts; it is also the beauty and excitement of observing the celestial bodies. Since many introductory astronomy courses do not provide adequate observing time to acquaint the student with the sky, the appendix on constellations is constructed so that the student, with minimal assistance from the instructor, can begin to observe and study the stars intelligently.

I am greatly indebted to several astronomers who have read parts of the manuscript and have made many valuable suggestions. I want to thank Dr. Billy A. Smith and Bruce Fitzpatrick for reading the entire manuscript and for making perceptive comments and helpful suggestions. Every effort has been made to eliminate typographical and factual errors; however, for those that I have overlooked, I alone am responsible. I am particularly grateful to those astronomers and institutions who have provided the excellent photographs, especially to Mike Donahoe, NASA Ames Research Center, and Paul L. Wenger, NASA Jet Propulsion Laboratory for their cooperation and patience in selecting and providing the many beautiful photographs of space exploration. I also am very grateful to Mrs. Electra Paulick for typing the manuscript, to Mrs. Miriam Paine for her indispensable aid in proofreading, and to my son Dean for his editorial help. I would also like to thank the Addison-Wesley production staff for their capable, generous, and enthusiastic assistance in making this book a reality. Finally, I want to thank my wife, Ethel, for her patience and understanding in maintaining a normal home in the midst of a chaos of notes and papers while the manuscript was being written. Without the generous and capable assistance of these many friends and colleagues, the completion and publication of the book would have been impossible.

Bakersfield, California N. A. P.
September 1972

CONTENTS

PREVIEW

Nearly all of us have had some exposure to astronomy in our daily life, even if we have not formally studied the subject. Science fiction stories, television shows, and movies convey images of alien planets, stars, and the dark emptiness of space. News stories report on discoveries about the sun, the planets, or more remote and exotic phenomena. The very pattern of our daily and yearly activities is defined by astronomical phenomena in that the alternation of day and night and the changing of the seasons are the result of celestial events. We will explore all of these topics as we progress through this book. First, however, let us briefly survey the kinds of objects to be found in the astronomical universe. This "preview" will help you to orient yourself to the topics we will be covering. More important, it will also help you to appreciate the magnitude of the universe you are about to begin exploring. For some of you, this exploration will be only the start of a lifelong interest in astronomy. The universe is always around you, inescapably, and merely awaits your curiosity. Some of you may participate literally in the exploration; your children, if not you yourself, may walk on Mars.

In the next few paragraphs, then, we will make a quick inventory of the basic types of astronomical objects, attempting to depict them in their relation to one another and to you. Because as a species we are still earth-bound, we will begin with our home planet, the earth. The earth is one of nine planets, all roughly similar objects circling our star, the sun. It is held together by the force of gravity, which is responsible also for shaping it into nearly a sphere. The rotation of the sphere on its axis causes the alternation of daylight and darkness at any given point on the surface of the earth, as that point slowly swings around into the light of the sun and then out of it again. The earth is large by human standards, but tiny by astronomical standards. It is approximately 8000 miles (12,000 kilometers) in diameter. You can gain some appreciation of its size by imagining how long it takes to drive across the United States and remembering that such a trip covers only about one-eighth of the earth's circumference. The earth has an atmosphere and a magnetic field, properties shared by many other astronomical objects.

Like some of the other planets, the earth is not a solitary traveler. It has a satellite companion, the moon. The moon is bound to the earth by gravity and orbits the earth in about one month. Unlike the earth, the moon has no atmosphere. It is also smaller than the earth, having a diameter only about one-fourth as large. Currently the outpost of direct human exploration, the moon lies about a quarter of a million miles from the earth. To visualize the relative sizes and the separation of the earth and the moon, imagine that the earth is a softball, the moon a golfball, and the distance between them about ten feet.

Just as the moon orbits the earth, the earth orbits the sun, bound to it by gravity. The sun, our star, is of course an object very different from either the earth or the moon. It is entirely gaseous, with no solid surface. More important for us, it is self-luminous. In its deep interior the sun generates immense quantities of energy, which slowly leak through to the solar surface and pour outward through the dark, cold reaches of the solar system, giving light and warmth to the earth and other nearby planets. Like the earth and the moon, the sun is shaped by gravity into a sphere, it rotates, and it possesses a magnetic field. But, as you might imagine, the sun is enormously larger than the earth both in size and in mass. The sun contains about 1000 times more material than all of the planets put together and is about 100 times the linear size of the earth. If the earth were represented by a pinhead, the sun would be larger than a basketball and would be located roughly 100 feet away. The actual distance from the earth to the sun is about 93 million miles (150 million kilometers). It is here that we first encounter the problem of trying to grasp the scale of astronomical objects. It is very easy to say the sun is 93 million miles away, but it is another matter to understand how large a distance that is. It would take you three years simply to *count* to 93 million, if you counted one number per second day and night.

Although there exist a few so-called giant and dwarf stars, most stars are similar to the sun in dimensions and mass. Like other astronomical objects, stars occur in various groupings. Many stars occur in pairs or triples called binary or multiple star systems. Others are grouped in star clusters (Fig. I.1). Even in such a grouping, typical distances between stars are of a magnitude that is nearly impossible for us to grasp. We saw that if

FIG. I-1
The Pleiades, NGC 1432, a typical galactic cluster in Taurus, which shows reflection nebulosities. (Courtesy of the Hale Observatories.)

FIG. I-2
The Large Magellanic Cloud, a satellite of our Milky Way. It is a typical small irregular type galaxy. (Courtesy of the Lick Observatory.)

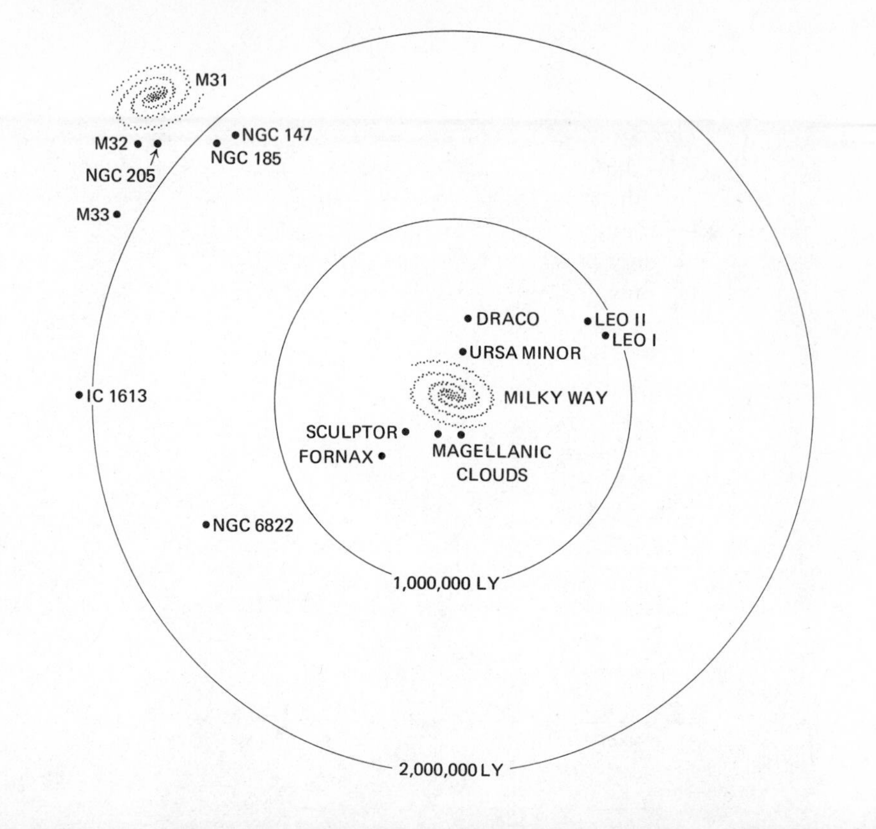

FIG. I-3
Two irregular galaxy clusters:
a sketch of the local group ►
and the large cluster in Hercules. ▼
(Photo courtesy of Kitt Peale
National Observatory.)

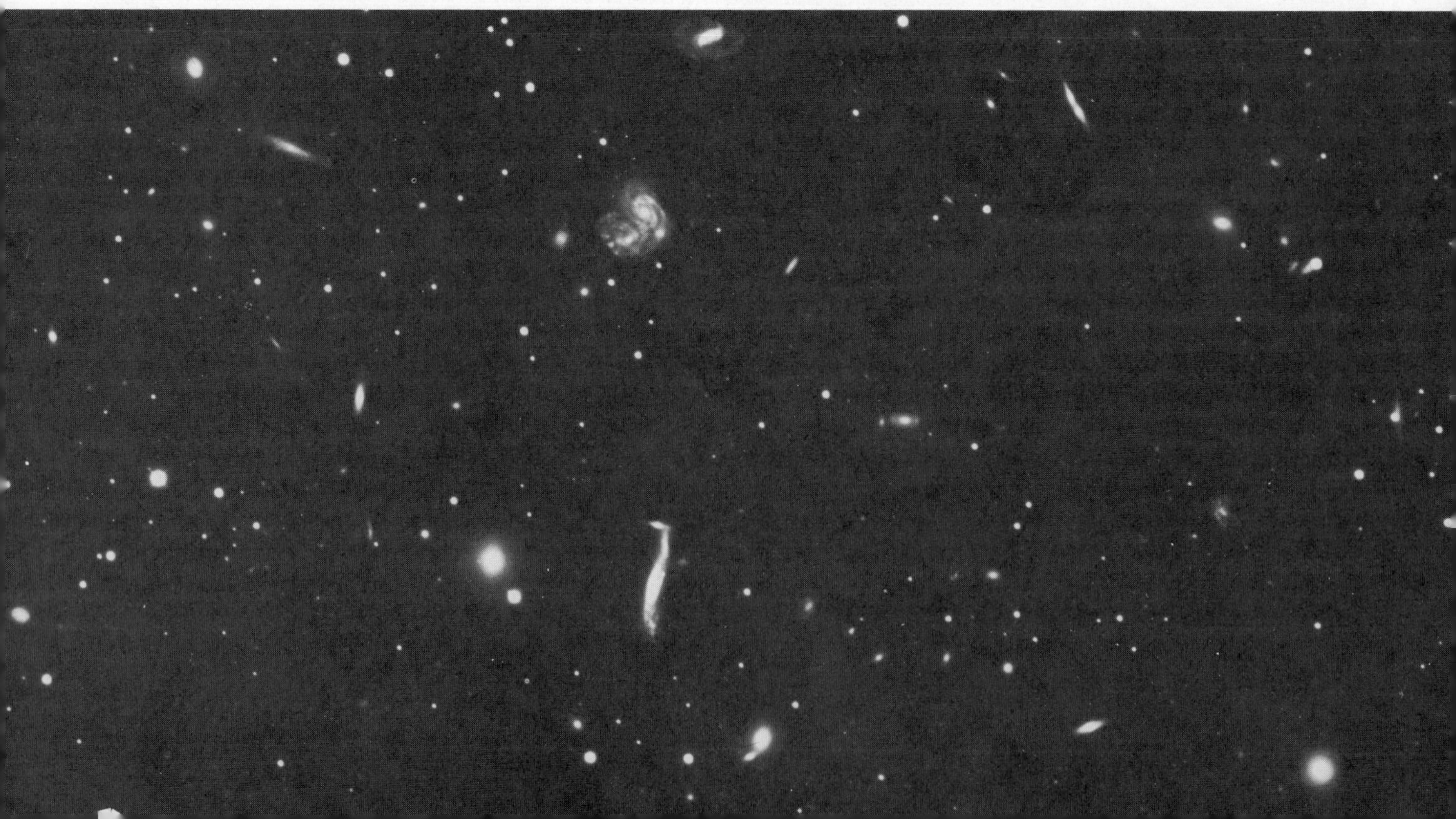

the earth were a pinhead, the sun would be a basketball 100 feet away. Now suppose that the sun itself were a pinhead; then the next nearest star—Alpha Centauri—would be *16 miles* away! Try to imagine some expanse of land (your county perhaps) with nothing in it except the heads of two pins. That may partially convey the enormity and emptiness of space.

In actuality, the distance from the earth to Alpha Centauri is more than 24 trillion miles—a number so vast that it would take nearly a million years to count to it, counting one number every second. Thus, if the first person who ever knew how to count had begun the job of counting off those miles and had handed the task on to his or her children, and they in turn had handed the task of counting on to their children, and so on up to the present, the number 24 trillion would still not have been reached.

Since distances between stars are so immense, astronomers find it helpful to use special units of distance to describe the scale of star systems. Just as you would not normally give the distance between New York and Los Angeles in terms of inches, so astronomers do not normally give the distances between stars in terms of miles. Instead they use a unit called the light year, which is simply the distance that light travels in one year. Light travels 186,000 miles in one second, or roughly seven times around the world in the time of a heartbeat. If you multiply $60 \times 60 \times 24 \times 365\frac{1}{4}$ to obtain the number of seconds in a year, then multiply the result by 186,000, you will see that a light year is nearly 6 trillion miles. Thus Alpha Centauri, for example, is located some 4.3 light years from the earth.

As we turn once again to look at the tiny, glowing points of light that are the stars, we see that beyond their groupings into clusters, the stars are grouped on a still larger scale into gigantic systems called galaxies (Fig. I.2). Our own galaxy is a flattened, disk-shaped system called the Milky Way. Galaxies occur in other shapes also; some, for example, are nearly spherical, with the stars distributed in a swarm.

Galaxies are like other astronomical systems in that they rotate and are held together by gravity. The distance across a typical galaxy is about 100,000 light years and, as you might imagine, the time required for it to complete one rotation is correspondingly huge. In all of human existence, our own galaxy has not completed one rotation. In fact, the sun in its motion around the galaxy is just about back to the place where it was when dinosaurs inhabited the earth. Thus astronomy deals not only with immense spans of distance, but also with immense spans of time.

You can probably guess the next step up the scale of size and outward in space. We have seen that planets are grouped around stars and that stars are grouped into clusters and then into galaxies. Galaxies themselves are grouped into what are called galaxy clusters (Fig. I.3). You can form a mental picture of a galaxy cluster by imagining our star system, the Milky Way, shrunk to the size of a dime, with other galaxies of similar size scattered about on the floor as if you had dropped a handful of coins. Distances between galaxies in a cluster are measured in millions of light years.

Whether there are further levels of structure in the universe is unclear. There is some evidence that galaxy clusters are themselves grouped into clusters of clusters, but the distances are so immense that so far it has not proved possible to verify this unambiguously. Does structure exist on a scale even larger than clusters of clusters of galaxies? That seems unlikely, because when we finally reach that scale we are nearly at the scale of the universe itself. However, this is but one of the many unanswered questions we will encounter on our trip across the universe. The solutions await the work of you and your children.

CHAPTER 1
EARLY ASTRONOMY

1.1 ANCIENT ASTRONOMY

The night sky, with its impenetrable darkness broken by small dots of flickering light, has had an unceasing fascination for human beings. Primitive people must have been perplexed and thrilled by the continual rising and setting of the sun, which marked the ending and the beginning of darkness, by the ever changing and yet unchanging faces of the moon, by the myriad of bright stars disappearing each day with the coming of the light, and by the beautiful Milky Way stretching across the sky like a great luminous river. Through fear, awe, and curiosity, early people gave meaning to these mysterious objects and events by weaving around them myths and superstitions. They worshipped the sun, moon, and planets as gods because they believed that the lives of these bodies mirrored human existence and exerted a persuasive influence and control over human destiny. As people pondered the night sky, astronomy, the oldest of the sciences was born.

The stars, which were considered to be the lesser gods, appeared to be fixed to the inside of a large inverted bowl. Later, this concept was extended to that of a sphere with only one-half of its surface visible at any one time. With the passing of time, the entire star sphere appeared to rotate about an axis that passed near the north pole star, which was the only star that remained nearly stationary in the sky. All the other stars followed an arc across the sky but appeared to remain stationary in relation to one another. Since they seemed to be arranged in definite groupings, primitive people tended to see them as outlines of the men, women, animals, and objects found in their religion. When they had thus identified a number of these groupings, which were called constellations, the sky became a friendly domain, and the stars were no longer terrifying lights.

As primitive people continued to observe the heavens, they began to acquire knowledge about the celestial objects, which they used for the practical purposes of determining direction, position, and time. Later, this information became the basis for the development of astronomical thought. Long before the invention of instruments for measuring the passing of time, the sun, moon, and stars served as clocks. By observing the sun's daily motion of rising in the east, reaching its highest position above the horizon, and setting in the west, it was

By permission of Johnny Hart and Field Enterprises, Inc.

possible to establish the moments of sunrise, noon and sunset and the intervals of day and night. The cycle of the seasons could be recognized by the fact that the sun's path differed from day to day (Fig. 1.1). In the winter, the sun appeared to rise in the southeast, follow a low path in the sky, and set in the southwest. At noon, when the sun was low in the sky, objects appeared to cast long shadows. In the summer, the sun appeared to rise in the northeast, follow a high path in the sky, and set in the northwest. At noon, when the sun was high in the sky, objects appeared to cast short shadows. On the first day of spring and autumn, the sun appeared to rise directly in the east and set directly in the west. When the sun first rose north of the east point, it was spring and time to prepare the land for planting; when the sun first rose south of the east point, it was autumn and time to harvest the crops. As primitive people recognized the continuous cycle of the moon's changing shape, they used its motion and that of the sun to count the passing of the days, months, and years.

1.2 CHINESE ASTRONOMY

As early as 4000 B.C., the Chinese ushered in the first important period in the history of astronomy, that of ancient astronomy. This period also included the astronomy developed by the great civilizations in the region of Mesopotamia and Greece.

Although the authenticity of some of the Chinese records has been questioned, most scholars recognize the Chinese as being among the first astronomers. As early as 2000 B.C., they made systematic observations of celestial bodies, determined the length of the year to be 365¼ days, established constellations that were used to guide them in their travels, and developed a calendar that enabled them to predict the beginning of the seasons. The Chinese were able to predict solar and lunar eclipses; however, on occasions they made mistakes or failed to predict an eclipse. One such occasion was recorded in the third century B.C., when the court astronomers, Hi and Ho, were put to death for neglecting to predict a solar eclipse which occurred while they were attending a garden party.

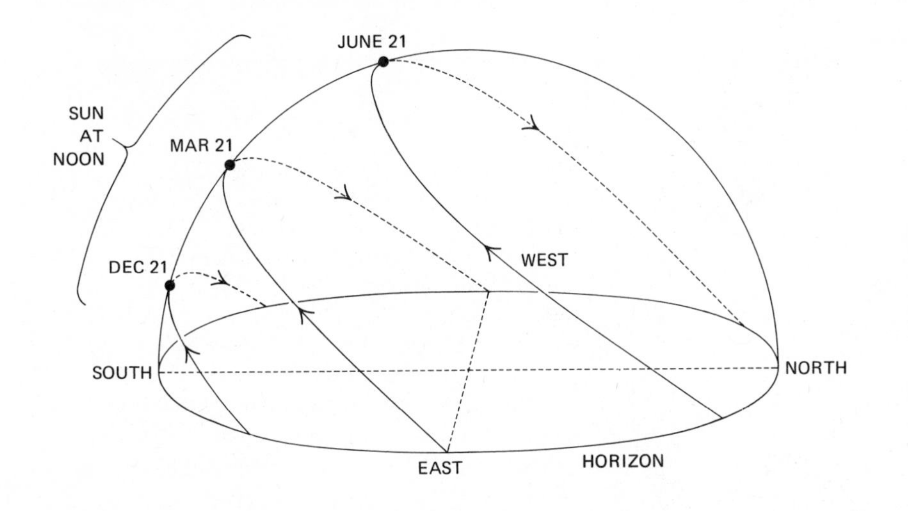

FIG. 1.1
Seasonal paths of the sun. In winter the sun rises in the southeast and sets in the southwest. In spring and autumn it rises in the east and sets in the west. In summer it rises in the northeast and sets in the northwest. At noon its position above the horizon is lowest in the winter and highest in the summer.

It appears that the Chinese were the first to recognize that the moon moves eastward in relation to the background stars and takes approximately 28 days to complete one revolution around the sky. This is revealed in one of their myths, which says that the moon has 28 wives and spends one night with each as it moves around the sky. To keep track of the moon's position and the passing of time, the Chinese divided the moon's path into 28 stations, each about 13 degrees in length, the distance that the moon travels each day.

Although the Chinese did not understand the nature of comets, they kept accurate records of their appearances, including Halley's comet of 467 B.C. They also kept records of the appearance of "guest stars," which we know as novae and supernovae. These stars suddenly and unexpectedly burst into brilliance thousands of times greater than normal so that some are clearly visible in the daytime sky. One such star recorded by the Chinese was the Supernova of 1054, which appeared in the constellation of Taurus the Bull. The remnant of this great explosion (visible during the day for three weeks) is believed to be the Crab Nebula (Chapter 14.7).

While Chinese astronomy was quite impressive, its influence on western astronomy was not great because of China's isolated geographical position.

1.3 MESOPOTAMIAN ASTRONOMY

Ancient astronomy also flourished in the region known as Mesopotamia, around the great fertile valleys formed by the Indus, Tigris, Euphrates, and Nile rivers. These were the home of the Indian, Babylonian-Chaldean, and Egyptian civilizations.

Indian astronomy has come down to us in legends. Their astronomical concepts were based primarily on imagination, religious beliefs, and preconceived ideas. One of the most interesting of these ideas is their concept of the universe. The Buddhists believed in a round earth that was falling continuously in space; however, no one was able to detect its motion—a fact which allowed the Hindus to believe that the earth was supported by four elephants that stood on the back of a large turtle resting on a coiled snake (Fig. 1.2). Since these animals were included in their religion and were considered sacred, the Hindus believed that they were involved in the support of the earth.

The great civilizations that flourished between the Tigris and Euphrates rivers were the Babylonian and the Chaldean. The Babylonian came first and was later absorbed by the Chaldean. The history of these civilizations is recorded on numerous clay tablets that have been

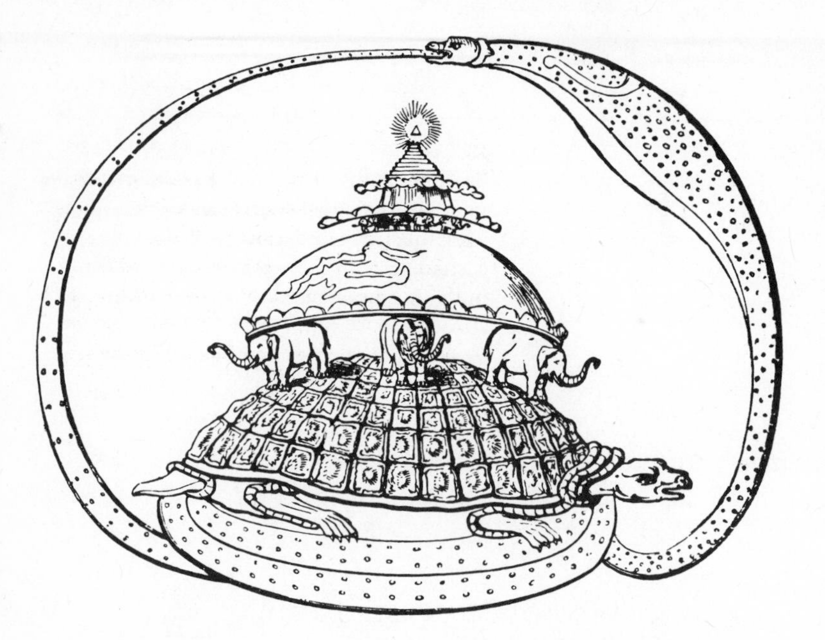

FIG. 1.2
The Hindu concept of the universe. Mount Meru, the earth, and the infernal regions were carried by the tortoise, the symbol of force and creative power. It rested on the great serpent, the emblem of eternity. The three worlds were the upper region, the residence of the gods; the intermediate region, the earth; and the lower, or infernal, region. At the summit of Mount Meru, which was supposed to cover and unite the three worlds, was the triangle, the symbol of creation. (Courtesy of The Bettmann Archive, Inc.)

discovered in many areas of Mesopotamia. On one Babylonian clay tablet (from about 2000 B.C.) were recorded the movements of Venus and the omens associated with it. On a Babylonian boundary stone (from about 1200 B.C.), three people are shown with the symbols of the sun, moon, and Venus over their heads. The sun represented a goddess; the moon, a king; and Venus, his daughter. The Babylonians were one of the few ancient peoples who gave the moon male attributes. According to the Babylonian concept of the universe (Fig. 1.3), the earth, which was enclosed by a wall and supported a dome where all the celestial bodies were located, rested on a chamber of water.

An interesting Chaldean clay tablet was engraved with three concentric circles divided into twelve sectors with thirty-six areas. In each area, the name of a constellation and a number was recorded. No one has been able to decipher these inscriptions; however, astronomers believe that the tablet represents some form of a calendar. From these records, we know that the Chaldeans gave the names that we still use today to many of the star group-

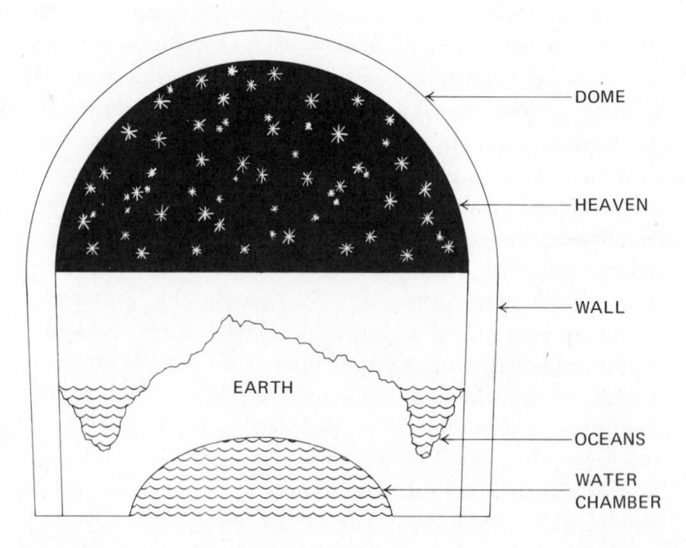

FIG. 1.3
The Babylonian concept of the universe. The earth, which rested on a water chamber, was enclosed by a wall supporting a dome where all the celestial bodies were located.

ings: Gemini the Twins, Scorpius the Scorpion, and Taurus the Bull.

The Chaldeans thought of the universe as a star-studded hemispherical dome in continuous motion above the earth. The dome and the earth were completely surrounded by an envelope of water which at times fell upon the earth.

The Chaldeans were able to predict solar and lunar eclipses, observe the position and motion of the planets, record the appearance of meteors and comets, and measure the passing of time. In spite of these accomplishments, their astronomy never reached a high state of development because the Chaldeans lacked the knowledge of geometry. Moreover, there was no practical need for astronomy because the nature and structure of the universe was fully explained in their religion. The Chaldean priests used their astronomical knowledge only as a means of establishing the dates of religious festivals and as the basis for astrology. It is from the Chaldean astronomers, in fact, that most current astrological traditions are derived. Presumably the notion that celestial occurrences influenced matters on earth was based on the observation that different star patterns became visible with the changing seasons. Ancient astronomers had no way of knowing that both the changing seasons and the visibility of different star groups were caused merely by the earth's motion around the sun. To them, it may have looked as if the stars *caused* seasonal changes. At any rate, since the motions of celestial bodies were predictable, the Chaldeans believed them to be supernatural and to exert great influence and control over human destinies. While many people today find entertainment in their horoscopes, there is no scientific evidence to suggest that an individual's personality or destiny is in any way affected by the positions of the stars or planets.

1.4 EGYPTIAN ASTRONOMY

Early Egyptian astronomy was primitive and naive. During their 2000-year history, the Egyptians developed several concepts of the universe. A mural from the tomb of Rameses VI, who reigned about 1000 B.C., shows that one of the early concepts was anthropomorphic, that is, the sky, earth, and air were given human attributes (Fig. 1.4). The Egyptian goddess of the heavens, Nut, arched her body over the earth, which was represented by a man. Between them was a child, which represented the air. According to one legend, sunset occurred when Nut swallowed the sun each evening, and sunrise occurred when she gave birth to it each morning. In another legend, a river which was traversed daily by the sun god Ra in his boat flowed over the arched body of Nut. At sunset, the sun god Ra disappeared below the earth into the realm of the dead, and at sunrise he was reborn above the earth.

A much later and more realistic concept pictured the universe in the shape of a rectangular box with Egypt located at the bottom center and surrounded by great

FIG. 1.4
The early Egyptian anthropomorphic concept of the universe. (Courtesy of the Yerkes Observatory.)

mountain ranges. The sky was represented by a flat ceiling with holes in it and was supported by four great mountain peaks; the stars were lamps of different sizes, shapes, and colors. The sky river that flowed through the mountain ranges was traversed daily by the sun god Ra. When his boat disappeared behind a mountain peak, it was sunset—the signal for the gods who stood on top of the ceiling to lower the lamps by means of cables through the holes in the ceiling. This is how the Egyptians explained the appearance and disappearance of the stars.

From the writings of the Greek historian Herodotus (about 450 B.C.), we learn that the astronomical accomplishments of the Egyptians were many and great. They recognized and named many of the bright stars and constellations. As early as 2500 B.C., they determined that the length of the year was 365 days, which they eventually divided into twelve equal months of thirty days each with the exception of the last month, which was allotted the five extra days. They also discovered that the year was short one day every four years, and to compensate for it they introduced the leap year. Although the days and nights were divided into twelve hours each, the length of the hour varied with the seasons to account for the differences in the length of the day. The Egyptians used the sun dial in the daytime and the water clock at night to mark the passing of time. They constructed calendars for use in predicting future astronomical events.

However, even though their geometry was advanced to a fairly high degree, they used it, not as a tool in the development of astronomy, but rather in the resurvey of the lands that were flooded periodically by the Nile river. Another deterrent in the development of Egyptian astronomy was that, since astronomical knowledge was considered sacred, all observations and information about it were kept secret within the priesthood. This was done to preserve and restrict to the priests the activity of making astrological predictions.

There is still controversy as to whether the pyramids have astronomical significance. They appear to be aligned so that an interior passage points to what was then the pole star. It seems doubtful, however, that they were used as "observatories" in the traditional sense.

1.5 GREEK ASTRONOMY

Our first knowledge of Greek astronomy appeared in the Homeric poems written during the ninth century B.C. According to Homer, the ancient Greeks believed that the earth was a flat, circular disk and the sky, a spherical dome. Since the five planets visible to the unaided eye—Mercury, Venus, Mars, Jupiter, and Saturn—appeared to wander in relation to the stars, they called them *Planetes*, which means "wandering stars." Mercury and Venus, which always appear near the sun, confused the Greeks, who believed that each consisted of two separate bodies: when they were following the sun, they appeared as evening stars; and when they were leading the sun, they appeared as morning stars. The Greeks also recognized meteors and comets. They called the meteors "falling stars" because they believed that the sporadic streaks of light were stars falling from the sky. They called the comets *Kometes*, which means "long-haired," because the comets' tails resembled long, flowing hair.

The ancient Greeks observed that the sun appeared to move eastward approximately one degree each day in relation to the stars, making one complete revolution around the sky every year, and to pass through the same constellation at the same time each year. They detected these motions when they observed the sun setting each evening with different stars in the background. They plotted its path, which they called *ecliptic*, among the stars. The band of constellations through which the sun passed was called the *zodiac* (Fig. 1.5). The band is about 16° wide, and its center line is the ecliptic. Since the sun appears to move eastward in relation to the stars, the star sphere appears to move westward faster than the sun, which causes the stars to appear to rise and set about one degree of arc, or four minutes of time, earlier each day (Chapter 6.7).

The Greeks became the first scientific astronomers when they separated their preconceived ideas and religious beliefs from science. They were interested in finding the solution to the problem of the nature of the universe and in their attempt to find it they applied reason and logic. Greek science, aided by geometry, became an intellectual discipline which attempted to coordinate and

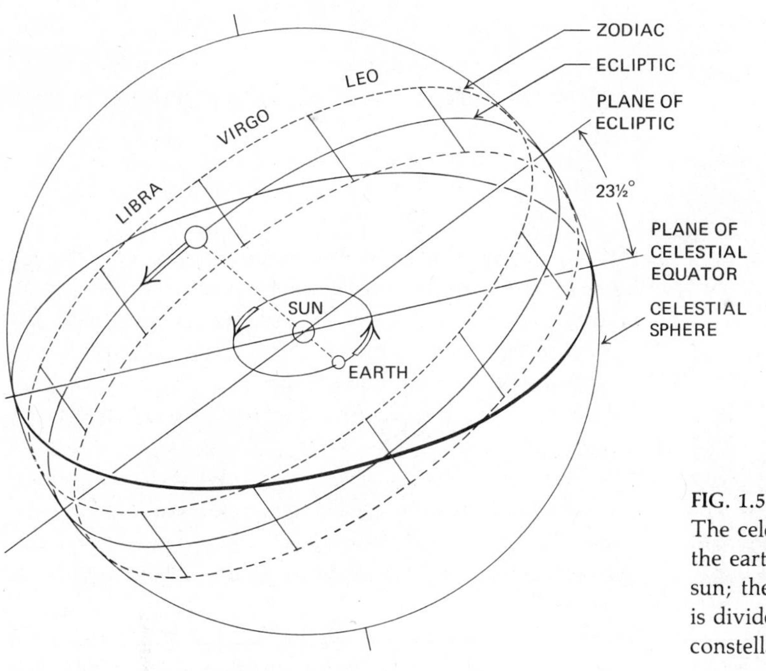

FIG. 1.5
The celestial sphere. The celestial equator is a projection of the earth's equator; the ecliptic is the apparent path of the sun; the zodiac is the band surrounding the sun's path, which is divided into 12 roughly equal parts, each named after a constellation.

understand the basic facts that it discovered. Modern science had its inception in the Greek mind. The generalizations that they developed from daily experiences gave satisfactory explanations of many natural phenomena.

1.6 THE IONIAN SCHOOL

Greek astronomy comprised several schools of learning. The four most famous in early Greek history were the Ionian, Pythagorean, Academy, and Lyceum. The fifth great school flourished in the Hellenistic period at the Alexandrian Museum. Let us take a brief look at the philosophy and science of each of these great schools.

The Ionian school, whose science was developed around geometry, was founded by Thales of Miletus (about 624–547 B.C.). His place in the history of astronomy rests on his great achievement of predicting the total solar eclipse that occurred on May 28, 585 B.C. during the battle between the Medes and the Lydians. This was an astonishing feat because it was difficult to predict accurately the occurrence of a solar eclipse, since the available knowledge of eclipses failed to take into consideration the concept of parallax (Chapter 11.1).

Thales proposed that water was the cause of all things in the universe; it was the primary element and the source from which the three basic elements—air, earth, and fire—were derived. The tiny particles in each of these three basic elements combined in different ways to form all the things in the universe. His theory was based on a false assumption (that water is the primary element) and a false generalization (that it is the source of all things). However, Thales was not altogether wrong, because his philosophy was based on the generalization that certain substances were the building blocks for other substances.

Anaximander (about 611–546 B.C.), who was one of Thales' pupils, disagreed with his teacher's belief that water was the primary element. He believed that all things in the universe were formed from an infinite mass,

By permission of Johnny Hart and Field Enterprises, Inc.

which he called the *apeiron*. Although the apeiron was never defined, it was considered to be in continuous motion, as pieces were being broken off to form the things in the universe. When the universe was formed, the apeiron separated into two parts, one consisting of hot particles and the other, cold. The cold particles became the earth, which was surrounded by the hot particles in the form of a sphere of flames. Eventually the flames were caught in a whirlpool of air which formed into circular tubes enclosing the flames and surrounding the earth. A hole in each tube permitted the flames to appear as the sun, moon, and stars. In this primitive theory, Anaximander proposed that the revolution of the celestial bodies around the earth was produced by the rotation of the tubes and that eclipses occurred when the holes of the sun and moon were either partially or completely closed.

1.7 THE PYTHAGOREAN SCHOOL

Pythagoras was born in Samos about 560 B.C. He left Greece for political reasons and emigrated to the Italian peninsula, where he established his school. Although the philosophy of the Pythagoreans was based on mysticism, they believed that knowledge was acquired through logical insight and that numbers were the substance of all things.

The astronomical philosophy of the Pythagoreans was established by three of its greatest members—Pythagoras, Philolaus, and Parmenides. They developed a geometrical concept of the universe that contained ten concentric spheres (Fig. 1.6). The center of the universe was occupied by the central fire. The sun, moon, earth, counter-earth, and five planets each occupied a sphere and revolved around the central fire. The fixed outer sphere was occupied by the stars. The central fire was invisible from the earth because of the presence of the counter-earth which, as it revolved around the central fire, always maintained a position between and in line with the earth and the central fire. The earth revolved around the central fire once every day and always had the same face turned toward the fire. With respect to the sun and the stars, the earth rotated and produced the intervals of day and night.

The Pythagorean spherical concept of the universe resulted from observations that gave astronomical

FIG. 1.6 ►
In the Pythagorean concept, the universe consisted of ten concentric spheres. The center was occupied by the central fire. Each celestial body occupied its own sphere. The counter-earth always occupied a position between the earth and the central fire.

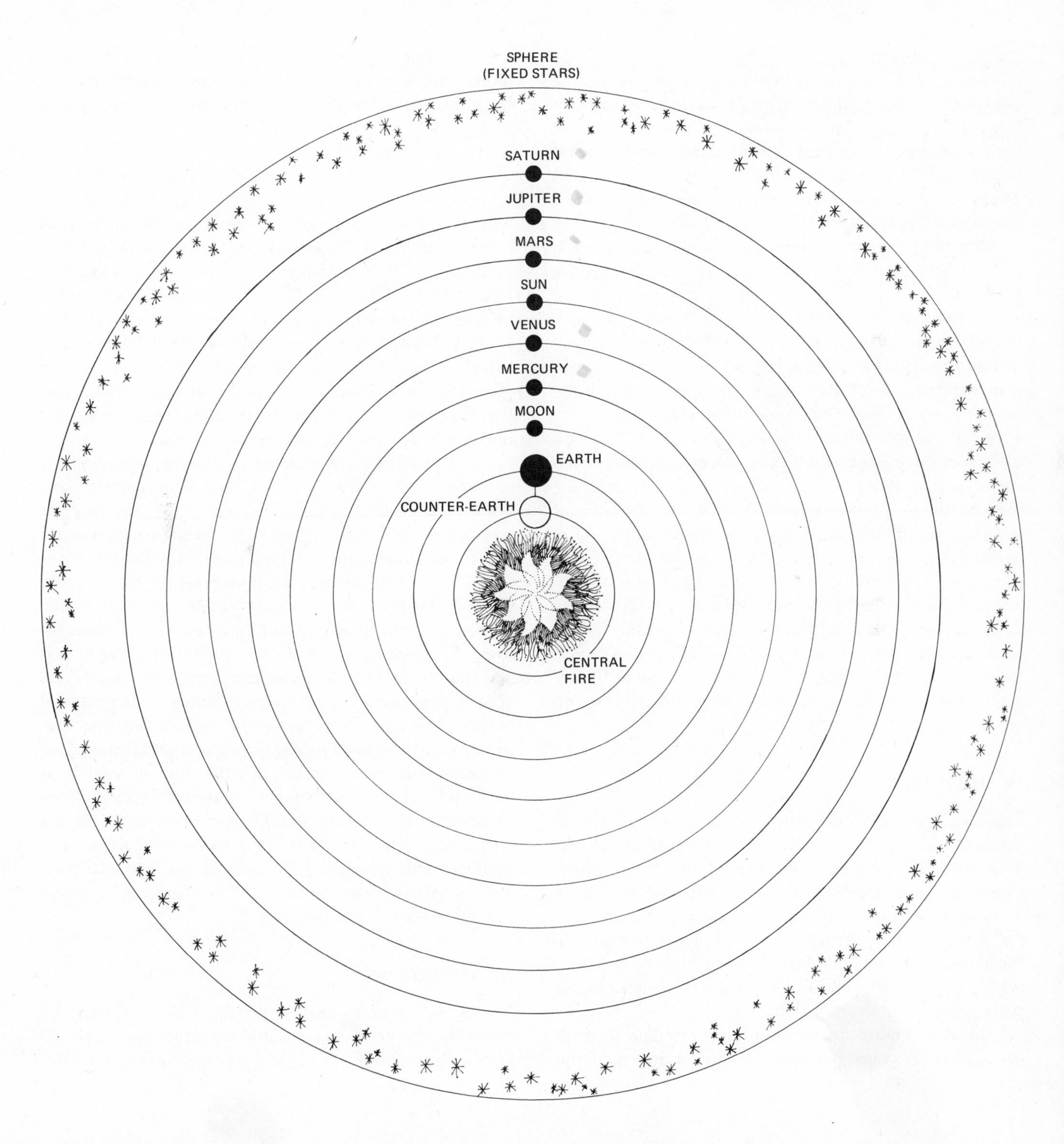
SPHERE
(FIXED STARS)
SATURN
JUPITER
MARS
SUN
VENUS
MERCURY
MOON
EARTH
COUNTER-EARTH
CENTRAL
FIRE

thought a tremendous push forward. The Greeks observed that in Greece the constellation of Ursa Major the Big Bear always remained above the horizon, whereas in Egypt it appeared to move below the horizon for short periods of time. From these observations, they theorized that the earth is a spherical body floating within the sky. They then concluded that the fundamental shape of the celestial bodies and the sky itself is spherical.

When Pythagoras discovered that, by subdivision with a movable bridge, the strings of a lyre could produce the third, fourth, and fifth octaves with ratios of $\frac{1}{2}$, $\frac{4}{3}$, and $\frac{3}{2}$, he noted that only the four basic numbers of 1, 2, 3, and 4 occur in the ratios and that their sum is the perfect number 10. He extended his discovery to the universe which, he believed, is in complete harmony with all the celestial bodies, each producing its own musical sound. This concept was called the *harmony of the spheres* and was probably influenced by the myth of Orpheus, which conveyed the idea that music exerts a magical influence over all things. Pythagoras tried to prove (but without success) that the musical sounds of the celestial bodies depended on their distances from the central fire. He reasoned that a body closer to the central fire moved slowly and produced a deeper sound, whereas a body farther away moved faster and produced a higher sound. Pythagoras believed that the distances from the central fire and the sounds produced by the bodies were in an arithmetical series, that is, any value could be obtained by adding a common term to the preceding value.

1.8 THE ACADEMY

The Academy was established in Athens in the fourth century B.C. by the Greek philosopher Plato (about 428–348 B.C.), who as an idealist rejected the observational and experimental approach to astronomy because he believed that the world which we experience through our senses is an apparition. The real world exists in the form of ideas, and the objects of the visible world are simply copies of the real objects that have existed in the perfect state from the beginning of time.

Most of Plato's astronomical concepts and theories are found in his dialogue *Timaeus* and are presented in a literary style of poetic imagery that makes it almost impossible to translate into definite factual information. In spite of this difficulty, his theories and philosophy continued to have a tremendous influence on scholars for nearly 2000 years.

As knowledge of the geography of the earth increased, the Pythagorean concept of the universe gave way to a theory proposed by Eudoxus (about 408–355 B.C.), a disciple of Plato, which is known as the *Spheres of Eudoxus*. In his attempt to express in mathematical terms Plato's ideas about the motions and positions of the planets, he succeeded in developing a new concept that expressed the apparent irregularity of planetary motions. Prior to this, no one had been able to explain why a planet might gradually stop its motion eastward across the sky, reverse its direction (become retrograde) and then gradually revert to its original motion.

Eudoxus' model consisted of a series of concentric spheres, with the sun, moon, and five planets each occupying its own sphere. To reproduce the irregular motions of the sun and moon required three spheres each and to reproduce those of the planets, four spheres each. The irregular planetary motions were reproduced by placing the concentric spheres within each other and revolving them at a uniform rate, but about different axes. By carefully selecting the orientation of each axis and the rotational velocity of each sphere, Eudoxus was able to reproduce the apparent motions of the celestial bodies. This model was consistent with all the cosmological models developed by the Greeks in that all attempted to explain the how rather than the why of what was observed. The Greeks simply reproduced the motions observed and made no attempt to reproduce the actual conditions. The concept of the Spheres of Eudoxus was modified and refined by Callipus (about 370–300 B.C.) and by Aristotle (about 384–322 B.C.) by adding more spheres to the celestial bodies.

1.9 THE LYCEUM

The Lyceum was established in Athens about 344 B.C. by Aristotle, the great logician and the most famous of all Greek philosophers. His greatest contribution to astronomy was his thorough and critical analysis of the concepts of previous philosophers and his presentation of personal views.

In opposition to Plato's beliefs, Aristotle accepted the validity of sensory data. The observable phenomena constituted the real world, whereas ideas and concepts were merely the essence of the phenomena. Although Aristotle employed logic as his key analytical tool, he used the observable world as his starting point. He wrote that everything on the earth was made from the four basic elements—water, air, earth, and fire—and that all the celestial bodies, including space itself, were made from the fifth element, which he called the *quintessence*. The earth elements were continuously changing from one form into another; whereas the celestial element always remained unchanged and perfect.

Aristotle rejected the Pythagorean concept of the universe and accepted that presented in the model of the Spheres of Eudoxus. He also sensed that the earth was spherical and presented two evidences to prove it. The first was that the shape of the earth's shadow always appeared as an arc on the face of the moon during a lunar eclipse. This evidence was not accepted by some of the philosophers, because they reasoned that a cylindrical earth pointed toward the moon during an eclipse would also produce an arc on the moon's face. Although this was a possibility, Aristotle rejected it because the cylindrical earth would have to be fixed in its orientation in space to always produce an arc shadow. The second evidence was that the entire star sphere appeared to be displaced as an observer moved north or south on the earth's surface. That is, to an observer at a particular latitude the north pole star appeared to be at a certain height above the horizon. When the observer moved to the north, the pole star would appear to move higher above the horizon; when the observer moved to the south, the pole star appeared to move closer to the horizon. Aristotle reasoned that this could happen only on a spherical earth. He also believed that the earth was located in the center of the universe and that it was farther from the sun than from the moon. According to Aristotle, this had to be true because during a total solar eclipse the moon completely obliterated the sun.

1.10 THE ALEXANDRIAN MUSEUM

During its later period Greek astronomy was centered in Alexandria, Egypt, which became the true center of western culture under the rule of several Greek Ptolemies. Many of the scholars of this period were members of the great school called the Alexandrian Museum. This school, with its famous library and observatory, reached its greatest heights in astronomy from 300 to 200 B.C. with such members as Aristarchus, Eratosthenes, and Hipparchus. The library, with its irreplaceable treasure of works by all the ancient classical writers, was destroyed during religious wars of the seventh century.

Aristarchus (about 300–250 B.C.) proposed a heliocentric theory of the universe, according to which the sun and the stars were considered to be fixed bodies, with the sun located in the center of the universe and the stars on the outer sphere. The earth and the planets revolved around the sun in circular orbits. This theory was rejected because there was no observable change in the apparent position of the stars, such as one would expect if the earth moved about the sun. Aristarchus said that the star sphere was at a tremendous distance from the earth, so great that as the earth revolved around the sun, it was impossible to observe any apparent motion in the stars. His theory of the universe, proposed as it was during the third century B.C., was amazingly accurate but was ignored by scholars until the fifteenth century A.D.

Eratosthenes (about 276–194 B.C.), a Greek geographer and the third head of the Alexandrian Museum, was the first person to measure the circumference of the earth. He accomplished this feat by an ingenious method based on simple logic and a sound principle of geometry. On June 21, the first day of summer (summer solstice, Chapter 6.2), Eratosthenes observed in the town of Syene, Egypt (the modern city of Aswân) that the noon sun appeared to be directly overhead, because its rays completely illuminated the entire floor of a deep, dry well (Fig. 1.7). On the same date in Alexandria, 5000 stadia to the north of Syene, the noon sun was not directly overhead, because objects in that city were casting shadows. He knew that the sun's distance from the earth was great; therefore, he reasoned correctly, the sun's rays reach the

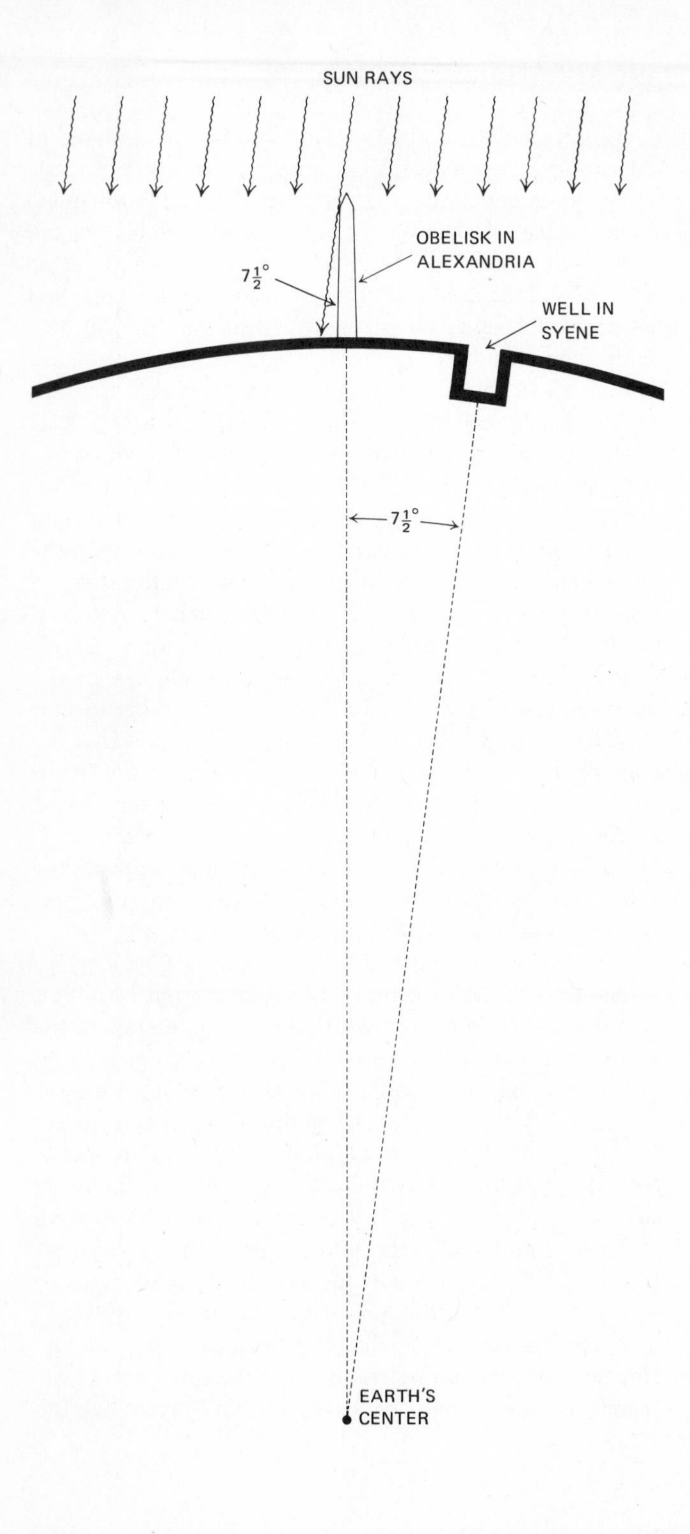

FIG. 1.7
Eratosthenes' method of determining the circumference of the earth. The shadow angle produced by the obelisk at Alexandria is about 7½°. The angle at the center of the earth is also 7½° because, when two parallel lines are cut by a straight line, the corresponding angles are equal. The distance between the two cities is about 5000 stadia. Since the shadow angle is about 1/48 of 360°, the distance between the two cities is about 1/48 of the earth's circumference. Therefore, the circumference of the earth can be found by multiplying 48 times 5000.

earth along parallel lines. He then measured the angle between the suns rays and an obelisk, using its shadow, and reasoned that its value of about 7½° was equal to the angle at the earth's center that is subtended by the distance between Syene and Alexandria. Since 7½° is 1/48 of 360° (the circumference of a circle) he deduced that the distance from Syene to Alexandria must be equal to 1/48 of the earth's circumference. Multiplying the distance between the two cities by 48 produced the correct value of the earth's circumference, within one percent. The error was introduced by three factors: the two cities do not lie on a north-south line; Syene is not exactly on the Tropic of Cancer, which means that the sun is not directly overhead on the first day of summer; and the *stadium*, the unit of distance that Eratosthenes used, was a rounded number which was determined by runners. Nevertheless, the measurement of the earth's circumference stands as a tremendous triumph of logic and reason.

Hipparchus, who was born in Nicaea about 175 B.C., probably was the greatest astronomer of antiquity. He is credited with the development of the eccentrics and epicycles that were used in several models of the universe to reproduce the apparent motions of the celestial bodies. He accurately determined the length of the seasons, and from this he developed a chart that gave the position of the sun on the ecliptic for each day of the year. He compared the position of the important stars of his day with those recorded by astronomers during the preceding 150 years and made the startling discovery that the earth's

By permission of Johnny Hart and Field Enterprises, Inc.

axis is not fixed in space but precesses gradually (Chapter 6.3). With this discovery, he recognized the importance of old astronomical records and proceeded to compile an accurate catalog of the positions of nearly 1000 stars to be used by future astronomers. His work was justified when later astronomers, while using his catalog, made several important discoveries. He also devised the magnitude scale for measuring stellar brightness that is still used today (See Chapter 11).

1.11 THE PTOLEMAIC SYSTEM

The last of the great Greek astronomers was Claudius Ptolemy, who lived in Alexandria about A.D. 150. His greatest contribution to astronomy was his famous book *Almagest*, in which he summarized the astronomical theories of his day and presented formally for the first time the earth-centered concept of the universe. This concept did not, as we have seen, originate with Ptolemy, and he gave full credit to Hipparchus for the theoretical and observational information that served as the foundation for the theory—including the idea that the earth is a sphere—and for providing the eccentric and epicycle theories.

According to his formulation, called the *Ptolemaic system*, each planet moved in a small circle (epicycle), the center of which moved on a larger circle (the deferent) around a stationary earth located in the center of the universe (Fig. 1.8). Since Mercury and Venus were always seen near the sun, Ptolemy placed the centers of their epicycles on a line between the earth and the sun, where they remained as the sun moved in its deferent. In developing this system, Ptolemy reasoned that if these bodies moved in circular orbits around the earth, an observer would always see them moving in one direction; this would not agree with the observed phenomena because at times the planets appeared to move in the opposite direction (retrograde motion). Therefore, to "save the phenomena," that is, to show geometrically the observed motions of the planets, Ptolemy had each planet move along the circumference of a small epicycle, whose center (C) moved along the circumference of a larger deferent with its center at A (Fig. 1.9). The center of the epicycle moved at a uniform

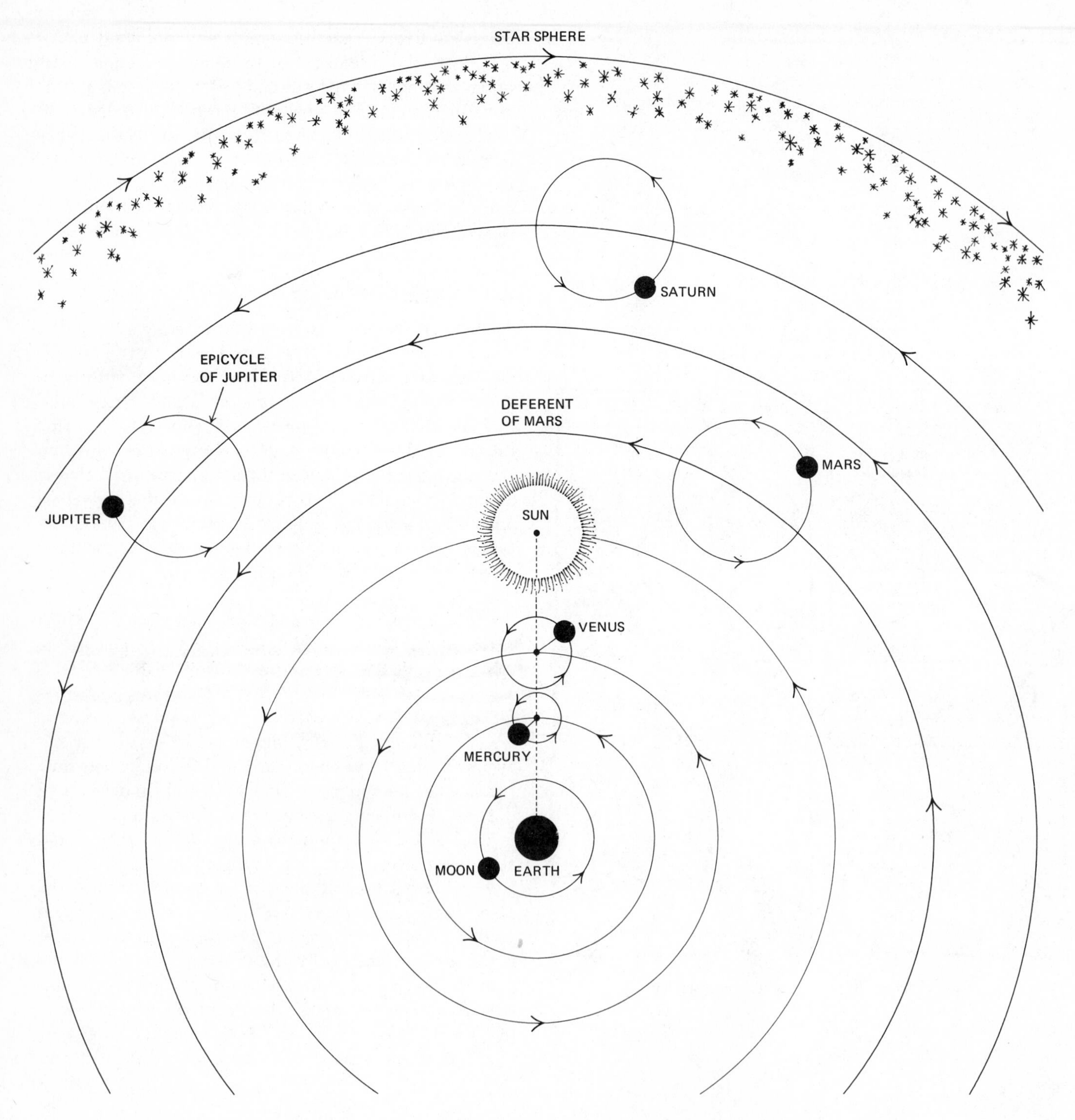
STAR SPHERE
SATURN
EPICYCLE
OF JUPITER
DEFERENT
OF MARS
MARS
JUPITER
SUN
VENUS
MERCURY
MOON
EARTH

◄ FIG.1.8
The Ptolemaic system of the universe. Each planet moves in an epicycle whose center revolves around the earth. The centers of the epicycles for Mercury and Venus are fixed in line with the sun and Earth, since these two bodies were always seen close to the sun.

speed around point Q, called the *equant*, which was located on the opposite side of the deferent's center from the earth. The epicycle produced the retrograde motion and also accounted for minor observable variations in a planet's motion. Retrograde motion was produced when the planet was inside the deferent.

Ptolemy never claimed that his cosmological model described the actual conditions. It simply reproduced geometrically the observed motions of the celestial bodies and provided the means by which their positions could be easily predicted for any particular time. For over fourteen centuries, the *Almagest* was accepted as the prime source for knowledge of the theories of Greek astronomy and was used as the basis for all astronomical work. When the sun-centered theory was proposed by Copernicus in the sixteenth century, many astronomers continued to use the Ptolemaic system to predict the positions and motions of the planets because its intricate system of epicycles provided them with more accurate values. Ironically, the old, totally incorrect theory gave better agreement with observations than the more conceptually correct explanation of Copernicus (Chapter 2.1).

1.12 ANCIENT MESO-AMERICAN ASTRONOMY

The great civilizations that flourished in South and Central America from about 500 B.C. to 1500 A.D. developed an extensive body of astronomical knowledge. Although, from our present information, these ancient civilizations exerted no influence on the development of astronomy because of their isolated geographical position, their astronomical achievements were so important that they deserve mention.

Evidence of the astronomical work pursued by the ancient Aztec, Mayan, and other civilizations comes in the form of carvings on fragmentary rock slabs. From

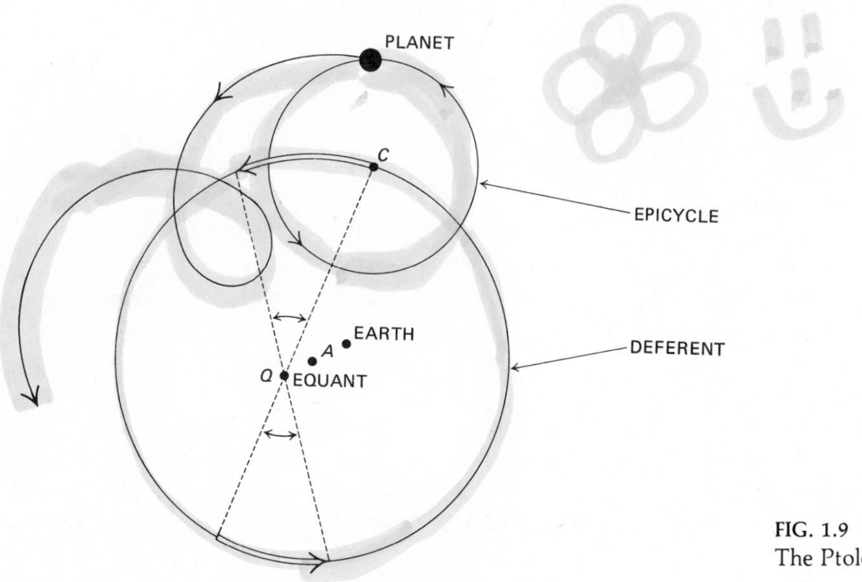

FIG. 1.9
The Ptolemaic epicyclic system.

these we learn that they recognized star groupings, which they associated with animals and objects in their religion. They also recognized the daily changing path of the sun in the sky and the apparent motion of the star sphere about the north celestial pole. Their main interest in the heavenly bodies appears to have centered on the sun, the moon, and the planets, especially Venus and Mars.

The ancient Mesoamericans also developed several elaborate calendars. The one that they used for agricultural purposes had 365 days, and every fourth year had 366 days. The calendar that they used for astronomical purposes was most complex because it was based on the synodic periods of the moon and the planets.

1.13 MEDIEVAL ASTRONOMY

Let us take a brief look at the period between the introduction of the Ptolemaic system about A.D. 150 and the Copernican system in 1543. The Huns started their westward movement during the third century because of the great pressure that was applied from the East by the Chinese and Mongols. As the Huns moved slowly westward during the next 200 years, they conquered and destroyed everything in their path. Their sacking of Rome in A.D. 455 marked the beginning of the slow decline of the Roman empire and the beginning of the rise of the Byzantine empire, which had been established by the Greek emperor Constantine when he moved the capital of the Roman empire to Constantinople. In 1453, the Byzantine empire collapsed when the Turks captured Constantinople.

During the period from A.D. 400 to 1453 (which is known as the medieval period, or the Middle Ages), the acquisition of knowledge declined steadily because of the hostility that existed between the pagans and the Christians. Since the great schools in Greece and the Alexandrian Museum were pagan, they and their students were greatly resented by the newly converted Christians. In their enthusiasm for orthodoxy, the Christians destroyed many of the pagan institutions, such as the great library at Serapis, and burned many books that represented the heritage of Greek knowledge and culture.

With the medieval period in Europe, astronomy went into a state of dormancy. During these years, the Arabs became the trustees of Greek astronomical thought; many Greek treatises, the most important being Ptolemy's *Almagest*, were translated into Arabic. The Arabs invented our present, simplified number system and introduced the algebraic approach to the solution of scientific problems as opposed to the geometric approach of the Greeks. Arab science began to filter back into Europe through Spain in the tenth century.

SUMMARY

For ancient peoples the changing aspects of the sky (motion of sun, moon, and stars) served as the only methods for determining the passage of time. All of the basic units of time such as the year, month, and day were known to preliterate people. By approximately 200 B.C. the Greeks had constructed the first "scientific" models of the astronomical universe, had correctly measured the circumference of the earth, and realized that the earth was a sphere. The complexities of planetary motion were not properly understood, but by 100 A.D. a model which allowed reasonably accurate predictions of planetary positions had been devised by Ptolemy.

REVIEW QUESTIONS

1. How did early peoples measure the passage of (a) the day, (b) the seasons, (c) the years?
2. How does the sun's apparent position in the sky change throughout the year?
3. What did Thales believe the basic units of the universe to be?
4. Explain how primitive people might have established the cycle of the seasons.
5. For what purpose was astronomical knowledge used in the early cultures?
6. What model of planetary motion was proposed by Aristarchus? Was he right?

7. What was the significance of the ecliptic, the zodiac, and the constellations to the ancient Greeks?
8. Cite several reasons why the ancient Greek astronomers believed that the shape of the earth was spherical.
9. List the principal differences the ancients recognized between the planets and the stars.
10. How did Aristotle defend his arguments against the heliocentric universe? Where did he err in his approach?
11. Hipparchus is considered by many to be the greatest astronomer of the pre-Christian era. What were his major contributions to astronomy which gained him this acclaim?
12. What is an epicycle? Why did Ptolemy's model of planetary motion require epicycles?
13. Make a sketch of Ptolemy's model of the solar system. Label the parts.
14. What is the name given to the zone of the sky through which the planets and sun appear to move?
15. Why do you think the zodiac was divided into 12 "houses"? [*Hint:* The moon requires about 30 days to go through a cycle of phases.]
16. If the twelve zodiacal constellations are each 30° in length and the sun is just entering the constellation of Aries, in what constellation will the sun be in 50 days?
17. Describe one of the mythical views of the universe discussed in the text. Do you find that more or less satisfying than the more mechanical model of Ptolemy? Why?

CHAPTER 2
COPERNICUS TO EINSTEIN

Toward the end of the medieval period, there was an awakening in the pursuit of knowledge throughout Europe because an atmosphere was developing in which scholars were free to think without religious or political interference or control. This awakening was a slow process which had developed from the failure of the crusades and the subsequent challenge to classical orthodoxy, from the translation of Arabic and Greek books into Latin, and from the founding and growth of secular universities. Modern astronomy had its birth during the sixteenth century when scholars began to think, question, and challenge the classical doctrines and principles of Aristotelian science.

It was in the early part of the sixteenth century that Nicolaus Copernicus proposed his heliocentric concept of the universe. Although this concept was essentially medieval, it marked the start of a movement away from Greek rationalism because it put a crack into the long sacrosanct Ptolemaic system. The rationalistic approach of the Greeks regarded mathematics and logic as the source of all knowledge and employed abstract reasoning in developing conceptual systems from simple observational material. For centuries, people believed that the earth-centered world (Ptolemaic system), the product of Greek science, was the only conceivable one, and this view was supported by the Church. The introduction of the heliocentric system shook the complacency of medieval dogma. Although the methods used by Copernicus were not revolutionary, they did infuse a new spirit and excitement into the developing science of astronomy.

2.1 THE COPERNICAN SYSTEM

Nicolaus Copernicus was born in Thorn, Poland, in 1473. He was brought up by a wealthy uncle whose sole desire was to have him pursue an ecclesiastical career. Copernicus studied secular and ecclesiastical law and received a doctor's degree in canon law. While at the University of Cracow, his interest in mathematics and astronomy was kindled. When he attended the Universities of Bologna, Padua, and Ferrara in Italy, he studied and mastered Greek, read practically all the works of the Greek astronomers, and kept up with the theoretical and observational aspects of astronomy. From these pursuits, he gained the incentive to simplify the complicated Ptolemaic system.

Copernicus' greatest contribution to astronomy was his development of the heliocentric concept of the universe, in which all planets, including the earth, revolved around the sun. Although this concept appeared in his book *On the Revolution of the Celestial Bodies* in 1543, the year of his death, his ideas on cosmology were already well known, since a letter written to a friend had been copied and widely circulated among interested people. About the end of the nineteenth century, there appeared two published copies of this letter, known as *The Commentariolus*, which allowed scientists to reconstruct the original work. In this letter, Copernicus wrote that all the ancient concepts of the universe, which employed spheres and eccentric circles to explain the motions of the celestial bodies, failed to show the conditions as they existed. His system was simpler and closer to reality, for it was based on the assumptions that the earth rotates daily about its axis; that all celestial bodies have different centers; that the earth's center is not the center of the universe, but simply the center of the earth's and the moon's orbit; that all celestial bodies appear to revolve around the sun, which is at or near the center of the universe; and that a body closer to the sun travels at a greater orbital velocity than one farther away.

Ptolemy assumed that the sky revolved around a stationary earth because he believed that if the earth rotated about its axis it would be torn apart by the force produced by this motion. Copernicus disagreed with this conclusion because, he reasoned, if it were true, then the sky, which is a much larger body than the earth, should have been torn apart many years ago; he therefore concluded that it is the earth rather than the sky that rotates.

In spite of Copernicus' radical cosmological ideas, the Ptolemaic influence on him was great—to the extent that he conceded the celestial bodies as moving in epicycles. Copernicus' system and the Ptolemaic were identical except for two major differences: (1) Copernicus interchanged the positions of the sun and the earth and discarded the equant point; and (2) in order to account for the variations in the orbits of the celestial bodies and to approach reality more closely, he assumed that the planets move in thirty-four epicycles—seven for Mercury, five for Venus, three for Earth, five each for Mars, Jupiter, and Saturn, and four for the Moon.

2.2 THE RETROGRADE MOTION OF THE PLANETS

Ptolemy observed that constellations that appeared on the western and eastern horizons at the same time would reappear later on the horizon in opposite directions. From this he concluded that the earth is at the center of the universe. Copernicus disagreed with this conclusion and assumed without proof that the earth revolves around the sun, as was first proposed by Aristarchus in the third century B.C. (Chapter 1.10).

Copernicus presented an esthetically simple explanation for the apparent looping of the planets in their motion through the sky. The actual motional direction of all planets around the sun, he reasoned, is always forward. When seen from the earth, they normally appear to move in an easterly direction in relation to the stars, a phenomenon called *direct motion*; however, they periodically appear to reverse their direction and to move westerly, that is, in *retrograde motion*. His explanation for retrograde motion was based on the planet's motion relative to the earth's and on the assumption that bodies closer to the sun travel at greater orbital velocities. For example, the earth, which is closer to the sun than Mars, takes one year to complete one revolution around the sun, while Mars takes nearly two years.

In Fig. 2.1, the inner circle represents the earth's orbit; the outer circle, Mars' orbit; and the looped line, the apparent path of Mars as seen from the earth against the background of the stars. Periodically, the earth overtakes Mars, which appears to remain stationary. As the earth passes, Mars appears to move in a retrograde motion, and as the earth moves away Mars appears to stop again, then move in a direct motion. Thus, by

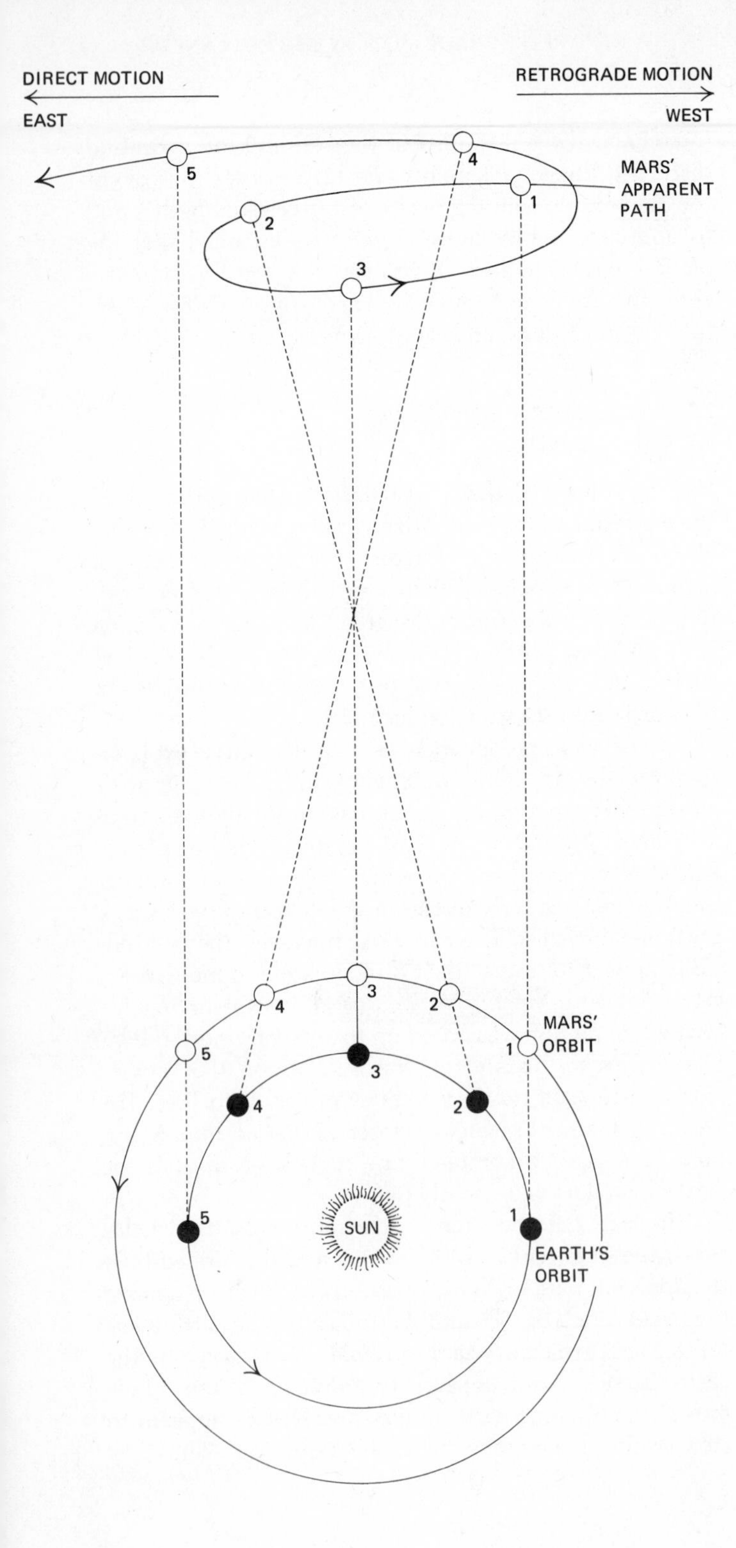

FIG. 2.1
The retrograde motion of Mars. As the earth overtakes, passes, and moves away from Mars, Mars appears to slow down, stop, retrograde, stop, and resume its direct motion in relation to the stars in the background.

assuming that the sun is at the center of the universe and the planets, including the earth, revolve around it, Copernicus was able to explain in a simple and natural way the mystery of the looping of the planets.

2.3 GIORDANO BRUNO

It was most unfortunate that the great work of Copernicus, *On the Revolution of the Celestial Bodies*, was too difficult for the average person to read and understand and was published at a time when its acceptance was almost impossible. One of Copernicus' champions was Giordano Bruno, a philosopher of great intellect, who firmly believed in the Copernican concept and extended it to include some of our present-day concepts of the universe. His mission in life appeared to be the presentation and defense of the Copernican system as well as his own cosmological ideas. He accomplished his mission by lecturing extensively and thus influencing many people throughout Europe.

By intuition and logical deduction, Bruno rejected the concept of a universe with a center. The universe, he felt, is an immense, limitless space in which all the celestial bodies move freely. He also considered the possibility that life inferior or superior to that on earth existed on these and other bodies. He maintained that the planets are cold bodies similar to the earth and that they shine by reflected sunlight. The stars, on the contrary, are very hot bodies and are visible because they give off their own light. The sun, which is but another star, is not located on the firmament, but is much closer to the earth than are the other stars. Also, the stars are at different and enormous distances from the earth—an idea which, for the first time, extended the concept of the universe beyond the star sphere to infinity. These ideas developed by Bruno were truly prophetic. For expressing them, the Inquisition found him guilty of heresy and burned him at the stake in Rome in 1600. Before he died, he said, "The time will come when all will see what I have seen."

2.4 TYCHO BRAHE

In 1546, three years after the death of Copernicus, Tycho Brahe was born. Of noble birth and in line to inherit wealth and property, he was brought up by a wealthy but childless uncle, although his parents were living. His uncle enrolled him in the Universities of Copenhagen and Leipzig, where his education centered on languages and law. While Brahe was at the University of Copenhagen, a predicted solar eclipse occurred, an event which changed the course of his life. He became intensely interested in mathematics and astronomy and devoted most of his time to acquiring as much knowledge as possible in these areas. His scientific pursuits were conducted without the permission of his uncle, who regarded such activities as merely hobbies for people of noble birth.

With the death of his uncle, Brahe was able to pursue his astronomical work without any interference. He built his first observatory in Augsburg, Germany, and had skilled artisans construct the largest and finest instruments of his day—a quadrant to measure the altitude of celestial bodies and a sextant to measure the angle between the bodies. These were the same instruments used by the Greeks except that his measuring scales were more refined and accurate. With these instruments Brahe observed and measured the position of a brilliant object that appeared in the constellation of Cassiopeia on the evening of November 11, 1572. Since the object was clearly visible in the daytime and its position was accurately measurable, Brahe was able to prove that it was a new star. It is known today as *Tycho's star;* in actuality it is not a new star, but a supernova (Chapter 14.6).

King Frederick II of Denmark was so impressed with Brahe and his accomplishments that he invited him to teach mathematics and astronomy at the University of Copenhagen and later to be his court mathematician. Frederick presented Brahe with the island of Hven off the coast of Denmark, near Copenhagen, and provided him with money to build an observatory and an annual income to maintain it. On Hven, Brahe built Uraniborg, "The Castle of the Heavens," the finest observatory in the world. It was truly a castle fit for a king. Its elaborate facilities included an observatory with the most accurate instruments of his day, a machine shop, a laboratory for making glass, living quarters for himself, assistants, servants, and guests, and a jail. At Uraniborg Brahe spent fifteen years of his life, observing and measuring the positions of the celestial bodies. He used his fabulous equatorial armillary to measure with great accuracy the angles between the stars and to establish the first new star catalog since Hipparchus. He also secured the most extensive and accurate data on the motions and positions of the planets prior to the invention of the telescope. Brahe also calculated the error due to refraction of light by the earth's atmosphere and determined that the equinoxes annually precess about 51 seconds (Chapter 6.6). In 1577 Brahe observed the great comet and made it the subject of one of his books, in which he proved that since the comet was much farther from the earth than was the moon, it was therefore an astronomical rather than a meteorological phenomenon.

Brahe ruled Hven with a firm, harsh hand, which was often quite ruthless. He made many enemies, so that when Frederik II died, he was forced to abandon Uraniborg and leave Denmark. In 1599 he arrived in Prague at the invitation of his new patron, Emperor Rudolph II, to serve as his court mathematician. Two years after his arrival, Brahe died.

2.5 THE TYCHONIC SYSTEM OF THE UNIVERSE

Throughout his life, Brahe strongly believed that our home, the earth, is at the center of the universe and that all celestial bodies revolve around it. He could not accept the Copernican system, for he was never able to observe the parallax in any star, that is, its apparent motion in relation to the other stars in the background. In his attempt to make the Copernican system more compatible with his beliefs, Brahe devised a simple solution that preserved his own beliefs and still explained the looping of

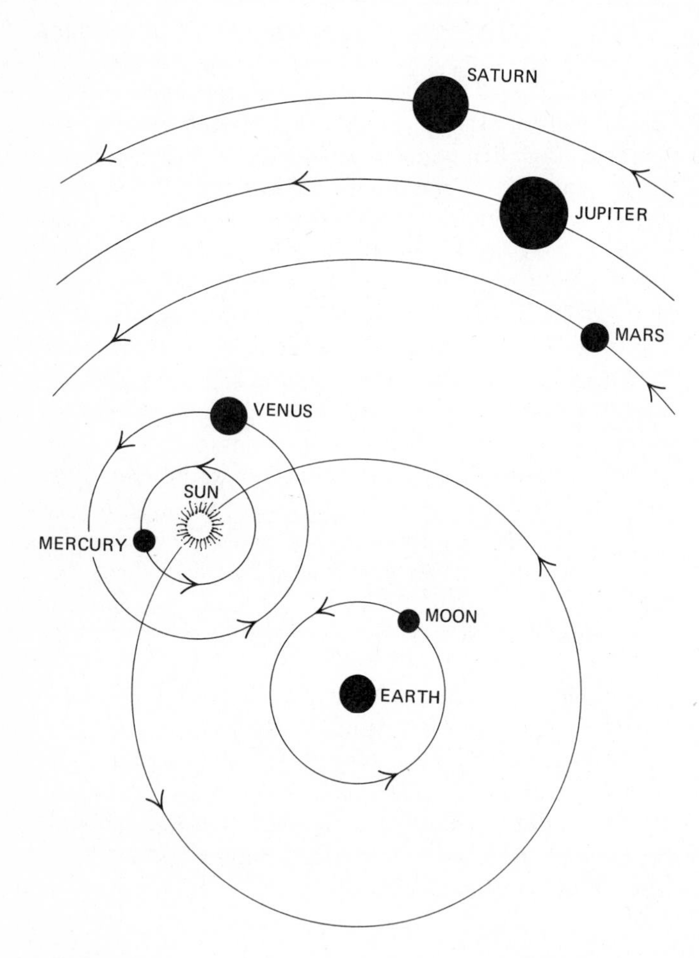

FIG. 2.2
The Tychonic system, a combination of the Ptolemaic and Copernican systems. The moon and the sun revolve around the earth; Mercury, Venus, Mars, Jupiter, and Saturn revolve around the sun.

the planets (Fig. 2.2). The Tychonic system was a combination of the Ptolemaic and Copernican systems. At the center of the universe was the earth; the sun revolved around the earth, and the planets revolved around the sun. To account for the irregular motions of the planets, Brahe reasoned that the orbits of Mercury and Venus were very small, with radii less than the distance between the earth and the sun. The orbits of Mars, Jupiter, and Saturn, on the other hand, were very large, with radii greater than the distance between the earth and the sun.

2.6 JOHANNES KEPLER

Johannes Kepler was born on December 27, 1571, in Weil Der Stadt, Germany. His ambition was to become a Lutheran pastor; however, while attending the University of Tubingen, he came under the influence of Father Michael Mastlin, a professor of mathematics and astronomy, who provided him with the interest and motivation to abandon theology and pursue the study of astronomy. Mastlin taught that the Ptolemaic system was correct, but Kepler disagreed with him and assumed that the Copernican system was correct.

Later, when Kepler was teaching mathematics and astronomy at the boys' school at Graz, he recalled that Euclid's last book demonstrated that only five regular shapes are possible—the cube, tetrahedron, octahedron, dodecahedron, and icosahedron. Kepler reasoned that if these shapes should fit between the spheres of the planets, then their relative distances from the sun could be easily calculated. The shape that fits between the spheres of two planets would establish the orbital limit of the inner planet. Thus, the orbital limit for Jupiter is established by the cube; for Mars, the tetrahedron; for Earth, the dodecahedron; for Venus, the icosahedron; and for Mercury, the octahedron. Kepler's findings were published in his first book, *Mysterium Cosmographicum*. Brahe was so impressed with the book's contents that he invited Kepler to join him in his work in Prague. In 1600, the year before Brahe's death, Kepler joined Brahe in Prague, and both worked on the *Rudolphian Tables*, which were completed and published in 1627. Based on the Copernican system and Kepler's three laws of planetary motion,

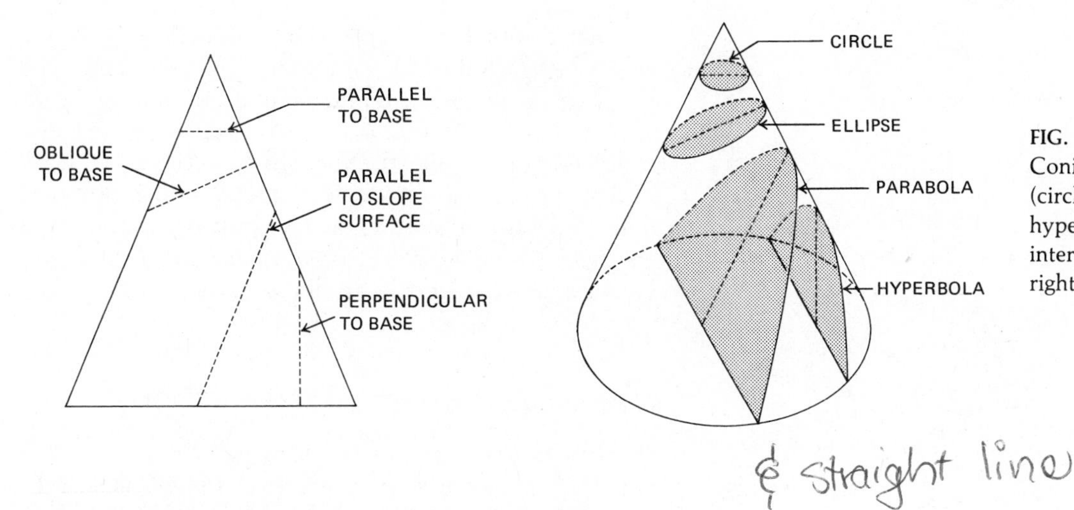

FIG. 2.3
Conic sections, the curves (circle, ellipse, parabola, and hyperbola) formed by the intersection of a plane and a right circular cone.

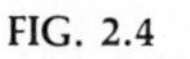

these tables listed the orbital positions of the planets. The meeting of Brahe and Kepler was of tremendous importance to astronomy; when Brahe died in 1601, he bequeathed his valuable cache of planetary data to Kepler, material that provided the foundation for Kepler's laws of planetary motion and ushered in the age of modern astronomy.

2.7 CONIC SECTIONS

Before we discuss Kepler's laws of planetary motion, it will be helpful to take a look at the properties of conic sections—curves formed by the intersection of a plane and a right circular cone (Fig. 2.3). A plane parallel to the cone's base cuts the cone along a curve which is a *circle.* A plane oblique to the cone's base produces an *ellipse.* A plane parallel to the sloping surface of the cone produces a *parabola,* a curve which is open at one end. A plane more inclined to the cone's base than a parabola produces another open-end curve, called a *hyperbola.*

The elements of an ellipse are the *major axis, minor axis, center,* and *foci* (Fig. 2.4). The maximum diameter (AB) is the major axis; the minimum diameter (DE) is the minor axis; the intersection of the two diameters (C) is the center; and the foci are the two fixed points (F and F'), located on the major axis. An important property of

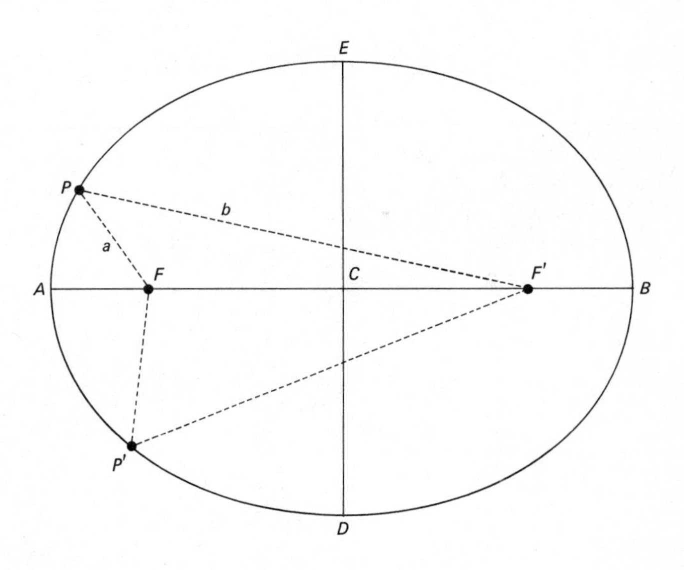

FIG. 2.4
The elements of an ellipse: major axis (AB), minor axis (DE), center of ellipse (C), and foci (F and F'). The sum of the distances between any point P on the ellipse and the two foci is always constant ($a+b$ is constant). The eccentricity of the ellipse is FF'/AB.

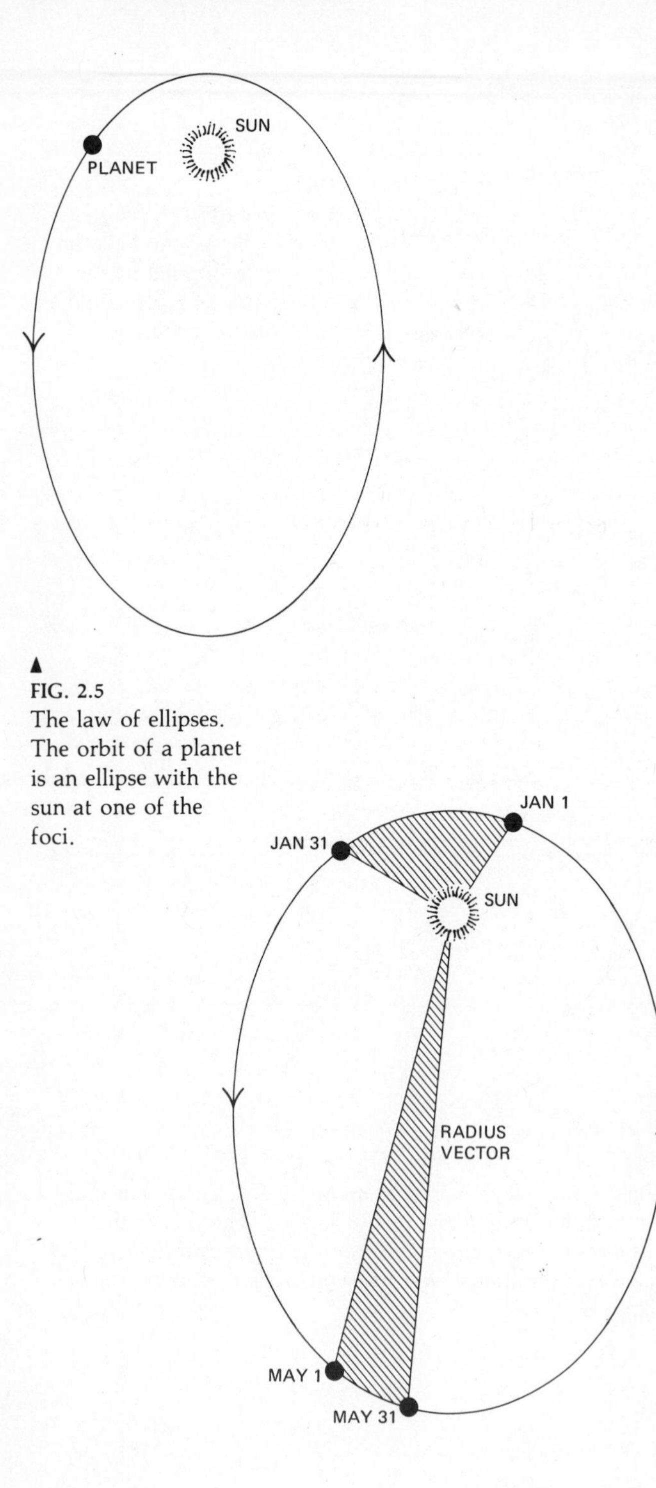

FIG. 2.5
The law of ellipses. The orbit of a planet is an ellipse with the sun at one of the foci.

an ellipse is that from any point (P) on the ellipse, the sum of its distances to the foci ($a+b$) is constant. The shape of an ellipse is determined by the distance between the foci. As this distance increases, the shape of the ellipse becomes more elongated. When the foci coincide with the center, the curve is a circle. The eccentricity of a curve is defined as the ratio of the distance between the foci and the major axis—zero for a circle, one for a parabola, between zero and one for an ellipse, and greater than one for a hyperbola.

2.8 KEPLER'S LAWS OF PLANETARY MOTION

After more than five years of tedious work with Brahe's observational data on the positions of Mars, Kepler concluded that the observational data could apply only to an elliptical orbit, and that therefore the orbits of all planets are elliptical. It also became apparent to him that the orbital velocity of Mars is variable, that is, faster when nearest the sun. He therefore concluded that the orbital velocity of a planet depends on its distance from the sun.

In 1609 Kepler published his two basic laws of planetary motion in his book *Commentaries on the Motions of Mars*. The first law, known as the *law of ellipses,* states that the orbit of a planet is an ellipse with the sun at one of the foci (Fig. 2.5). The second law, known as the *law of areas,* states that the radius vector, (the straight line that joins the sun and the planet) sweeps equal areas in space in equal intervals of time. This means that as the planet moves in its orbit, the area produced by its radius vector is proportional to the time. In Fig. 2.6, the time intervals and the areas are equal; therefore, the orbital velocity of the planet is greatest when it is nearest the sun.

Kepler's third law of planetary motion was published in his book *The Harmony of the World*. He

FIG. 2.6
The law of areas. The radius vector of a planet sweeps equal areas in space in equal intervals of time.

believed that a divine harmonic relationship existed between the minimum and maximum velocities of the planets, which he demonstrated by a mathematical relationship between the planet's period of revolution and its distance from the sun. He chose the earth's period of revolution (one year) as a time unit so that the periods for the planets Mercury, Venus, Earth, Mars, Jupiter, and Saturn are 0.24, 0.62, 1.00, 1.88, 11.86, and 29.46. He defined the earth's distance from the sun to be one unit (referred to now as an astronomical unit); therefore, the distances of the planets from the sun are 0.39, 0.72, 1.00, 1.52, 5.20, and 9.54. By squaring the period and cubing the distance, Kepler established his third law, the *harmonic law,* which states that the square of the sidereal period of a planet is directly proportional to the cube of its mean distance from the sun.

The *sidereal period* is the true period of revolution about the sun with respect to a fixed star or as seen by a hypothetical observer on the sun. The harmonic law is expressed by the formula

$$p^2 = ka^3,$$

where P is the sidereal period in years, a is the mean distance from the sun in astronomical units, and k is a constant whose value varies with the type of units used. Since the earth's mean distance from the sun is one astronomical unit and its sidereal period is one year, when a planet's period and distance are compared to the earth's the proportion is reduced to the simple relationship of

$$p^2 = a^3.$$

A planet's sidereal period, or distance from the sun, may be determined from this relationship. For example, what is the sidereal period of a planet if its mean distance is four astronomical units?

$$p^2 = a^3$$

$$p^2 = (4)^3 = 64$$

$$p = \sqrt{64} = 8 \text{ years}$$

Kepler's third law was entirely empirical because it was based completely on observations without the theoretical foundation provided later on by Newton's laws.

2.9 GALILEO GALILEI

Galileo Galilei was born in Pisa, Italy, on February 15, 1564. At the age of 16, he enrolled at the University of Pisa to obtain a medical degree. He left the university without a degree because his primary interest was in mathematics. About six years later, he returned to the University of Pisa as a professor of mathematics; at the age of 28, he was appointed professor of mathematics at the University of Padua, where he remained for the next 18 years.

In 1610 Galileo published his book *Sidereus Nuncius,* in which he reported the important astronomical observations and discoveries that he had made with the use of his new telescopes. He was the first man to observe the surface features of a celestial body. He saw the mountains, valleys, and craters on the moon's surface and calculated from the length of the shadows they cast that its mountains were much higher than those on the earth. He observed the "rings" around Saturn as two small fixed disks on either side of the planet because, unfortunately, his telescope was not able to resolve them. He was amazed to discover the presence of many dim stars within the Milky Way and to realize that the light from these stars produced the Milky Way. When he observed the image of the sun projected on a white paper, he recognized that there were dark blotches on the sun (sunspots) which varied in number and location over a period of time. He also recognized that the sunspots move across the sun's image, disappear behind its limb, and then reappear from behind the other limb. From this observation he concluded that the sun rotates about its axis. His discoveries of the four satellites revolving around Jupiter, which revealed the presence of a miniature solar system in the universe, and the phases of Venus, which could happen only if the planet revolves around the sun, convinced him beyond the shadow of a doubt that the Copernican model of the universe was correct. Prior to these observations, Galileo had been an advocate of the Ptolemaic system.

In 1632 Galileo published his *Dialogue Concerning The Two Chief World Systems*, a book in which he presented the Ptolemaic and the Copernican systems as hypotheses. The book was published under the auspices of the Catholic Church. Moreover, the material was given the form of a dialogue among three people, one of whom represented Galileo's own views on cosmology. In spite of these precautions, the book brought him into direct conflict with the Catholic Church and almost immediately after its publication, the Sacred Congregation of the Index placed the book on the forbidden list. It was kept there until 1835.

Galileo was tried under the Inquisition, found guilty of heresy, and prohibited from doing anything further in astronomy. Under the circumstances, the sentence was relatively light. He was not imprisoned but was allowed to return to his villa near Florence. Moreover, he was allowed to continue his scientific studies. In 1636 the results of his labor appeared in the book entitled *Discourse and Mathematical Demonstrations Concerning Two New Sciences*, in which he defined velocity and uniform and accelerated motions in terms of his own experimental evidence, and refuted the Aristotelian mechanics by demonstrating experimentally that it failed to explain the observable phenomena. To acquire this knowledge, Galileo developed and used the quantitative experimental method, one of his major contributions to science.

Galileo's heresy trial raised many interesting questions. His troubles appear to have been caused by his own actions. Pope Urban VIII was upset with Galileo's subterfuge in obtaining permission to publish his treatise from the Florentine inquisitor after he had been requested to make certain changes. The Pope, it was also rumored, felt that Simplicius, the bumbling character Galileo created to espouse the Ptolemaic system, was a caricature of himself. Even with these considerations in mind, the basic question remains: does a society have the right to restrict the free flow of ideas? Such restrictions in the field of science can set back progress of knowledge for years.

2.10 ISAAC NEWTON

Isaac Newton was born on a farm at Woolsthorpe, near Grantham, England, on December 25, 1642, the year of Galileo's death. His mother wanted him to become a farmer, but his ability in mathematics and mechanics gave him the opportunity to attend Trinity College, where he received his bachelor of arts degree and was elected a fellow of the college. When the Great Plague of 1665 closed Trinity College, Newton was forced to return to his farm for two years, a move which proved the most productive of his life. It was during this period that he developed the binomial theorem, invented the calculus, and laid the foundation for his discoveries in light and optics.

At the age of 26, Newton was appointed the Lucasian Professor of Mathematics at Trinity College. A year later in 1668, he invented the reflecting telescope (which used mirrors) as a way of avoiding the color fringes produced by the lenses then available. Further attempts to solve this problem led Newton to discover the spectrum of light (Chapter 3.3). In 1672, when Newton was appointed a fellow of the Royal Society, he read a paper explaining the construction of the reflecting telescope and the work that he had completed on light and optics. The ensuing controversy over its contents caused Newton to lose his interest in science, but that interest was revived several years later and resulted in his monumental *Principia*, in which he organized the research he had performed and recorded over a period of 20 years on the motions of bodies.

To understand the motions of the celestial bodies, we must consider how motion is produced, maintained, and changed. Newton's treatment of motion appeared in the *Principia*, which was published in 1687. In formulating his three laws of motion and the universal law of gravitation, Newton took into account Kepler's three laws of planetary motion, Galileo's law of inertia (which states that a body at rest will remain at rest unless it is acted upon by an external force), and Galileo's idea that all bodies fall at the same rate regardless of size.

To understand motion, we must first define the following basic concepts. The *volume* of a body is a measure of the physical space that it occupies, and the *mass* is the amount of matter it contains, the mass of a body remains constant whether it is on the earth, in space, or on the moon and can be changed only by adding matter to it or subtracting matter from it. The amount of mass contained in a given volume is the *density* of the body, which is usually expressed in units of grams per cubic centimeter.

Every material body has the fundamental property of resisting a change in its state of motion; that is, a body at rest resists being set in motion, and a body in motion resists being stopped. Although *inertia* is usually used as a synonym for momentum, it is also identified with mass—the extent to which a body has inertia is the extent to which the body has mass; therefore, the mass of a body can be measured by its inertia.

The most important characteristic of a moving body is its speed. *Average speed* is the ratio of the total distance covered to the total time it took. *Instantaneous speed* is speed at any instant of time, such as that recorded by the speedometer of an automobile. The *velocity* of a body is not its speed alone, but rather speed in a given direction. One thousand miles (1600 km) per hour represents the speed of a body, whereas 1000 miles per hour in a southeasterly direction represents its velocity. Any change in the velocity, e.g., when a body stops or starts, increases or decreases its speed, or changes its direction, is called *acceleration* and is expressed in units of distance per unit of time per unit of time, for example in miles (or kilometers) per hour per second.

2.11 NEWTON'S FIRST LAW OF MOTION

The basic tendency of a material body to resist any change in its state of motion (if it is at rest, it remains so indefinitely; if it is in motion, it continues in motion in one direction at uniform velocity unless a force alters its state of motion) was formulated by Newton in his first law of motion. This law, known as the *law of inertia*, states that "every body tends to remain at rest or in motion with uniform velocity in a straight line unless it is acted upon by an external force." Since mass is related to inertia, an increase in a body's mass produces an increase in its inertia. For example, it is easier to get an empty shopping cart moving than a full one.

2.12 NEWTON'S SECOND LAW OF MOTION

The second law of motion, known as the *law of force and acceleration*, states that "if a force acts upon a body, the body accelerates in the direction of the force." For example, a space ship that is moving at a uniform speed will accelerate when one of its rocket engines is ignited. Acceleration has already been defined as an increase or decrease in speed or a change in the direction of motion; thus, anything that accelerates a body is a force. The degree of acceleration depends on the mass of the body and the force applied. When the force is increased, the acceleration is increased; when the mass is increased, the acceleration is decreased. Therefore, acceleration of a body varies directly as the force applied and inversely as the mass of the body. In other words, force can be expressed as the product of the mass of the body and its acceleration: force = mass × acceleration.

2.13 NEWTON'S THIRD LAW OF MOTION

Although the first two laws of motion involve only the force that acts on a single body, in actual practice the force involves the interaction of two bodies. Newton's third law of motion, known as the *law of reacting forces*, states that "for every force that acts on a body, there is a second force equal in magnitude but opposite in direction that acts on another body." The acting force is exerted by the first body on the second body, and the reacting force is exerted by the second body on the first. Although these forces are equal in magnitude, they are opposite in direction and can never neutralize each other because they act on two different bodies. A well-known example of action and reaction occurs when a man steps from a small boat to a dock. The force that he exerts on the boat (acting force) causes the boat to move away from the dock. The boat exerts a force equal in magnitude and opposite in direction (reacting force) on the man, causing him to move toward the dock.

From Newton's laws of motion, we derive the concept of *momentum*. A body's momentum is a measure of its inertia and is expressed as the product of the body's mass and velocity: momentum = mass × velocity.

2.14 CIRCULAR MOTION

Newton's second law of motion states that an object in motion with no force acting on it will continue to move in a straight line at constant velocity (or will remain at rest). If instead of moving in a straight line, an object moves in

By permission of Johnny Hart and Field Enterprises, Inc.

a circle (or along any curved path) at constant speed, it can only be because there is a force acting to constantly change the object's direction. For the familiar case of a weight tied on a string swung around one's head, it is the string which supplies the force.

For a planet moving about the sun, inertia continuously causes the planet to try to move in a straight line; however, the force of gravity (see below) constantly pulls it off a straight-line path. A planet's orbit can be determined by balancing the effects of inertia and the gravitational force. For any object moving along a curved trajectory, the magnitude of the force keeping it on this trajectory—called the *centripetal force*—is mv^2/r, where m is the object's mass, v its orbital speed, and r the distance from the central mass. (This expression was independently derived by the Dutch physicist Christian Huygens and Isaac Newton.)

The effect of inertia is sometimes referred to as a *centrifugal force*. The centrifugal force gives the illusion of being directed along the radius of the object's curve, balancing the centripetal force (Fig. 2.7). However, if the centripetal force were suddenly turned off (in the case of a stone swung on a string, if the string were to break) the object would *not* move radially outward. It would merely continue moving in a straight line in whatever direction it had been traveling at the instant the force was turned off.

2.15 NEWTON'S UNIVERSAL LAW OF GRAVITATION

Through ingenious geometrical proofs, Newton deduced from Kepler's three laws of planetary motion and the centripetal force formula a great deal more information about the forces that act on the planets. From Kepler's laws, he deduced that between the sun and a planet an attractive force exists that acts entirely in the plane of the planet's orbit and along the straight line that connects the two bodies. This force is proportional to the product of their masses and inversely proportional to the square of the distance between their centers. Although Newton's inquiry was limited to the solar system, he concluded that this force is universal. Therefore, he made his famous

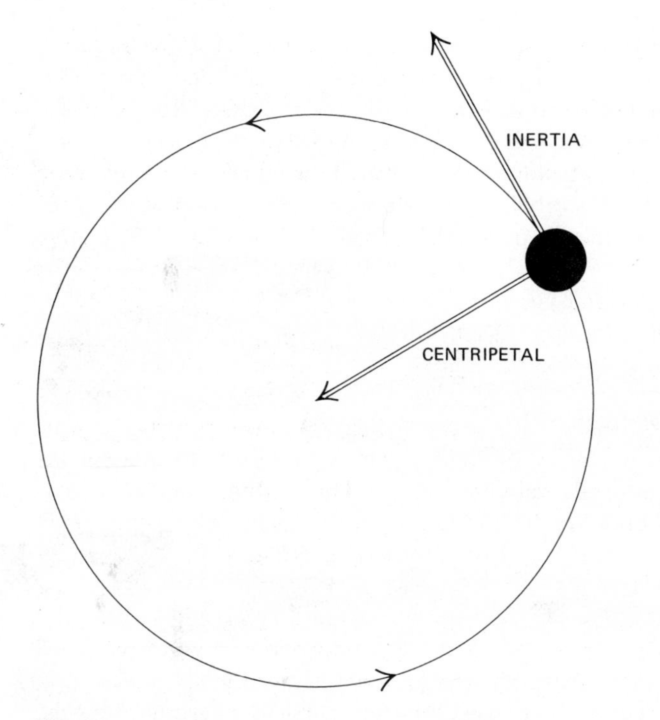

FIG. 2.7
A body in circular motion. Inertia moves a body in a straight line. Centripetal force acts toward the center, producing the acceleration which moves the body in circular motion.

generalization that every mass attracts every other mass by a force that he called *gravity*. He stated this generalization formally in his universal law of gravitation: "Every particle in the universe attracts every other particle with a force that is directly proportional to the product of the masses of the two particles and inversely proportional to the square of the distance between them." He expressed this relationship by the formula

$$F=G(Mm)/d^2,$$

where F is the gravitational force between the two bodies, G is the gravitational constant, M and m are the masses of the two bodies, and d is the distance between their centers.

In confirming Kepler's third law of planetary motion,

$$p^2=ka^3,$$

Newton restated it more precisely to include the masses of the bodies. For example, the product of the total mass of two mutually revolving bodies, such as the planet Mars and its satellite Deimos, and their period of revolution is proportional to the cube of the mean distance between them. He expressed this relationship by the formula

$$(M+m)P^2=4\pi^2Ga^3,$$

where $M+m$ represents the total mass of Mars and its satellite, a is the satellite's distance from Mars, P is the satellite's sidereal period, and G is the gravitational constant. When Mars is compared to the earth, and its satellite's distance and sidereal period are expressed in terms of the moon's mean distance from the earth and its sidereal period, the mass of Mars can be determined in terms of the earth's mass (Chapter 8.2). This is possible when we assume that the mass of a body is considerably greater than that of its satellite.

2.16 WEIGHT

Although they are two different things, mass and weight are often confused. The terms can, however, be used interchangeably without introducing a serious error as long as the body to which they refer remains on the earth. However, when the body leaves the earth, the terms cannot be used interchangeably. The weight of a body is a measure of the force of gravity exerted by the earth on the body. According to Newton's second law of motion, when a force F is exerted on a body of mass m, the body will accelerate according to the relationship $F=ma$. When a body of mass m falls freely on the earth with a constant acceleration of g (32 feet, or 9.81 meters per sec^2), the force exerted by gravity on the body is its weight and may be expressed by the relationship $W=mg$.

Since this force decreases as the distance from the center of the earth increases, the weight of a body decreases as its distance from the center of the earth increases.

2.17 EINSTEIN'S THEORY OF RELATIVITY

Until the twentieth century, it was believed that Newton's laws of motion and gravitation applied to all natural phenomena. In Newtonian mechanics, the universe was considered to be permeated with an invisible medium through which the celestial bodies moved, and the immovable space was used as a fixed reference. This medium was also used in the propagation of light, hence it was called the *luminiferous ether*. The ether was accepted as a fact even though it was impossible to prove or disprove its existence in space. It is interesting to note that the nineteenth century Scottish physicist James C. Maxwell, who philosophically accepted the existence of a substance in space because he could not get himself to accept the fact that there could be waves without a medium, formulated his electromagnetic theory without any reference to the ether (Chapter 3.1).

In 1887 A. Michelson and E. Morley conducted a very simple experiment to discover whether a medium existed. Their specific problem was to determine whether the earth was at rest or moving in space. If a medium existed, there would be a recordable difference in the velocity of light moving toward the earth's direction of motion and away from it. This difference would be equal to the velocity of the earth in space. To accomplish this, they accepted that the velocity of light is 186,284 miles per second (299,787 km/sec), and that the earth's orbital velocity is 18 miles (30 km) per second. Then, they reasoned that light which travels in the same direction as the earth would have a velocity of 186,266 miles (299,757 km) per second; in the opposite direction its velocity would be 186,302 miles (299,817 km) per second. To detect these small differences, they constructed an instrument which they called the *interferometer*.

The results of their experiment were astounding. No apparent motion of the earth through space was discernible. Although the experiment has been repeated many times, no one has been able to detect the presence of a medium in space. Many explanations for the negative results of the Michelson-Morley experiment have been presented in an attempt to preserve the existence of the ether. According to one such explanation, when a body moves at a high speed through the ether, its size might be altered so that its relative speed with respect to the ether could not be detected.

In 1905, at the age of 26, Albert Einstein proposed a solution to the problem of the ether in his *Special Theory of Relativity*. Einstein rejected the theories that a medium exists in space and attempted to show that the principle of relativity is valid for mechanics as well as for Maxwell's equations on electromagnetism. Einstein assumed that the velocity of light in free space is the same for all uniformly moving bodies. This formed the theoretical foundation for Einstein's special theory of relativity. This theory is based on two basic postulates: (1) the velocity of light is always constant at any point in space, and (2) a universal medium in space cannot be detected. Since the medium is undiscernible, an absolute frame of reference cannot exist; therefore, space is meaningless unless it is occupied by bodies. It is impossible to determine whether or not a body is in motion or in what direction it is moving without some reference in space. Therefore, in a limitless space, the positions and motions of bodies can be described only with reference to one another.

In 1916 Einstein announced his *General Theory of Relativity*, an extension of his special theory to include those bodies whose motion is produced by acceleration. In the general theory, Einstein predicted new laws of motion and a new law of gravitation. He explained gravity as the acceleration of a body toward the earth produced by the action of the earth's gravitational field on the body. Acceleration is not affected by either the material or physical state of the body; therefore, the earth's gravitational field affects a stone and a feather in the same manner and accelerates both at exactly the same rate toward the earth. Einstein also determined that a ray of light that moves in a gravitational field reacts in the same way; therefore, he concluded, gravity and acceleration are equivalent. This conclusion was stated in his famous *Principle of Equivalence*: "A gravitational field of force in space is equivalent in every way to an artificial field of force which results from acceleration and there is no way possible to distinguish between them."

Relativistic astronomy does not contradict, but rather supplements, classical astronomy. The laws of classical physics work in excellent fashion for most situations. However, in regions of strong gravitational fields, or where velocities are near the speed of light, it is necessary to either make "relativistic corrections" or apply the more general laws of special and general relativity.

2.18 THE SCIENTIFIC METHOD

In order to give science a philosophical basis, many textbooks have accepted the concept of the scientific method, which has been given a cookbook formulation. We are told that in the solution of a problem by the scientific method, certain prerequisite steps are usually taken. The existence of the problem is recognized, and an objective is formulated for its solution. From observations or planned experiments, data are accumulated, and a hypothesis is presented to explain the observable phenomena or experimental results. Deductions, formulated from the hypothesis, are tested by predicting new phenomena from the observable data or new results from the experiments. Finally, if the tests are successful, the hypothesis is accepted; if they are unsuccessful, the hypothesis is either modified or rejected.

Such a post-mortem examination of the process of discovery is both simplistic and myopic. Scientific discoveries have almost never been made by following a recipe and almost never begin with a specific problem. Scientists are not computers; they do not sit in their laboratories waiting for an order from the main office to solve problem 432. The great advancements in science have been made by those who, because of some intense creative drive within them, have attempted to discover the order of nature. Usually, their search begins from a disquietude in their psyche. Thus a traditional explanation for a phenomenon in nature does not "set right" with them. Refusing to accept dogma, they begin to look at the facts with their own eyes. Through a leap of the imagination, they may visualize a new order—a theory has dawned. How this leap occurs is little understood. The psychology of the creative process is still a mystery, and the sources of inspiration are myriad and often strange. Charles Darwin first realized the possibility of evolution as a theoretical concept after reading Thomas Robert Malthus' *Essay on Population* "for amusement"!

Although exhibiting the same creative impulse as artists, scientists are concerned with a different aspect of life. They attempt to discern and systematize the pattern and order of physical nature. Scientists must work within the framework of facts. Facts are the basis of science, and all concepts must be tested against the hard reality of facts. For this reason, orderly experimentation and minute analysis can never be overemphasized. Scientists must in the end show that these concepts fit the observable facts. But the concept itself is a product of the mind. Nature lays before us a kaleidoscope of facts and phenomena. The scientist attempts to give these disorganized facts a unity. As Blake writes, "Where man is not, nature is barren."

One crucial point must be made: *science is not static*. No conceptual explanation of nature can ever be considered the ultimate truth. Each new discovery is a stepping stone, a bridge, that makes the next step possible. Science, like life, is a dynamic process, continuously evolving. Isaac Newton, recognizing the debt he owed to those who preceded him, said "If I have seen further than other men it is because I stand on the shoulders of giants."

SUMMARY

During the sixteenth and early seventeenth centuries astronomy, like many other areas of human endeavor in the Western world, underwent a major revitalization. In about 1500, Copernicus proposed that the structure of the solar system made better sense with the sun (rather than the earth) at the center, and thus revived the idea of a heliocentric solar system which had earlier been recorded by Aristarchus. Observational evidence accumulated by Tycho Brahe and analyzed by his student Kepler around 1600 led to the first mathematical expressions for the motions of the planets. Kepler's three laws of planetary motion (especially the first and third) will reappear many times in the rest of this text. The first law

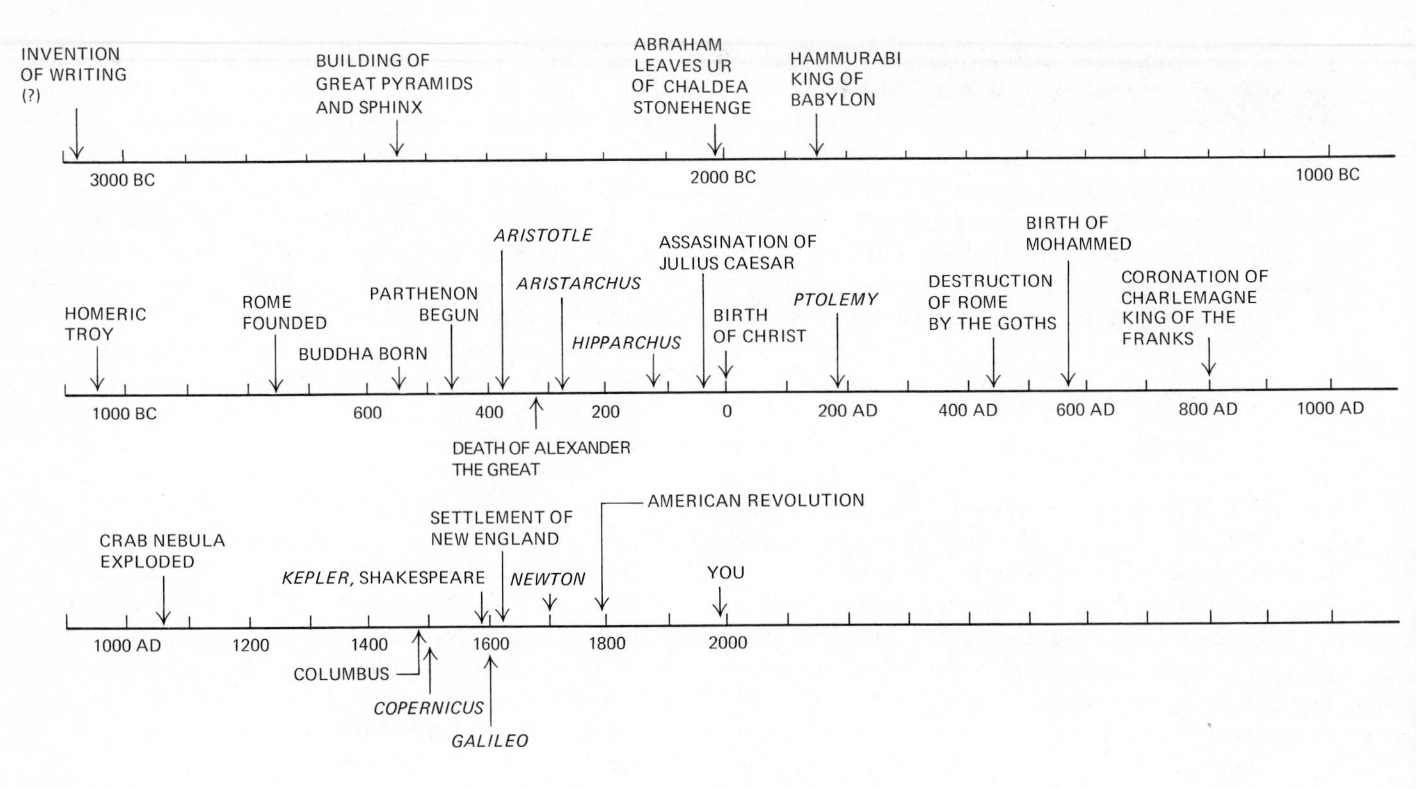

FIGURE 2.8

states that planets move in elliptical orbits; the third law relates the time required to complete an orbit to the size of the orbit. The final accomplishment of this era in astronomy was the work of Newton, who showed mathematically that Kepler's laws were natural consequences of the law of gravity and the laws of motion, which apply to all bodies.

In Chapters 1 and 2 we have sketched the history of astronomy from its beginnings up to recent times. Figure 2.8 is a time line that relates some of the important people mentioned to other significant events in the history of Western civilization.

REVIEW QUESTIONS

1. What was Copernicus' greatest contribution to astronomy? Were his cosmological ideas considered radical?
2. Discuss the similarities and differences between the geocentric (Ptolemaic) and the heliocentric (Copernican) concepts of the universe.
3. How did Copernicus explain the retrograde motion of the planets?
4. List some of Giordano Bruno's astronomical ideas that were truly prophetic.
5. List the important works of Tycho Brahe and explain how they contributed to the development of astronomy.
6. Why are circles, ellipses, and parabolas called conic sections?
7. What are Kepler's three laws of planetary motion? Draw a simple picture to illustrate what each means.
8. What is the difference in a planet's speed when at perihelion and aphelion? Explain your answer by Kepler's second law.

9. Jupiter is about five astronomical units from the sun. What is its orbital period?
10. Pluto takes about 240 years to circle the sun. How far away from the sun is it?
11. What important astronomical discoveries were made by Galileo? What major contributions to science in general were made by Galileo?
12. What discovery did Galileo make which convinced him that the Copernican concept of the universe was correct?
13. List the important astronomical contributions made by Newton.
14. According to Newton's universal law of gravitation, what happens to the gravitational force between two bodies when the distance between them is decreased?
15. How did Newton restate in a more precise manner Kepler's third law of planetary motion?
16. What is the difference between mass and weight?
17. State the two postulates of Einstein's special theory of relativity and briefly discuss the meaning of each.
18. State Einstein's "principle of equivalence" in his general theory of relativity and briefly discuss its meaning.

CHAPTER 3
LIGHT AND THE ATOM

3.1 THE NATURE OF LIGHT

Some knowledge of the properties of light is extremely important to the study of astronomy. While there are a few types of astronomical objects for which astronomers have first-hand evidence (meteorites and moon rocks), nearly all other information about the astronomical universe has been obtained by the analysis of the radiation flowing to earth across the immense reaches of space. To understand the messages thus conveyed, we need to understand a little of the nature and origin of light.

Light travels faster than anything else in the universe. Its velocity in empty space is 186,000 miles (300,000 km) per second. In one second a beam of light could travel seven times around the earth. The distance light travels in a year (a light year) is a useful measure of the enormous distances we will encounter in describing stars and galaxies. A light year is about 6 trillion miles (10 trillion km).

The study of light emitted and reflected by celestial bodies has revealed their distance, size, mass, temperature, luminosity, composition, age, history, and velocity with respect to the earth. Although we have learned a great deal about the way light behaves, its true nature and why it acts as it does are still not fully understood. Today, after three centuries of research, several theories of light have been formulated, but no single theory has proved capable of explaining all of the properties of light.

Isaac Newton proposed the *corpuscular theory*, according to which light consists of material particles called corpuscles that move in a beam from the source to the eye of the observer. This theory was accepted because it offered the most efficient explanation for the phenomena of the reflection and refraction of light. In 1815 the French physicist Augustin Fresnel demonstrated that under certain conditions it was possible for two beams of visible light to cancel each other. This phenomenon could not be explained by the corpuscular theory; therefore, the Dutch physicist Christian Huygens proposed the *wave theory*, according to which light is the disturbance caused by the movement of the medium in space. This medium was never defined, but its behavior was compared to the motion of water waves. In fact, for much of our subsequent discussion,it will be useful to think of light waves as behaving like the ripples that spread out on the surface of a pond into which a stone is tossed.

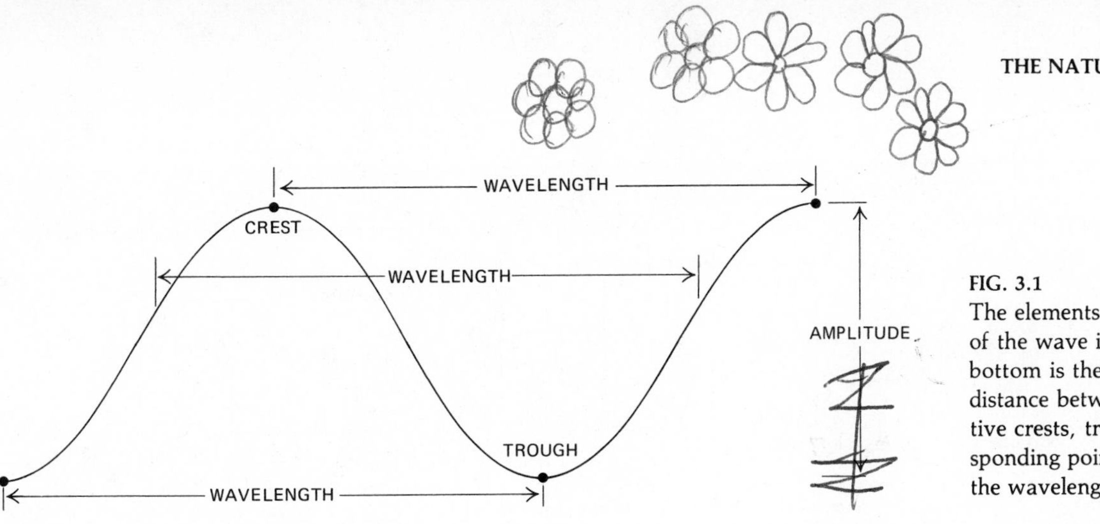

FIG. 3.1
The elements of a wave. The top of the wave is the crest; the bottom is the trough. The distance between two consecutive crests, troughs, or corresponding points on the wave is the wavelength.

The principal elements of a wave of visible light are shown in Fig. 3.1. The top of a wave is the crest, the bottom is the trough and the distance between two consecutive crests, troughs, or corresponding points on the wave is the *wavelength* (λ). Wavelengths are generally expressed in centimeters; however, for the shorter wavelengths, the more convenient units of micron (or micrometer) and angstrom are used. One micrometer (μ) equals 10^{-4} centimeters (one ten-thousandth of a centimeter), and one angstrom (Å) equals 10^{-8} centimeters (one hundred-millionth of a centimeter). The number of wave crests that pass a given point in one second is the *frequency* (f) of the light. The product of the frequency and the wavelength determines the velocity of the wave (c), which is expressed symbolically,

$$c = f \times \lambda .$$

In terms of the water waves analogy, the wavelength is the space between the crests; the frequency is the number of waves passing per second; and c is the wave's speed toward shore.

Our knowledge of the nature of light took a tremendous step forward in 1864, when James Maxwell presented his electromagnetic theory. He mathematically predicted that waves of electromagnetic energy should move through space. In 1888 Heinrich Hertz verified Maxwell's theory and demonstrated that heat, light, and radio waves differ only because their wavelengths are different. We use the term electromagnetic to describe the radiation because it consists of both electric and magnetic fields, which continually grow and decay. Physically the wavelength measures the distance between the regions of maximum electric field strength.

The wavelengths of electromagnetic radiation range from the long radio waves, which are over 10^5 centimeters in length, to the gamma rays, which are less than 10^{-12} centimeters (Fig 3.2). Between these two extremes are radar, infrared, visible light, ultraviolet, and x-rays, in decreasing order of their wavelengths. The entire class of electromagnetic waves is referred to as the electromagnetic spectrum. Visible light is one form of energy and represents a very small portion of the total spectrum. Its wavelengths range from about 4000 Å (violet) to about 7000 Å (red).

We are most familiar with the visible part of the electromagnetic spectrum because that is the wavelength range to which our eyes are sensitive. We can feel infrared energy as heat on our skin, and with relatively simple devices we can detect radio energy. Up until the early part of the twentieth century, astronomers could only study objects in the visible region of the electromagnetic spectrum, mainly because the necessary technology for exploring other spectral regions was not available, but also because the earth's atmosphere prevents energy from most of the other regions from getting through to the surface (Chapter 5.6). As technology developed and

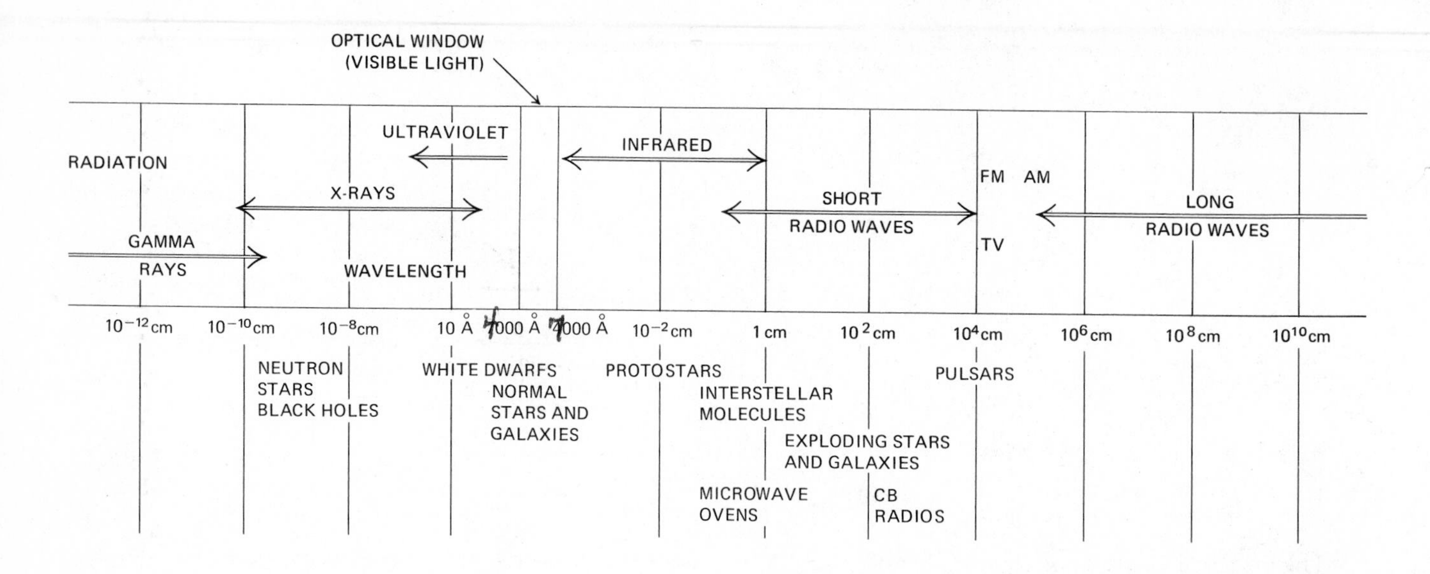

FIG. 3.2
The electromagnetic spectrum. The wavelength of visible light determines its color—4000 Å is violet, and 7000 Å is red. Shown also are some astronomical and more familiar objects and the spectral range at which they are most readily observed.

astronomers began to explore the sky at previously unstudied wavelengths, new kinds of objects were discovered and many previously studied celestial bodies were found to have a whole new range of properties. Newly born stars that are still relatively cool radiate primarily at infrared wavelengths. The cold gas clouds from which the stars form radiate at radio wavelengths. The remnants of dying stars radiate strongly at radio and x-ray wavelengths. Thus astronomy is no longer a matter of studying only the visible light of stars and planets. It involves nearly the entire electromagnetic spectrum.

Experiments have demonstrated that radiation is emitted or absorbed in discrete packets of energy, called *quanta,* that occupy a definite amount of space (roughly given by the wavelength) and consist of energy units called *photons.* Since a photon is indivisible, the quanta appear in whole numbers of photons.

One of the most important properties of light is that it carries energy. This energy is what gives you a sunburn, makes you feel warm in the sunlight, and fuels plant growth, among other things. The energy of a photon, which is proportional to its frequency, may be expressed by the formula

$$E=hf$$

or

$$E=hc/\lambda=\text{constant}/\lambda,$$

where E is energy, h is Planck's constant,* f is frequency, λ is wavelength, and c is the velocity of light. These formulas indicate that radiations of the same frequency (or wavelength) have equal amounts of energy. Also, since the wave velocity is equal to the product of its wavelength and frequency, and since the energy of a photon is inversely proportional to its wavelength, the energy of a photon of violet light (short wavelength) is greater than that of a photon of red light (long wavelength). This is

*See Appendix 5 for numerical values.

basically why ultraviolet light will give you a sunburn while ordinary visible light will not.

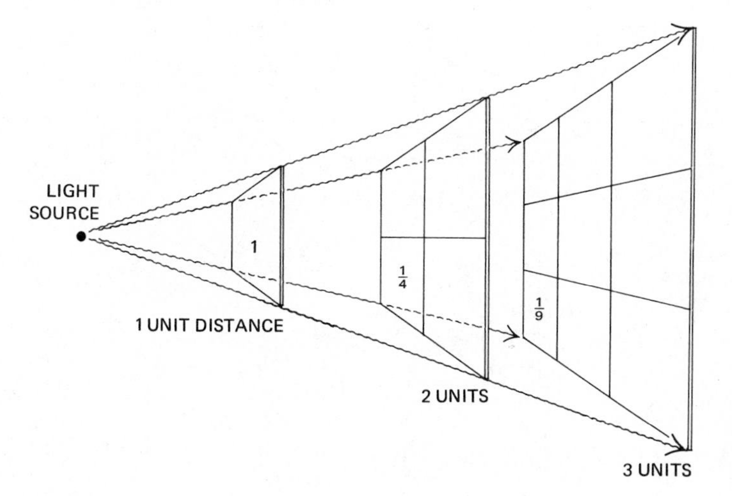

FIG. 3.3
The inverse square law. The amount of visible light that passes through a unit area decreases with the square of its distance from the light source.

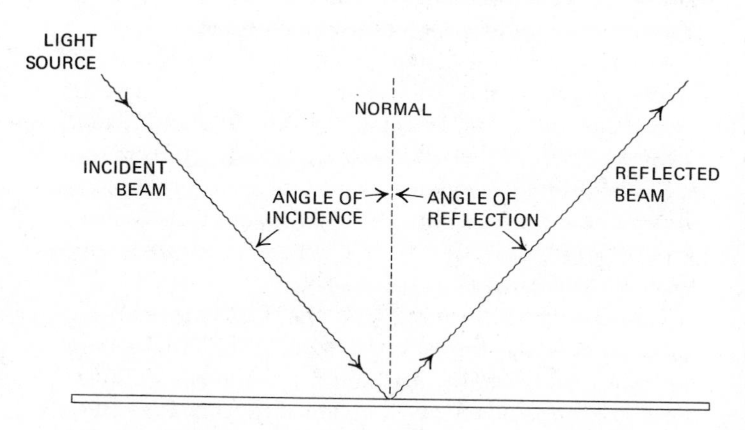

FIG. 3.4
Reflection of light. When light is reflected from a smooth surface, the angle of incidence equals the angle of reflection of light, and both beams lie in the same plane with the normal.

3.2 THE INVERSE SQUARE LAW

Intuitively, one can easily recognize that the amount of light received by an object decreases as its distance from the light source is increased (Fig. 3.3). This important property in the propagation of light is indicated by the *inverse square law*, which states that the amount of light decreases as the square of the distance.

Anyone who has driven at night has intuitively used the inverse square law to judge distances to intersections, street lights, and traffic lights. Astronomers also use the inverse square law to measure distances, which is one of the reasons it is so important for us to understand. For example, if a telescope receives a certain amount of light from a light source that is one unit distance away, it will receive only one-quarter the amount of light when the distance is doubled, one-ninth when the distance is tripled, and one-sixteenth when quadrupled.

3.3 THE PROPERTIES OF LIGHT

The properties of light that are basic in the design and construction of astronomical instruments are *reflection*, *refraction*, and *dispersion*.

Objects become visible when light from them is reflected to our eyes. When a beam of light strikes a rough surface light is reflected in all directions, and when it strikes a smooth surface the light is reflected in a single direction. As shown in Fig. 3.4, the perpendicular to the reflecting surface is called the *normal*. The angle between the incoming ray (incident beam) and the normal is called the *angle of incidence*, and the angle between the reflected ray (reflected beam) and the normal is called the *angle of reflection*. When light is reflected from a smooth surface, the angle of incidence equals the angle of reflection, and both beams lie in the same plane with the normal.

When a light beam passes from one medium into another of different density, it is refracted (bent). A pencil placed in a glass of water shows this effect clearly.

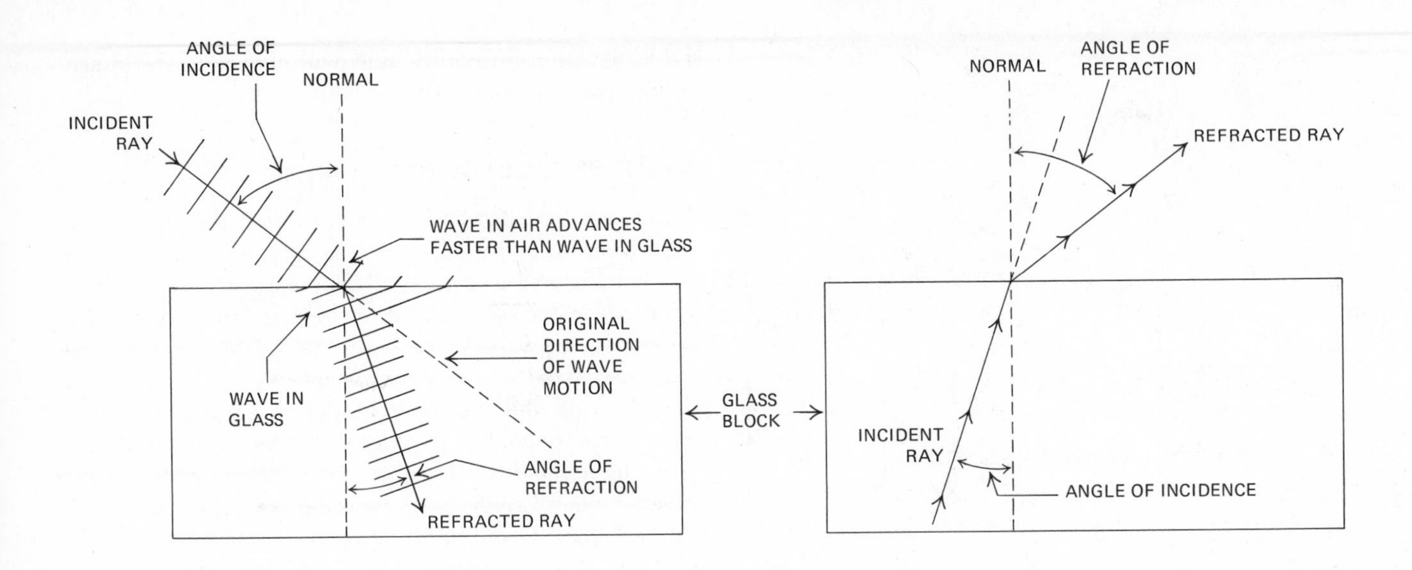

FIG. 3.5
Refraction of light. When a light ray passes from one medium into another of different density, the angle that it makes with the normal is always less in the medium of the higher density. The slowing down of the waves is indicated schematically for the left-hand light ray. The part of the wave still in the air advances faster than the wave in the denser medium. This causes the bending.

The pencil will appear bent at the water surface. The refraction is toward the normal when the second medium is more dense than the first and away from the normal when the second medium is less dense than the first. Refraction occurs because light waves travel more slowly in the denser medium. As shown in Fig. 3.5, a light beam that passes from air into glass is refracted toward the normal because the glass is the denser medium; when it passes from glass to air, it is refracted away from the normal. Another way of stating this property is that when a light beam passes from one medium into another of different density, the angle it makes with the normal is always less in the medium of the higher density.

Early telescopes using crude lenses often showed colored fringes around the images of stars and planets. In his attempt to eliminate these "chromatic aberrations," Newton observed that when visible light passes from one medium into another, such as air into glass, it disperses into its component colors (wavelengths)—red, orange, yellow, green, blue, and violet; he concluded, therefore, that color is a basic property of light.

Visible light is a mixture of the wavelengths in the visible range of the electromagnetic spectrum. In a vacuum, their velocities are uniform, while in a medium, such as glass or even air, they are different. Therefore, when a beam of visible light passes through a material substance, since each color travels at a slightly different velocity, each wavelength is refracted by a slightly different amount. Blue light travels more slowly than red in

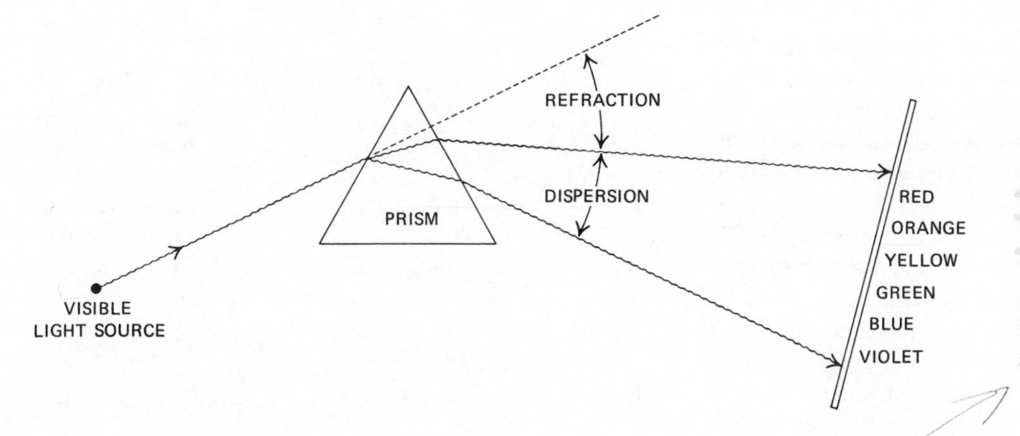

FIG. 3.6
The spectrum. When a beam of visible light passes through a prism, its component wavelengths are refracted differently. When the dispersed light beam falls on a screen, the colors of the different wavelengths are spread out and made visible.

most media. Thus when light is passed through a prism, as shown in Fig. 3.6, its component wavelengths are refracted differently so that when the beam emerges from the prism, it is dispersed in a fan-shaped beam; when focused on a screen, the colors of the different wavelengths are arranged in the order of their frequencies from red to violet. This display of colors is called the spectrum. *Spectroscopy*, the study and analysis of spectra, provides very important data about the celestial bodies—their chemical composition, temperature, pressure, radial velocity, and rotation.

3.4 SPECTRA

With the invention of the spectrograph (Chapter 4.18), it was learned that spectra of different light sources may have different general appearances. A spectrum may be a continuous band of different colors (continuous spectrum), a series of bright-colored lines (emission or bright-line spectrum), or a continuous spectrum with fine, dark lines superimposed on it (absorption or dark-line spectrum).

The *continuous spectrum* (Plate 1*), which shows that all visible wavelengths are present, is produced by a hot solid or very dense gas. The electrons in the atoms are continuously interacting with one another to produce the blending of the colors of all wavelengths.

The *bright-line spectrum* (Plate 2), which shows only a few bright lines, is produced by a rarefied gas under low pressure. Since the atoms are relatively far apart and the electrons are free to move in their usual orbits (see Chapter 3.10), its spectrum displays their discrete characteristic lines.

The *absorption spectrum* (Plate 3), is produced when light from a continuous-spectrum source passes through a cooler gas that is under low pressure. The continuous spectrum is produced by the light source, and the dark lines are produced by the energy absorbed from the light source by the cooler gas. Each dark line occupies the exact position of the bright-line spectrum that would be produced if the cooler gas alone were heated to incandescence and emitted its own radiation.

*Color plates 1–13 appear following p. 56.

3.5 THE BLACK BODY

A radiating body would eventually deplete its energy and reach a temperature of absolute zero if it were unable to absorb energy from outside sources. Actually, in the surrounding region there are hotter bodies that are radiating energy as well as cooler bodies that are absorbing energy; therefore, a body will absorb energy if it is cooler than the surrounding region and will radiate energy if it is hotter. All bodies are radiators as well as absorbers, and a good radiator is also a good absorber. The so-called *black body* absorbs all the radiation that falls on its surface. Although there is no such body, a familiar approximation is a cavity hollowed out of a piece of charcoal. Since stars radiate energy in all wavelengths and behave almost like black bodies, a quantitative analysis of a black body can furnish important information about the stars' emission and absorption of radiation.

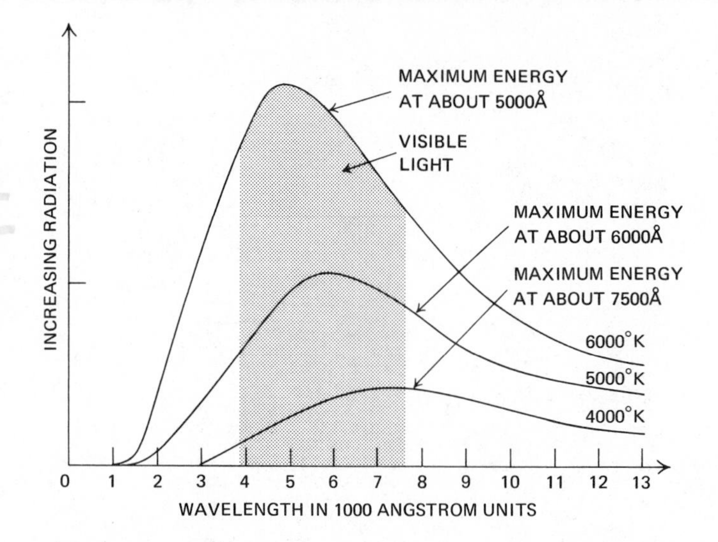

FIG. 3.7
The energy distribution curve of a perfect radiator (black body) shows energy emitted at three different temperatures. An increase in temperature increases the total radiation (represented by the area under the curve) and shifts the wavelength of maximum radiation (represented by the peak of the curve) toward the shorter wavelengths. The area under the curve represents the total energy emitted by a black body every second from one square centimeter of its surface.

3.6 THE ENERGY DISTRIBUTION CHART

When heated, a black body emits energy in all wavelengths. The distribution of its energy at any particular temperature can be indicated by a smooth, solid curve on an energy distribution chart, where energy is plotted against wavelength. In Fig. 3.7, the relative intensity of the energy emitted over the entire electromagnetic radiation range is shown for three temperatures. Note that the curves are smooth without any bumps or dips. A black body is said, therefore, to emit a continuous spectrum. Figure 3.8 shows for comparison the charts of continuous emission, and absorption line spectra based on those illustrated in Plates 1 to 3. The maximum height of a curve occurs at the wavelength at which maximum energy is radiated. Since the curve indicates the amount of energy radiated at a given wavelength, the maximum height of the curve occurs at the point where the maximum energy is emitted. A black body at a temperature of 6000 °K* radiates its maximum energy at about 5000 Å. When the temperature is increased, the energy emitted is increased in all wavelengths, and the maximum energy occurs at progressively shorter wavelengths.

This principle is illustrated by an electric-stove burner or by a log in a fireplace. At a low temperature, the general color is dull red because maximum energy is being emitted in the long red wavelengths. As the temperature increases, the maximum energy being emitted shifts toward the shorter wavelengths, and the general color changes progressively from red to orange to yellow. Since temperature is a measure of energy, there is a rela-

*See Appendix 4 for a discussion of the Kelvin (K) temperature scale.

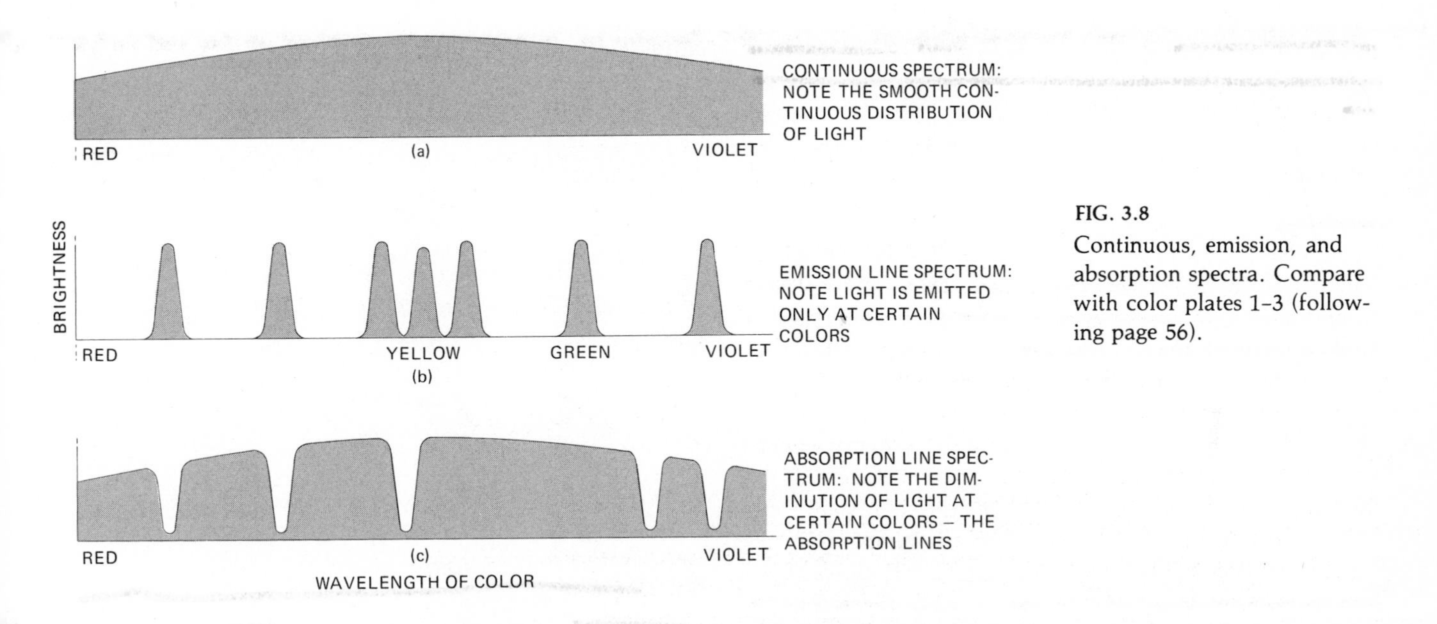

FIG. 3.8
Continuous, emission, and absorption spectra. Compare with color plates 1-3 (following page 56).

tion between an object's temperature and the color of the emitted light. And since the energy carried by light depends inversely on its wavelength, it is not surprising that the hotter, more energetic source emits more energetic and hence bluer light. An electric burner never gets hot enough to emit blue light because it would melt at such temperatures. However, if you have ever watched someone welding, you no doubt recall the colors seen in the smoke and flame when the material was being welded. There the light does appear blue, and the temperature is high enough to melt metal.

3.7 THE RADIATION LAWS

The wavelength of the maximum energy radiated by a black body at a given temperature is the wavelength of the maximum height of the curve on the distribution chart. This value can be obtained from the relationship

$$\lambda_{max} = 2.898 \times 10^7/T,$$

known as Wien's law, where λ is the wavelength in angstroms and T is the absolute temperature in degrees Kelvin or °K. This relationship was determined when it was recognized that the wavelength of maximum emission is inversely proportional to the absolute temperature. Hence, under most circumstances, a star's color is a direct measure of its temperature.

For instance, using Wien's law, we can determine the surface temperature of the sun by observing the wavelength at which it emits its maximum energy. The sun's radiation is observed with a *spectrograph*, and its maximum radiation is measured with a *radiometer* (Chapter 4.13). Since the maximum energy is emitted at about 4700 Å, we substitute this value in Wien's formula to determine the sun's effective surface temperature.

$$4700\ \text{Å} = 4.7 \times 10^3\ \text{Å}$$

$$\lambda_{max} = 2.898 \times 10^7/T$$

$$T = 2.898 \times 10^7/\lambda_{max}$$

$$= 2.898 \times 10^7/4.7 \times 10^3$$

$$= 6164\,°\text{K}.$$

The total energy that a black body emits every second from one square centimeter of its surface is represented by the area under the curve on the energy distribution chart. This value can also be determined from the relationship

$$E=\sigma T^4,$$

which is known as *Stefan's law,* where E is the total energy at a given temperature, T is the absolute temperature, and σ is Stefan's constant. In this relationship the total energy emitted is proportional to the fourth power of the absolute temperature. A body with a temperature of 1000°K radiates a certain amount of energy every second from one square centimeter of its surface. When the temperature is doubled to 2000°K, the energy emitted is increased 16 times (2^4), and when the temperature is tripled to 3000°K, its energy is increased 81 times (3^4). Using Stefan's law, we can determine the surface temperature of the sun by substituting the total energy that it emits from one square centimeter of its surface in one second.

A star's radius, in terms of the sun's radius, can also be determined from Stefan's law. In the following example, the subscripts $_\odot$ and $_*$ are used to designate the sun and star that have been observed, with the following data recorded: the surface temperature of the sun, $T_\odot$, is 6000°K; the surface temperature of the star, T_*, is 3000°K; and the star's luminosity, L_*, is 400 times brighter than the sun's, $L_\odot$.

The total energy emitted every second from one square centimeter of the star's surface is obtained from Stefan's law

$$E_*=\sigma T_*^4.$$

When this value is multiplied by the star's surface area, $4\pi r_*^2$, its luminosity, L_*, is obtained. By comparing the star's luminosity with that of the sun, $L_\odot$, the star's radius, in terms of the sun's radius, is obtained. The following procedure is presented as an illustration of a situation which at first appears to be complicated but which, by recalling elementary procedures of algebra, becomes quite simple. Its solution is based on simple proportion.

$$\frac{(\sigma T_*^4)\,(4\pi r_*^2)}{(\sigma T_\odot^4)\,(4\pi r_\odot^2)} = \frac{L_*}{L_\odot}, \qquad \frac{(T_*^4)\,(r_*^2)}{(T_\odot^4)\,(r_\odot^2)} = \frac{L_*}{L_\odot}$$

$$\frac{r_*^2}{r_\odot^2} = \frac{L_*(T_\odot^4)}{L_\odot(T_*^4)}, \qquad \frac{r_*}{r_\odot} = \frac{\sqrt{L_*}\,(T_\odot^2)}{\sqrt{L_\odot}\,(T_*^2)}.$$

Since the ratio $T_\odot^2/T_*^2$ is 4, and the ratio $\sqrt{L_*}/\sqrt{L_\odot}$ is $\sqrt{400}$, when these values are substituted in the preceding equation, we produce the ratio

$$\frac{r_*}{r_\odot} = (\sqrt{400})\,4=80.$$

This means that the radius of the star is 80 times greater than the radius of the sun. This is one way in which the radii of stars may be found.

The energy emitted every second from one square centimeter of the black body's surface at a particular temperature and wavelength is represented by the ordinate to the curve on the energy distribution chart. This value as well as Wien's Law can also be determined from a somewhat complicated relationship, derived by the German physicist Max Planck, which is known as *Planck's law.*

3.8 THE STRUCTURE OF THE ATOM

To understand the process of the emission and absorption of light by celestial bodies, we must first understand the structure and behavior of the atom. About 450 B.C., the Greek philosopher Democritus proposed that matter consists of discrete particles—an idea that was rejected by both Plato and Aristotle. Centuries later, Newton and many other scientists accepted Democritus' idea solely on the basis of intuition. In 1808 the idea became well established on experimental data when the English scientist John Dalton proposed the atomic theory. The present modern theory of the structure of the atom was given in 1911 by the English scientist Ernest Rutherford. His atomic model consisted of a small, extremely concentrated nucleus of positively charged particles that were surrounded by a relatively large, tenuous cloud of negatively charged electrons (Fig. 3.9).

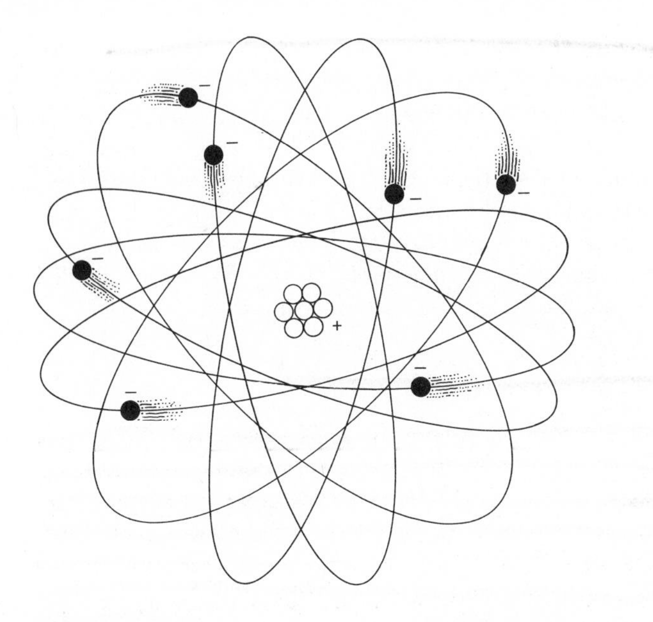

FIG. 3.9
The Rutherford model of the atom—a small, extremely concentrated nucleus of positively charged particles surrounded by a large, tenuous cloud of negatively charged electrons.

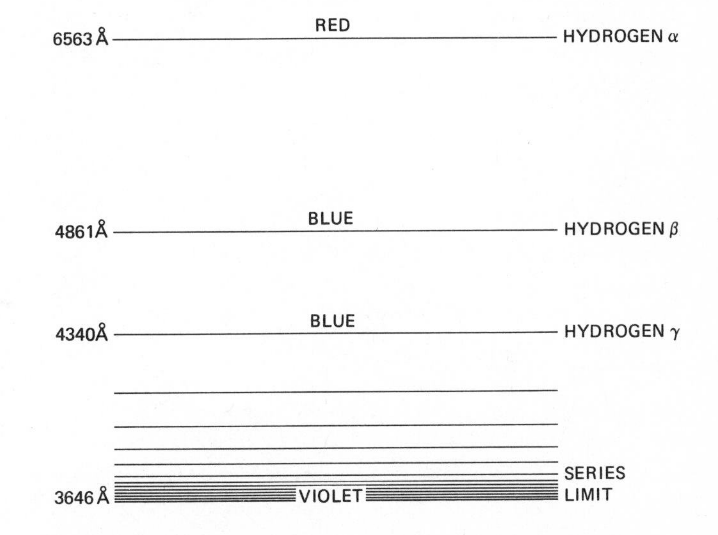

FIG. 3.10
The Balmer series of hydrogen. The spectral lines are in the visible region of the spectrum and are closer as the wavelengths become shorter.

A study of the spectrum of the simplest atom, hydrogen, has revealed the important fact that its spectral lines are distributed in definite and regular patterns called *spectral series*. The first series was discovered by J. Balmer, a Swiss high school mathematics teacher, who recognized the regularity of the hydrogen spectral lines in the visible range of the spectrum. The Balmer series is shown in Fig. 3.10. The line with the longest wavelength and strongest intensity is designated hydrogen-alpha, $H\alpha$, with a wavelength of 6563 Å. As the wavelengths decrease, the lines appear less intense and closer together. The limit of the series has a wavelength of about 3646 Å.

In 1913 the Danish physicist Niels Bohr presented his famous model of the hydrogen atom and a theoretical interpretation of its spectrum based on the concept of the *quantum*. The model consisted of one proton in the nucleus with one electron orbiting around it. Many developments since this model was presented reveal that the model and Bohr's interpretation do not adequately describe the structure or the behavior of the atom; however, it represented a great step forward in the development and understanding of the atom and quantum mechanics. Some of Bohr's concepts still serve as a link between *classical physics*, which is based on the laws postulated by Newton, and *quantum physics*, which is based on the laws governing the motions of extremely small bodies.

An explanation for the emission spectrum of hydrogen is provided in quantum physics. Atoms exist only with discrete energy levels, each characterized by a definite amount of energy. These levels are considered to be concentric shells that are centered around the nucleus

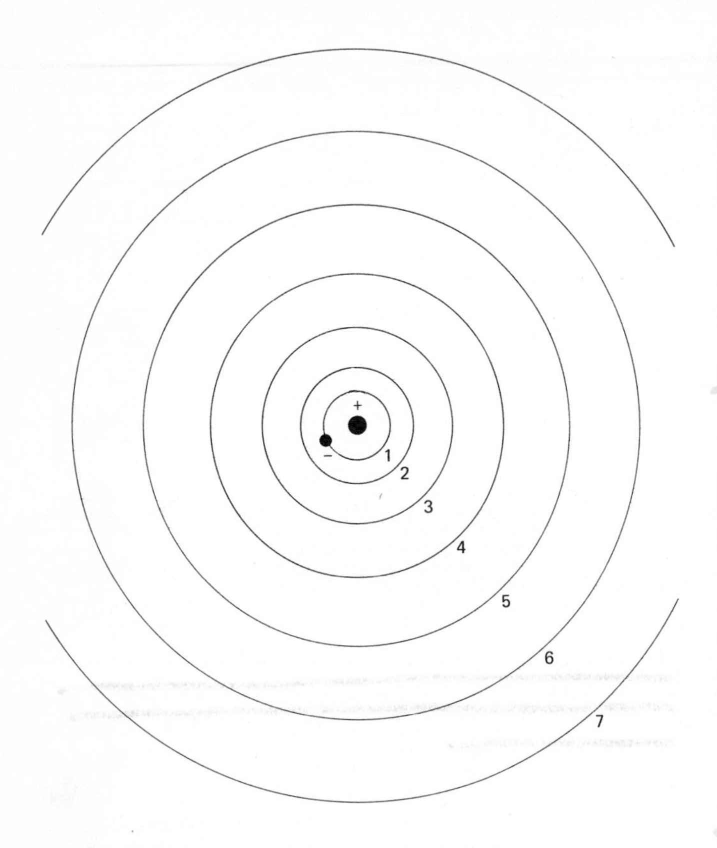

FIG. 3.11
The main energy levels in the model of an atom are designated by the quantum numbers 1, 2, 3, . . . , etc.

(Fig. 3.11). They represent the region in space where the electrons in each level are most frequently found. The main energy levels are designated from the nucleus outward, by the quantum numbers 1, 2, 3, 4. . . , etc. For more complex atoms, the maximum number of electrons that can exist in each main energy level is determined by the formula

$2n^2$,

where n is the quantum number of the energy level. This means that the first level can hold only 2 electrons; the second level, 8; the third level, 18; and so on. Normally, the electrons fill the lower energy level to its maximum capacity before they enter the next higher level; however, there can never be more than eight electrons in the outermost level. The main energy level may have as many as four sublevels, and the sublevel may be further divided into additional levels of two electrons each.

3.9 THE PARTS OF THE ATOM

The atom is the smallest particle into which a chemical element can be divided and still retain its chemical identity. An atom consists of three basic elementary particles: *protons, neutrons,* and *electrons.* The electron, with a mass of 9.1096×10^{28} grams, is the least massive and is negatively charged. The proton is 1836.11 times more massive than the electron and is positively charged. The neutron has about the same mass as the proton, but is electrically neutral.*

The atom has a negatively charged cloud of electrons (whose numbers range from 1 in the hydrogen atom to 103 in the lawrencium atom) that revolve around a dense and positively charged nucleus. The nucleus consists of one proton in the hydrogen atom and a combination of protons and neutrons in the heavier atoms. The protons and neutrons are held together by a strong attraction, called the *nuclear force.* The electrons are held in shells, or orbits, by an *electrostatic force,* which exists between the positively charged protons and the negatively charged electrons—a force that, like gravity, varies inversely as the square of the distance between the particles. The number of protons in the nucleus determines the atomic number of the atom, which is 1 for hydrogen, 2 for helium, and 92 for uranium. The mass of the atom is its *atomic weight.* Since almost all of the atoms have atomic weights that are fractional decimals and most of their mass is in the nucleus, the atomic weight is rounded off to a whole number to establish the mass number of the atom.

*Recent experiments show that protons and neutrons may themselves be composed of still more fundamental particles called "quarks."

An atom with the same atomic number as another atom but with a different mass, that is, with a different number of neutrons, is called an *isotope.* A hydrogen atom with one proton and one neutron in its nucleus has a mass number of 2, is an isotope of the normal hydrogen atom, and is called *deuterium.* A hydrogen atom with one proton and two neutrons has a mass number of 3, is also an isotope of hydrogen, and is called *tritium.* The symbolic representation of deuterium is $_1H^2$. The subscript 1 indicates that there is one proton in the nucleus. The superscript 2 indicates that there are two particles in the nucleus. Since the nucleus contains protons and neutrons the number of neutrons can be found by subtracting the subscript from the superscript.

3.10 EMISSION AND ABSORPTION

When the normal hydrogen atom is unexcited, the electron occupies the atom's lowest energy level (referred to as its ground state), which is the one closest to the nucleus. In this condition, the electron is in the *ground state* and is stable. When the electron absorbs a discrete amount of energy, it becomes *excited* and moves to the next higher energy level. When it has absorbed a sufficient amount of energy to pass through all of the atom's permissible energy levels and escape from the atom, it is said to be *ionized,* and the nucleus in this condition is called an *ion.* Excitation can also be produced by the collision of one particle with another, causing part of the energy of the collision to be absorbed by the atom.

An atom always seeks its lowest energy level; therefore, it remains in the excited state for only a very brief period of time—less than one hundred-millionth of a second—before it returns to the ground state. As an electron moves from one energy level to the next lower level, it emits a photon of energy of definite wavelength. If e represents the energy of the atom in the excited state and e' represents the energy at the next lower level, the electron emits energy equal to $e - e'$ as it moves from the higher to the lower level. The movement of an electron to either a higher or a lower energy level is called a *transition.*

According to the present theory of the structure of the atom, a distinctive wave pattern with properties that can be expressed mathematically is associated with each energy level. A transition from a higher to a lower level is a transformation from one wave pattern to another and produces a spectral line. For the hydrogen atom (Fig. 3.12), all transitions from any energy level to the ground state produce a series of spectral lines in the ultraviolet region of the spectrum, which is called the *Lyman series.* All transitions from any higher level to the second energy level produce a series of spectral lines in the visible region of the spectrum, which is called the *Balmer series.* All transitions from any higher level to the third energy level produce the *Paschen series,* and those to the fourth energy level produce the *Brackett series.*

Since the Balmer series of spectral lines is in the visible region of the spectrum, it will be analyzed more thoroughly. A transition from the third to the second energy level produces the first line in the series, which is called H (for hydrogen) alpha, with a wavelength of 6563 Å and a red color. A transition from the fourth to the second energy level produces the second line, which is called H-beta, with a wavelength of 4861 Å and a blue color. A transition from the fifth to the second energy level produces the third line, which is called H-gamma, with a wavelength of 4340 Å and a blue-violet color. Although there are many more lines in the Balmer series, as well as in the other series, only the first few are considered because the spectral lines produced by transitions from the outer energy levels are progressively closer together and are thus more difficult to isolate. These transitions explain how and why an emission spectrum is produced.

Before we explain how the absorption spectrum is produced, let us quickly review the excitation of the atom. We know that when light from a source which produces a continuous spectrum is allowed to pass through a cooler, low-pressure gas, it produces an absorption spectrum. The light source which produces the continuous spectrum emits photons in all wavelengths; therefore, when the light passes through the cooler, low-pressure gas, these photons bombard the atoms of the gas and are absorbed by their electrons. The energy of the photons

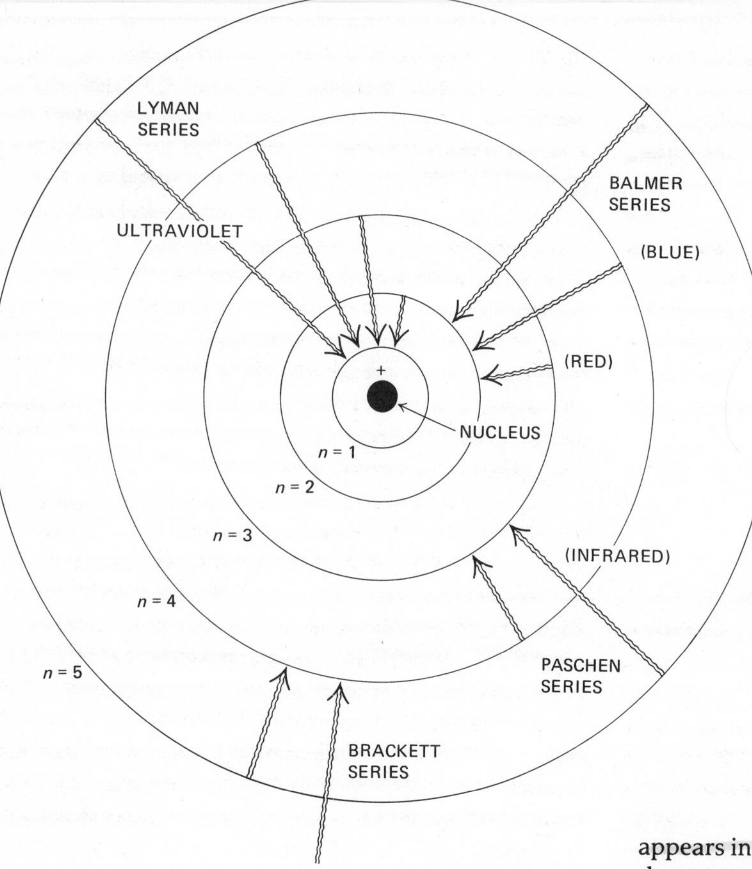

FIG. 3.12
Spectral lines produced in the emission of light by the hydrogen atom as an electron moves from a higher to a lower energy level in the Bohr model.

appears in the electrons as increased energy, which causes them to make an upward transition. The electrons remain at the higher levels for about one hundred-millionth of a second, then fall back to their ground state. In the downward transition, they emit energy in the form of photons, which may or may not be of the same wavelength as those that were absorbed. If the transition was made directly to the original level, they are the same; if the transition was made in steps—in one or more energy levels at a time, or cascading—the wavelengths are different.

The absorption spectrum is produced in the following manner. When the cooler, low-pressure gas is not present, all the wavelengths in the beam from the light source will reach the spectrograph and be recorded as a continuous spectrum. When the gas is present, its atoms will absorb the energy from a light beam, produce an

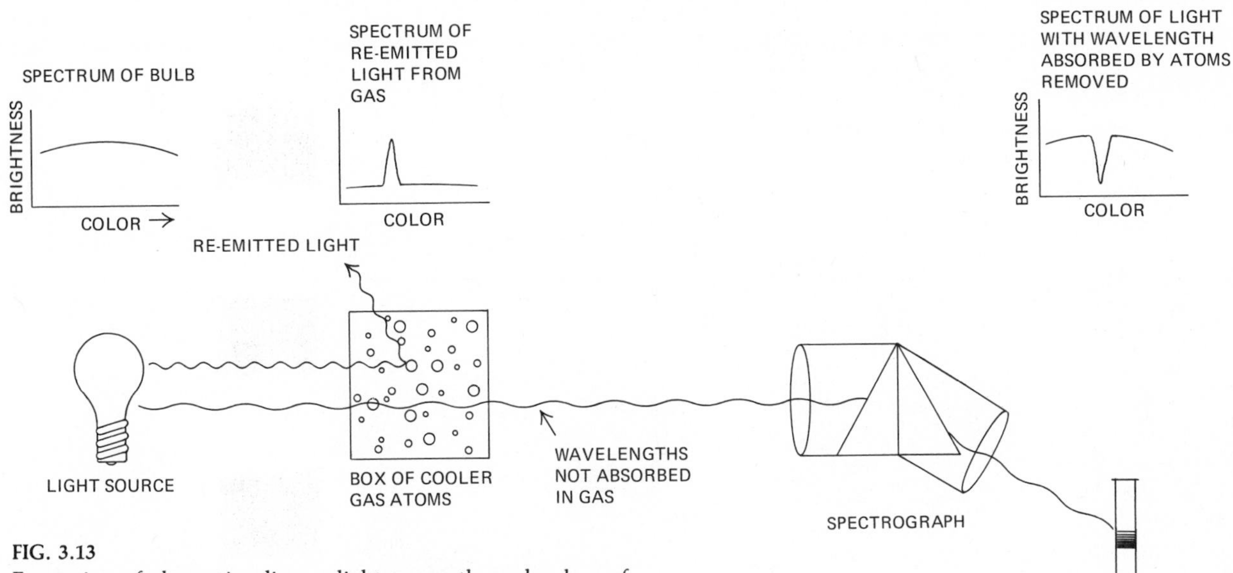

FIG. 3.13
Formation of absorption line as light passes through a box of cooler gas. Atoms in the box absorb light and re-emit it in directions that do not reach the spectrograph.

upward transition, followed by a downward transition in which the gas atoms re-emit energy. Since the energy is re-emitted in all directions, very little of it will reach the spectrograph; therefore, an absorption spectrum will be produced—a continuous spectrum with those wavelengths that did not reach the spectrograph recorded as dark lines (Fig. 3.13).

3.11 THE DOPPLER EFFECT

Christian Doppler, an Austrian physicist, demonstrated in 1842 that the frequency (and therefore the pitch) of sound increases as its source approaches the observer and decreases as it recedes. The rise and fall of the pitch of a locomotive's whistle or automobile horn as it approaches and passes an observer is evidence of this phenomenon, which is called the *Doppler effect*. The effect occurs whether the source of the sound, the observer, or both are in motion.

The Doppler effect also applies to light. In Fig. 3.14, the observer is stationary in all three views; the star is stationary in the top view, approaching the observer in the middle view, and receding from the observer in the bottom view. When the star is stationary, its light waves reach the observer at normal frequency. As the star approaches the observer, the wavelengths are shortened and the frequency is increased; that is, more waves reach the observer in a given interval of time. This effect is indicated by a shift of the spectral lines toward the violet end of the spectrum. The reverse is true when the star is moving away from the observer—the wavelengths are increased, the frequency decreased, and the effect indicated by a shift of the spectral lines toward the red end of the spectrum. Both the shift of the spectral lines and the

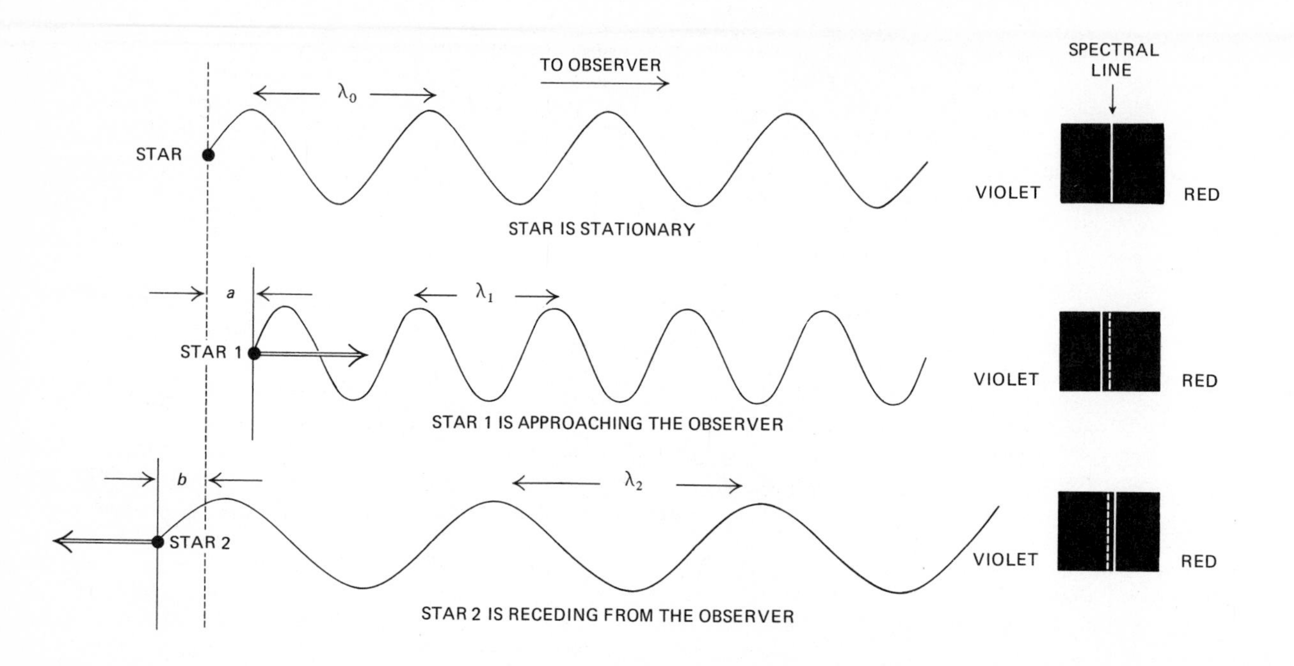

FIG. 3.14
The Doppler effect. When a star is stationary, its light waves reach the observer at normal frequency. When the star is approaching the observer, the wavelength is decreased because during the emission of one wavelength, the star has moved the distance a. Since the wavelength has been decreased, the spectral line shifts toward the violet. When the star is receding from the observer, the wavelength is increased because during the emission of one wavelength, the star has moved the distance b away from the observer. Since the wavelength has been increased, the spectral line shifts toward the red.

rate at which the body is approaching, receding, or rotating about its axis can be established by comparing the spectrum of a body with a known source whose spectral lines have already been accurately measured.

The velocity of a moving object can be found by observing the Doppler shift of spectral lines that it emits. The shift in wavelength divided by the original wavelength is the object's velocity toward or away from the observer divided by the speed of light. This can be expressed in the formula

$$\frac{\lambda - \lambda_0}{\lambda_0} = \frac{v}{c},$$

where λ_0 is the wavelength known to be emitted when the object is at rest; λ is the wavelength actually observed; v is the object's velocity along the line from it to the observer, and c is the speed of light. Note that the shift described above will not occur unless the object's distance from the observer is changing.

Using the Doppler shift for example, we can find the velocity of a distant galaxy. The galaxy spectrum has a

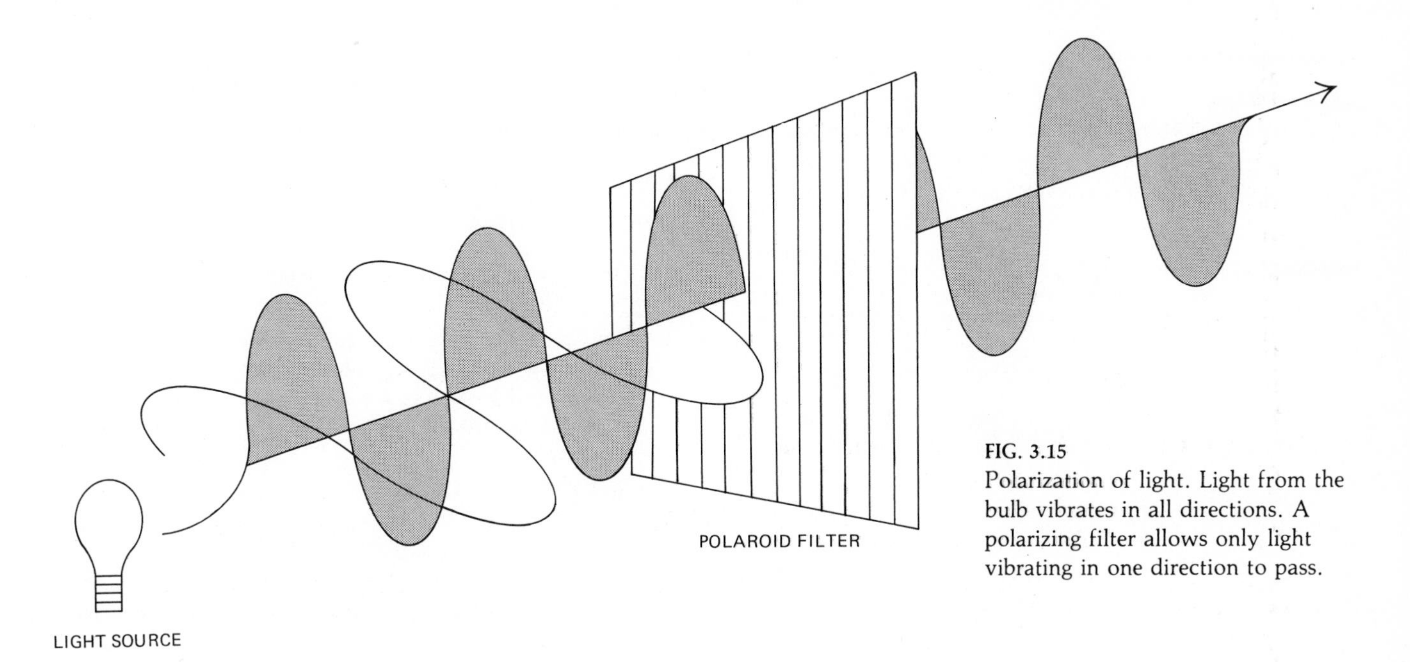

FIG. 3.15
Polarization of light. Light from the bulb vibrates in all directions. A polarizing filter allows only light vibrating in one direction to pass.

line that we identify as H-α (λ_0 = 6563 Å), but we observe it shifted to λ = 7219 Å. How fast is the galaxy moving? Is it approaching us or receding?

$$\frac{\lambda - \lambda_0}{\lambda_0} = \frac{v}{c}$$

$$\frac{v}{c} = \frac{7219 - 6563}{6563}$$

$$\frac{v}{c} = \frac{656}{6563}$$

$$\frac{v}{c} = 0.1$$

The velocity is therefore 0.1 c, or 30,000 km/sec. The galaxy is moving away from us because the wavelength is lengthened from 6563 Å to 7219 Å.

3.12 NONTHERMAL RADIATION

Most of the electromagnetic radiation that we perceive in the world around us is produced by the motion of electrons between atomic energy levels. Because such radiation is usually associated with a heated substance, it is often referred to as *thermal radiation.* In many astronomical objects, electromagnetic radiation is produced by nonthermal processes. The radiation in these instances arises from the motion of high-speed charged particles (electrons, etc.) in a magnetic field. Nonthermal radiation arises on the earth in atomic accelerators, especially in a type of accelerator called a *synchrotron.* Thus nonthermal radiation is often called *synchrotron radiation.*

Nonthermal radiation, unlike that produced from thermal sources, is often strongly polarized; that is, the electromagnetic waves vibrate in a preferred direction. In unpolarized light the vibrations are in random directions (Fig. 3.15). Nonthermal radiation from astronomical sources, in addition to being polarized, has a continuous spectrum of a very different shape than that associated with a hot gas or solid (Fig. 3.7). These properties allow astronomers to distinguish between thermal and nonthermal radiation. The kind of astronomical objects that emit nonthermal radiation most strongly are exploding stars and peculiar galaxies, which we will study in detail in Chapters 14 and 15, respectively.

SUMMARY

Our knowledge of the astronomical universe comes almost exclusively from the study of radiation emitted or reflected by astronomical objects. Radiation is a form of electromagnetic energy and travels through space as a wave. The wavelength characterizes the type of radiation. The regions of the electromagnetic spectrum that are of particular importance in astronomy are—in order of increasing wavelength—x-ray, ultraviolet, visible, infrared, and radio. The energy carried by an electromagnetic wave increases as its wavelength decreases. The amount of radiation received by a distant observer depends inversely on the square of the distance to the source and the intrinsic brightness of the source (inverse square law). Light waves may interact with matter in a variety of ways. Refraction is the bending of light as it passes from one substance into a different one. This principle is used in the operation of lenses. Light can also be dispersed (spread out) into its component colors (wavelengths) to form a spectrum.

Many astronomical objects are approximately perfect radiators and are called black bodies. Black bodies are important because their radiation obeys well-defined laws. In particular, radiation from a black body depends only on the object's temperature. Hotter objects radiate more strongly at shorter wavelengths, and the total amount of energy radiated is proportional to the fourth power of the absolute temperature.

Radiation may be emitted or absorbed. Both processes can be understood conveniently in terms of the motion of electrons in their orbits about the nuclei of atoms. Emission corresponds to the dropping of an electron to a lower orbit; absorption corresponds to the lifting of an electron to a higher orbit.

The wavelength of radiation from a moving source is altered by the Doppler shift. Approaching objects experience a blue shift, receding ones a red shift. The Doppler shift allows an object's velocity of approach or recession to be measured.

REVIEW QUESTIONS

1. Explain what is meant by wavelength, frequency, and the speed of a wave.
2. The light from a point source is allowed to pass through an opening of one square unit area placed at one unit distance from the source. How large an area will it cover when the distance is tripled? What law applies in this situation?
3. Jupiter is about five times as far from the sun as is the earth. How much brighter (or fainter) will the sun appear from Jupiter than from the Earth?
4. The stars in the Big Dipper appear about one hundred times brighter than those in the star cluster "The Beehive." If they have the same intrinsic brightness, how much farther away from us is the star cluster than the Big Dipper stars?
5. Put a pencil in a glass half full of water. Does the pencil look straight when seen from different angles? What property of light does this demonstrate?
6. Put a penny at the bottom of a tin can and arrange it on the table in front of you so the penny is just hidden by the top edge of the can. Now have a friend pour water into the can. Is the penny still invisible? Suggest how this example of refraction is related to the apparent position of the setting sun.
7. Which kind of electromagnetic radiation has the longer wavelength, infrared or x-ray? What are typical values for the wavelength of visible light and radio waves?
8. Arrange the following in order of decreasing wavelength: x-rays, yellow light, ultraviolet, violet light, infrared, and gamma rays.
9. What are the three types of spectra and how is each produced?
10. What information about the light source can be deduced from its spectrum? Explain your answer.
11. What is meant by the Doppler effect as it applies to light? What information about the light source can be deduced from the Doppler effect? Give examples.

12. What is a perfect radiator? Is there such an object? Why is a perfect radiator called a black body? What important information can be obtained from a quantitative analysis of a black body?

13. The mean temperature of the earth's surface is about 300 °K (roughly 50 °F). Use Wien's law to find out at what wavelength the Earth radiates most of its energy. In what region of the spectrum is this?

14. Suggest how you could find out how hot the filament in a lamp is by using (a) Wien's law, b) Stefan's law.

15. What sort of spectrum would you expect from the burner on an electric stove? On a gas stove?

16. List the basic particles that compose the nuclei of atoms. What are their characteristics? How do nuclei differ? What is the term given to two or more elements whose atoms have the same atomic number but different atomic mass? Discuss the stability of the nuclei of such atoms.

17. Draw a sketch of what an atom may look like, labeling the parts. How would you redraw the atom to indicate the emission of light? How would it differ from the picture for absorption of light?

18. If a tube of cold sodium atoms absorbs light at 5890 Å, at what wavelength might you expect it to emit light when heated? What color is this, approximately?

19. Draw sketches of a hydrogen, a helium, and a carbon atom. Put in all the relevant fundamental particles.

20. Explain what takes place when an atom: (a) absorbs and (b) emits radiation. Why do atoms emit light in definite wavelengths?

21. Explain what is meant by: (a) normal, (b) neutral, (c) excited and (d) ionized atoms.

CHAPTER 4
THE TOOLS OF THE ASTRONOMER

Astronomy is not simply a body of theories that has come into existence fully grown like Athena from the head of Zeus. Astronomy is closely associated with technology. The invention of a new instrument or the improvement of an old one has often produced a tremendous impact on the development of astronomy. The Ptolemaic system was seriously challenged when Galileo, using the newly invented telescope, discovered four of Jupiter's satellites and realized that they were bodies that did not revolve around the earth. The use of the photographic plate in astronomical research allowed astronomers to observe objects that were not visible through a telescope. The invention of the radio telescope greatly extended the astronomical horizons because it gave astronomers the ability to penetrate interstellar dust, as well as allowing the detection of radiation from cold interstellar gas, young stars, and high-energy particles produced by exploding stars and galaxies. Similarly, x-ray telescopes in orbit have made it possible to observe the extremely energetic phenomena believed to be associated with the final collapse and death of stars.

4.1 EARLY INSTRUMENTS

One of the most important and frequently used instruments that originated in antiquity was the *astrolabe*—a simple device consisting of a metallic disk whose outer edge was divided into 360° with a pointer that rotated about the disk's center. When the astrolabe was held vertically, the height of a celestial body above the horizon was measured; a second measurement with it held horizontally allowed the position of a celestial body to be determined. The astrolabe was also used to tell time by observing the direction of a shadow cast by an object. Other early astronomical instruments were the wall quadrant for measuring the angle of a celestial body as it crossed the observer's meridian, the armillary sphere for determining the coordinates of a celestial body, and the sundial (Plate 4) and water clock for determining time.

4.2 ELEMENTS OF THE TELESCOPE

The optical telescope, which was invented in 1608 by the Dutch lens grinder Hans Lippershey, was until this cen-

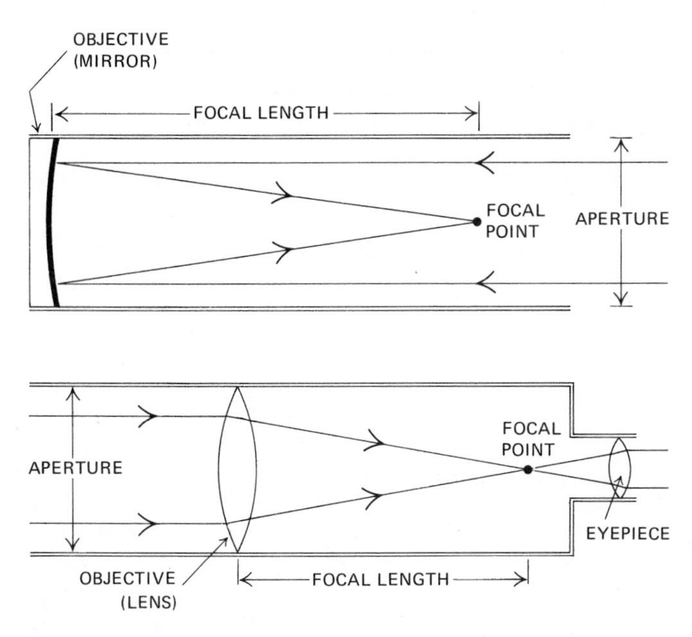

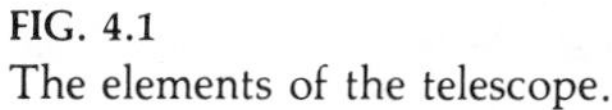

FIG. 4.1
The elements of the telescope.

tury the astronomer's basic tool. It serves the deceptively simple function of resolving the images of celestial objects and bringing new and more distant ones into view. The telescope has either a lens or a mirror, called the *objective*, whose purpose is to collect the light from an object and form its image (Fig. 4.1). This is accomplished by allowing the light to pass through the lens or to be reflected by the surface of the mirror. The point at which the light is brought to a focus and the image is formed is called the *focal point*. Its distance from the center of the lens or the surface of the mirror is the *focal length*. The ratio of the focal length to the diameter of the objective is called the *focal ratio*, or the f-ratio. Thus a 10-inch, f-9 telescope has an objective with a diameter of 10 inches and a focal length of 90 inches. The f-ratio is important because it is a measure of how rapidly a faint object may be photographed. For example, with an ordinary camera a picture may be taken in dim light at f/4 in 1/50 of a second, while in full sunshine the shutter speed might be 1/1000 of a second for the same f-ratio. The image formed at the focal point is examined with an *eyepiece*, which is simply a magnifying glass of short focal length.

4.3 PROPERTIES OF THE IMAGE

The image that is formed by a telescope has three important properties—*size*, *brightness*, and *resolution*. The size of the image depends on the focal length of the objective—it increases as the focal length increases. This is the purpose of a telephoto lens on a camera. The longer focal length lens (say 300 mm) gives a larger image than the 50-mm focal length lens commonly found in some cameras.

The amount of light collected by a telescope is called its *light-gathering power* and is proportional to the area, the square of the radius, or the square of the diameter of the objective. If the light-gathering power of a 2-inch telescope is 4, that of a 4-inch telescope is 16. That is, when the diameter of the objective is doubled, the light-gathering power is increased four times. For a point source of light, such as a star, the brightness of its image depends on the amount of light collected by the objective. As the size of the objective is increased, the image becomes brighter, and more of the dimmer stars become visible. For an object that appears as a disk, such as the moon, the brightness of its image varies inversely as the square of the focal length of the telescope. Two telescopes with objectives of the same diameter collect the same amount of light, regardless of their focal lengths. If the focal length of one telescope is twice that of the other, it will form an image whose linear dimension and area are twice and four times as large, respectively, as the image formed by the other telescope. Since both telescopes receive the same amount of light, the brightness of one square unit of surface of the image formed by the telescope with the

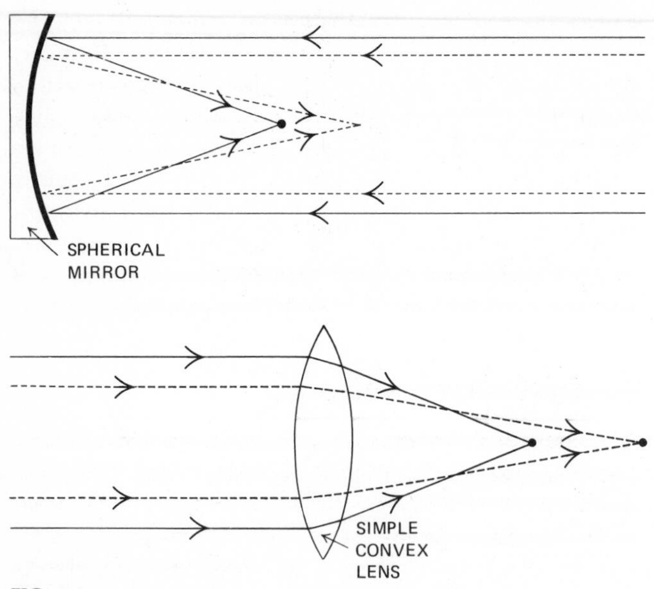

FIG. 4.2
Spherical aberration. A spherical mirror and a simple convex lens do not bring parallel rays of light to a focus at a single point.

longer focal length is one-fourth as great as the brightness of one square unit of the image formed by the telescope with the shorter focal length.

A telescope does not reproduce the geometrical shape of a star; it simply produces an image called a *diffraction pattern*, which consists of a bright central spot, containing about 85% of the total light, surrounded by faint concentric rings. The size of the diffraction pattern depends on the diameter of the objective and the wavelength of the light and has no relation to the brightness of the star. The ability of a telescope to separate the angular distance between two neighboring stars so that two distinct diffraction patterns are visible is called the *resolving power* of the telescope. If the telescope cannot resolve this angular distance, the two stars will appear as a single image of two overlapping patterns. The resolving power increases as the size of the diffraction patterns decreases. This is accomplished by using a telescope with a larger objective. The resolving power is expressed in simplified form by the relationship

$$d=4.56/a,$$

where d is the smallest angular distance in seconds of arc between two stars which the telescope can just barely separate, and a is the diameter of the objective in inches. A 10-inch telescope can resolve two stars that are 0.456 seconds apart, while the 200-inch Hale telescope can resolve two stars that are 0.023 seconds apart. These are theoretical values. Actually, the real resolving power is always lower, primarily due to atmospheric conditions.

4.4 ABERRATIONS OF THE TELESCOPE

A convex lens and a spherical mirror have imperfections (aberrations) that make it impossible for them to produce a perfect image. *Spherical aberration* results when light is brought to a focus at a greater distance in front of either the lens or the mirror when it passes near its axis rather than near its periphery (Fig. 4.2). This effect can be corrected in a lens by using a second lens with a different index of refraction, thereby canceling the error produced by the first lens. This second lens is also important for correcting chromatic aberration as described below. Spherical aberration in a mirror can be eliminated by changing its spherical surface to a parabola.

Chromatic aberration results when visible light passes through a lens; the short wavelengths are refracted more and are focused closer to the lens than the long wavelengths (Fig. 4.3). Before a lens is corrected, its purpose must be established so that those wavelengths important to its purpose are corrected. For visual observation, the lens is corrected for the yellow and green wavelengths because these are more sensitive to the eyes. For photographic work, the blue and violet wavelengths are corrected because most photographic plates are sensitive to these wavelengths. The corrections are accomplished by placing a second lens of different index of refraction close to the convex lens. The two lenses bring the two colors to a common focal point. Chromatic aberration can also be reduced by increasing the focal length of the lens; however, this is expensive because it

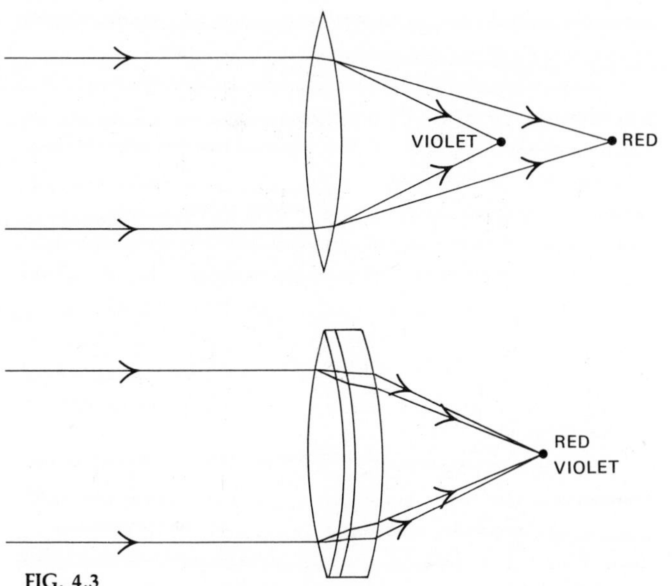

FIG. 4.3
Chromatic aberration. A simple convex lens brings blue and red light rays to a different focus. Two lenses of different index of refraction bring the two rays to a single focal point.

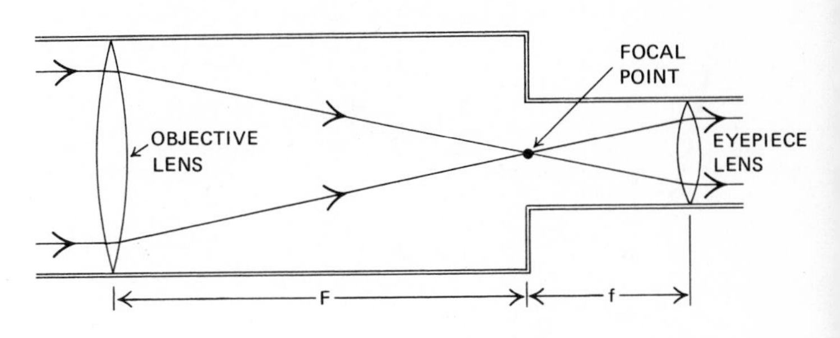

FIG. 4.4
The refracting telescope. For visual use, the eyepiece is used to refract the light into parallel rays. For photographic work, the photographic plate is placed at the focal point.

increases the size and weight of the telescope mount. One reason mirrors are so widely used in telescopes is that they have no chromatic aberration.

4.5 KINDS OF OPTICAL TELESCOPES

The two general types of optical telescopes in use today are *refractors* and *reflectors*. In both types, the objective, which produces the image, is a fixed part of the telescope and cannot be changed without changing the instrument completely. In a refractor, the objective is a lens; in a reflector, a mirror. The eyepiece, which is a magnifying lens of short focal length, is easily interchangeable.

4.6 REFRACTING TELESCOPES

The essential features of a refracting telescope are two lenses separated by a distance equal to the sum of their focal lengths (Fig. 4.4). The larger lens serves as the objective and the smaller as the eyepiece of the telescope. The fixed objective collects the light from an object, which it receives in parallel rays, focuses it, and forms its image at the focal point. The movable eyepiece receives the light from the focal point, magnifies it, and transmits it to the eye in parallel rays. Actual eyepieces usually use two or more lenses to reduce aberrations.

4.7 REFLECTING TELESCOPES

The objective of a reflecting telescope is typically a paraboloidal mirror, which reflects the parallel rays of light it collects from an object and focuses them at the focal point where the image is formed. When the diameter of the objective is very large, as in the 200-inch Hale telescope, the observer and the photographic equipment may be located at the focal point (Fig. 4.5a). This system, sel-

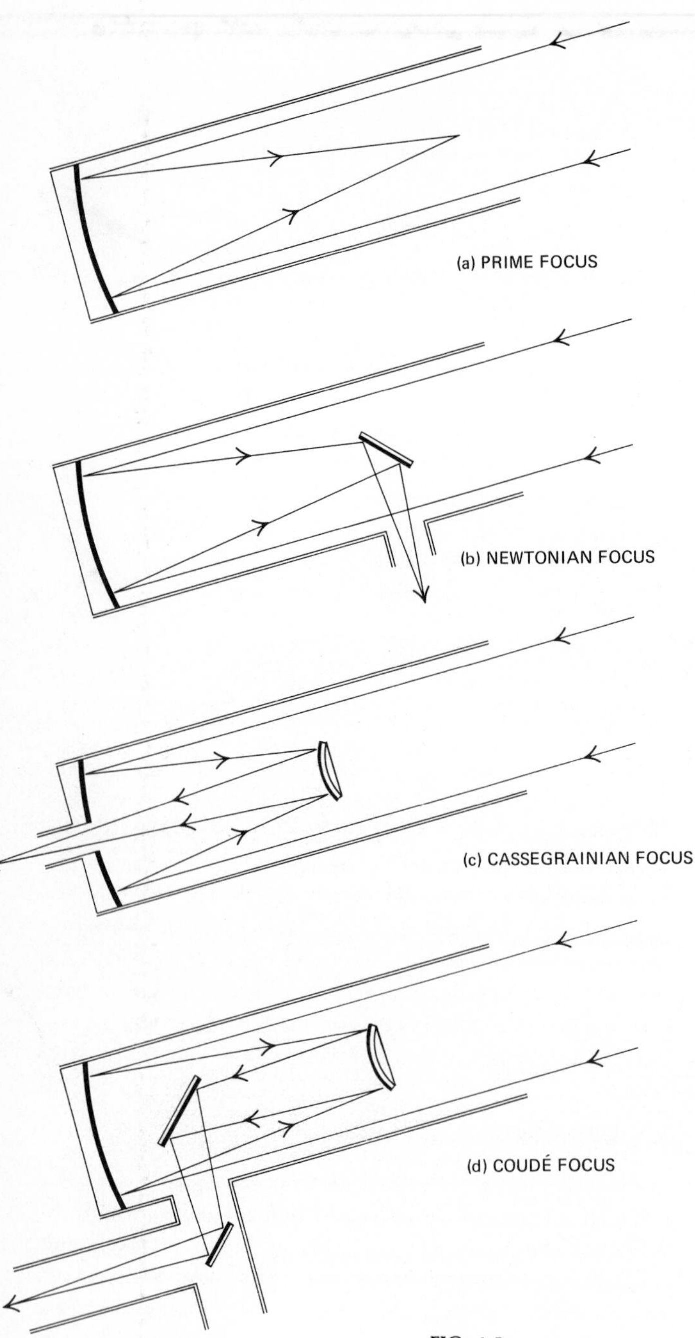

FIG. 4.5
Reflector-type telescopes.

dom used for visual observation but used extensively in photography and spectroscopy, is called the *prime focus*.

The *Newtonian focus*, invented by Newton, is a popular system for visual observation, especially in the smaller telescopes (Fig. 4.5b). It consists of a secondary flat miror located in the axis of the telescope at an angle of 45°, which diverts the light rays to the side of the tube. The image is formed and observed through an eyepiece located on the side of the tube. The diagonal consists of two crossed vanes supporting the secondary mirror in the tube and blocks out a small amount of the incoming light. This slight loss is more than compensated for by making the focal point and the eyepiece accessible to the observer.

Another paraboloidal mirror system, called the *Cassegrainian focus*, uses a secondary convex hyperbolic mirror located on the axis of the telescope at 90° to the axis (Fig. 4.5c). The secondary mirror diverts the light rays back through a hole in the objective and forms the image behind the objective where it can be observed visually or photographed. The Cassegrainian focus produces a more convenient instrument than does the Newtonian focus (because the focal point is located at the lower end of the telescope) and a larger image, with more magnification (because its focal length is usually three times longer than the Newtonian focus). Its disadvantage is a smaller field of view.

A four-mirror system, the *Coudé focus*, uses a secondary hyperbolic mirror similar to that in a Cassegrainian focus and a flat mirror located between the objective and the secondary mirror, which diverts the light down through a hole in the polar axis of the telescope to a stationary observation station (Fig. 4.5d). Since the image rotates with the telescope, a dove prism is used to maintain the image in a fixed position by rotating the image in the opposite direction. As a fixed-focus system, the Coudé telescope is used for spectroscopic analysis with large and complicated auxiliary equipment.

4.8 THE 200-INCH HALE TELESCOPE

The 200-inch (5 meter) Hale telescope at Hale Observatory (Mt. Palomar) in southern California is one of the

PLATE 1 (above, top) Continuous spectrum

PLATE 2 (above, center) Bright-line spectrum

PLATE 3 (above, bottom) Absorption spectrum

PLATE 4 (below, left) The giant Intihuatana at Machu Picchu, Peru, is probably the world's largest sundial. Carved out of rock, it was used to mark the sun's maximum movement to the north in its annual pilgrimage in the sky. (Photograph by J. Paul Freed)

PLATE 5 (below, right) The distance between the twin 90-foot paraboloids at the Owens Valley Radio Observatory north of Los Angeles can be varied to provide the desired base-line interferometer. (Photograph from Owens Valley Radio Observatory, California Institute of Technology)

PLATE 6 (left) The earth as seen by the Apollo 10 astronauts at a distance of one-quarter million miles. The west coast of North America is clearly visible through a break in the cloud cover which appears to obscure the rest of the earth's land masses. (NASA photograph)

PLATE 7 (above) The rising earth as seen by the Apollo 8 astronauts about 5° above the lunar horizon. The earth is 240,000 miles away, and its sunset terminator appears to bisect Africa. (NASA photograph)

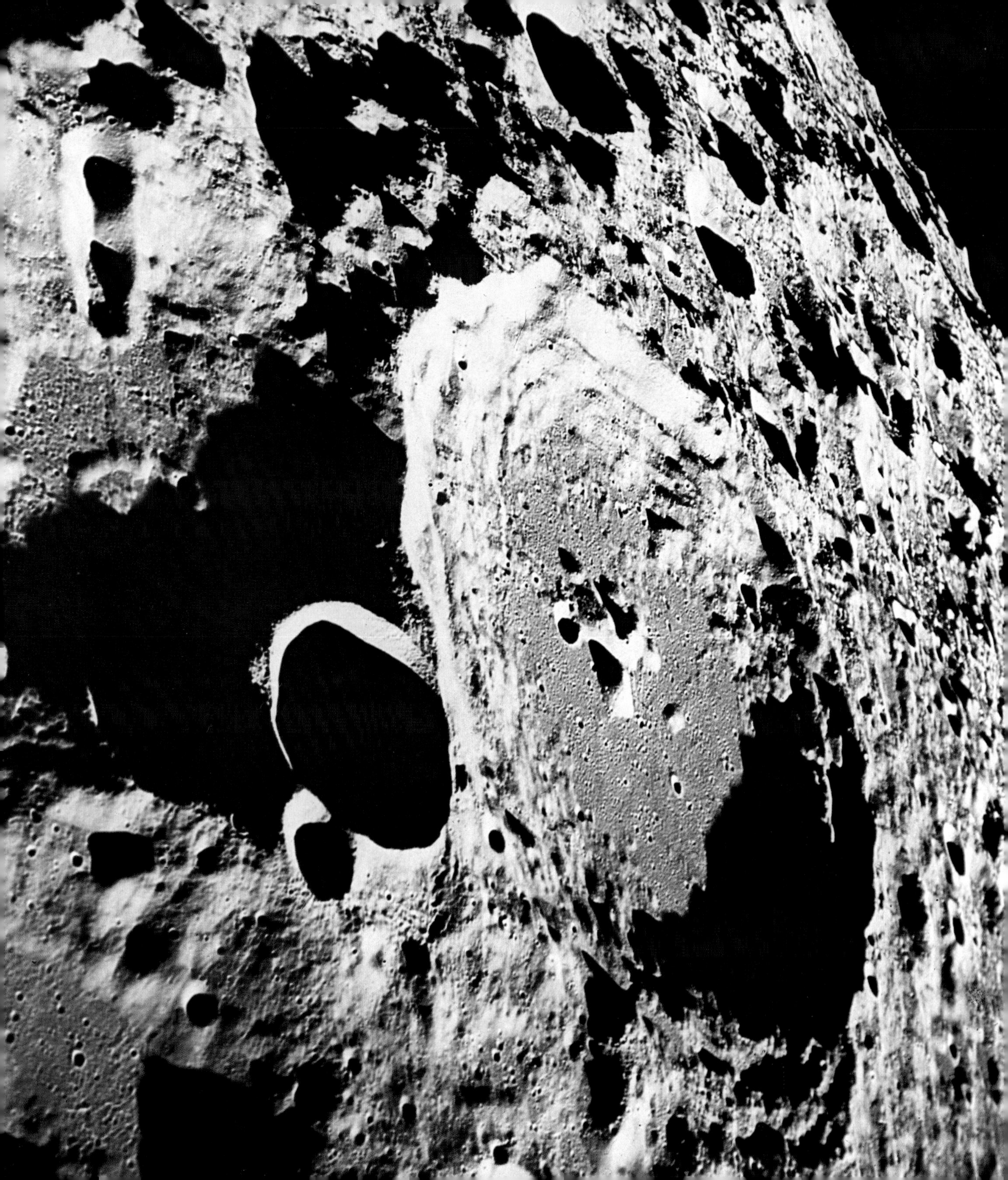

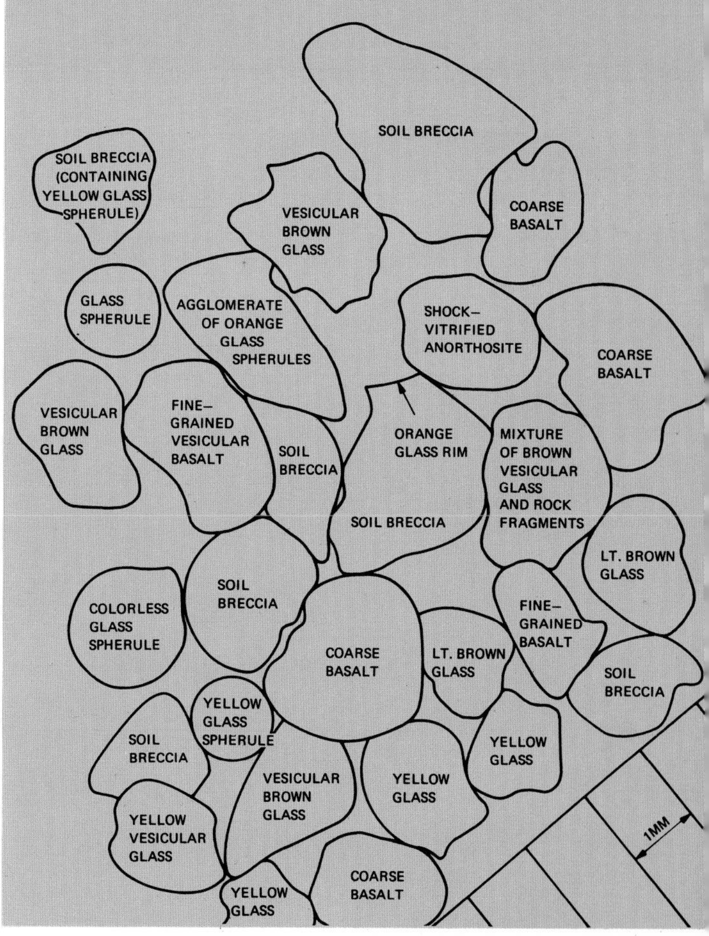

PLATE 8 (left) Oblique view of the lunar far side photographed from Apollo 11. The large crater is the 50-mile diameter International Astronomical Union #308. The lunar far side appears to be more rugged, with fewer and smaller maria. (NASA photograph)

PLATE 9 (above) A representative selection, with identifying key, of particles from the Smithsonian Astrophysical Observatory's Apollo 11 lunar soil sample collected on Mare Tranquillitatis by Neil Armstrong. (This photograph appeared in the August 1970 issue of *Scientific American;* key courtesy John A. Wood, Smithsonian Astrophysical Observatory)

PLATE 10 (right) Glass spherules of various colors and crystals from Apollo 11 rock samples. The diameter of the largest spherule is 0.4 millimeters. (NASA photograph)

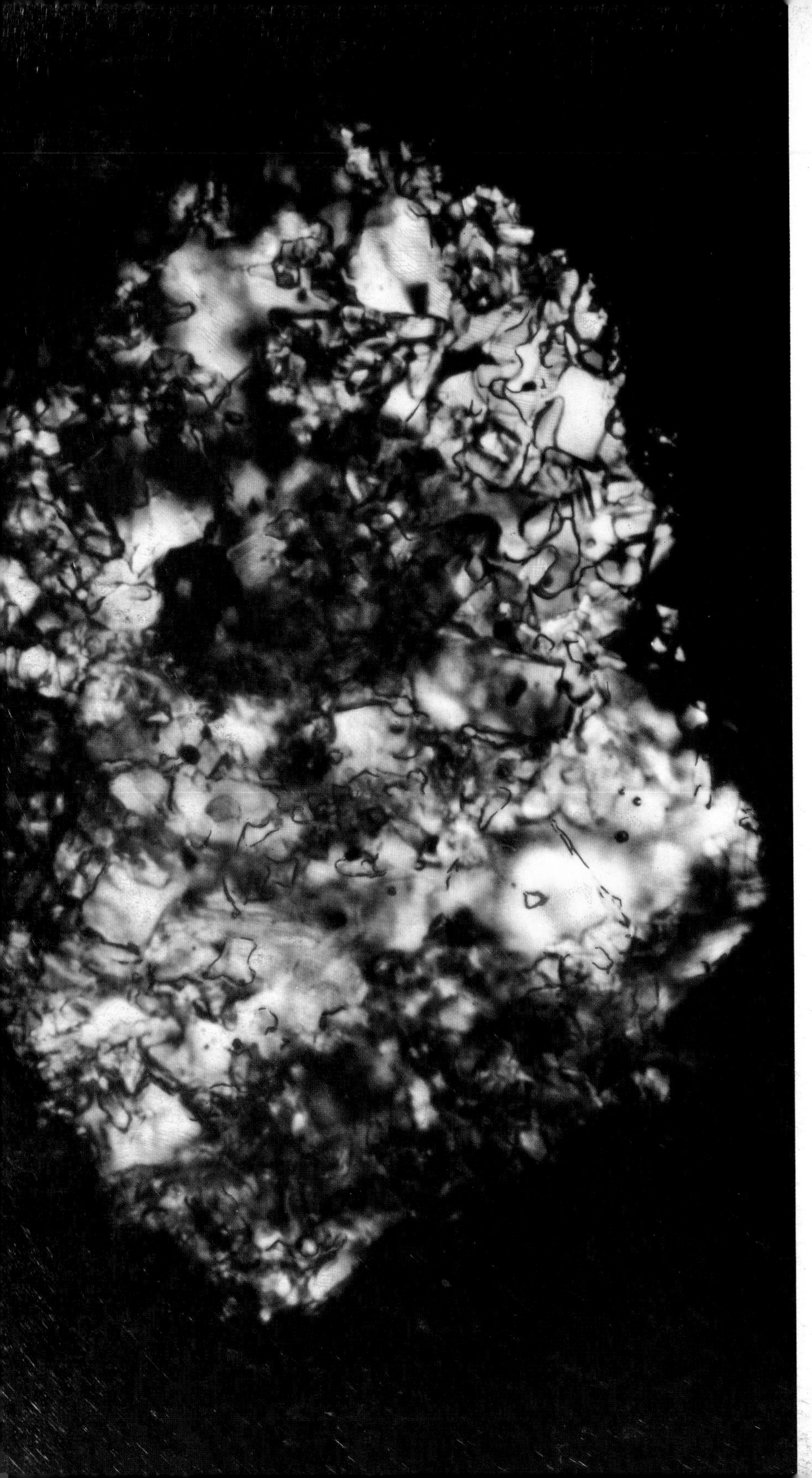

PLATE 11 (left) Photomicrograph in polarized light of a rock fragment of anorthositic gabbro. Magnification 300. (NASA photograph)

PLATE 12 (right) This beautiful and remarkable solar eclipse was photographed by the Apollo 12 astronauts. The black disk of the earth is in front of the sun just before totality, producing the striking "diamond ring" effect. (NASA photograph)

PLATE 13 Surface of Mars as viewed by the Viking lander, showing that Mars is truly a red planet. (NASA photograph)

FIG. 4.6
The 200-inch Hale telescope as seen from the east, is pointing to zenith. (Courtesy of the Hale Observatories.)

technological wonders of our century. Its objective is a paraboloidal mirror, 200 inches in diameter, 24 inches in thickness at the edge, and about 29,000 pounds (13,000 kilograms) in weight. (The mirror is large enough to park two automobiles on it side by side.) The construction of this giant telescope, which was conceived in 1928 by the American astronomer George Ellery Hale, required many years of painstaking work. The major problem was in casting the disk; the technical problems that had to be overcome were staggering. Two years and over half a million dollars were lost at the outset in an attempt to cast the disk out of quartz. After several setbacks, a perfect disk using Pyrex was cast in 1934. Seven hours were required to cast, ten months to cool, and eleven years to polish the disk. The Hale telescope was completed and placed in operation in 1949 (Fig. 4.6).

This telescope was the first one to use a diameter large enough to permit a cage to be placed at the prime focus for the observer and auxiliary equipment. The amount of the incoming light intercepted by the cage is no more than that intercepted by the secondary mirror of the Newtonian focus. The cage is six feet in diameter and

FIG. 4.7
This view of the 200-inch Hale telescope shows the observer in the prime-focus case and the reflecting surface of the 200-inch mirror. (Courtesy of the Hale Observatories.)

is divided into two sections—the upper section provides working area for the observer, while the lower section houses the photoelectric equipment, spectrograph for low dispersions, and the secondary mirrors for the Coudé and the Cassegrainian focus systems. Most of the photographic work is done at the primary focus (Fig. 4.7).

4.9 THE 236-INCH SOVIET TELESCOPE

The world's largest optical telescope, the 236-inch Soviet reflector located near the village of Zelenchukskaya in the Caucasus mountains northwest of Tiflis, has recently begun operation. The mirror, which is nearly 20 feet in diameter, is housed in an 82-foot tube. The prime-focus cage is located at the upper end of the tube, and two observation stations are on each side of the tube's horizontal axis.

Because of its tremendous size and weight—the telescope weighs 850 metric tons and the moving parts weigh

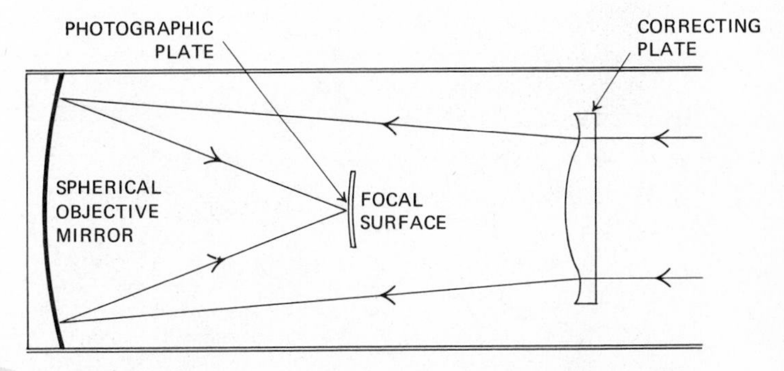

FIG. 4.8
The optical elements of the 48-inch Schmidt telescope at Mount Palomar Observatory.

800 metric tons—its mounting is *altazimuth*, which permits the telescope to be rotated easily about a vertical axis for altitude and about a horizontal axis for azimuth. This mounting has three advantages: the horizontal axis always remains horizontal for all observations; the loading conditions remain the same for all azimuth positions; and the mirror-support system is simpler than the equatorially mounted telescope (see below). It has two disadvantages: in tracking a star, the azimuth and altitude speeds are nonuniform, and the field of view must be adjusted for rotation. To compensate for these disadvantages, a digital computer is used. The design and construction of this telescope are major accomplishments for the Soviet Union, and when the telescope becomes fully operational it will be one of the world's great astronomical research tools.

4.10 THE SCHMIDT TELESCOPE

In 1932 Bernhard Schmidt, an optician at the Hamburg Observatory, invented a lens-mirror system that permitted the use of a concave spherical mirror as the objective of a telescope. He accomplished this by placing at the center of curvature of the objective a thin aspherical lens, which provided the necessary correction for the spherical aberration of the objective (Fig. 4.8). Even though the correcting lens was difficult and expensive to produce, its use was justified because it permitted the photographing of a large field of view that was relatively free from aberrations. The placement of the photographic plate along a curved surface produces an image that is nearly perfect in its entire field of view. With such features, the Schmidt telescope is used almost exclusively for photographing large areas of the sky. The size of the Schmidt telescope is determined by the diameter of the correcting lens, which is two-thirds of the objective. The large Schmidt telescope at Palomar has a 48-inch correcting lens and a 72-inch objective (Fig. 4.9).

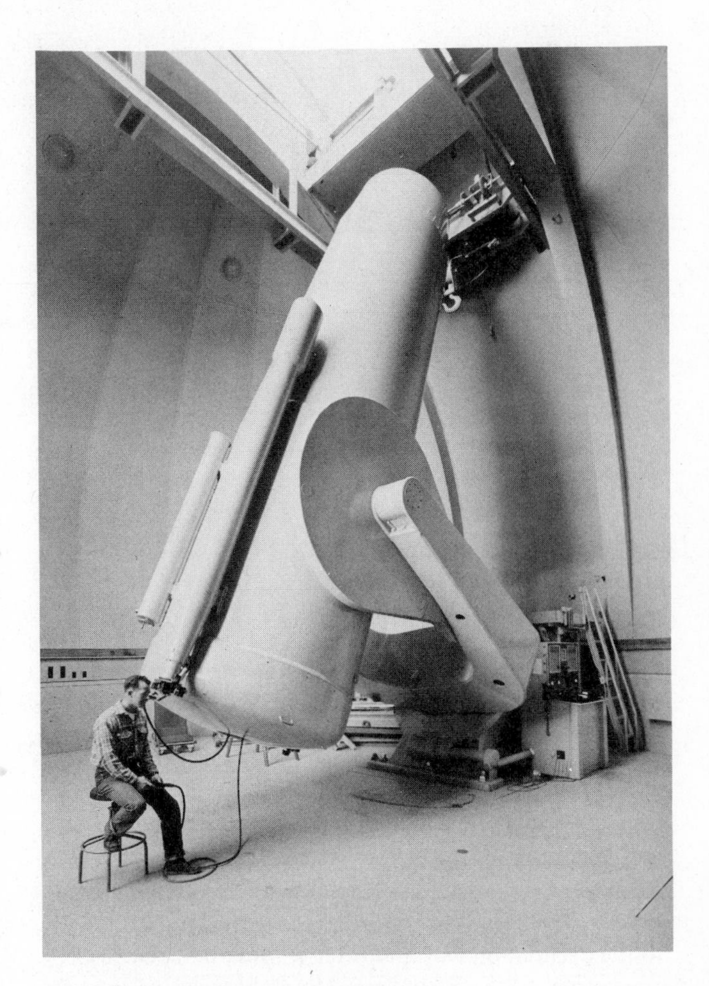

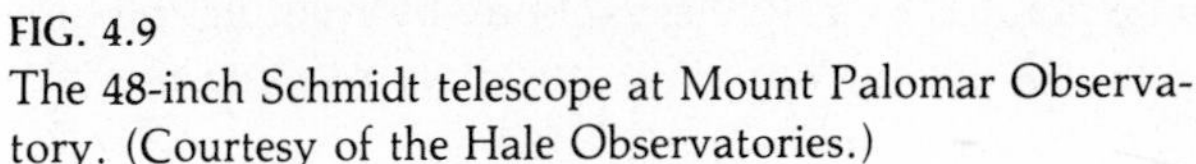

FIG. 4.9
The 48-inch Schmidt telescope at Mount Palomar Observatory. (Courtesy of the Hale Observatories.)

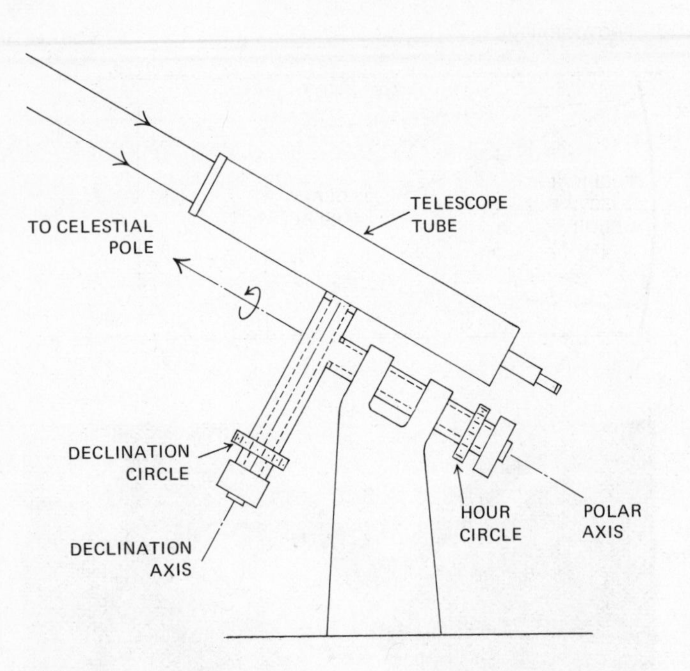

FIG. 4.10
An equatorial telescope mounting. The polar axis supports the declination axis and the hour circle. The declination axis supports the telescope tube and the declination circle.

4.11 TELESCOPE MOUNTINGS

Most telescopes are supported by an *equatorial mount*, which rotates the instrument about its *polar axis*, which is parallel to the earth's axis (Fig. 4.10), thus permitting the telescope to follow a celestial body in its daily motion across the sky. The polar axis is inclined to the horizontal by an angle equal to the latitude of the observer and is directed toward the north celestial pole near Polaris. Attached and at right angles to the polar axis is the *declination axis*. The telescope tube is mounted at right angles to the declination axis and rotates about it. The telescope can be pointed to any celestial body by rotating the tube around the declination axis. By simply rotating the telescope about its polar axis at the proper uniform rate, the telescope will follow the body's daily motion.

4.12 THE MULTI-MIRROR TELESCOPE (MMT)

When any large optical telescope, either reflector or refractor, is pointed to different positions in the sky, its mirror or lens flexes slightly. These distortions, while tiny by normal standards, are sufficient to seriously degrade the quality of the image. One of the advantages that reflecting telescopes have is that the mirror can be supported over its entire back surface, which helps to limit the flexing. A lens, by contrast, can only be supported at its edges, otherwise light cannot pass through it. Even mirrors distort slightly as they are tipped to receive light from different parts of the sky.

One technique that is now being developed to cope with these distortions involves the use of a number of small mirrors that can be individually pointed and aligned with a computer. The multi-mirror telescope (MMT) uses six small (72-inch) mirrors instead of one large piece of glass. The smaller mirrors are also easier to fabricate than a single large one, but still have a collecting area and resolution comparable to that of a single mirror 175 inches in diameter. Since they are mounted independently, the distortions produced by tilting can be partially corrected by making minute changes in the orientation of each mirror separately. (Correcting the shape of a mirror has been carried even further with a new telescope, designed to offset the blurring produced by the atmosphere. Its "mirror" is made of rubber, which can be poked and tweaked by a computer controlled device to balance out distortion in images produced by air turbulence.) The MMT was built jointly by the University of Arizona and the Center for Astrophysics in Cambridge, Massachusetts, and is located on Mt. Hopkins, about 50 miles south of Tucson, Arizona (Fig. 4.11).

4.13 THE MAGNIFICATION OF THE TELESCOPE

Magnification is defined as the number of times the diameter of a body is increased when viewed through a telescope. The maximum useful magnification depends

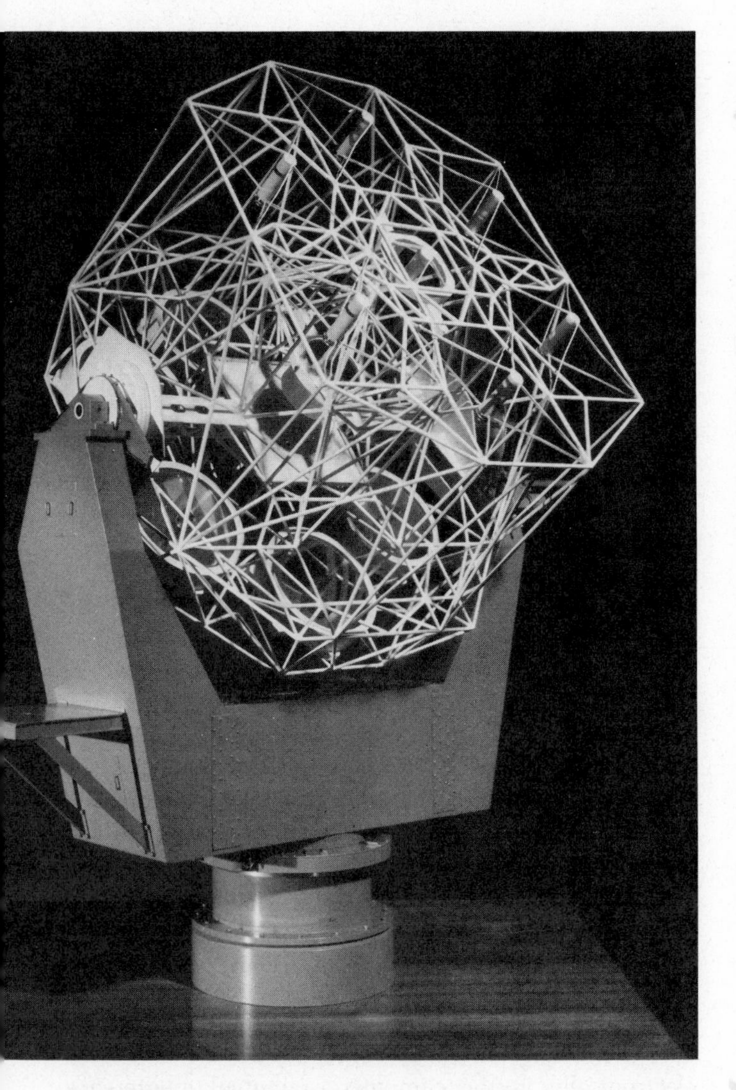

FIG. 4.11
Multi-mirror telescope. (Courtesy of Smithsonian Astrophysical Observatory.)

on the brightness of the body and the steadiness of the atmospheric conditions. An increase in the brightness permits an increase in the magnification. An increase in the unsteadiness of the atmospheric air does not permit an increase in the magnification, because when the body is magnified, the air disturbance will also be magnified, producing a blurred image.

The magnifying power of a telescope is determined by the ratio of the focal length of the objective to the focal length of the eyepiece. It is expressed by the relationship

$$M = F/f,$$

where M is the magnification, F is the focal length of the objective, and f is the focal length of the eyepiece. If the focal length of the objective is 10 feet and the focal length of the eyepiece is ½ inch, the magnifying power of the telescope is

$$M = \frac{10\,(12)}{1/2} = 240.$$

Since the objective and its focal length are fixed in a telescope, the magnification of the image can be changed only by using eyepieces of different focal lengths. However, the magnifying power of a telescope cannot be increased effectively beyond a certain limit even under the best observational conditions. This limit is about 50 times per inch of the objective's diameter. An 8-inch telescope, which is a typical size for student use, has a maximum effective magnification of about 400.

4.14 RADIO TELESCOPES

Although celestial bodies emit radiation in all wavelengths, much of the radiation does not reach the earth's surface because it is absorbed by the molecules in the atmosphere. Prior to 1931, the optical telescope limited our view of the universe to the visible wavelengths of the electromagnetic spectrum; however, in 1931, Karl Jansky at Bell Telephone Laboratories accidently discovered radio waves coming from the region of the Milky Way. He realized that they were being emitted by celestial

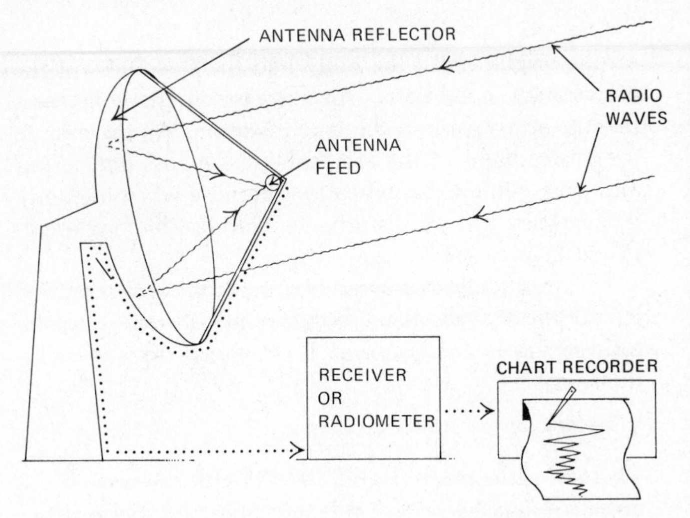

FIG. 4.12
The elements of a radio telescope.

bodies because their sources moved westward at the same daily rate as the stars. This discovery opened up the new branch of astronomy called *radio astronomy.* The construction of the radio telescope greatly extended our astronomical horizons. It opened up a new "window" in the electromagnetic spectrum by permitting the use of the radio region for the observations of stars and galaxies. Most of the radio observations are made in the range from 1-millimeter to 15-meter wavelengths, wavelengths for which the atmosphere is nearly transparent.

The principle of the radio telescope is based on several properties of radio waves. They can be refracted and reflected in the same manner as light waves; and they can be amplified by electronic techniques. In addition, except for the shortest radio waves, they are not influenced by the earth's atmosphere. The essential features of a radio telescope are the antenna and the receiver (radiometer). In Fig. 4.12, the antenna collects and concentrates the radio energy at its focus, where it is picked up by a "feed" that conveys it to the radiometer, where the energy is measured and its value recorded. The radiometer is a specially constructed receiver that eliminates the noise of a regular receiver by using special circuitry. This is necessary because the internal noise of a regular receiver is far greater than the signal created by the radio energy.

The resolving power of a radio telescope is its ability to separate the angular distance between two radio sources. This ability varies inversely with the wavelength of the radio energy and directly with the diameter of the antenna dish, and is similar to the resolving power of an optical telescope. Since the wavelengths of radio radiation are considerably greater than the wavelengths of visible light, the resolving power of a paraboloidal radio telescope is much less than that of an optical telescope of the same diameter, that is, the radio telescope is less effective in distinguishing details.

To offset the moderate angular resolving power of a single radio telescope, the *radio interferometer* was developed. It is simply two or more antennas connected to a single receiver. The antennas may be placed several miles apart, as are the two California Institute of Technology 90-foot steerable reflectors at Owens Valley, California, which establishes a long-baseline interferometer, or several thousand miles apart, using antennas in different parts of the world. In the latter case, the two observatories observe the same region of the sky simultaneously. The signal is recorded on tape along with an accurately timed series of pulses provided by an atomic clock. Afterwards the two tapes are brought to one computer and played back, using the timing pulse to synchronize them. When the taped signals are combined electronically in the computer, a mutual interference pattern results, which can be mathematically analyzed to resolve the radio source. Experiments with very long-baseline interferometers (VLBI) have achieved resolving powers better than those obtainable with optical telescopes, whose resolution is limited by atmospheric effects.

Because radio telescopes must be large in order to have a high resolving power, restrictions arise on the physical structure of the antenna. It is difficult to fabricate a structure that is both large (for maximum sensi-

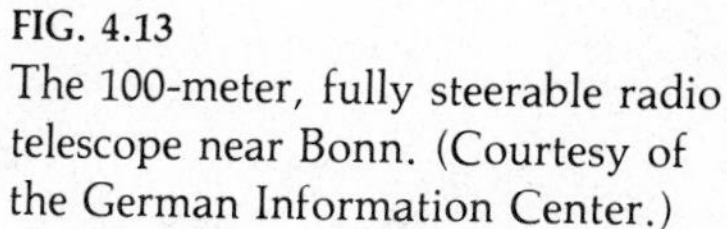

FIG. 4.13
The 100-meter, fully steerable radio telescope near Bonn. (Courtesy of the German Information Center.)

tivity) and maneuverable so that it can be pointed at a given region of the sky for tracking a celestial object. An excellent example of a steerable antenna is the 250-foot paraboloidal reflector at the Jodrell Bank Experimental Station at the University of Manchester, England, By being mounted on a double circular rail track, 350 feet in diameter, it can be easily pointed to any position in the sky. The instrument, designed for both radio and radar research, uses a wide range of frequencies and band widths. The world's largest fully steerable radio telescope is the new Bonn 328-foot at the Max Planck Institute for Radio Astronomy, Bonn, West Germany (Fig. 4.13). It is altazimuth-mounted on a circular track that is only 210 feet in diameter. The reflector dish has a very short focal length of about 100 feet and will operate down to a wavelength of 1.2 centimeters.

In some radio telescopes the antenna is fixed and the rotation of the earth is used to direct the antenna to different areas of the sky. This procedure, however, limits the time an object can be observed. Some antennas have a fixed reflector with a movable "feed" (the location where the signal is picked up and sent to a receiver). This allows a small amount of tracking and eliminates the necessity of tipping and steering a huge metal reflector. The Arecibo antenna (Fig. 4.14) is the most famous example of this type of telescope. Located near the port of Arecibo, Puerto Rico, and operated by Cornell University, the Arecibo radio telescope fills a natural bowl that is over 1000 feet (300 meters) in diameter. Its feed antenna is located 435 feet (130 meters) above the ground. The reflector dish was originally made with chicken wire, but recently it has been resurfaced to allow operation at shorter wavelengths.

Some antennas are composed of numerous small reflectors, each of which can be independently pointed at a radio source. The signals are then electronically added together. This technique is called *aperture synthesis* (that is, a single large aperture is created by adding the contri-

FIG. 4.14
The great radio-radar telescope at the National Astronomy and Ionosphere Center in Arecibo, Puerto Rico. The feed system is suspended at the focus by cables from the three towers. The National Astronomy and Ionosphere Center is a national research center operated by Cornell University under contract with the National Science Foundation. Courtesy of NAIC/NSF.

butions from many small ones). The two most noted examples of this type of antenna array are the five-mile Cambridge Interferometer in England and the aptly named VLA (very large array) located in New Mexico. When completed the VLA will extend along a Y-shaped set of railroad tracks for a distance of about 15 miles (21 km) along each arm of the Y (Fig. 4.15). It will have an effective collecting area about 17 miles in diameter, but, because it will consist of a number of fully pointable smaller antennas, it too can be directed to track any source in the sky. With the VLA it should be possible to make maps of radio sources that show detail on a scale better than that obtainable with optical telescopes.

4.15 LIGHT DETECTORS

The refracting and reflecting telescopes described in earlier sections all serve to collect light and bring it to a focus. When most people think of an astronomer at work they picture someone with one eyeball pressed to a telescope. In fact, while many important discoveries have been made with simple direct visual observations, nearly all astronomical work today is done with detectors that are specially designed to analyze the light and make a permanent record of the data accumulated. Probably the simplest of these detectors is the *photographic plate.* Plates are special films mounted on a glass sheet. The glass prevents the photosensitive surface (emulsion) from curling and distorting with age. Photographic plates are used directly or in conjunction with image tubes or image-intensifying systems to record not only direct photographs of galaxies (or whatever), but also to record spectra.

Photographic plates have the limitation that more than a single photon must fall on the film at each point to produce an image. For faint objects, to obtain a photograph may require exposing the film for many hours. Recently considerable ingenuity has gone into developing intensification systems that can reduce the time required to obtain a photograph from hours to minutes and that allow the recording of much fainter images than was previously possible. Image tubes take advantage of the photoelectric effect, whereby an incoming photon strikes

FIG. 4.15
The VLA. Located near Socorro, New Mexico, it extends across a nearly flat plain. The individual antennas move on railroad tracks. (Courtesy of the National Radio Astronomy Observatory under contract with the National Science Foundation.)

a special surface and ejects an electron. The electron is caught in an electric field, which accelerates it down the tube toward a phosphor surface at the other end (similar to the screen of a television set). Having gained energy from the electric field, the electron causes the phosphor to emit more light than was originally responsible for creating the electron. The electrons can be focused on the screen, thereby creating a picture far brighter than the original light falling on the tube.

4.16 THE STELLAR INTERFEROMETER

Scientists have recently developed new and exciting auxiliary instruments such as the image tube, which increases the effective light-gathering power of a telescope, and the laser beam, which permits the distance to the moon to be measured with great precision. Here we shall discuss a few basic auxiliaries to show how a star's angular diameter and brightness can be measured and how light emitted by bodies can be observed and analyzed.

The stellar interferometer, which was invented by the American physicist A. A. Michelson, is used to measure the angular diameter of stars. Its operation is based on the property of light called *interference*. As Fig. 4.16 illustrates, an amplification in the wave is produced when the crest of one wave meets the crest of another wave or when the trough of one wave meets the trough of another wave. When the crest of one wave meets the trough of another wave, they cancel each other. This property is also illustrated in Fig. 4.16. When parallel light passes through two slits, the light which emerges from each slit moves outward in spherical waves as though the light source were at each slit. Along lines 1, 2, and 3, the crests of the two waves meet, producing *intensification*, which is visible as bright regions; between

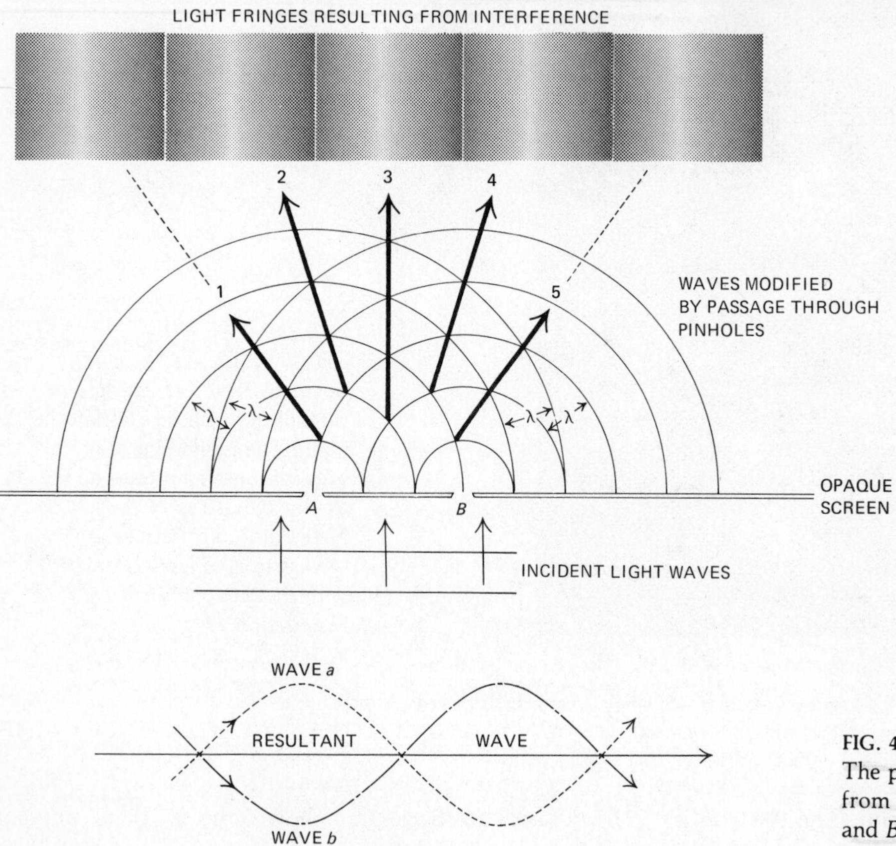

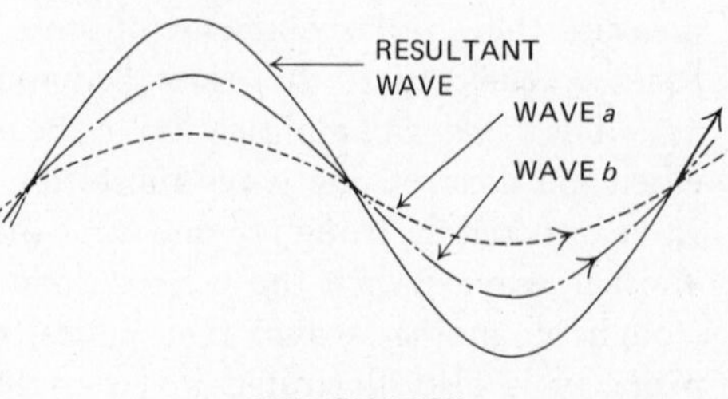

FIG. 4.16
The principle of the interferometer. Light from a single source passes through slits (*A* and *B*) in an opaque screen. Amplification results along lines 1, 2, 3, 4, and 5 when the crests of two waves meet and produce bright fringes. Cancellation results between these lines when the troughs of the two waves meet and produce dark fringes.

these lines, the troughs and the crests meet and produce *cancellation,* which is visible as dark regions.

Basically, the interferometer consists of a steel beam with a movable mirror at each end. As shown in Fig. 4.17, the light from a single star falls on the surface of the two movable mirrors and is directed to an eyepiece. The image appears as a fringe pattern of alternating bright and dark parallel bands. To obtain the angular diameter of the star, the two movable mirrors are moved until the fringe pattern disappears. The distance between the mirrors is related to the angular diameter of the star.

The classical methods of stellar interferometry have been further extended to two new techniques: *Hanbury-Brown & Twiss* and *Speckle* interferometry. The Hanbury-Brown & Twiss interferometer consists of two large mirrors mounted on a circular track. They can be pointed at a star and the light beams are then recorded and combined electronically to produce interference patterns. The spacing between the mirrors is typically 500 feet. With this interferometer it has been possible to measure the angular diameters of about fifty stars.

Speckle interferometry is a very new technique, made possible only with the advent of image-intensifying systems. When you look at a star with a normal telescope, what you see is a large blurry image composed of the so-called diffraction-limited image (set by the size of the telescope) made fuzzy by the many turbulent elements in the atmosphere (Fig. 4.18). Under very high magnification and using very short exposures, it is possible to see the many small separate diffraction-limited images of the telescope. Once these are recorded on film they can then be added together, either with a special laser device or using a mathematical technique, to produce a crude picture of the star's surface. With Speckle interferometry it has been possible to see blotches on the surface of some giant stars and also to detect double stars previously unresolvable as separate objects.

4.17 THE PHOTOELECTRIC PHOTOMETER

The photoelectric photometer is used to measure the brightness of a body to an accuracy of several thousandths of a magnitude. Its operation is based on the

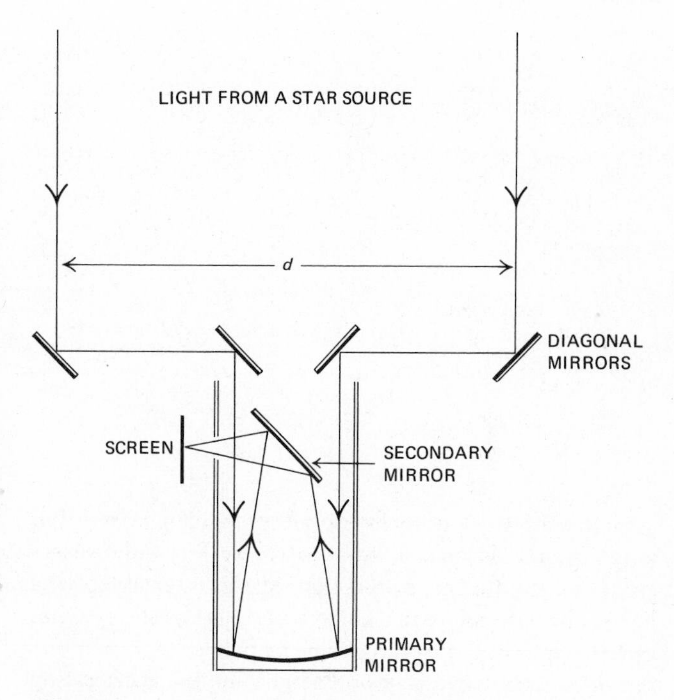

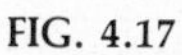
FIG. 4.17
The stellar interferometer. The four small mirrors are mounted on a steel beam; the inside ones are fixed, and outside ones are movable.

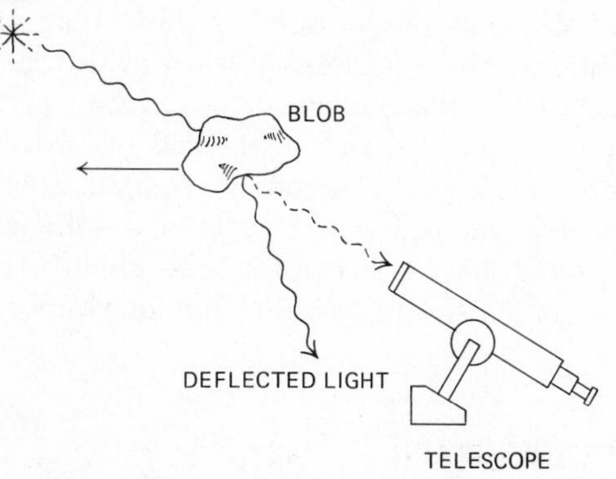

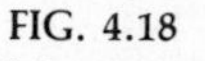
FIG. 4.18
Moving atmospheric irregularities shift image of star by deflecting light passing through them.

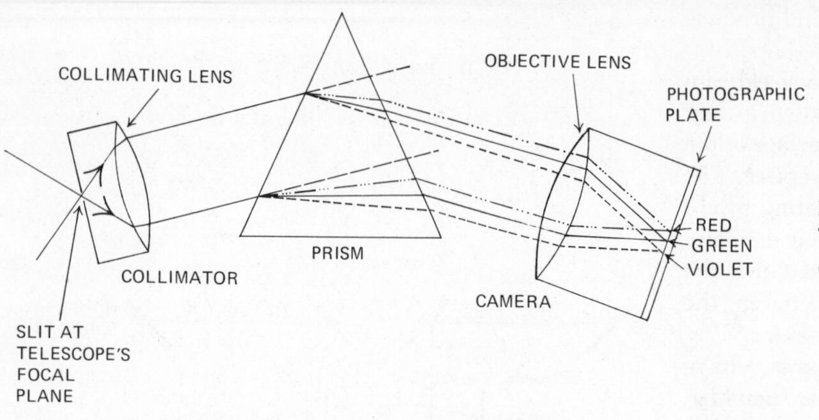

FIG. 4.19
The optical elements and principles of a spectrograph.

principle of the photoelectric effect, which states that when light falls on a light-sensitive surface, photoelectrons are emitted in proportion to the intensity of the light. The light meter in a camera and the electric eye that opens a door operate on the same principle.

When the frequency of the light and the accelerating potential, i.e., the voltage between the emitting and collecting surfaces, are both held constant, the photoelectric current is a linear function of the intensity of the light. The flow of current is measured by a microammeter and is a measure of the body's brightness.

When the intensity of the light is extremely weak or the required measurement is most sensitive, a photoelectric cell (photomultiplier tube), in which there are several amplification stages, is used. When light from a celestial body falls on the photosensitive surface, one or more electrons are emitted and accelerated toward the second surface, where more electrons are emitted. This is repeated on each successive surface until a sufficient number of electrons have been emitted to be amplified by an electronic amplifier and counted by an electronic counter.

4.18 THE SPECTROGRAPH

The spectrograph, one of the most important instruments of astronomical research, permits the observation and analysis of the light emitted by a celestial body. Its basic function is to identify the wavelengths, intensities, and characteristics of the light emitted. From this information, the astronomer can derive the physical characteristics and behavior of the celestial body.

The essential features of the spectrograph are a narrow slit, collimator, prism, and camera (Fig. 4.19). The light rays from a celestial body converge at the focal point of the instrument where the narrow slit is located. They pass through the slit, diverge, then pass through the collimator lens, and emerge in parallel rays. When the rays pass through the prism, each wavelength is deviated by a different angle to produce the different colors in the spectrum. When the spectrum is photographed, the instrument is called a *spectrograph*.

A diffraction grating can be used in place of the prism to produce the spectrum. It consists of a flat surface that is ruled with thousands of very fine, parallel and equidistant lines, which disperse the light beam into the spectrum by diffraction. In the transmission grating the surface is transparent and the light is refracted, while in the reflection grating the surface is opaque and the light is reflected. The lines in both gratings can be either ruled in the surface or formed by alternating clear and opaque strips.

Both the prism and the diffraction grating are widely used in spectroscopic work; however, they differ considerably in the spectra that they produce. In a prism, the red wavelengths are deviated the least and the violet the most, whereas the opposite is true in a diffraction grating. This means that the spectrum produced by the grating is the reverse of that produced by the prism. The dispersion of the light beam is greater toward the violet end of the spectrum produced by a prism, while the grating disperses the wavelengths more uniformly. Also, the angle of deviation of any wavelength can be accurately determined from the line spacing of a grating and can be approximated from the type of material used in the construction of a prism.

A spectrograph is designed so that the spectrum of a known source (comparison spectrum) can be placed on both sides of the spectrum of the celestial body. For high-dispersion spectra in the ultraviolet region, an iron-arc comparison is used; in the red, yellow, and infrared regions, a neon-discharge tube is used. For low dispersions, mercury and helium-discharge tubes are used. The light from the arc or discharge tube is transmitted through the slit of the spectrograph by right-angle quartz prisms located in front of the slit. The spectral lines in the arc and discharge tube spectra have been measured very carefully and are used to identify and analyze the spectral lines in the spectrum of the celestial body.

4.19 LIMITATIONS

Astronomers, armed with an array of elaborate and precise observing and recording instruments that are capable of obtaining an impressive amount of astronomical data, nevertheless cannot escape the fact that both they and their instruments are hampered by inherent limitations. As a human being, the astronomer is limited by sense inadequacies, which may lead to errors. The light rays from an object produce different effects on individuals because of eye defects, perspective, personal emotions, and bias. The astronomer's ability to observe and record what actually occurs in the universe is made more difficult by natural-light distortions produced by atmospheric disturbances, by night-sky radiation, and by other sources. And, although the instruments used have extended our ability to see farther out in space with greater clarity and definition, they also have their own built-in errors and limitations.

SUMMARY

The optical telescope remains one of the major tools for gathering astronomical information. The basic purpose of the telescope is to gather light and bring it to a focus so that an image may be formed. The image may then be photographed or analyzed in other ways (spread into a spectrum, have its intensity measured, etc.). The main light-collecting and image-forming part of the telescope is the objective. Either a lens or a curved mirror will serve this purpose. Telescopes using lenses for the objective are called refractors, while those using mirrors are called reflectors. In general, reflectors are used in today's telescopes because mirrors are easier to make, do not introduce color distortion (chromatic aberration), and are easier to mount in the telescope.

A wide variety of auxiliary equipment can be attached to a telescope to analyze and record the light. Photometers measure the brightness of a source. Spectrographs disperse the light into a spectrum and record it. Image intensifiers make images brighter and thus help reduce the time necessary to obtain the information.

Astronomers use many other regions of the electromagnetic spectrum in addition to that of visible light. Radio telescopes collect radio waves and focus them at a feed antenna which is in turn attached to a receiver. The radio signal strength from different locations in the sky can then be measured and a "map" of the radio source constructed. This may then be compared with a photograph of the object to see where the radio emitting regions lie (at the center? on either side? etc.).

Other regions of the spectrum, particularly the x-ray and infrared regions, may be studied by means of special-purpose telescopes to yield x-ray and infrared maps of astronomical objects, analogous to the radio maps described above.

REVIEW QUESTIONS

1. When was the telescope invented? What purpose do the objective and the eyepiece of a telescope serve?
2. What determines the size of the image formed by a telescope? How can the size of the image be increased?

3. What is meant by the light-gathering power of a telescope? How does the light-gathering power of a 12-inch compare with that of a 6-inch telescope?
4. What is meant by resolving power? A general formula for resolving power, d, in seconds of arc of a telescope with a diameter D operating at a wavelength λ is $d = 2{,}000{,}000\,\lambda/D$. The wavelength and telescope diameter must be given in the same units, such as inches or centimeters. For visible light, $\lambda = 0.0002$ inches (approximately), which then gives the formula in Chapter 4.3. Given this information, what is the solving power (a) of a 6 inch telescope? b) of a 10-meter diameter radio telescope operating at a wavelength of 1 millimeter? c) For what size radio telescope would the resolving power at 10 centimeters be equal to that of a 6-inch optical telescope with visible light?
5. What is meant by spherical aberration? Is it present in both a lens and a mirror? Explain how this defect can be corrected in a mirror.
6. What is chromatic aberration? Explain why it is not present in a mirror. How can this defect be corrected in a lens?
7. What are the essential features of a refracting telescope? Explain the purpose of each.
8. An image may be viewed from several different positions in a large reflecting telescope. Where is each position located and what is its optical system called?
9. List the advantages and disadvantages of both reflector and refractor telescopes.
10. Describe the Coudé-focus type of reflecting telescope. For what type of astronomical work is this focus system used?
11. Describe the principle of the Schmidt telescope. Why is it suitable for photographic surveys of large areas in the sky?
12. When an eyepiece of focal length ⅛ inch is used in a 10-inch telescope of focal length 60 inches, what is the telescope's magnifying power? How can the magnifying power of a telescope be increased? What is its practical limit of magnification?
13. Describe the essential elements of a radio telescope and explain their function.
14. What is meant by an "equatorial mount"? Give its important advantages. Why are equatorial mounts not generally used for a radio telescope?
15. Compare (in terms of advantages and disadvantages) the radio telescope and the optical telescope.
16. List the advantages of photographic over visual observation.
17. What is an interferometer? Why do astronomers use it? What special advantage do interferometers offer to radio astronomers?
18. What is an image intensifier? How does it differ from a photomultiplier?
19. What is the basic function of a spectrograph? What are its essential features? What is a diffraction grating? Discuss the differences in the spectra produced by a prism and a diffraction grating.

CHAPTER 5
THE EARTH: A PHYSICAL BODY

To most of us, the earth means solid ground beneath our feet. We see no evidence that the earth is really a ball of rock spinning at roughly 1000 miles per hour at the equator and hurtling through space around the sun at more than 60,000 miles per hour. It is perhaps just as well, considering what happens to many of us on a roller coaster or loop-the-loop.

There are many reasons for studying the earth. As an astronomical object, the earth shares many general properties with the other planets. For example, it is shaped by gravity, it rotates, it has an atmosphere, etc. Thus, by understanding our home planet, we can better understand other planets. Furthermore, we will see that despite similarities to some of the other planets, our world, which we take so much for granted, is a very special place. Of all the members of the solar system, the earth alone possesses an environment in which we can supply our needs and live without elaborate devices protecting us.

Let us begin by taking ourselves to Mars, which provides an excellent observation platform for viewing the earth. From this vantage point, we can observe that the shape of the earth is nearly spherical. By noting the motion of its permanent surface features, we learn that it rotates about its axis approximately once every 24 hours. By noting the change in its position from leading to following and back to leading the sun, we also learn that it revolves around the sun approximately once every 365 days. These observations from Mars are simple and direct; however, from the earth they must be determined by more subtle methods.

5.1 THE SHAPE OF THE EARTH

The ancient Greeks believed that the earth was round, and they presented several empirical arguments to prove their belief. Anaximander observed that the constellation of Ursa Major the Great Bear always remained above the horizon in Greece, while in Egypt a portion of the constellation appeared to go below the horizon each evening. Since this would have been impossible to observe on a flat surface, he concluded that the earth was spherical.

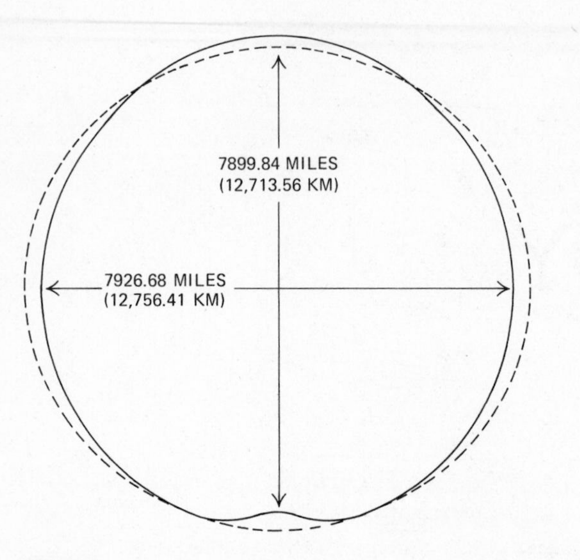

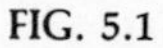
FIG. 5.1
The shape of the earth. The equatorial diameter is 7926.68 miles, and the polar diameter is 7899.98 miles. (The drawing is greatly exaggerated.)

Herodotus, who noted the apparent shifting of the stars as an observer travels north or south on the earth's surface, argued that the earth was spherical. He observed that the north pole star always remained at the same height above the horizon when seen from a ship in the Mediterranean, which is an east-west sea. However, when the ship left the Mediterranean and headed north for the British Isles, the north pole star appeared to rise to a higher point above the horizon. He concluded that this could be observed only on a spherical earth.

Aristotle, observing that the earth's shadow always appeared curved on the face of the moon during a lunar eclipse, inferred that the earth was spherical. Today we have additional evidence—the high-altitude photographs of the earth taken from spacecraft clearly show the earth to be nearly spherical.

5.2 THE EARTH AS AN OBLATE SPHEROID

The earth, however, is not a perfect sphere. In 1745 two French scientific expeditions, both sponsored by the French Academy of Sciences, produced evidence that the shape of the earth is an oblate spheroid, that is, a sphere whose equatorial diameter is slightly longer than its polar diameter (Fig. 5.1). The expeditions measured three equivalent arcs of latitude (a portion of a great circle that passes through the north and south poles): one in Peru (which is near the equator), another in Paris, and a third in Lapland (which is near the north pole). Since the measurements were all different, they concluded that the length of an arc of latitude changes from the equator to the poles. The distance on the earth's surface that corresponds to one degree of latitude is greater near the poles than near the equator. This flattening of the earth's poles, which is called the *oblateness of the spheroid*, is defined as the ratio of the difference of the two diameters to the equatorial diameter and is expressed as 26.70/7926.68, or approximately 1/298.

A study of the motions of orbiting artificial satellites revealed that the earth is an oblate spheroid with a slightly indented south pole and a slightly extended north pole.

5.3 THE MASS OF THE EARTH

There are several methods by which the earth's mass can be determined. A simple and accurate method was developed by P. von Jolly, who used a large and precise analytical balance. This experiment is also important because, once we know the mass of the earth, we can find the value of G, the gravitational constant. Jolly's method is showin in Fig. 5.2, in which the gravitational attraction between the earth and a known mass is compared to the gravitational attraction between two other known masses.

When equal known masses M and M' are placed in the two pans, the analytical balance remains in equilibrium, that is, the balance remains in a horizontal position because the two masses are equidistant from the earth's center and both are attracted equally by the earth's gravitational force. In terms of Newton's universal law of gravitation, these two forces can be expressed mathematically as $(GMM_e)/R^2$ and $(GM'M_e)/R^2$, where G is the gravitational constant, M and M' are the two equal masses, M_e is the earth's mass, and R is the distance of each mass from the earth's center. When a large sphere of mass A is placed below the left pan at a distance d between the centers of the sphere and mass M, the balance will move downward to the left because of the unequal attraction between the sphere and the two equal masses. The attraction between the sphere and mass M is greater because the distance between their centers is small, whereas the attraction between the sphere and mass M' is negligible because the distance between their centers is large. Equilibrium is restored when mass B is placed in the right pan. Since the balance is in equilibrium, the forces that tend to move the balance downward to the left must equal those that tend to move it downward to the right. This equilibrium can be expressed by the mathematical relationship

$$\frac{GMM_e}{R^2} + \frac{GMA}{d^2} = \frac{GM'M_e}{R^2} + \frac{GBM_e}{R^2}.$$

Since

$$\frac{GMM_e}{R^2} = \frac{GM'M_e}{R^2},$$

the relationship is reduced to

$$\frac{GMA}{d^2} = \frac{GBM_e}{R^2}.$$

When this relationship is solved for the earth's mass, it becomes

$$M_e = \frac{MAR^2}{Bd^2}.$$

Since the values of all the quantities are known, when they are substituted in the relationship, the earth's mass is found to be approximately 5.98×10^{24} kg (6.6×10^{21} tons).

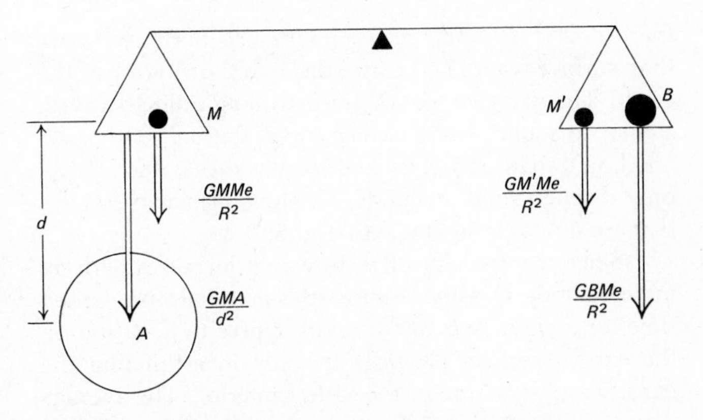

FIG. 5.2
The measurement of the earth's mass by Jolly's balance. With only M and M' (equal masses) on the scales, the earth pulls down on both sides equally. The presence of A under the left pan adds a weak extra gravitational force downward. To bring the scales back to balance, an additional mass, B, must be added to the right pan.

5.4 THE INTERIOR OF THE EARTH

Even though civilization has existed on the earth for several thousand years, our knowledge of the earth's structure and composition is still quite limited. We have been able to penetrate its surface to a depth of only about 5 miles (8 km), and the material recovered is similar to that at the earth's surface. Our knowledge of the earth's interior has been derived from the study of the movement of seismic waves produced by earthquakes or subterranean explosions.

The slowest waves are the *surface waves*, which travel like an ocean wave near the earth's surface and follow its curvature. The other two kinds of waves are the *primary* (*P*) and the *secondary* (*S*). Both are faster than surface waves and travel through the interior of the earth. The primary waves are fast, longitudinal waves, similar to sound waves, which travel through both solid and liquid materials. The secondary waves, which travel only through solid materials, are slow, transverse waves that are similar to light waves (Fig. 5.3).

Since the velocity of these waves increases with an increase in the density of the earth's material, the elapsed time for a wave to travel from its source to a station on the earth's surface provides the key in estimating the density and structure of the earth's interior. The average density of the earth is 5.5 grams per cubic centimeter, and the average density of the material at or near the earth's surface is 2.7 grams per cubic centimeter, a fact which indicates that the interior is considerably denser than the surface. (To give you a feeling for what these densities mean, water has a density of 1 gram per cubic centimeter. Ordinary rock and iron have densities of about 3 and 8, respectively.)

The solid part of the earth consists of three main layers of material: *crust, mantle,* and *core.* The crust is the outer, most familiar layer; the mantle is below the crust; and the core is the layer at the earth's center. The crust is not uniform in either thickness or composition. The thickness varies from about 3 miles (5 km) under the oceans bottoms to over 30 miles or (48 km) below the mountainous parts of continents. An analysis of the composition of the crust reveals that most of the elements are in scarce supply. The approximate proportions of the most abundant elements are: 46.6% oxygen, 27.7% silicon, 8.1% aluminum, 5.0% iron, 3.6% calcium, 2.8% sodium, 2.6% potassium, and 2.1% magnesium.

The mantle extends to a depth of about 1800 miles (2900 km) and consists of the basic silicates that are rich in magnesium and iron. Seismographic studies indicate that the core is divided into two parts, the outer and the inner. The *outer core* is a shell slightly over 1000 miles (1600 km) thick, while the *inner core* has a radius of slightly less than 1000 miles. Direct information about the composition of the core is not available; however, scientists believe that the outer core consists mainly of nickel-iron in the liquid state, because the secondary waves do not pass through this layer, and that the inner core consists of nickel-iron in the solid state.

From the study of the temperatures that have been recorded in deep mine shafts and wells, geologists have estimated that the temperature of the crust near the earth's surface increases at the rate of about 1°F for each 150 feet of depth; however, it is very unlikely that this rate continues to the earth's core. Since the heat in the inner core escapes through the outer core, the mantle, and the crust, and since the rate of conductivity of the earth's material is low, the inner core must cool at a relatively slow rate.

The source of heat is believed to be the radioactive decay of elements such as uranium, thorium, and radioactive potassium. Each time an atom decays it adds a tiny amount of heat to the interior. Over millions of years this has raised the temperature high enough to melt rock.

Radioactive material can be used to establish the earth's age. The decay of a given element takes place at a constant and uniform rate, which is expressed as the element's *half-life.* This term refers to the length of time it takes for half of its atoms to decay to a stable state. Uranium 238 has a half-life of 4.5 billion years, which means that in this period of time one-half of its atoms will decay. Since this is a continuing process, one-half of the remaining atoms will decay during each successive 4.5 billion year period. By comparing the present amounts of radioactive elements to the decay products found in the surface rocks, one can obtain a fairly accurate estimate of the earth's age. The age of the oldest earth rocks has been estimated to be between 3.7 and 3.9 billion years, and the age of the earth at about 4.6 billion years.

5.5 MOTIONS IN THE EARTH'S INTERIOR

We tend to think of the earth beneath our feet as solid and immovable. Occasionally, we are jolted by an earthquake, and the ground is shaken and rearranged locally before returning to its seemingly dormant state. However, if we could watch for millions of years, we would see that there is a steady movement of the crust that totally dwarfs the effects even of large quakes. This motion is called *continental drift,* or *plate tectonics.*

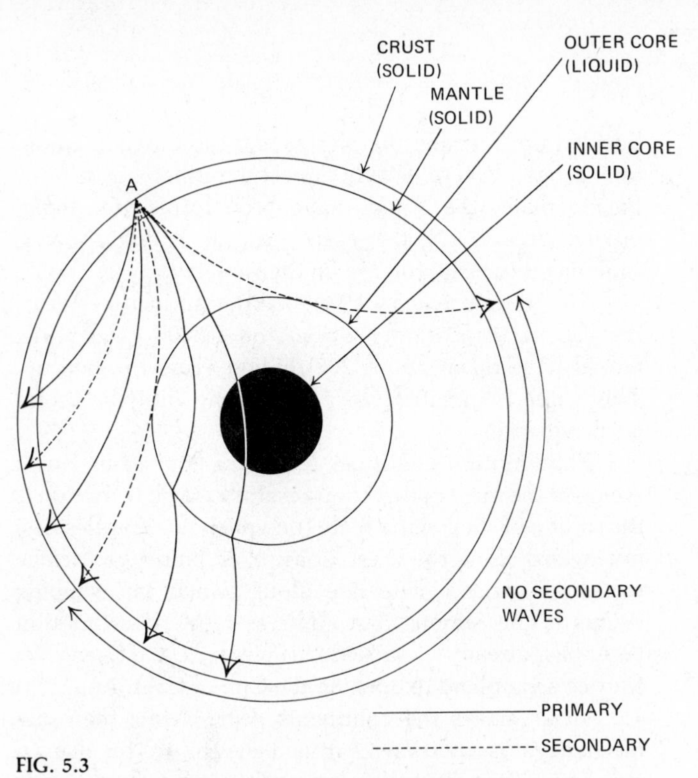

FIG. 5.3
The interior structure of the earth, and the paths of the primary and secondary waves through the earth.

FIG. 5.4
The earth, showing the coastlines of South America and Africa.

An examination of a map of the world (Fig. 5.4) shows that the eastern coast of South America matches extraordinarily well to the western coast of Africa. Similar matches of coastlines can be made for many other parts of the world. Not only do the coastline shapes match, but fossils, radioactive ages, and mineral deposits seem to match up as well. A species of fossil animal found in a small region of southern Africa is also found in a tiny region of South America at the corresponding point on the opposite side of the Atlantic Ocean.

The general agreement of coastline shapes of Africa and South America was noted by Alfred Wegener, a German meteorologist, in 1912. His suggestion that the two were originally one land mass, which had subsequently cracked into two pieces and drifted apart, was met with much debate. Despite the strong circumstantial evidence that Wegener amassed, most geologists remained unconvinced. Since that time the discovery of a ridge running along the middle of the Atlantic Ocean (with similar ridges found in many other areas of the world's oceans) and the discovery that the age of the rock became progressively greater at increasing distances from the ridge, as well as many other pieces of evidence, have led to the almost universal acceptance of the motion of crustal plates over the earth's surface.

There appear to be about a dozen active plates now (Fig. 5.5) and wherever one plate is found to be pushing against its neighbor, the crust of the earth shows signs of geological activity. Where North America nudges against the Pacific plate, the edge of the North American plate is

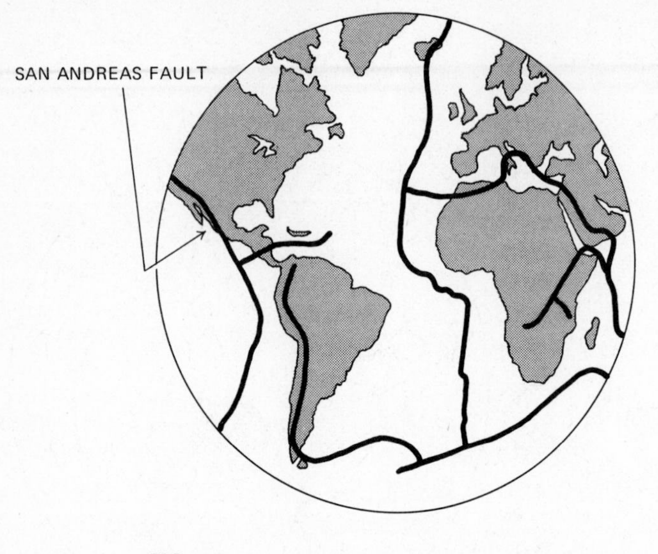

FIG. 5.5
Plate boundaries on the earth.

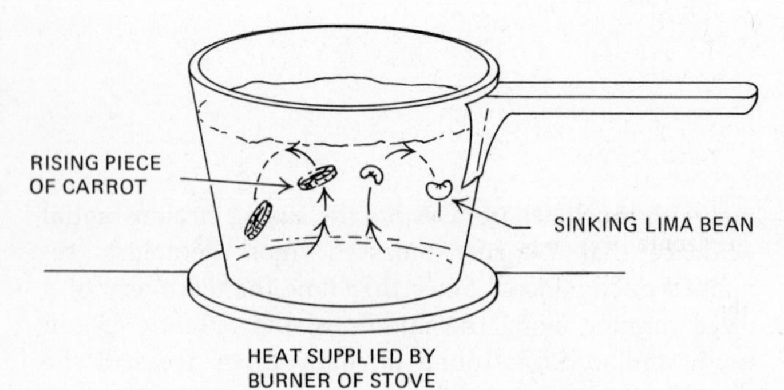

FIG. 5.6
Convective motion in a pan of vegetable soup.

buckled up, forming the Sierras and the Cascade mountain range. Where South America pushes against the Pacific plate, the Andes have been forced up. India, having broken off the east coast of Africa, now is ramming into Asia, forcing up the Himalayas.

The motion is very slow,—about an inch per year; the gap of the Atlantic Ocean, now 3000 miles wide, would have taken about 200 million years to open up. This value is in satisfactory agreement with the geological evidence.

Plate motion continues today, of course. In North America the most conspicuous example of plate motion is the sliding of the Pacific plate (on which Los Angeles lies) northward along the West Coast from Northern Mexico to San Francisco. The line along which this slipping occurs is the famous San Andreas fault. Plate motion here has already "cracked" a piece of continent off Mexico's mainland to open up the Gulf of California.

What makes the continents drift? While the exact mechanism is unknown, it is believed to be due to *convective motions* in the earth's interior. Convection is a process we will encounter numerous times in our study of astronomy. It is the motion of material as a result of heating and is a way in which heat is transported. A familar example is the rising and sinking motions of pieces of vegetable in a pan of boiling soup on a stove (Fig. 5.6). Any fluid or gaseous substance may experience convection if heated sufficiently from the bottom. While normally one does not think of the interior of the earth as behaving like a fluid, the mantle is sufficiently hot that under the influence of a large enough force, it will flow very slowly, a bit like extremely cold motor oil. It is thought that the heating that drives convective motions in the earth's interior is produced by radioactive decay. Matter heated near the bottom of the mantle rises, reaches the crust, and then spreads out to flow along just under the crust. The crust is dragged along by this motion (Fig. 5.7). There is still much to be learned about how plate motion occurs and if, in fact, the mechanism described above is the correct one. Since earthquakes are thought to occur as the crust slips, there are great practical reasons for the study of the earth's interior. It is thus partly in the hope of better understanding our home planet that so much interest is attached to study of the interiors of other planets (Chapter 8).

5.6 THE ATMOSPHERE OF THE EARTH

The earth's atmosphere is a complex, invisible, and vital shell which completely encircles the earth. It protects us from the deadly ultraviolet radiation of the sun and helps to maintain a fairly uniform temperature on the earth by means of the "greenhouse effect" (Chapter 5.8). Although the atmosphere is vital to the existence of life on the earth, it creates problems for the astronomer, because it seriously hampers viewing of the celestial bodies (Fig. 5.8).

The atmosphere is a mixture of gases (78% nitrogen, 21% oxygen, and 1% water vapor, carbon dioxide, and the inert gases) and of pollutants (sulphur dioxide, nitrous oxide, and ammonia) of human origin. Although at the present time these pollutants have not affected the composition of the atmosphere, they have begun to affect the ecology, our health, and possibly our very existence. Many projects have been initiated by scientists to determine how these pollutants are affecting the environment. One interesting project, "Project Astra," was launched by the astronomer Paul W. Hodge and his associates at the Univeristy of Washington to study the changes produced in the earth's atmosphere by pollutants. They compared the brightness of the zenith stars recorded at the Mount Wilson Observatory during a 50-year period and discovered that there has been a dimming of about 0.3 magnitude in the ultraviolet light and 0.1 in the visible light.

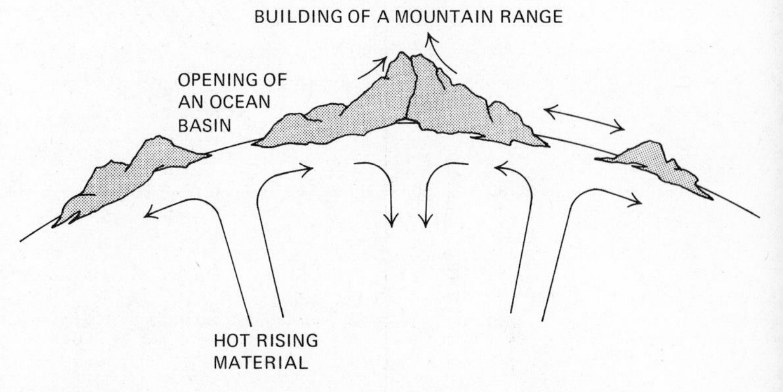

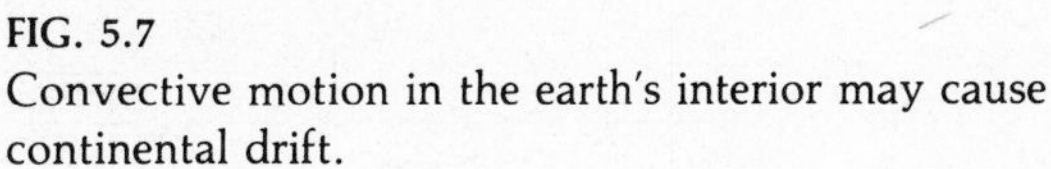

FIG. 5.7
Convective motion in the earth's interior may cause continental drift.

FIG. 5.8
Transmission curve of the earth's atmosphere (schematic). Horizontal scale is stretched in the visible and infrared. Atmosphere is opaque to long-wavelength radio waves due to their reflection off the ionosphere.

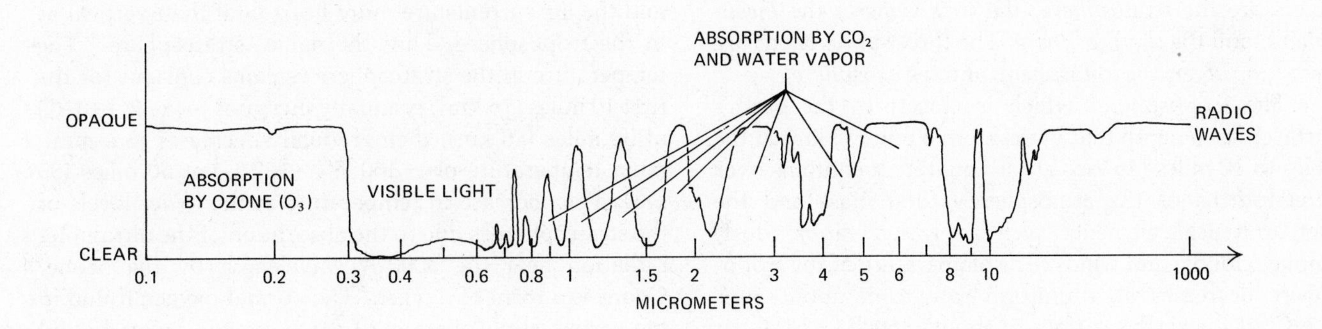

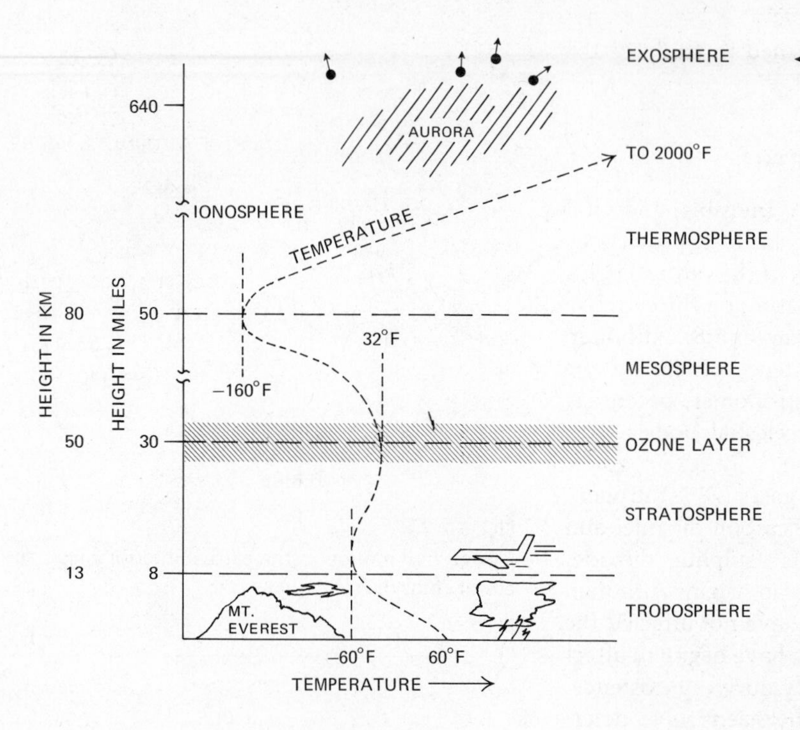

◀ FIG. 5.9
The vertical distribution of the earth's atmosphere. Note how the ozone layer heats the atmosphere at about 30 km.

FIG. 5.10 ▶
The effect of atmospheric refraction on viewing is maximum at the horizon and zero at the zenith, causing celestial bodies to appear higher above the horizon than they really are. To an observer unaware that the starlight has followed a curved path, the light appears to come from direction B. (The effect is exaggerated in the drawing.)

The atmosphere's density decreases rapidly with an increase in altitude. Calculations indicate that about 50% of the atmosphere's total mass is located within three miles above the earth's surface. The data obtained by artificial satellites clearly show that the density at elevations above 100 miles varies considerably from day to day as the result of solar activity (Chapter 10).

The atmosphere is composed of four basic layers with three important subdivisions (Fig. 5.9). The major layers are the *troposphere*, the *stratosphere*, the *mesosphere*, and the *thermosphere*. The three subdivisions are the *ozone layer*, the *ionosphere*, and the *exosphere*.

The troposphere, which is closest to the earth's surface, has a depth that varies from 5 miles (8 km) at the poles to 10 miles (16 km) at the equator. It contains over three-fourths of the atmosphere's total mass and the meteorological elements such as water vapor, dust, smoke, clouds, and winds. The temperature of the troposphere decreases at a uniform rate from about 56°F (13°C) at the earth's surface to about −60°F (−51°C) at the top of the layer. The troposphere is continuously being mixed by air rising at one location and sinking back at another. This circulation in the lower layer is what gives the layer its name and is partly responsible for the varieties of weather we experience.

Above the troposphere is the stratosphere, which has a depth of 40 miles (64 km) and extends to 50 miles (80 km) above the earth's surface. There are very few clouds and basically no weather changes in this layer, and the air currents are more horizontal than vertical as in the troposphere. Thus the name "stratosphere." The temperature in the stratosphere remains constant for the first 10 miles (16 km), gradually increases to +32°F (0°C) at 30 miles (48 km), then gradually decreases to a minimum temperature of −160°F (−107°C) at 50 miles (80 km). This increase in temperature at the lower levels of the stratosphere is due to the absorption of the ultraviolet radiation near the 3000-Å wavelength by the ozone. Ozone is a form of oxygen. The normal oxygen found in the earth's atmosphere is O_2 (two oxygen atoms bound

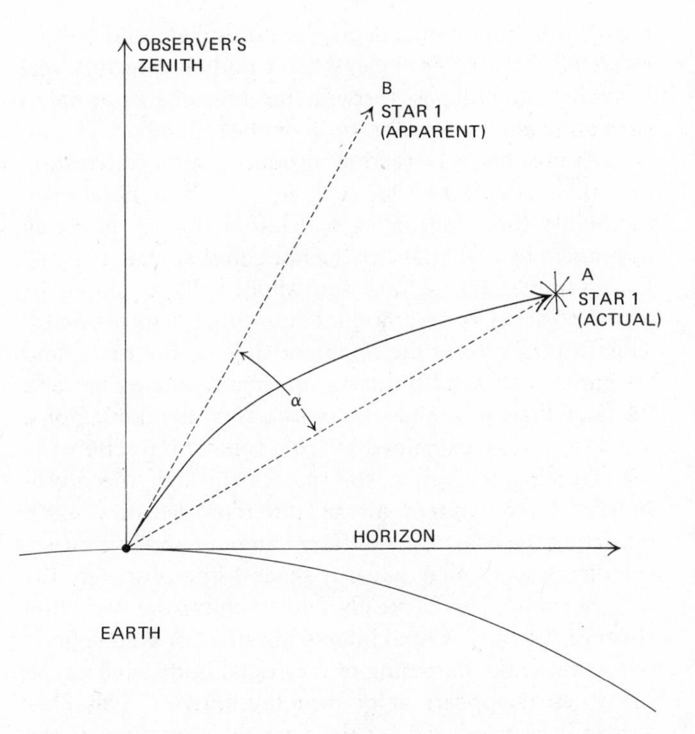

together). Ozone is three oxygen atoms bound together and is thus denoted O_3. Ozone is produced when O_2 is dissociated into atomic oxygen (O), which then recombines with a normal O_2 to yield O_3.

The ozone layer is believed to play an extremely important role in making life on earth possible. Without the attenuation of ultraviolet rays that the layer provides, most animals would suffer severe sunburn merely by venturing outside. Even many plants would be harmed by the intense short-wavelength energy. Unfortunately, our understanding of the ozone layer is very incomplete. The thickness of the layer seems to change from year to year and even from season to season, becoming thinner in the winter months. There has been considerable concern expressed about the effect of such pollutants as some aerosol can propellants and exhaust from jet aircraft, on the ozone layer. All of these artificial contaminants are believed to initiate chemical reactions that destroy ozone. If the ozone layer were destroyed, life on earth might be seriously affected. There are, however, many natural processes that also destroy ozone. Chlorine atoms from volcanic gases seem to be one such culprit. So far it has proved very difficult to assess exactly what the dangers are to the ozone layer. Until atmospheric scientists have a better idea of the natural changes that occur in this layer, it would be wise to be careful of it.

Above the stratosphere lie the mesosphere (middle layer) and the thermosphere, the hot, tenuous, extreme upper region of the atmosphere. Within the thermosphere and extending to about 350 miles (563 km) above the earth's surface is the ionosphere, so named because many of the gas molecules have been broken up into atoms, ions, and electrons by intense ultraviolet radiation and high-speed particles. The ionized particles appear in layers at different heights based on the solar radiation that produced the ionization. These layers reflect radio wavelengths longer than 15 meters, therefore making long-range radio transmission possible. The burning of meteoroids and the display of auroras occur primarily in the ionosphere.

Even though the density of the atmosphere at the upper level of the ionosphere is extremely tenuous, the atmosphere extends beyond the ionosphere, where it gradually thins out to practically nothing. This outer layer is called the exosphere because atoms here leak slowly out into space and are lost.

5.7 ATMOSPHERIC REFRACTION

The presence of an atmosphere around the earth produces several astronomical effects through the processes of refraction, scattering, and absorption.

The density of the earth's atmosphere decreases with an increase in elevation. For simplicity, the earth's atmosphere is assumed to be stratified in layers of decreasing densities. When a beam of light from a star passes through the atmosphere obliquely, the increasing atmospheric density causes it to be progressively refracted by a greater amount as it approaches the earth's surface. When the light beam reaches the observer, its direction is more vertical than when it first entered the earth's atmosphere (Fig. 5.10). Refraction is maximum for a body that is at or near the horizon and zero when a body is directly overhead. When a body is near the horizon, its light will

By permission of Johnny Hart and Field Enterprises, Inc.

pass through a greater depth of atmosphere and will be refracted at a greater angle; when a body is directly overhead, its light will pass through the atmosphere normal to each atmospheric layer and will not be refracted.

Atmospheric refraction produces some interesting astronomical effects. One such effect is the apparent elevation of the position of a celestial body—the body appears to be higher above the horizon than it actually is. For example, the setting sun appears lifted above its actual position by an amount about equal to its diameter. This in turn affects the apparent time of the rising and setting of a celestial body—it appears to rise earlier and set later than it would otherwise. The "twinkling" of a star can also be explained by atmospheric refraction. As the light beam from a star passes through the atmosphere, it encounters air of different densities, each refracting the beam by a different amount and in a different direction so that when it reaches the observer, the light beam appears unsteady and produces the twinkling effect of the star. A final interesting effect of atmospheric refraction is the flattening of a celestial body such as the sun when it appears at or near the horizon. This phenomenon is produced by the unequal refraction of the light from the lower and upper limbs (edges) of the body. The light from the lower limb is refracted more than that from the upper limb because it is closer to the horizon and passes through a greater depth of atmosphere. The bottom edge of the sun thus appears lifted, causing the sun to appear "squashed."

Many people erroneously believe that the apparent increase in the size of the moon at or near the horizon is also caused by atmospheric refraction. Actually, when the moon appears near the horizon, its distance from the earth is about one earth-radius farther away than when it is observed higher in the sky; therefore, its actual size is smaller and is further reduced by atmospheric refraction. In spite of these two facts, the moon appears to be noticeably larger when it is near the horizon than at a higher altitude. The explanation is provided by psychologists, who maintain that it is an optical illusion. When the moon is observed near the horizon, it is seen against the earth's surface features visible on the horizon, which causes the moon's image to appear larger. When the

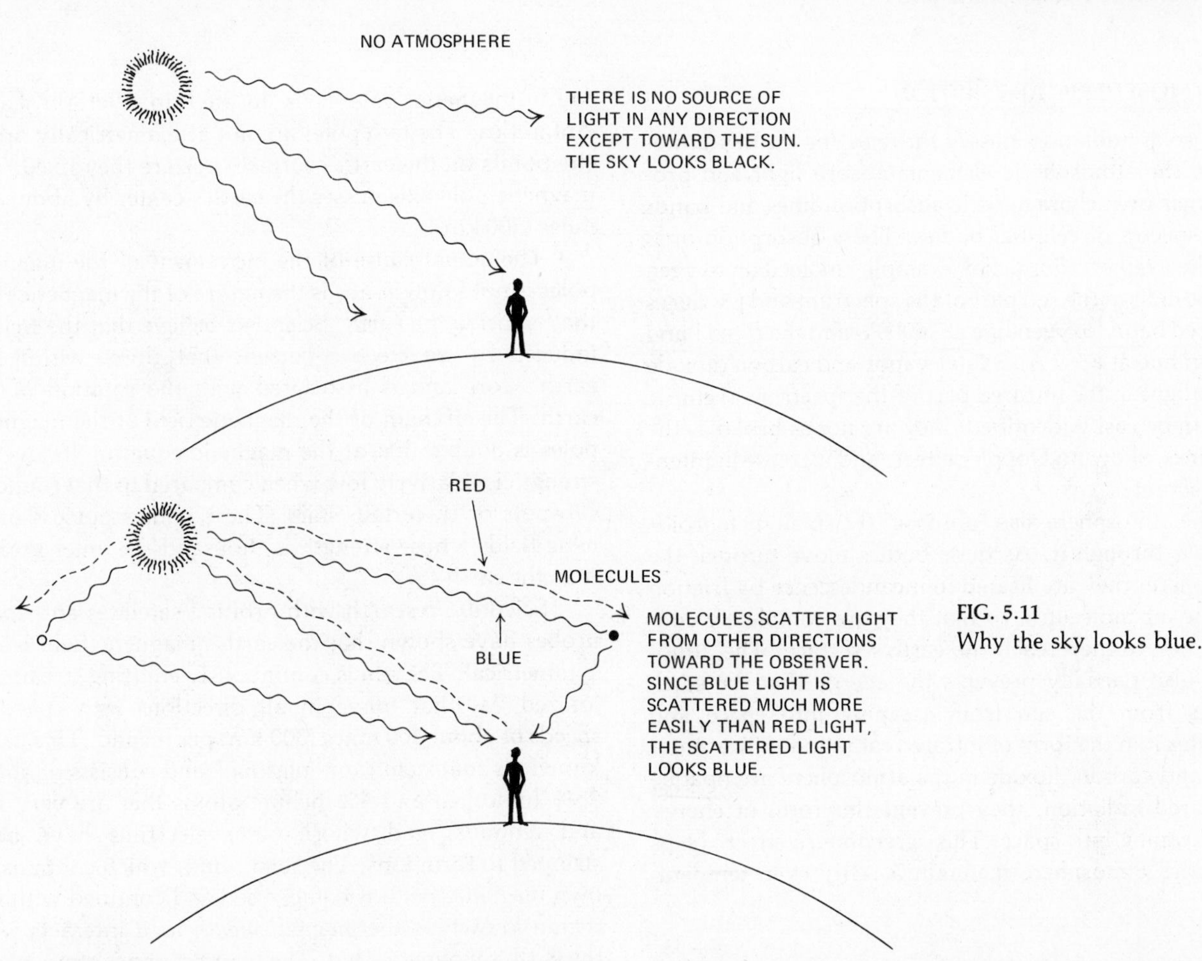

FIG. 5.11
Why the sky looks blue.

moon is observed high in the sky, it appears smaller because it is seen by itself.

5.8 ATMOSPHERIC SCATTERING

The air molecules and the extremely small solid particles in the atmosphere scatter the shorter wavelengths of visible light more effectively than the longer wavelengths. Scattering occurs when the light interacts with the individual air molecules and solid particles. In the encounters, the light beam is deflected in random directions by the interfering substances.

Atmospheric scattering produces the reddening of rising and setting sun and the blue color of the sky. The sun on the horizon appears redder than the noonday sun, because the direct light beam to the observer passes through a greater depth of atmosphere and more of the blue light is scattered. In other words, the direct beam of light does not contain any more red light—it simply has less blue light; therefore, it appears redder. Since sunlight is scattered innumerable times by the molecules in the atmosphere, it becomes progressively bluer as the distance from the direct light beam increases, which causes the sky to appear blue. If the earth had no atmosphere, the daytime sky would appear black, and the stars and planets would be clearly visible. Normally, the stars and planets are not visible in the daytime, because scattered sunlight is much brighter than starlight or the reflected light on the planets (Fig. 5.11).

5.9 ATMOSPHERIC ABSORPTION

When solar radiation passes through the earth's atmosphere, the atmospheric elements absorb light and produce their own characteristic absorption lines and bands in the spectra of celestial bodies. These absorption lines are called *telluric lines*. For example, molecular oxygen absorbs light in the red part of the spectrum and produces the *A* red band (oxygen line at 7600 Å) and the *B* red band (oxygen line at 6867 Å). Water vapor and carbon dioxide absorb light in the infrared part of the spectrum. Telluric lines can be easily identified; they are not as broad as the solar lines, show no Doppler effect, and increase in intensity at sunset.

The atmosphere also "absorbs" the small meteoroids that pass through it. As these bodies move through the atmosphere, they are heated to incandescence by friction with the air molecules, so that they are completely consumed before they reach the earth's surface. The atmosphere also partially prevents the energy that the earth receives from the sun from escaping into space and reradiates it in the form of infrared radiation. Since water vapor and carbon dioxide in the atmosphere are opaque to infrared radiation, they prevent this form of energy from escaping into space. This "greenhouse effect" helps the earth's atmosphere maintain a fairly even temperature.

5.10 THE EARTH'S MAGNETIC FIELD

The earth's magnetic properties were known to the Chinese as early as 1300 B.C. For centuries sailors have used these properties to set and maintain their course at sea. When a magnetized needle that is free to rotate in a horizontal plane is placed in the earth's magnetic field, it will orient itself parallel to the magnetic lines of force. Therefore, its orientation indicates the direction of the force at that particular point on the earth's surface. The points where the needle is parallel to the earth's surface mark the *magnetic equator*, and the two points where the needle is perpendicular to the earth's surface mark the *north* and *south magnetic poles*. The north magnetic pole *N* is located in the Hudson Bay area of Canada, and the south magnetic pole *S* is located in Victoria Land, Antarctica. The two poles are not at diametrically opposite points on the earth's surface, nor are they fixed. The magnetic pole axis misses the earth's center by about 300 miles (500 km).

The actual cause of the movement of the magnetic poles is not known; nor is the nature of the magnetic field that is inside the earth. Scientists believe that the field is induced by an electric current that flows within the earth's core and is associated with the rotation of the earth. The strength of the magnetic field at the magnetic poles is double that at the magnetic equator. Its overall strength is relatively low when compared to that found in sunspots or in certain stars. These bodies possess magnetic fields whose strength is thousands of times greater than the earth's.

Scientific research with orbiting satellites and space probes have shown that the earth's magnetic field is not symmetrical. The sun is continuously emitting streams of ionized gas that move in all directions into space at speeds of about 300 miles (500 km) per second. This gas is known as *solar wind*, or "plasma," and consists of about 95% hydrogen and 5% helium atoms that are very hot and tenuous, and whose outer electrons have been stripped to form ions. The solar wind, which contains its own magnetic field, is compressed and confined within a region known as the *magnetosheath* as it interacts with the earth's magnetic field. The tenuous upper atmosphere of the earth, where magnetic fields control the structure, is called the *magnetosphere* (Fig. 5.12).

The shape of the earth's magnetic field is altered considerably by its interaction with the solar wind. The side facing the sun is compressed into an elliptical shape whose outer limit is about 50,000 miles (80,000 km) from the earth's center. The side away from the sun is extended into the shape of a comet-like tail whose diameter and length are each about 160,000 miles (257,000 km)—more than half the distance to the moon. The presence of the tail was revealed in 1961 from measurements taken by the artificial satellite Explorer 10. The field within the tail is divided into two regions by a neutral plane that is parallel to the magnetic equator. North of the neutral plane,

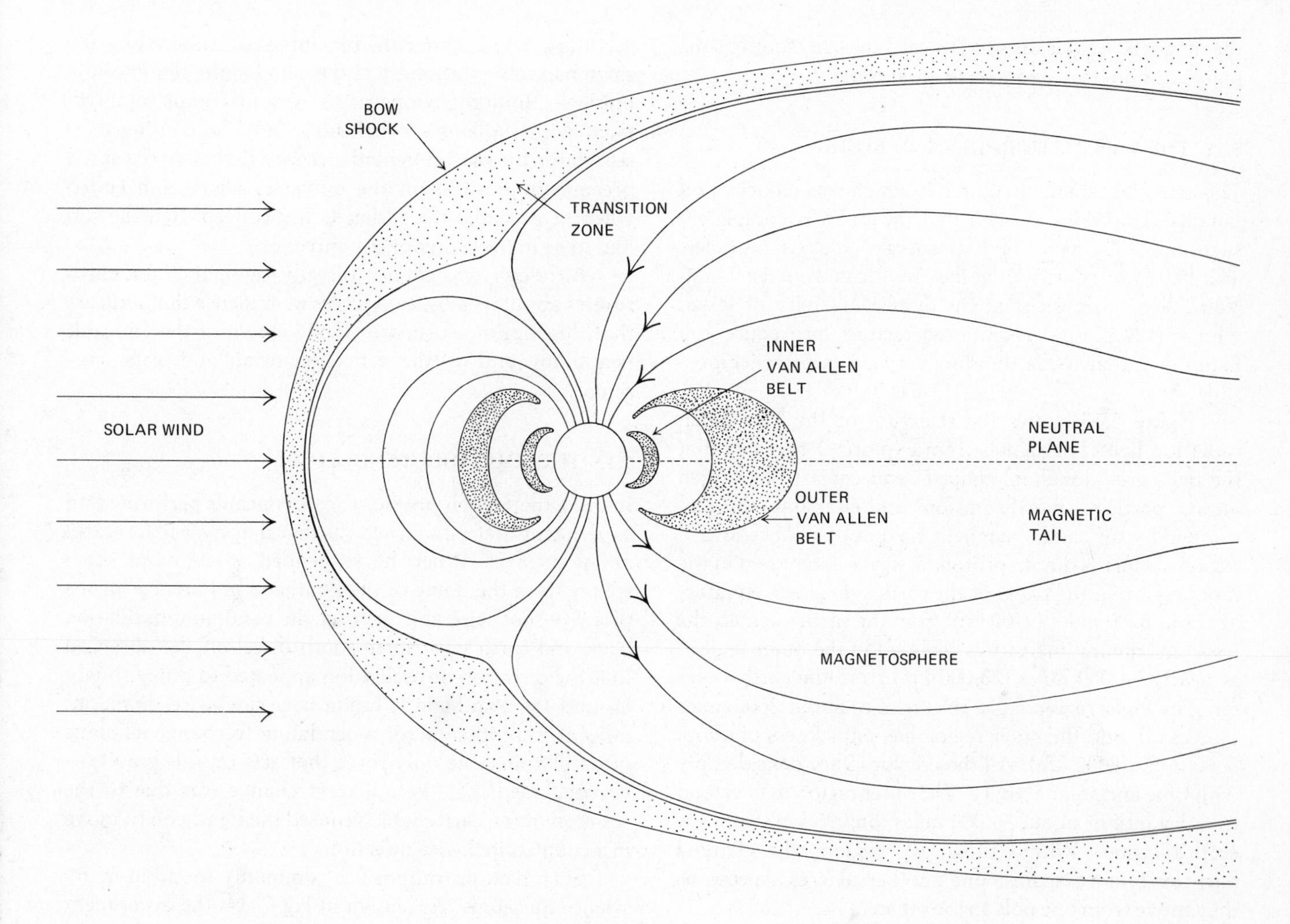

FIG. 5.12
The interaction of the earth's magnetic field with the solar wind.

the magnetic field is directed toward the sun; south of the plane, it is directed away from the sun.

5.11 THE VAN ALLEN RADIATION BELTS

The artificial satellite Explorer 1, which was launched on January 31, 1958, revealed that the earth is completely surrounded by belts of high-energy charged particles. The belts are named after the American scientist James Van Allen, a physicist at the State University of Iowa, who devised the radiation-detecting procedure on Explorer 1 that made the discovery of its presence possible.

Figure 5.13 shows the structure of the Van Allen radiation belt. The signals from Explorer 1 revealed that the belts are "doughnut-shaped" and consist of charged atomic particles, mostly protons and electrons, that are trapped by the earth's magnetic field in two concentrated regions. The maximum proton intensity is centered in the inner region in the plane of the earth's magnetic equator, at about 6000 miles (9700 km) from the earth's center; the lesser maximum intensity is centered in the outer region, at nearly 14,000 miles (23,000 km) from the earth's center. The inner region has a thickness of about 3000 miles (5000 km), and the outer region has a thickness of about 5000 miles (8000 km). All these values vary considerably with time and solar activity. The outer region may extend to a distance of about 36,000 miles (58,000 km) from the earth's center. Within these two regions the charged particles bounce against one another at great speeds as they move from one pole to the other.

5.12 THE ROTATION OF THE EARTH

The idea that the earth rotates about its axis had its roots in antiquity. Cicero wrote that some of the philosophers of his period believed that all the celestial bodies, except the earth, were fixed in space and that the motions of the other bodies were produced by the earth's rotation. Others believed that this theory was nonsensical. One of these was Ptolemy, who reasoned that if the earth rotates, the force of its rotational speed would tear the earth apart and scatter the bits into space; therefore, the earth had to be stationary and located in the center of the universe. Initially, Copernicus was in complete agreement with Ptolemy's conclusion. Still, he continued to search for the reason behind a theory that gave the sun a preeminent position in the universe. His search ended when he accepted the earlier Greek concept that the sun had to be in the center of the universe.

Although one cannot directly sense that the earth rotates about its axis, two pieces of evidence that indicate that this motion exists are (1) the action of the Foucault pendulum, and (2) the deflection of air and water currents.

5.13 THE FOUCAULT PENDULUM

In 1851 the French physicist Jean Foucault performed an experiment that established the fact that the earth rotates about its axis. When he suspended a 62-pound brass sphere from the dome of the Pantheon in Paris by means of a 219-foot wire and then set the pendulum oscillating back and forth in a north-south direction, he observed that the direction of oscillation appeared to move slowly around the Pantheon rotunda in a clockwise direction. Since it is impossible for a pendulum to change its plane of oscillation—the only force that acts on it is gravity—he concluded that the apparent change was due to the rotation of the earth, which caused the Pantheon to move in a counterclockwise direction.

Such demonstrations are commonly found in many science museums. As shown in Fig. 5.14, the experiment is more obvious when the pendulum is located at the north pole, because the apparent period of the earth's rotation is equal to its actual period of nearly 24 hours. At the equator, the apparent period of the earth's rotation is infinity, because there is no apparent change in the direction of oscillation. If the pendulum is oscillating in a north-south direction, it will always move perpendicular to the equator and therefore show no apparent change in its direction of oscillation. Between the poles and the equator, the apparent period of rotation will vary from nearly 24 hours at the poles to infinity at the equator.

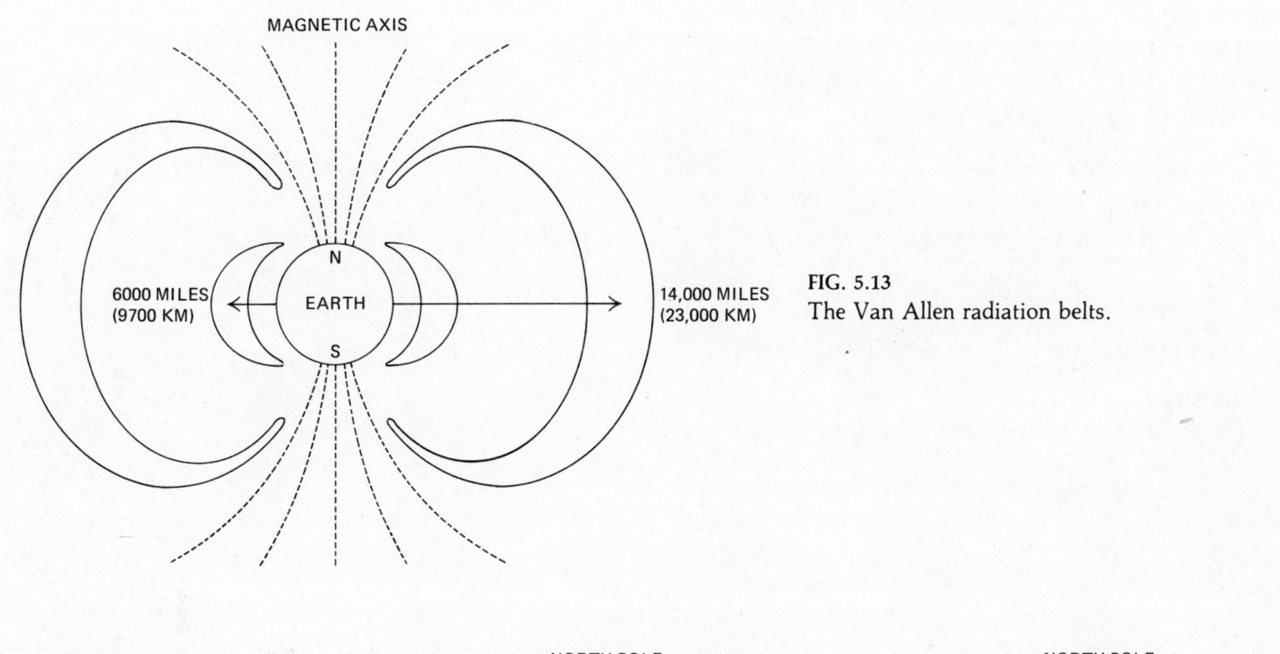

FIG. 5.13
The Van Allen radiation belts.

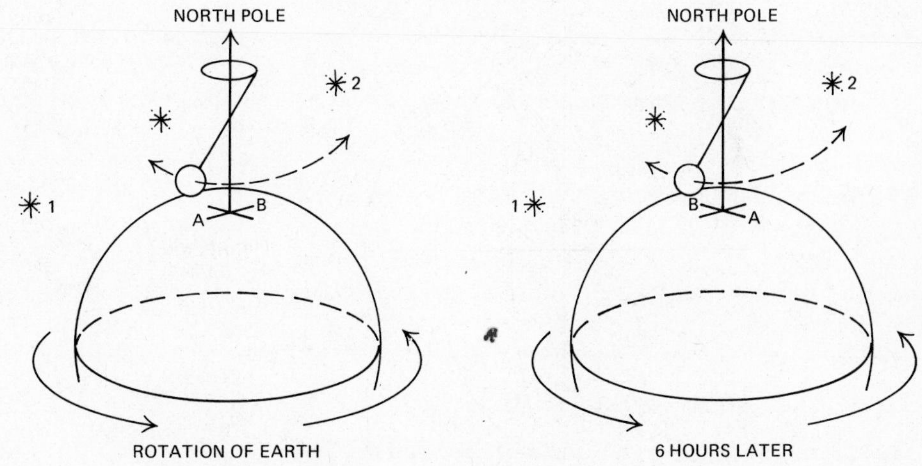

FIG. 5.14
The Foucault pendulum. The pendulum swings in a plane oriented toward stars 1 and 2. The observers at B see it swinging back and forth in front of them. As the earth turns, the observers are carried around until the pendulum is swinging at them. Thus the pendulum appears to have changed its plane of motion.

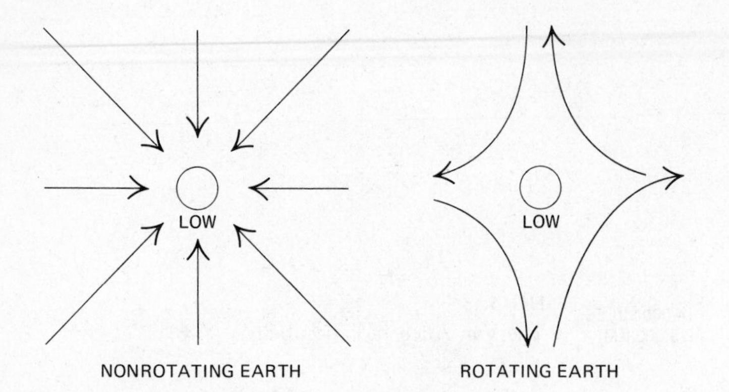

FIG. 5.15
The deflection of air and water currents by a rotating body.

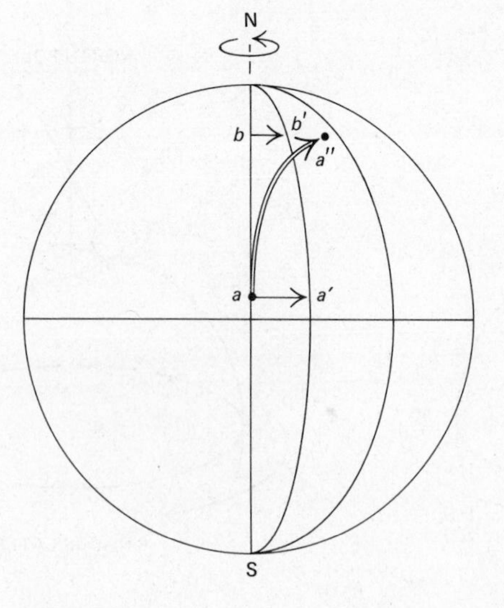

FIG. 5.16
The Coriolis effect. In the northern hemisphere, a body (*a*) moving northward is deflected to the right. An observer at *a* is unaware that the body when pushed toward *b* already has a component of motion toward *a′* due to the earth's rotation.

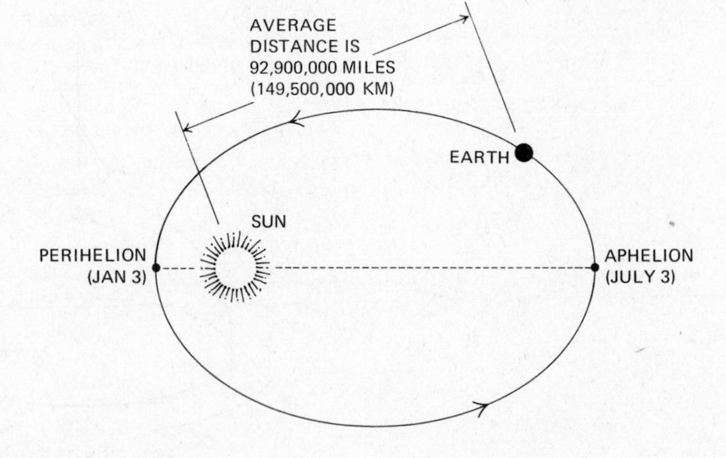

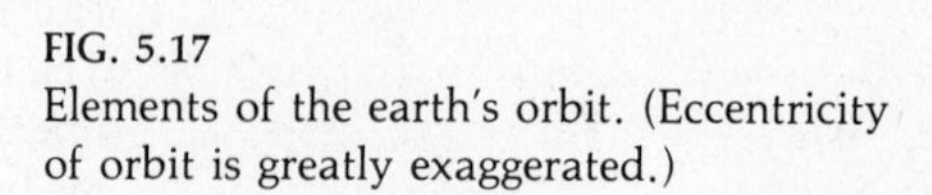
FIG. 5.17
Elements of the earth's orbit. (Eccentricity of orbit is greatly exaggerated.)

5.14 THE DEFLECTION OF AIR AND WATER CURRENTS

A second proof that the earth rotates is the deflection of air and water currents. As shown in Fig. 5.15, the currents on a stationary earth converge radially toward the low-pressure area, whereas on a rotating earth they move in a counter clockwise direction in the northern hemisphere and clockwise in the southern. As the currents move from the equator to the poles, they are deflected to the right in the northern hemisphere and to the left in the southern. This occurs in the following manner. All objects on the earth's surface rotate around the earth's axis about once every 24 hours at different speeds, depending on their location. If two objects are located on the same north-south line in the northern hemisphere, the one that is closer to the equator travels at a faster speed. If the two objects a and b (Fig. 5.16) are to complete one rotation in 24 hours, object a, which is closer to the equator, must move faster because it has a greater distance to travel. Therefore, in a given interval of time, objects a and b have moved to positions a' and b', respectively. If object a moves to the north, it retains its higher eastward speed; therefore, at the end of this interval of time at the higher latitude, it will be in position a'', ahead and to the east of position b'. Consequently, as object a moves northward in a given interval of time, it will be deflected to the right. This is known as the *Coriolis effect*. By analogy, ask yourself what would happen if you opened the door of a car moving at 20 mph and attempted to stroll away from the car. Would your path be a straight line away from the car?

5.15 THE REVOLUTION OF THE EARTH

The earth revolves eastward around the sun in an elliptical orbit once every year. As shown in Fig. 5.17, the sun is located at one of the ellipse's foci; therefore, its distance from the earth varies during the year. The earth is at perihelion, or nearest point to the sun, about January 3 and at aphelion, or farthest point from the sun, about July 3. The earth's mean distance from the sun, which is half the length of the major axis, is 92,900,000 miles (149,500,000 km).

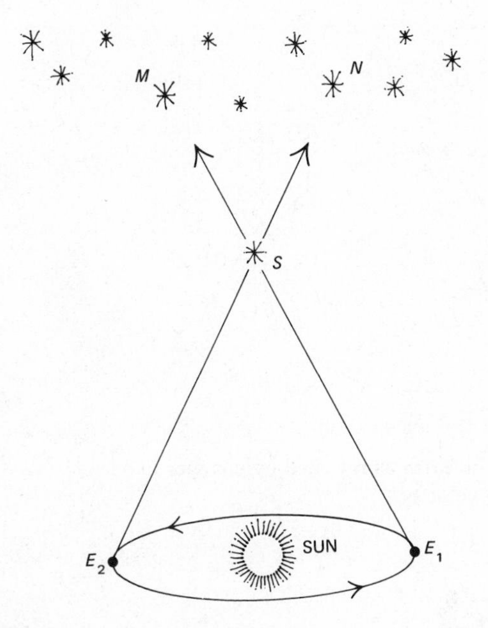

FIG. 5.18
The parallactic motion of the stars. As the earth revolves around the sun, the nearer star S appears to oscillate annually from M to N with respect to the more distant stars.

Although we accept the fact that the earth revolves around the sun, there are many who would find it difficult to present evidence of its motion. Three important proofs can be given: the *parallactic motion of the stars*, the *variation in the radial velocities of the stars*, and the *aberration of starlight*.

5.16 THE PARALLACTIC MOTION OF THE STARS

The parallactic motion of a star, which is its apparent motion with respect to the more distant stars, is one evidence that the earth revolves around the sun. When the earth is in position E_1 (Fig. 5.18), the star S is observed to be in line with the more distant star M. Six months later, when the earth is in position E_2, star S is observed to be in

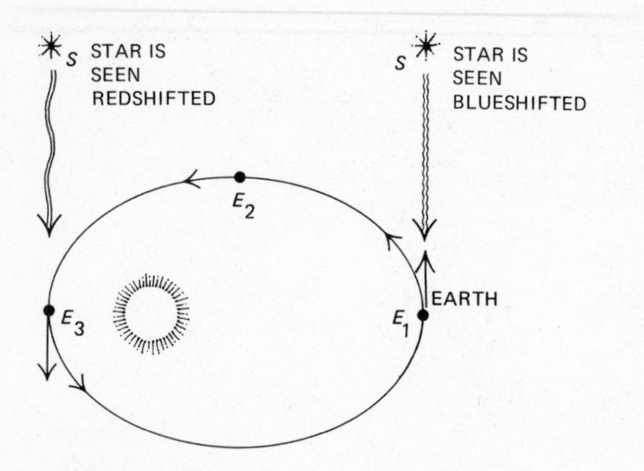

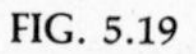
FIG. 5.19
Motion of the earth as revealed by changes in a star's radial velocity.

line with the more distant star N; twelve months later, when the earth has returned to position E_1, star S is again observed to be in line with star M. In one year, as the earth completes one turn around the sun, star S has appeared to have made one complete cycle in the sky. Its path is a straight line when the star is located in the plane of the ecliptic, a small circle when it is at 90° from the plane of the ecliptic, and an ellipse when it is between these two extremes. We will see in Chapter 11 that this shift in a star's position due to the earth's motion plays a crucial role in measuring distances to stars.

5.17 THE VARIATION IN THE RADIAL VELOCITY OF A STAR

Evidence that the earth revolves around the sun can also be gained by observing a star's radial velocity, that is, its velocity toward or away from the observer. If we look at a star that is at rest with respect to the sun, its measured radial velocity as seen from the earth will change as the earth moves in its orbit. (See Fig. 5.19.) At E_1 the radial velocity will be most negative (a blue shift). At E_2 there will be no measured radial velocity. At E_3 the radial velocity will be most positive (a red shift). This changing radial velocity is evidence that the earth moves around the sun and allows us to measure the earth's orbital velocity.

The radial velocity of a star is determined from the Doppler effect in the star's spectrum. The star is approaching the earth when the spectral lines are displaced toward the violet end of the spectrum and is receding from the earth when the displacement is toward the red end of the spectrum.

5.18 THE ABERRATION OF STARLIGHT

The aberration of starlight was first announced in 1727 (the year of Newton's death) by James Bradley, the Astronomer Royal at the Greenwich Observatory. The aberration of starlight, which is the apparent displacement of a star in the direction of the earth's motion, can be illustrated by observing the motion of raindrops on the side window of an automobile. On a calm day, when the raindrops are falling vertically, they will appear to move in a vertical direction downward on the window of a stationary automobile. When the automobile is moving forward, the raindrops appear to move in a slanting direction to the rear, and the degree of slant is determined by the speed of both the automobile and the falling raindrops.

A similar effect occurs with starlight reaching the earth. A telescope mounted on the earth is moving forward at the earth's orbital velocity, and the starlight is moving toward the earth at the speed of light. In Fig. 5.20, the solid line represents the true direction of the starlight, and the dotted line represents its apparent direction. When the earth is in position 1 and moving to the right, the telescope must be pointed forward along the dotted line if the star is to be seen. In this position, the angle of displacement is maximum at 20.5 seconds. When the earth is in position 3 and moving to the left, the angle of displacement is again maximum and in the direction in which the earth is moving. In position 2, the earth is moving toward the star; in position 4, it is moving away from the star. Therefore, in these two positions the displacement is zero. This means that as the earth revolves around the sun, the star appears to move in an ellipse 20.5 seconds of arc across centered on the star's true position.

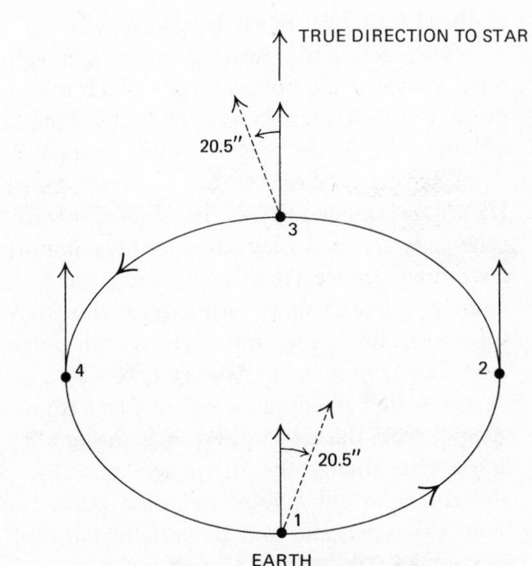

FIG. 5.20
The aberration of starlight. There is no aberration when the earth is moving toward the star (2) or away from the star (4). Maximum aberration of 20.5″ occurs when the earth is moving at right angles to the star, as in positions (1) and (3).

5.19 TWILIGHT

The instant when the sun's upper limb disappears below the horizon is called *sunset*. Even though the sun is below the horizon after sunset, the earth still receives some of its light by refraction, reflection, and scattering, and this light is visible above the western horizon.

Astronomical twilight is defined as the interval of time between sunset and when the sun's center is 18° below the horizon. This value represents the maximum angle at which light can appreciably illuminate the atmosphere. *Civil twilight* is defined as the interval of time between sunset and when the sun's center is 6° below the horizon. This value was selected arbitrarily, because when the sun is in this position, it is believed that the amount of light that the earth receives is insufficient for the continuation of ordinary outdoor activities. The duration of twilight, as shown in Fig. 5.21, varies with the latitude—it depends on the time required for the sun to travel the 18° below the horizon. The sun's path with respect to the horizon also varies with the latitude. At the equator, the sun sets vertically with the horizon; as the latitude increases, the angle becomes more oblique. The duration of twilight at the equator is about one hour, whereas in the northern midlatitudes it is several hours.

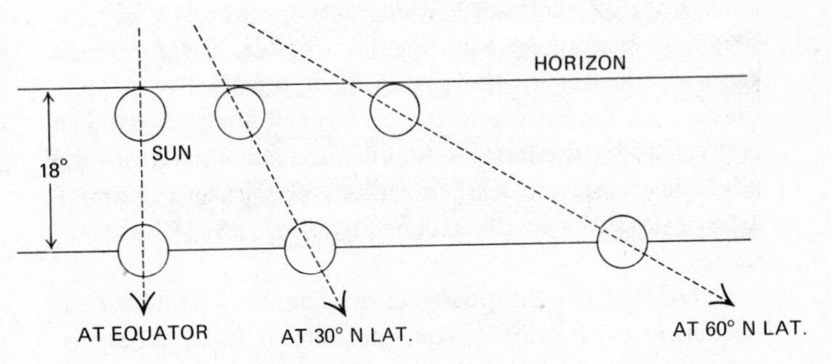

FIG. 5.21
The duration of astronomical twilight at different latitudes on the first day of spring. As the latitude increases, the duration of twilight increases because the sun's path becomes more inclined to the horizon. The sun moves along the dashed line. It can be seen from the figure that the sun has about twice as far to go between the lines at 60° as it does at the equator.

SUMMARY

The earth, our home planet, is representative of many of the other planets. It is basically spherical, although it bulges slightly at the equator due to its rotation. The interior, beneath the thin crust, can be divided into three main regions: a hot but solid zone of rock called the mantle; a liquid core of nickel-iron; and a solid inner core, also of nickel-iron. The interior is heated by the decay of radioactive elements. The interior regions are studied by analysis of earthquake waves.

The radioactive heating causes convective motions (circulation of the hot material) which ultimately causes motion of the crustal layers. The slow, long-term motion of large plates in the crust is called continental drift.

The atmosphere of the earth is mostly nitrogen (78%) and oxygen (21%), but contains traces of several other gases which play an extremely important role in governing the earth's temperature and habitability, namely ozone, water, and carbon dioxide. The atmosphere can be divided into regions with different properties. The lowest is the troposphere, the zone in which most weather phenomena occur. Next comes the stratosphere, then the mesosphere, which contains the ozone layer, and finally the thermosphere. The ozone layer absorbs solar ultraviolet radiation, thus protecting us from that energetic and potentially harmful light. The ionosphere reflects radio signals.

The atmosphere bends (refracts) light and scatters it. Refraction is responsible for the twinkling of stars and for the apparent distortion in the shape of the sun at sunrise or sunset. The scattering of light is what makes the sky appear blue.

The earth possesses two basic motions: its orbital motion about the sun and its rotational motion about its axis. Rotation manifests itself in the slow turning of a Foucault pendulum and in the Coriolis force, which causes deflection of the motion of air, water, or projectiles across the earth's surface. The earth's orbital motion is revealed by the parallactic shift in a star's position, the periodic changes in a star's radial velocity, and the periodic shift in the displacement of a star's image (aberration).

The earth also possesses a magnetic field, which influences the motion of some particles in its upper atmosphere and which is manifested at the surface by the orientation of a compass needle.

REVIEW QUESTIONS

1. List several evidences to show that the earth's shape is spherical.
2. Define the oblateness of the earth's spheroid. What is its value?
3. Describe a method that is used to measure the earth's mass. What is the approximate value for the earth's mass?
4. Discuss the probable structure of the earth's interior. How was information about the structure of the earth's core obtained?
5. What is meant by the half-life of a radioactive element? Explain how a fairly accurate estimate of the earth's age can be determined.
6. Why is the center of the earth hot?
7. What is meant by continental drift?
8. What is believed responsible for the motion of the crustal plates?
9. The temperature of the sea floor is found to be highest along the mid-oceanic ridges where new crust is forming. Can you think of a reason why this might occur?
10. How does the temperature vary as one moves up through the earth's atmosphere? Why does the temperature increase in the ozone layer?
11. What is the most abundant element in the earth? What is the most abundant element in the earth's atmosphere?
12. What is atmospheric refraction? Describe some of the interesting effects produced by atmospheric refraction.
13. Explain why the planets and stars are not normally visible in the daytime sky.
14. What are the "telluric lines"? How can they be identified from the absorption lines produced by celestial bodies?
15. Does a compass point to true north? Why?
16. Describe the structure and composition of the Van Allén radiation belts.
17. Describe and explain the changes in the shape and color of the setting sun.
18. Give two observable proofs that the earth both rotates and revolves.
19. At what time of the year at your own latitude will the longest period of twilight occur? Use a diagram to explain your answer. Explain by means of a diagram why the duration of twilight decreases with a decrease in latitude.
20. Since the earth rotates in a counterclockwise direction, explain why the stars appear to move in the opposite direction.

CHAPTER 6
THE EARTH: A CELESTIAL BODY

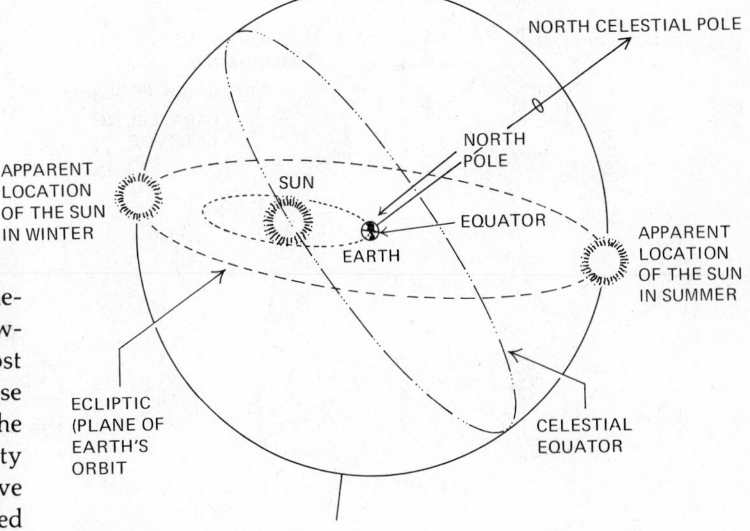

FIG. 6.1
The celestial sphere, an imaginary sphere surrounding the earth, on which the stars and planets appear to lie. The celestial sphere appears to rotate around the earth once each day. The earth is shown with its orbit about the sun for the winter season. Note that the sun appears south of the celestial equator.

There is a temptation to suppose that astronomical phenomena have little or no influence on our daily life. However, a little reflection shows that some of the most important and obvious changes in our environment arise from astronomical causes, such as day and night and the cycle of the seasons. Because of the extreme regularity with which these events repeat, even the most primitive people, as well as plants and lower animals, have used them as indicators of the passage of time. Astronomical measures of time are still used today and can be easily understood in terms of the position and motions of the earth. One must constantly bear in mind in the subsequent discussion that the earth has a variety of motions and that many of these occur in a cyclic and highly regular fashion.

6.1 THE ORIENTATION OF THE EARTH IN SPACE

If an observer stands on a rise and looks at the sky it appears to form a giant dome. At night this imaginary dome is dark and covered with stars. Astronomers call this imaginary dome the *celestial sphere*. It is represented as a giant sphere that surrounds the earth and the other objects in the solar system (Fig. 6.1). As the earth rotates, the celestial sphere appears to turn about the earth. It is, therefore, convenient to imagine that the celestial sphere

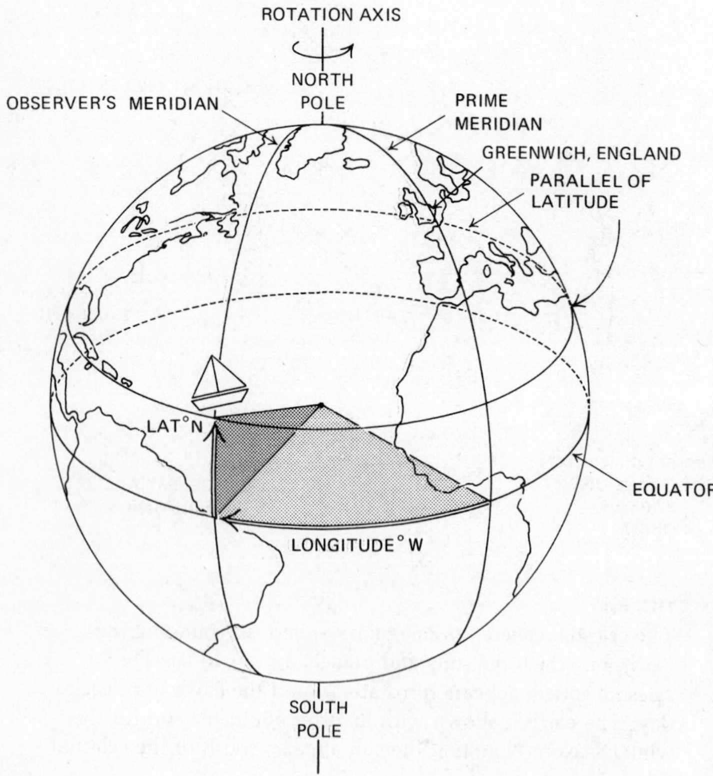

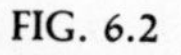
FIG. 6.2
The terrestrial sphere. The coordinates latitude and longitude are used to locate the position of places on the earth's surface.

moves, making one turn about the earth every 24 hours, while the earth remains at rest. Because the earth maintains the tilt of its rotation axis with respect to its orbital plane at a nearly constant angle of 23.5° (and because the direction of the earth's rotation axis is nearly constant), the celestial sphere can be imagined to keep nearly exactly the same orientation from year to year. The motion of the earth about the sun (its revolution) means that if an observer imagines the celestial sphere oriented with respect to the earth, the sun will appear to move across the surface of the sphere, completing one trip around the celestial sphere each year (Fig. 6.1). The apparent path of the sun along the celestial sphere is called the *ecliptic*. Its name derives from the fact that eclipses can occur only when the moon crosses this apparent path.

6.2 THE TERRESTRIAL SPHERE

An understanding of the relationship that exists between the earth and the celestial bodies is essential for navigating by land, water, or air, for establishing the position of the celestial bodies in relation to the observer's position on the earth, and for understanding the phenomenon of time. To determine the observer's position on the earth's surface, we first assume that the earth is a sphere and that a coordinate system (grid) based on the earth's axis of rotation (Fig. 6.2) is established. This axis is an excellent, convenient, and natural reference. Its ends are designated as the north and south poles of the grid. The great circle (formed by a plane passed through the center of the sphere) that is halfway between the poles is the equator. The great circles that pass through both poles and intersect the equator at right angles are *meridians*. The meridian that passes through the Greenwich Observatory, England, is the *reference meridian*, or *zero meridian*. The meridian that passes through the observer's position (the boat in Fig. 6.2) is the *observer's meridian*. The small circles (formed by a plane that does not pass through the center of the sphere) that are parallel to the equator are *latitude circles*.

The observer's position on the earth's surface is determined by two coordinates, *longitude* and *latitude*.

Longitude is the angle measured from the reference meridian, east or west along the equator, to the meridian that passes through the place. It varies from 0° to 180° east or west. Latitude is the angle measured from the equator, north or south along a meridian, to the latitude circle that passes through the place. It varies from 0° to 90° north or south. Since the earth is an oblate spheroid, the length of one degree of latitude at the poles is about 0.7 statute miles (1.13 km) longer than at the equator. At the poles it is about 69.4 statute miles (111.69 km), while at the equator it is about 68.7 statute miles (110.56 km).

The longitude and latitude of a position vary slightly over a period of years. The variation is produced by the movement of the land with respect to the earth's axis of rotation and involves the complex movement of the poles along two paths that are almost circular. One is slightly less than 10 feet, while the other varies from 10 to 20 feet (3.05 to 6.10 meters) in diameter. The period of motion along the first circle is about one year and is believed to be caused by the seasonal shifting of snow, ice and atmospheric masses; the period along the second circle is about 14 months and is believed to be caused by a natural oscillation due to the earth's mass not being symmetrical about its axis of rotation. The oscillations of an unbalanced rotating object are one reason why tires on an automobile should be balanced.

6.3 THE CELESTIAL SPHERE

We have seen in the previous section how a position may be denoted on the surface of the earth. A similar coordinate system may be used to locate celestial objects. The celestial sphere's grid is an extension of the terrestrial grid and is called the *equator system* (Fig. 6.3). The extension of the earth's axis to the celestial sphere is the *celestial axis*, and its ends mark the north and south celestial poles. The projection of the earth's equator is the *celestial equator*. There are also circles on the celestial sphere, analogous to the earth's meridians called *hour circles*. The observer's *celestial meridian* labeled in Fig. 6.3, is the projection of the observer's *terrestrial meridian*. The projection of a latitude circle is a *diurnal circle* and represents the daily apparent path of a celestial body.

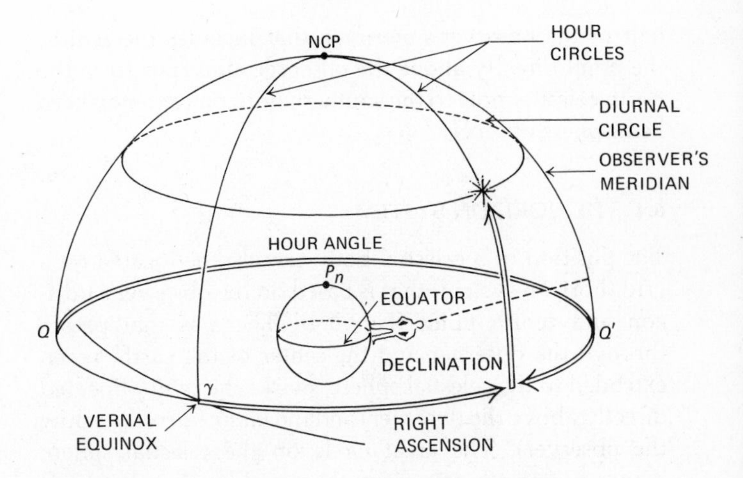

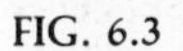
FIG. 6.3
The celestial sphere. The coordinates right ascension or hour angle and declination are used to locate the position of celestial bodies in the sky.

The position of a celestial body on the celestial sphere is determined by the coordinates *right ascension* (R.A.) and *declination* (Dec), which are analogous to longitude and latitude, respectively, on the terrestrial sphere.

Right ascension is the angle measured from the hour circle that passes through the vernal equinox (the intersection of the celestial equator and the ecliptic) eastward along the celestial equator to the hour circle that passes through the body. This angle varies from 0° to 360° but is measured in hours from 0–24 hours.

Declination is the angle measured from the celestial equator, north or south along an hour circle, to the diurnal circle that passes through the body. This angle varies from 0° to 90° north or south. The hour angle is measured from the upper branch of the observer's celestial meridian, westward along the equator, to the hour circle that passes through the body. The upper branch is the half of the observer's meridian that includes the zenith, the point directly above the observer, and runs from the north celestial pole to the southern horizon for a northern hemisphere observer.

6.4 THE HORIZON SYSTEM

The position of a celestial body can also be located on a grid (horizon system) that is based on the observer's horizon and zenith point (Fig. 6.4). The axis that passes through the observer and the center of the earth, when extended to the celestial sphere, marks the *zenith* (located directly above the observer) and the *nadir* (directly below the observer). The great circle on the celestial sphere whose points are 90° from the zenith is the observer's horizon, and the great circles that pass through the zenith and nadir and intersect the horizon at right angles are *vertical circles*. The small circles that are parallel to the horizon are *altitude circles*.

The two coordinates that determine the position of a body in the horizon system are *azimuth* and *altitude*. Azimuth is the angle measured from the north point on the horizon clockwise along the horizon to the vertical circle that passes through the body. The angle varies from 0° to 360°. Altitude is the angle measured from the horizon up along a vertical circle to the altitude circle that passes through the body.

6.5 THE SEASONS

Many people hold to the fallacious belief that the seasons are related to the earth's distance from the sun—for example, the earth is closer to the sun in summer than in winter. Actually, the northern hemisphere of the earth is nearest the sun in the winter (about January 3) and farthest from the sun in the summer (about July 3). During the year its distance from the sun varies by about 3 million miles (5 million km), and this difference produces only a small temperature change. Therefore, a change in the earth's distance from the sun does not produce a change in the earth's seasons. The two factors that produce the earth's seasons are the inclination of the earth's equator to the plane of the ecliptic and the revolution of the earth around the sun.

The rotation of the earth about its axis causes the earth to behave like a giant gyroscope. Any object set spinning tries to preserve its orientation in space, which, for example, is why a spinning coin will stand upright on a table until friction slows it down. The earth is so immense that its tendency to maintain its orientation keeps its rotation axis pointing in very nearly exactly the same direction for hundreds of years. As the earth moves about the sun, its north pole is always tilted at 23.5° with respect to its orbital plane. During half the year the north pole is tipped toward the sun; during the other half of the year it is tipped away from the sun (Fig. 6.5). When the north pole is tipped as close to the sun as it can be (so that the sun shines most directly down on the northern hemisphere), the *summer solstice* occurs. This defines the first day of summer. At this time of year the sun will be the highest above the horizon that it can be in the northern hemisphere.

The word *solstice* means "sun standing still" and is used to describe those times the sun's motion north and south across the sky seems to pause. This change in solar motion is easily noted and was used by many early cultures to signal the changing seasons. When the north pole is tipped as far away from the sun as possible, the *winter solstice* occurs. This defines the first day of winter. At this time of the year the sun will be lower in the sky at noon in the northern hemisphere than it is at any other time of year. At a latitude of 40°N (which is used here as a representative latitude for the United States), the sun at noon will be 73.5° above the horizon on the local meridian during the summer solstice and only 26.5° above the horizon on the local meridian at noon during the winter solstice. The sun's path on the celestial sphere (the ecliptic) forms a curve which swings north in the summer and south in the winter. Reference to Fig. 6.6, showing the sun's motion across the sky, also gives the dates corresponding to the times when the sun crosses the celestial equator moving northward (March 21, or the first day of spring) and southward (September 21, or the first day of autumn). The spring equinox is often referred to as the *vernal equinox* (vernal comes from the Latin word meaning green).

On the first day of summer the sun will be directly overhead as seen by an observer at latitude 23.5°N. Like-

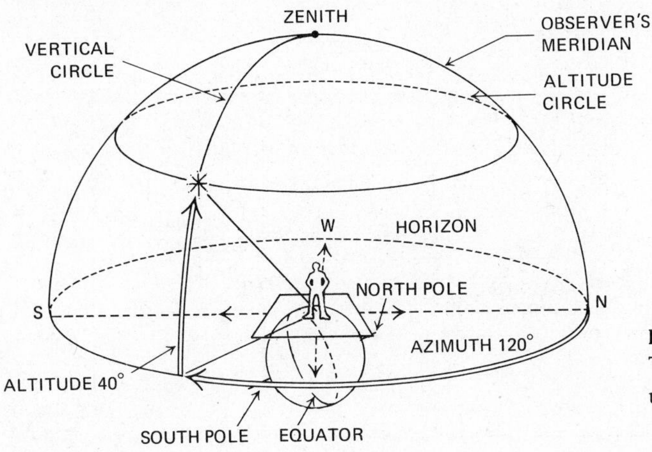

FIG. 6.4
The horizon system. The coordinates altitude and azimuth are used to locate the position of celestial bodies in the sky.

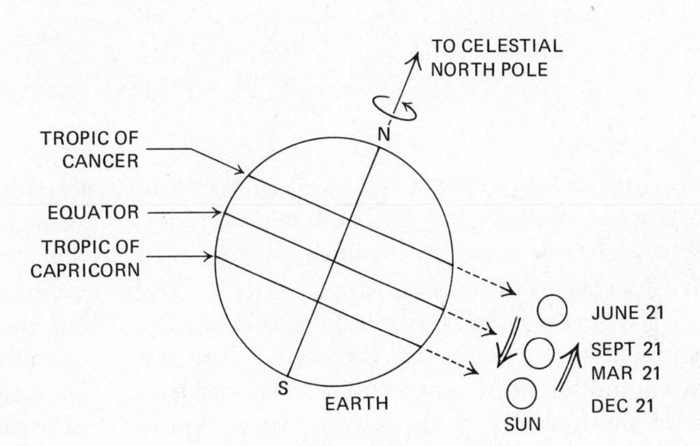

FIG. 6.5
Orientation of the earth's axis in space. The earth maintains its equator at an angle of 23.5° to the plane of the ecliptic. The north pole of the earth's axis of rotation points to the north celestial pole, which is within one degree of Polaris.

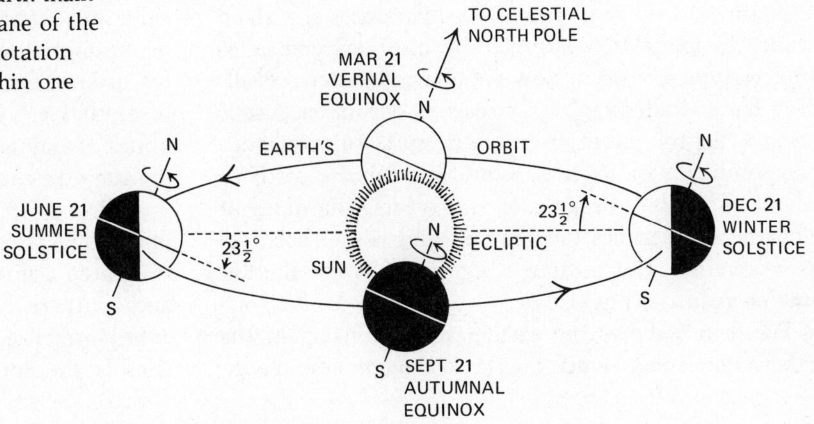

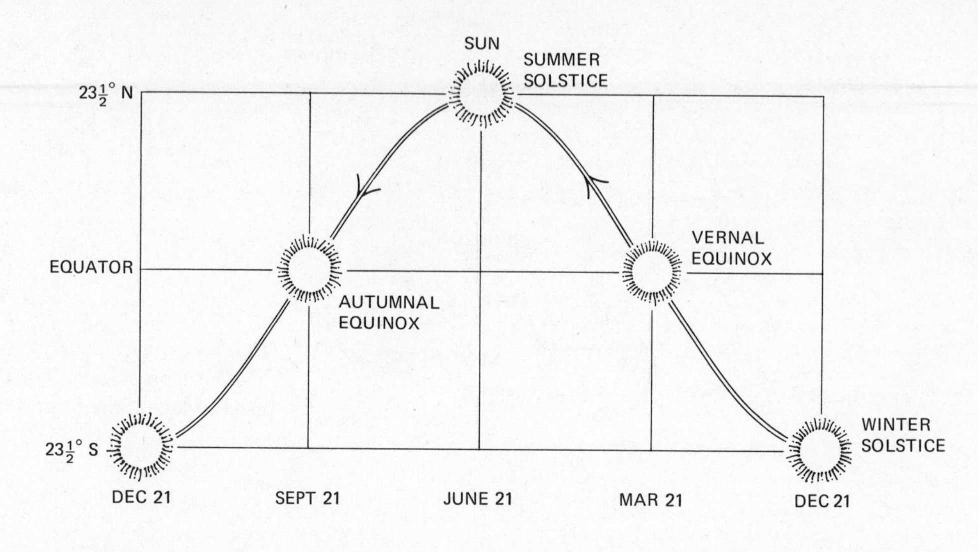

FIG. 6.6
As the earth revolves around the sun, the sun appears to move annually from 23.5° below the celestial equator (winter solstice) to 23.5° above the celestial equator (summer solstice), and back to 23.5° below.

wise, on the first day of winter the sun will be overhead for an observer at latitude 23.5°S. These two latitudes, which define a zone centered on the equator, are called the *Tropic of Cancer* and the *Tropic of Capricorn*. The word tropic is from the Greek word meaning to turn. The tropics represent that position on the earth where the sun's apparent motion north and south in the sky ceases. The dates of the first day of the seasons vary slightly from year to year because of leap years, occurring roughly at the twenty-first of March, June, September, and December, respectively. However, these dates are not the times of the most extreme temperatures at a given latitude. In temperate latitudes the most extreme temperatures typically occur about four weeks after the solstices. The so-called lag of the seasons occurs because it takes a while for the earth to warm up in summer and, correspondingly, it takes it some time for the earth to cool off in winter. Because of leap years, the dates at which the sun reaches the summer and winter solstices and the vernal and autumnal equinoxes vary slightly from one year to the next.

Figure 6.7 shows the earth-sun relationship in the northern hemisphere when the sun is at summer and winter solstices. At summer solstice, one-half of the earth's surface, including the north pole, is illuminated because the earth's polar axis is inclined toward the sun. The earth's surface from latitude 66.5° north to the north pole has the sun above the horizon for 24 hours each day, a condition called the *midnight sun*. The 66.5° north parallel is called the *Arctic circle*. At winter solstice, one-half of the earth's surface, including the south pole, is illuminated because the earth's polar axis is inclined away from the sun. The midnight sun is visible from the south pole to 66.5° south latitide. The 66.5° parallel of latitude is called the *Antarctic circle*. When the earth is at the vernal or autumnal equinox, the earth's axis points neither toward nor away from the sun; on these two dates, therefore, the entire earth's surface receives exactly twelve hours of daylight and twelve hours of darkness because the sun's diurnal path coincides with the celestial equator.

Figure 6.8 shows the sun's daily path on these four important dates for an observer in San Francisco, which is located approximately halfway between the northern and southern borders of the United States on about the 38th parallel of latitude north. At winter solstice, the sun rises in the southeast, reaches a height of about 28.5°

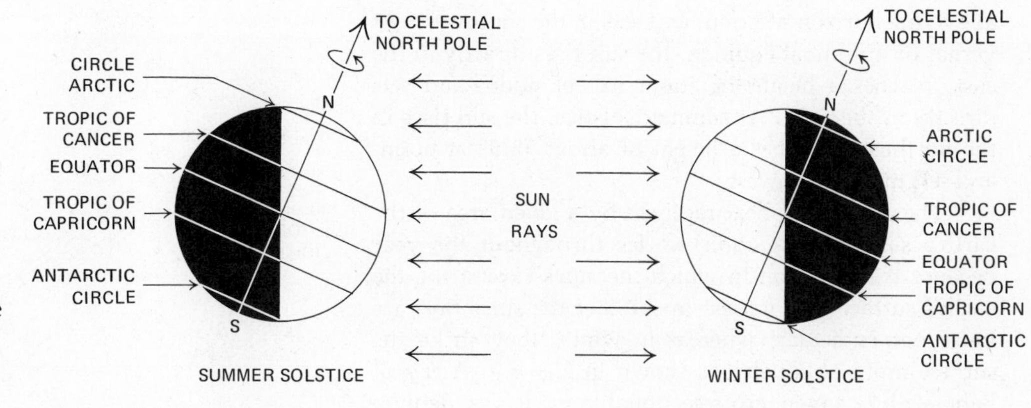

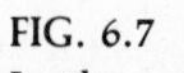

FIG. 6.7
In the northern hemisphere the longest day and shortest night occur at summer solstice. At winter solstice, the opposite occurs.

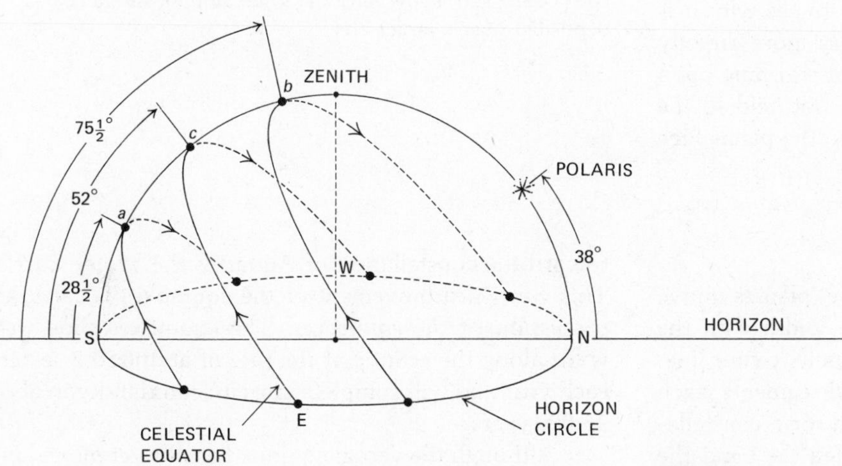

FIG. 6.8
The sun's daily path in the city of San Francisco at winter solstice (*a*), summer solstice (*b*), and vernal and autumnal equinoxes (*c*).

above the horizon at noon, and sets in the southwest. At vernal or autumnal equinox, the sun rises directly in the east, reaches a height of about 52° at noon, and sets directly in the west. At summer solstice, the sun rises in the northeast, reaches a height of about 75.5° at noon, and sets in the northwest.

The amount of heat received by a given area of the earth's surface (*insolation*) varies throughout the year because the direction in which the sun's rays strike the earth's surface also varies. In summer the sun's rays are more perpendicular, whereas in winter they strike the surface more obliquely, as shown in Fig. 6.9. A ray of light with a given cross-sectional area has a definite amount of heat. As this ray of light strikes the earth's surface at San Francisco at summer solstice, its heat will be deposited over a given area of its surface. At winter solstice, a similar ray of light with the same amount of heat strikes the earth's surface more obliquely and deposits the heat over almost twice the surface area; therefore, the temperature of a unit area on the earth's surface will be subsequently lower in winter. It is easy to see why it is warmer in summer, when the sun shines more directly down on the ground, if one thinks about warming one's hands in front of a fire. The hands are not held so the palms face each other; they are tipped so the palms face directly at the fire for maximum warmth.

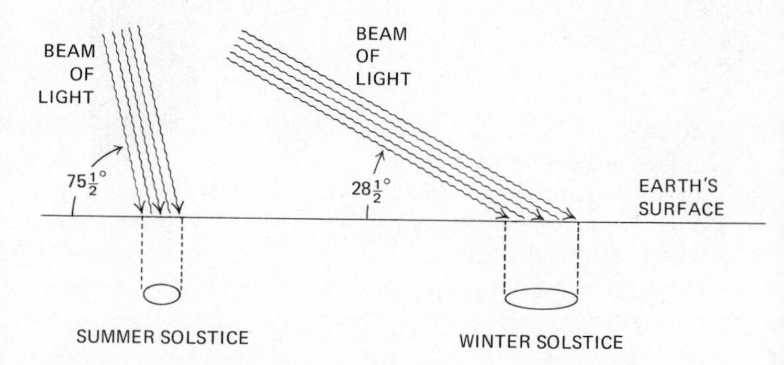

FIG. 6.9
At summer solstice in San Francisco, the sun is at its highest position above the horizon, and maximum energy is deposited on a unit area of the earth's surface in one minute. At winter solstice the sun is low, and the same amount of energy is deposited over a larger area.

6.6 THE PRECESSION OF THE EARTH

The ancient Greeks observed that the planets move within a band in the sky about 18° in width with the ecliptic, the apparent path of the sun, as its center line. They divided this band into twelve equal segments, each 30° in length, with each segment named for a constellation that lies within it. The Greeks called the band the *Zodiac* because many of the constellations represented animals. About 2000 B.C. the vernal equinox was located in the constellation of Aries the Ram. During the period that Christ lived, the vernal equinox had moved into the constellation of Pisces the Fishes. It is interesting to note that the early symbol of Christ was the fish and that the period was known as the Piscean Age. Today, the vernal equinox is still in the constellation of Pisces and is moving toward the constellation of Aquarius the Water Carrier. This westward movement of the equinoxes is called the *precession of the equinoxes.* The equinoxes slide westward along the ecliptic at the rate of about 50.2 seconds each year and will complete one circle in the sky in about 26,000 years.

Although the vernal equinox has not yet moved into the constellation of Aquarius, some people believe that the Age of Aquarius has already started. The misunderstanding appears to stem from the fact that the early astrologers had assigned only 24,000 years for the period of the earth's precession, or 2000 years for each age.

The precession of the equinoxes is produced by the earth's third principal motion, also called precession, which is defined as the slow circular motion of the earth's

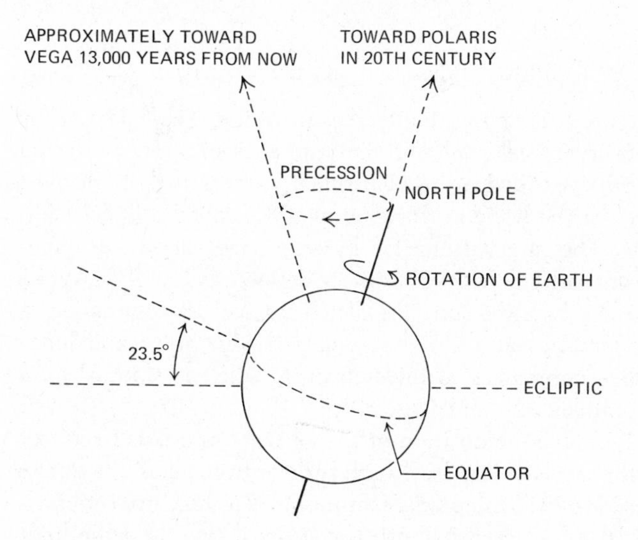

FIG. 6.10
The slow change in the direction of the earth's rotation axis (precession) and the consequent change in the pole star.

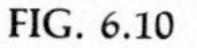

axis about a line that is perpendicular to the plane of its orbit (Fig. 6.10). This motion is caused by the gravitational attraction of the sun and moon for the earth. The sun's gravitational force tends to pull the earth's equatorial bulge toward the plane of the ecliptic, and the moon's differential gravitational force tends to pull the earth's bulge toward the plane of the moon's orbit. Since the moon's gravitational attraction for the earth is greater than the sun's, only the moon's effect on the earth will be illustrated.

The earth as a rotating body acts like a gyroscope and resists the moon's efforts to move the bulge to the plane of the moon's orbit. The earth's equator maintains its angle of 23.5° to the plane of the ecliptic, and the earth's axis moves in a direction opposite to the earth's rotation, describing a circle of 23.5° radius about the ecliptic pole, which is a point 90° from the ecliptic plane. When the earth is observed from above its north pole, the preccessional motion is clockwise. This motion is similar to that of a spinning top, whose axis of rotation is inclined toward the vertical. At the present time, the earth's axis points toward the star Polaris. Nearly 4000 years ago, the earth's axis pointed toward the star Thuban in the constellation of Draco the Dragon. About 12,000 years from now, it will be pointing toward the star Vega in the constellation of Lyra the Harp.

The precession of the equinoxes does not affect the seasons or their sequence. It simply causes a given season to occur when the earth is in a different position in its orbit around the sun with respect to the stars.

6.7 ARC AND TIME UNITS

The circumference of a circle is divided into 360 equal units (degrees). The degree is subdivided into 60 equal units (minutes), and the minute is subdivided into 60 equal units (seconds). A portion of the circumference is usually expressed as an angle or as an arc, because an angle of 12°17′12″ subtends, or corresponds, to an arc of the same value.

Astronomical angles are often expressed as units of time. On the first day of spring the sun follows a path in the sky which coincides with the celestial equator. During that 24-hour period, the sun travels along 360 degrees of arc. Therefore, in one hour the sun travels 15°; in four minutes, it travels 1°; and in one minute, it travels four seconds of arc.

Arc units		*Time units*
15 degrees	=	1 hour, or 60 minutes
1 degree	=	4 minutes
1 minute	=	4 seconds

6.8 TIMEKEEPERS

Although accurate records are not available, we can imagine what early attempts were made to understand the apparent motion of the celestial bodies and to measure the flow of time. The apparent motion of the sun and the moon presented two natural and simple means of measuring time. The recurring astronomical phenomenon of the rising sun was the interval of time defined by early peoples as the *day*. The interval of time between two consecutive new moons was defined as the *lunar month*. The interval of time called the *year* was determined by the apparent position of the rising sun. In the summer, it appeared to rise in the northeast. With each passing day, its rising position shifted slowly to the south so that by winter, it appeared to rise in the southeast. Then, its rising position shifted slowly to the north until it appeared to rise once again in the northeast in the summer.

One of the earliest devices used in determining the sun's position and the length of the year was the shadow clock, or *gnomon*, developed by the Egyptians. In its simplest form it was just a straight stick placed vertically in the ground. The length of the stick's shadow indicated the season. The longest shadow occurred in the winter, when the noon sun's position was low in the sky, and the shortest shadow occurred in the summer, when the noon sun's position was high in the sky. The counting of the sunrises during the interval of time between two consecutive appearances of the shortest shadow established the number of days in the year.

Since the gnomon could be used only during the day, when the sun was visible, the water clock, or *clepsydrae*, which could be used both day and night, was developed in Egypt about 1500 B.C. It operated on the principle of water entering or leaving a container at a regular rate of flow. In the middle of the seventeenth century, the Dutch physicist Christian Huygens invented the *pendulum*, which revolutionized the manufacture of timekeepers and made all previous methods and devices obsolete. With the invention of the pendulum and the discovery by Galileo that its period of oscillation is independent of its amplitude, the construction of a pendulum-operated clock was made possible. The pendulum clock has in turn been superseded by the electric clock. For the most precise timekeeping, however, atomic clocks, which count the vibrations of atomic systems, are used.

6.9 SIDEREAL TIME

Figure 6.11 shows that a crossing of the visible part of the observer's celestial meridian (upper branch) by a celestial body is called an *upper transit*; a crossing of the part below the horizon (lower branch) is called a *lower transit*. The interval of time between two successive upper transits by a celestial body is defined as one day. When the body is the sun, it is called a *solar day*; the moon, a *lunar day*; and a star, a *sidereal day*. Solar and lunar days commence at lower transit, whereas sidereal days commence at upper transit.

Since the position of the vernal equinox is fixed (for all practical purposes) with respect to the stars, it can be used to determine sidereal time. When the vernal equinox is at upper transit, both the sidereal time and the hour angle of the vernal equinox are zero hours.

The sidereal day is approximately 3 minutes 56 seconds shorter than the solar day. The difference can be easily understood from Fig. 6.12. When the earth has revolved around the sun from position 1 to 2, it has completed one rotation with respect to the vernal equinox and one sidereal day. Before the earth can complete one rotation with respect to the sun and one solar day, it must rotate about one more degree at the end of the sidereal day. To accomplish this, the earth must move in its orbit around the sun approximately 3 minutes 56 seconds. Since the sidereal day is about four minutes shorter than the solar day, the stars appear to rise and set about four minutes earlier each day by solar time; thus, at any given time each evening, the star sphere appears to have moved slightly to the west.

6.10 APPARENT SOLAR TIME

Time measured by reference to the actual sun is called *apparent solar time*. The apparent solar day begins at midnight, when the sun is at lower transit. The interval of time between midnight and noon (when the sun is at upper transit) is designated A.M. (ante meridian), which means "before the meridian," and the interval of time between noon and midnight is designated P.M. (post

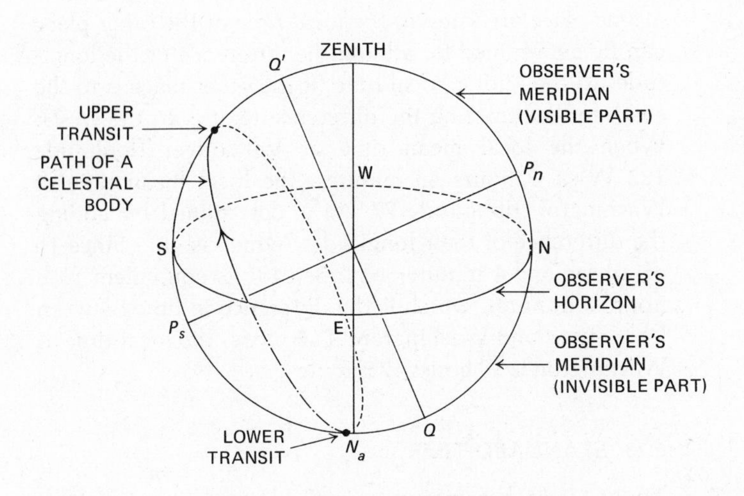

FIG. 6.11
The upper and lower transits of a celestial body in its daily path in the sky. When a body crosses the visible part of the observer's meridian, it is called upper transit. When a body crosses the observer's meridian below the horizon, it is called lower transit.

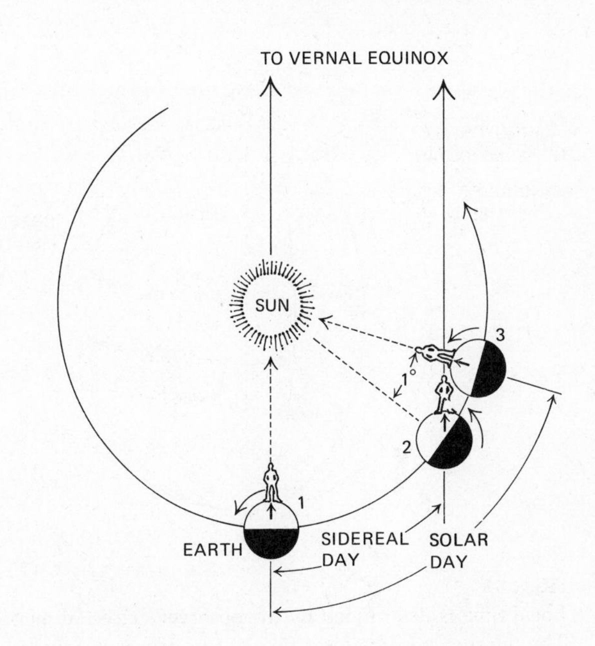

FIG. 6.12
The duration of time for the earth to complete one rotation with respect to the vernal equinox or a star is a sidereal day. A sidereal day is approximately 3 minutes 56 seconds shorter than a solar day, the time in which the earth completes one rotation with respect to the sun.

meridian), which means "past the meridian." Apparent solar time is variable for two important reasons: (1) according to Kepler's second law of motion, the earth's orbital motion is faster when the earth is closer to the sun, and (2) since the sun appears to move along the ecliptic rather than the celestial equator, it also moves either northward or southward as it moves eastward.

6.11 MEAN SOLAR TIME

Since the actual sun is not a uniform timekeeper, a fictitious sun was invented to move along the celestial equator at a uniform rate equal to the average rate of the actual sun's motion along the ecliptic. The time kept by the fictitious sun is called *mean solar time*. The difference between mean solar time and apparent solar time at any instant is called the *equation of time*. This difference is never more than 16.5 minutes and is tabulated for every day of the year in the *American Ephemeris (Nautical Almanac)*. This practice, however, was discontinued in 1965. The equation of time is designated (+) when the apparent time is faster and (−) when it is slower than the mean time. About four times each year, the apparent and mean solar times are equal, and then the equation of time is zero.

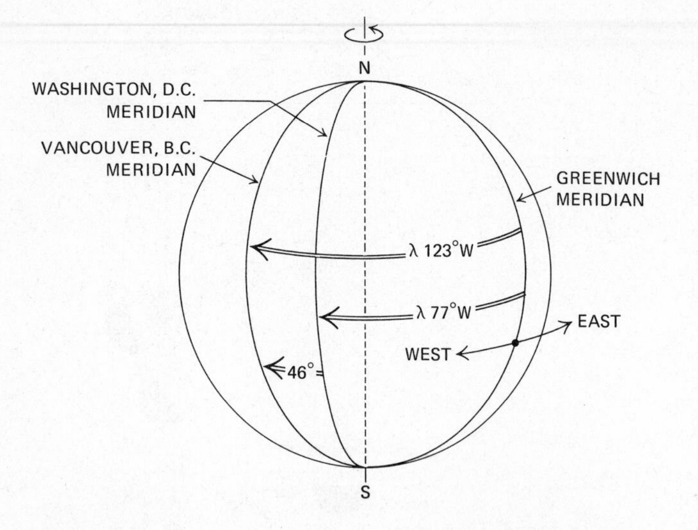

FIG. 6.13
Local time is determined by the observer's celestial meridian. The difference in the local time at two places is equal to the difference of their longitudes.

6.12 LOCAL TIME

Time determined by reference to the observer's celestial meridian is called *local time*. By international agreement, the local mean solar time at the meridian that passes through the Greenwich Observatory at Greenwich, England, is called *universal time*. Astronomical and navigational data are usually recorded in this time. Since the earth's rotational motion is to the east, the sun appears to move across the sky to the west; therefore, local times at places to the east of the observer are always later. For example, the local time at Vancouver, British Columbia, is earlier than at Washington, D.C., and the time at both places is earlier than at Greenwich, England, because the sun passes over the eastern meridians first. In Fig. 6.13 the difference in the local time at two places is equal to the difference of their longitudes.

When the longitude of both places and the local time at one place are known, the local time at the other place can be determined by adding the difference of the longitudes to the known local time (if the other place is to the east) and subtracting the difference (if it is to the west). When the local mean time at Vancouver (longitude 123°W) is 7 hours 43 minutes, the local mean time at Washington (longitude 77°W) is determined by adding the difference of their longitudes, which is 46°. Since 1° of arc equals 4 minutes of time, 46° is equivalent to 3 hours 4 minutes, which is the difference in time between Vancouver and Washington. Therefore, the local time at Washington is 10 hours 47 minutes.

6.13 STANDARD TIME

Today, if each community in the United States kept its own local mean time, as was the case nearly 80 years ago, the situation would be chaotic, and only places on the same meridian would have the same time. To avoid this situation, a world-wide system of time zones was established by dividing the earth into 24 zones, each 15° wide (Fig. 6.14). The time for all places in each zone is established by the zone's central meridian, which is designated as the *standard meridian*. The longitudes of the standard meridians are increments of 15 degrees, starting with the standard meridian which passes through the Greenwich Observatory (longitude 0°). The standard meridians to the west of Greenwich are designated positive; those to the east, negative.

The zone in which a place is located can be determined by dividing its longitude by 15. Chicago, with a longitude of 88° west, is in zone +6. This designation makes it simple to convert the observer's standard time to universal time by simply applying the zone number with its appropriate sign. The +6 designation means that the time in that zone is 6 hours earlier than universal time. An easy rule to remember is that "when the longitude is west, Greenwich is best, and when the longitude is east, Greenwich is least." "Best" implies later time, and "least," earlier time. What is the universal time when the standard time is 10:32 A.M. at longitude 79°42′ west? The

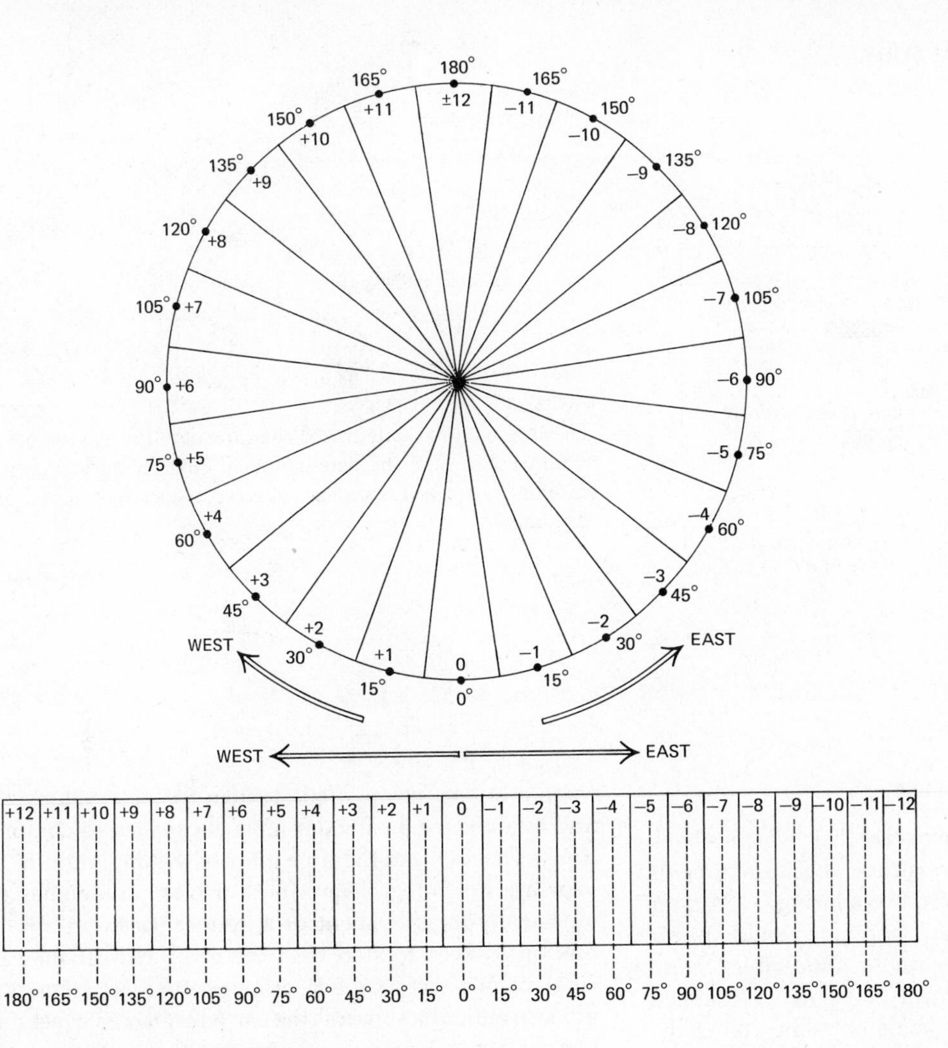

FIG. 6.14
Diagram of the earth's time zones as viewed from above the north pole. The degrees indicate the longitudes of the standard meridians of the zones from the Greenwich meridian. The inside numbers indicate the corrections in hours applied to universal time to obtain the standard time of the zones, where + means earlier and − means later.

standard meridian for longitude 79°42′ west is 75°, and its zone designation is +5.

Standard time	10:32
Zone number	+ 5:00
Universal time	15:32

Zone boundaries over water areas are regular, whereas those over land areas are most irregular. This difference was instituted for the convenience of the people in regional and local areas. There are places in the established time zones of the United States where the time differs by one hour from the standard time. The United States is divided into four time zones: *Eastern, Central, Mountain,* and *Pacific.* The longitudes of the standard meridians of each zone are: 75°, 90°, 105°, and 120°. The zone designations are: +5, +6, +7, and +8, which means that their standard times are from 5 to 8 hours earlier than universal time.

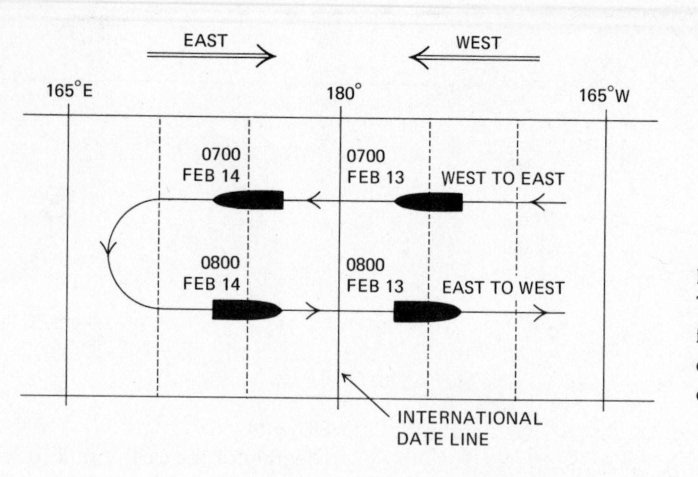

FIG. 6.15
The international date line. When the date line is crossed from west to east, the date is moved one day back. When the date line is crossed from east to west, the date is moved one day ahead.

6.14 INTERNATIONAL DATE LINE

Travelers heading westward, must set their watches back one hour for each 15° of longitude, because time is always later to the east; therefore, during a 24-hour period, they would lose 24 hours, and time would appear to be standing still. This, however, is impossible; therefore, by international agreement the date is changed when the traveler crosses an imaginary line called the *international date line*. The line generally runs along the 180° longitude meridian, except for the land areas, where it moves to the west around the Aleutian Islands and to the east around Fiji, New Zealand, and Siberia. When the international date line is crossed from east to west, the date is moved one day ahead, and when it is crossed from west to east, it is moved back one day (see Fig. 6.15).

6.15 THE EARLY HISTORY OF CALENDARS

The earth's rotation, the apparent motion of the sun, and the apparent motion of the moon serve as the foundation for determining time; however, all attempts to combine these motions into a single system for determining time and establishing a calendar have always led to complications. The fact that the time units of the day, month, and year are not integral multiples of one another has prevented the establishment of a system based on any one motion to keep in step with the other two. If the earth rotated an exact number of times for each revolution made around the sun or if the earth rotated an exact number of times for each revolution of the moon around the earth, there would be considerably less difficulty in developing an accurate calendar.

The earliest known calendar used by most of the ancient people was the *lunar calendar,* which is still used by the Mohammedans. Of the simplest type, this calendar is based on the observation that the new moon occurs regularly about every 29.5 days. The lunar year has 354 days and consists of 12 months. Odd months have 30 days, and even months have 29 days.

The Egyptians were the first to develop and use a *solar calendar* based on the annual apparent motion of the sun and the recurring cycle of the seasons. At first, there were 360 days in the year, but this was later revised to 365 days. This calendar had 12 months of 30 days each, except for the last month, which had 35 days. Later, the Egyptians discovered that the true length of the year was about 365.25 days; however, nothing was done about it until 238 B.C., when Ptolemy Evergetes I introduced the leap year, i.e., every fourth year had 366 instead of 365 days.

The ancient Jewish 12-month *lunisolar calendar*, a complex compromise of the lunar and solar calendars, is still in use today. Although it was adopted in the third centruy A.D., it had its roots in antiquity. It is probably the only calendar in which the counting of years starts not from any particular event in their history, but from the day of creation, which the Jews established as having occurred in the fall of 3761 B.C., according to our present calendar.

In the Jewish calendar, each month is based on the motion of the moon around the earth, that is, 29.5 days. This makes the lunar year of 12 months about 11 days shorter than the solar year. Since three lunar years are about 33 days shorter than three solar years, every third lunar year has 13 lunar months instead of 12. This correction is inadequate, because there are 1092.63 days in 37 lunar months and 1095.75 days in 36 months, which means that there are about four days less in three lunar years than there are in three solar years. To correct this discrepancy, the 19-year period was established in which 19 solar years contain 6939.60 days, while 19 lunar years (12 years of 12 months each and 7 years of 13 months each) contain 6939.69 days.

The ancient Roman calendar was first a lunar type of 10 months and later a lunisolar type of 12 months. The passing of the years was counted from the legendary date of the founding of the city of Rome. The year was designated by the symbol A.U.C., the first letters of the Latin phrase *ab urbe condita,* which means "from the founding of the city." Since the Roman calendar was partly lunar, it was necessary to periodically add days to it so that the religious holidays could be observed at the proper time. This task, the prerogative of the priest, was so poorly administered and abused that the calendar soon fell out of step with the occurrence of natural events—the first day of spring was occurring in December rather than in March.

6.16 THE JULIAN CALENDAR

The history of the calendar is a record of compromises and reforms. The first great reform was made on the Roman calendar in 46 B.C. by Julius Caesar on the advice and recommendation of the Greek astronomer Sosigenes of Alexandria. Julius Caesar decreed that the year 46 B.C., which became known as the "year of confusion," would have 445 rather than 354 days. This was done to correct the nearly 90-day difference that existed between the calendar and the occurence of the seasons. Caesar also decreed that the new calendar would commence on the first day of the new moon following the winter solstice in the year 45 B.C. In the new calendar, 31 days were assigned to the odd months, which were considered to be lucky, and 30 days to the even months, which were considered to be unlucky. February, however, was assigned 29 days because it was the "month of the dead." The Egyptian leap year was also added to the calendar. The new calendar is known as the *Julian calendar*, named in honor of Julius Caesar, who instituted its reforms. Julius Caesar also changed the name of the seventh month, Quintilis, to July, in honor of himself.

Minor changes were made in the Julian calendar in 8 B.C. by Augustus Caesar. He changed the eighth month, Sextilis, to August, in honor of himself, and made the month 31 days long by taking one day from February. This change produced three consecutive months of 31 days each, which was considered unlucky; to rectify this, September and November were changed to 30 days, October and December to 31 days.

Another minor reform was made in the Julian calendar in the early part of the fourth century A.D. by the Greek emperor Constantine, who introduced the so-called Christian seven-day week into the calendar by decree and made it legal throughout the Roman empire. Some scholars believe that the Christian seven-day week originated with the Babylonians rather than with the early Christians and that it was adopted by the Jews during their captivity. It is also believed that the Sabbath was adopted by the Jews from the Babylonian Sabbatu.

In the sixth century (about 528 A.D.) the Scythian monk Dionysius Exiguus introduced the convention of counting years from the date of Christ's birth. Thus A.D. (Anno Domini, meaning the year of [our] Lord) is sometimes written after the year.

6.17 THE GREGORIAN CALENDAR

Even though the Julian calendar eliminated much of the confusion in the Roman lunisolar calendar, it was far from perfect. By 1582 the Julian calendar had accumulated an error of 10 days, so that the first day of spring occurred on March 11 instead of March 21. To correct this error, which made the Julian year slightly longer than the tropical year, the second great reform on the Julian calendar was made in 1582. Authorized by the Council of Trent and instituted by Pope Gregory XIII, it dropped 10 days from the calendar and adopted a unique rule, suggested by the Vatican librarian, Aloysuis Giglio, to keep the calendar more closely in step with the length of the tropical year. All years divisible by 4 were designated leap years, except century years not divisible by 400, starting with the year 1700. This rule eliminated three days every 400 years. The new calendar was known as the *Gregorian calendar.*

Unfortunately, since the Gregorian calendar was instituted by a Catholic pope shortly after the Reformation, many princes who had become Protestants would not accept the papal bull directing them to use the new calendar. The Catholic world adopted it in 1582, and the first Protestant country adopted it in 1700. The last four adoptions occurred in the twentieth century: China in 1912, Turkey in 1917, Soviet Russia in 1918, and Greece in 1923. Despite the fact that the Gregorian calendar is a great improvement over the Julian calendar, the Greek and the Russian Orthodox churches still use the Julian calendar.

6.18 PROPOSED CALENDAR REFORMS

Some people consider the Gregorian calendar to be imperfect because during a period of 1000 years, the date of the first day of spring will change by one day. This error is negligible for all practical purposes and can be easily corrected; however, these people have proposed further reforms to the calendar. The three most interesting of the proposed reforms are the *13-month calendar,* the *World calendar,* and the *Jubilee calendar.*

In the 13-month calendar, a new month, "Sol," is placed between June and July. This year has 364 days, and each month has exactly four weeks of 28 days. The 365th day, which is placed at the end of the year, is considered an extra day and not part of a week. A 366th day is placed at the end of every fourth year, which is a leap year. It, too, is considered an extra day. The distinguishing feature of the 13-month calendar is that all the months are identical in length; therefore, only a one-month calendar would be required.

The World calendar year is divided into four equal quarters of 91 days. Each quarter is divided into three months of 31, 30, and 30 days, respectively. Since the year has 364 days, the extra day is placed at the end of the year, and the 366th day is placed at the end of June in the leap year. The distinguishing feature of this calendar is that any given date will fall on the same weekday every year.

Many people oppose the adoption of these new calendars on the ground that they violate the Christian seven-day week. The basic major calendar reforms in the past have been astronomically inspired, whereas the proposed calendars are based on a change in the traditional, seven-day week. The Jubilee calendar was proposed by several religious groups in an attempt to preserve the seven-day week. A most interesting and unique calendar, it contains 12 months and 52 weeks. Every fifth year, except those divisible by 400 or ending in 25 or 75, is a leap year of 53 weeks.

6.19 THE JULIAN-DAY CALENDAR

Many attempts have been made to determine from the Bible and other sources the exact date of creation. Medieval Jewish scholars established the date as 3761 B.C. In A.D. 1650 James Ussher, an Anglican archbishop, placed the date of creation at 4004 B.C. The oldest date of creation, 5508 B.C., was established by Greek Orthodox theologians.

In 1585 Julius Scaliger proposed a calendar according to which days are numbered consecutively from noon, universal time, January 1, 4713 B.C., the date of creation he had established. This is a very practical calendar because it has no weeks, months, or years. To obtain the interval of time between any two events or to determine the time of an event requires a simple addition or subtraction of two numbers. The day on which the event occurred is called the *Julian day.* This calendar is used by astronomers in predicting the date of the occurrence of a celestial event and the times of maxima and minima in the period of variable stars. The Julian-day numbers for each year are tabulated in the *American Ephemeris and Nautical Almanac* and the *Handbook of the Royal Astronomical Society of Canada.* The Julian day for January 1, 1979, for example, is designated as 2,443,875 J.D.

SUMMARY

The earth's rotation axis is tilted with respect to its orbital plane around the sun. Thus, as the earth moves in its orbit the northern and southern hemispheres are tipped alternately toward and away from the sun. When the northern hemisphere is tipped toward the sun, the sun shines on it more directly and that hemisphere receives its maximum heating effects. Thus it is summer in the northern hemisphere. When the northern hemisphere is tipped away from the sun, it receives less sunlight in that hemisphere and it is winter.

The direction in which the earth's rotation axis points slowly changes, a phenomenon called precession. Precessional motion of the rotation axis is very slow; one cycle requires 26,000 years.

The motions of the earth in space are used to establish time-keeping systems. The rotation of the earth defines one day; the revolution of the earth around the sun defines one year. Because the earth does not make an even number of full rotations in one orbit about the sun (it makes about 365¼ rotations), it is periodically necessary to adjust the calendar to make up for the accumulated extra quarter days. This is why there are leap years. Because of the earth's orbital motion, the sun does not move across the sky at the same rate throughout the year. Thus, for time-keeping purposes, it is the mean (average) motion that is used as a reference.

Various coordinate systems are used to identify locations on the earth and in the sky. On the earth, the reference points are taken as the north and south poles (the points where the earth's rotation axis passes through the surface). Lines drawn from pole to pole (segmenting the earth's surface like an orange) are called lines of longitude. By tradition the zero reference longitude (prime meridian) passes through Greenwich, England. The equator encircles the earth exactly halfway between the poles. Lines of latitude are drawn parallel to the equator and run around the earth. A location on the earth's surface can thus be specified by giving its latitude and longitude.

A similar coordinate system has been devised for the sky (celestial sphere), using the celestial pole and celestial equator as the basic reference marks. The celestial poles are located by extending the earth's rotation axis outward so as to pass through the celestial sphere. Positions on the sky are then labeled according to right ascension and declination, these terms being analogous to terrestrial longitude and latitude, respectively.

Alternatively, a horizon system is sometimes used to locate objects in the sky. In this coordinate system positions are specified by height above the horizon (altitude) and distance around the horizon from north (azimuth).

REVIEW QUESTIONS

1. Where is Polaris located as seen from (a) the north pole, (b) the earth's equator, (c) the south pole?
2. Why do all celestial objects appear to rise in the east?
3. Explain the location of the vernal equinox, autumnal equinox, summer solstice, and winter solstice on the ecliptic with respect to the celestial equator.
4. Is the sun ever directly overhead as seen from Los Angeles or New York? On which days of the year is the sun directly overhead as seen from the equator?
5. Discuss the validity of the statement that since summer in the northern hemisphere is warmer than winter, the sun is closer to the earth at that time.

6. What names are given to the days when the sun is directly overhead as seen from the Tropic of Cancer and the Tropic of Capricorn?
7. Suppose the earth were tilted at 45° instead of at 23.5°. Would the sun ever be directly overhead as seen from Los Angeles?
8. Will Polaris always be the pole star? Why?
9. Where would you expect the vernal equinox to be in the sky at midnight local time on the first day of autumn?
10. What is the precession of the equinoxes? What causes this motion? What is its period?
11. Suppose that the observer is at the equator. (a) Where is his/her zenith point located? (b) Where is the north celestial pole located? (c) Describe the daily motion of the celestial bodies.
12. Stars that never set during the course of the night because of their nearness on the sky to Polaris are called circumpolar stars. What is the minimum declination for a circumpolar star at your latitude?
13. What are the right ascension and declination of the (a) vernal equinox, (b) autumnal equinox, (c) summer solstice, and (d) winter solstice?
14. Convert 2 hours 18 minutes to arc units. Convert 79°30′ to time units.
15. Define a sidereal and a solar day. What is the difference between these two time intervals? Which is longer? Why? How does this difference affect the time that the stars rise on successive days as based on solar time?
16. Why was the concept of mean solar time invented? How does it differ from apparent solar time? Is the sun on the meridian at noon apparent solar time?
17. What timekeepers are used to determine apparent solar time and mean solar time? What is the equation of time? How is it used?
18. What is the difference in the local standard time between two places whose difference in longitude is 75°?
19. In what time zone do you live? What is the longitude of its central meridian? What is the difference between your standard time and local civil time?
20. Are the time zones spaced uniformly on the earth's surface? Explain.
21. What is universal time? Who uses it? Why?
22. Why was the international date line established? A ship moving in an easterly direction crosses the international date line at 2:30 A.M. standard time on February 13. What will be the time and date when the ship has crossed the line?
23. When was the Julian calendar introduced? Explain the reforms that were made to establish the Julian calendar. Who instituted these reforms?
24. On what calendar were the Gregorian reforms made? Why? What were these reforms? How did the people react to the Gregorian calendar? Why?
25. Will there be a need to reform the Gregorian calendar? What is your reaction to the 13-month calendar and the World calendar that have been proposed? Should they be adopted? Why?
26. Explain the principle on which the Julian-day calendar is based. What is the Julian day for an event that will occur on November 19, 1983?

CHAPTER 7
THE MOON: EARTH'S NEAREST NEIGHBOR

After the sun, the moon is the most conspicuous object in the sky. At a mean distance of 238,857 miles (384,401 km), it is earth's nearest neighbor and only natural satellite. Before the invention of the telescope, people saw the moon only as a beautiful and mysterious silvery sphere patched with irregular dark areas. Most of the pictures of the moon drawn in manuscripts and executed on medieval stained-glass windows depicted the dark areas as the outline of a human face. After the invention of the telescope, Galileo sketched rather crudely the surface features of the moon from telescopic observations and presented them with a complete description in his book *Sidereus Nuncius.* He named the dark areas *maria,* believing them to be seas, and considered the light areas as irregular land forms.

7.1 TRUE MOTIONS

The two major motions of the moon are *revolution* and *rotation.* When the sun is used as a reference point, the moon revolves around the earth in an elliptical orbit in about 29.5 days. This is called a *synodic month* and is the interval of time from one full moon to the next. When a star is used as a reference point, the moon completes one revolution around the earth in about 27.33 days. This interval of time is called a *sidereal month.* The difference between these two periods is approximately two days (Fig. 7.1).

As the earth revolves around the sun and the moon revolves around the earth during a synodic month, the moon continuously changes its position with respect to the sun and follows a wave-like path around the sun (Fig. 7.2). Since the sun's distance from the earth is nearly 400 times greater than the moon's, the moon's path is considerably flatter than that shown in the figure. Also, the gravitational force exerted on the moon by the sun is more than twice that exerted by the earth; therefore, the moon's greatest acceleration is always in the direction of the sun, which causes the moon's path to be concave toward the sun.

The moon's orbit is inclined to the orbital plane of the earth at a mean angle of 5°09′. This angle varies about 18 minutes because of the moon's perturbations. The moon's equator is inclined about 6.5° to its orbital

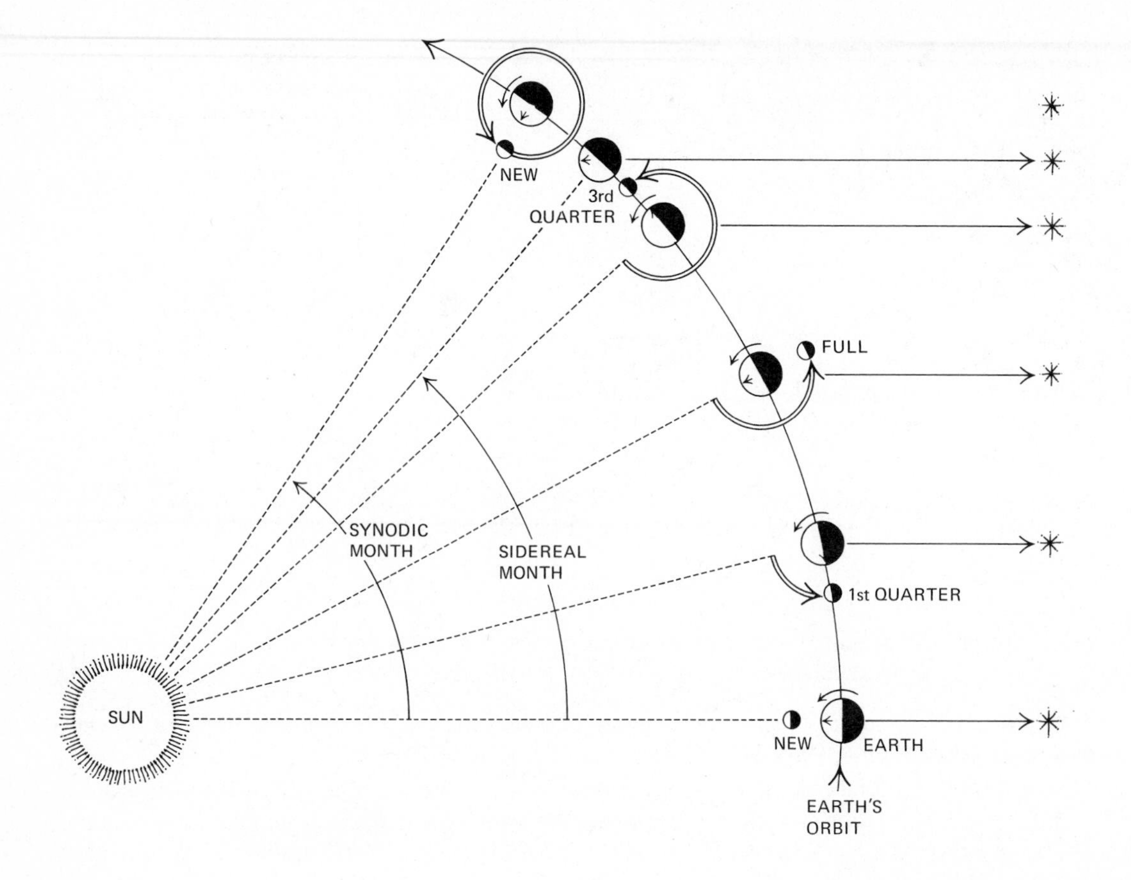

FIG. 7.1
The moon completes one revolution with respect to the stars in 27.33 days, the sidereal month. One revolution with respect to the sun is 29.5 days, the synodic month.

plane. The point in the moon's orbit that is closest to the earth is called *perigee,* and the point farthest from the earth is called *apogee* (Fig. 7.3). The perigee distance is about 221,457 miles (356,400 km), and the apogee distance is about 252,712 miles (406,700 km).

As the moon revolves around the earth, it rotates about its axis; its period of rotation is equal to its period of revolution. Therefore, the moon always presents the same face to the earth and does not appear to rotate (Fig. 7.4); with respect to the sun, the moon does rotate.

To illustrate this effect, put a wastepaper basket in the middle of the room and walk around it so as to always face the basket. Note that after completing a circle around the basket you have made a complete rotation yourself.

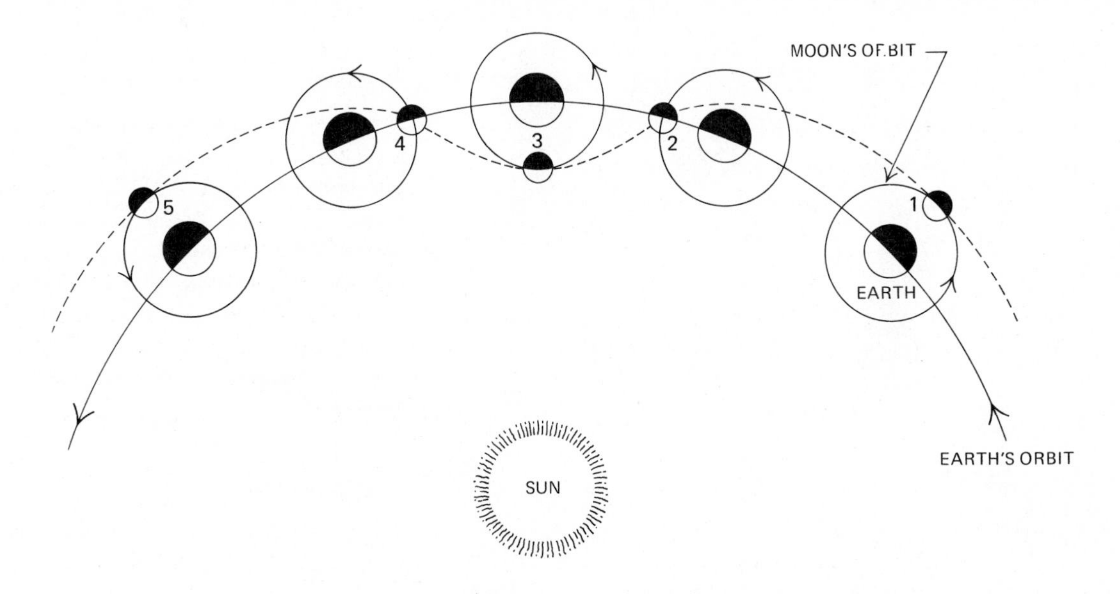

▲
FIG. 7.2
As the moon revolves around the earth, it moves along a wavy path which is always concave toward the sun, as indicated by the dotted line. Its path is greatly exaggerated here.

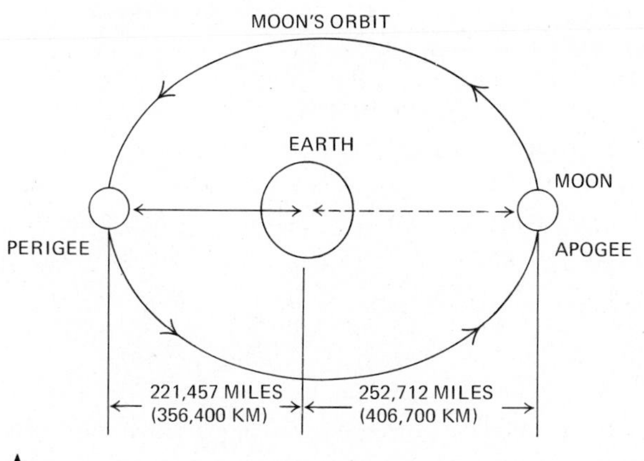

▲
FIG. 7.3
The lunar orbit around the earth. The nearest distance to the earth is called perigee; the greatest distance is called apogee. (Eccentricity is greatly exaggerated.)

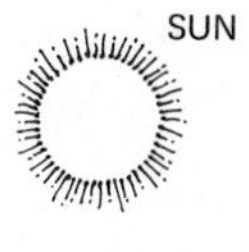

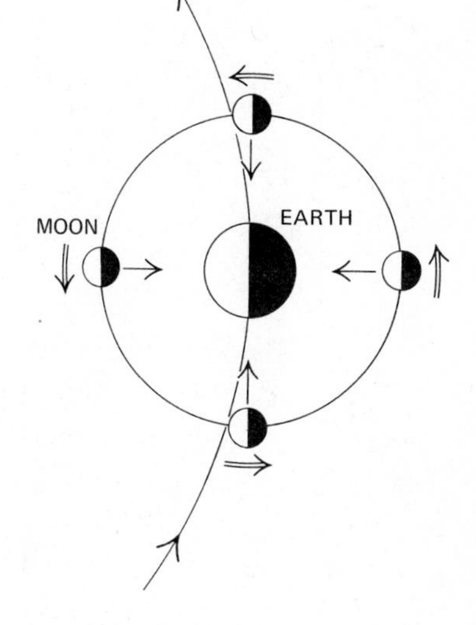

FIG. 7.4 ►
With respect to the earth, the moon does not rotate. With respect to the sun, the moon completes one rotation in 29.5 days.

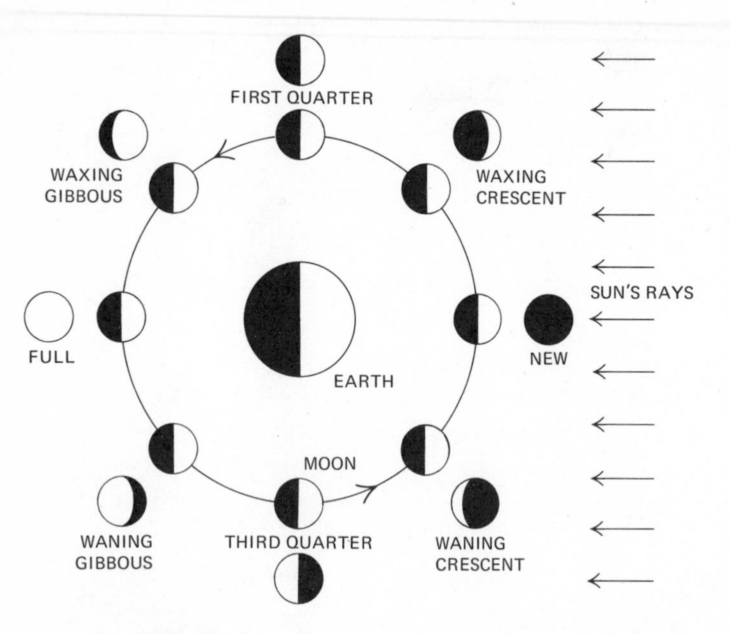

FIG. 7.5
Phases of the moon. The outer figures show the phases of the moon as seen from the earth.

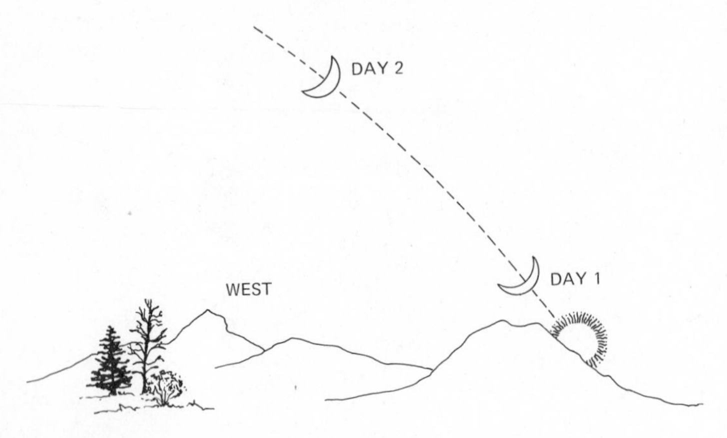

FIG. 7.6
Changing aspect of the moon at sunset for a series of days.

7.2 APPARENT MOTIONS

The moon appears to rise daily in the east, move across the sky, and set in the west. This daily apparent westward motion is caused by the rotation of the earth about its axis. Every day the moon and the sun appear to travel eastward with respect to the stars—the moon at the rate of about 13°, and the sun at the rate of about 1°. Therefore, since the moon travels eastward faster (due to its orbital motion about the earth) than the sun by about 12° each day, it appears to loop about the sun once every 29.5 days.

7.3 PHASES

The daily apparent eastward motion of the moon produces changes in the shape of its illuminated disk. These changes are called the *phases of the moon.* When the moon is between the earth and the sun, its disk is dark and its phase is new. The new moon rises and sets with the sun. For the next 7.5 days, the moon's phase is a waxing crescent (Fig. 7.5 and 7.6) and is seen rising higher in the western sky each day after sunset. When the moon has traveled about one-fourth of its path around the earth and is about 90° east of the sun, it is in the first-quarter phase, and one-half of its disk appears illuminated. The first-quarter moon rises at noon and sets at midnight. During the second 7.5 day period, the phase is waxing gibbous, and the moon rises between noon and sunset and sets between midnight and sunrise. When the moon has traveled one-half of its path around the earth and is about 180° from the sun—on the opposite side of the earth—it is in the full phase, and the entire disk appears illuminated. The full moon rises at sunset and sets at sunrise. During the last half of the cycle, the phases change from full, to waning gibbous, to third-quarter, to waning crescent, and back to new. A summary for the time of the rising and setting of the phases of the moon follows.

Phases	*Rises*	*Sets*
New	Sunrise	Sunset
First-quarter	Noon	Midnight
Full	Sunset	Sunrise
Third-quarter	Midnight	Noon

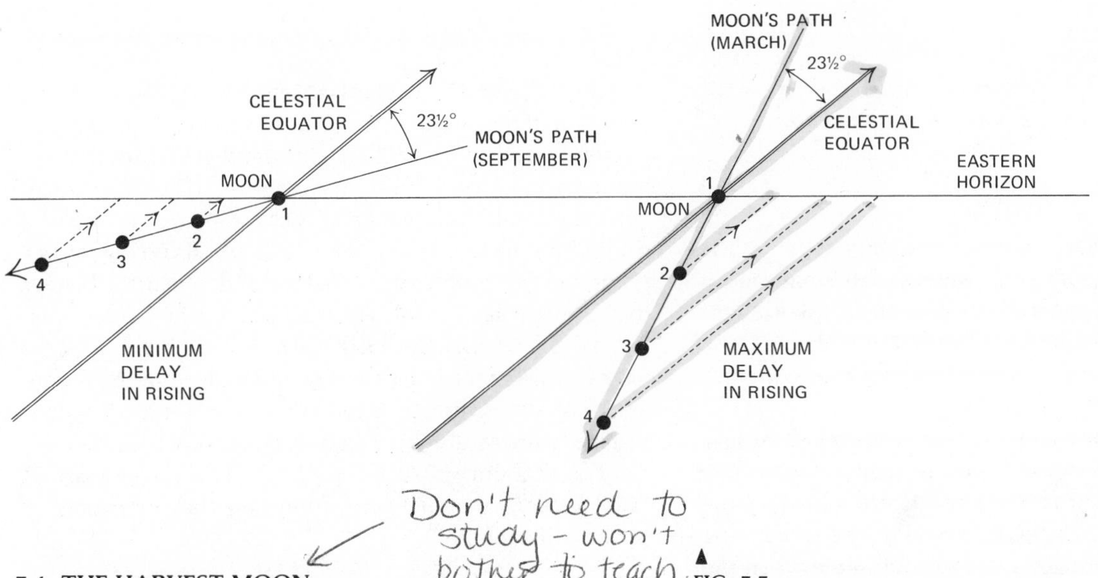

Don't need to study - won't bother to teach

7.4 THE HARVEST MOON

▲
FIG. 7.7
When the sun is at the autumnal equinox and is setting, the harvest moon is at the vernal equinox (1) and is rising. For several evenings in September, the moon rises only a few minutes later than it did the previous evening because its daily path is more nearly parallel to the horizon. The greatest delay in the rising of the moon occurs in March, when its path makes the greatest angle with the horizon.

The moon's apparent eastward motion produces an average daily delay of about 51 minutes in the time of moonrise and moonset. Thus at most times of year the nearly full moon is visible in the early evening on only one or two days each month. However, in the middle latitudes of the northern hemisphere in September, instead of rising 51 minutes later each night, the moon may rise only 10 minutes later if the sun is near the autumnal equinox. Thus the nearly full moon will appear shortly after sunset for several days. The full moon appearing at this time of year is called the *harvest moon* because by rising shortly after sunset it permits the farmer to extend the working day into the evening hours.

A graphic explanation of the harvest moon is shown in Fig. 7.7. In September at sunset, when the sun is at or near the autumnal equinox, the full moon is at or near the vernal equinox and is rising on the eastern horizon directly opposite the sun. At this time of the year, the ecliptic is below the celestial equator and makes a minimum angle with the horizon. Since the moon's orbital plane is inclined only 5°09′ to the plane of the ecliptic, we will assume for simplicity that both coincide and that the moon moves in the plane of the ecliptic. On the following evening at sunset, the moon is in position 2, having traveled 13° eastward along the ecliptic. Due to the earth's rotation, the moon's daily motion is along lines parallel to the celestial equator; therefore, since the length of this path to the horizon is considerably less than 13°, the harvest moon appears to rise much earlier than the average 51-minute delay.

A similar situation occurs during the next several days as the moon travels to positions 3 and 4. In October, the situation has changed only slightly; the angle that the ecliptic makes with the horizon has increased by a small amount so that the full moon, or *hunter's moon,* rises later than the harvest moon, but still much earlier than the average 51-minute delay. In March, the situation is reversed. The autumnal equinox is on the eastern horizon and the ecliptic, which is above the celestial equator, makes a maximum angle with the horizon; therefore, the full moon rises later than the average 51-minute delay.

7.5 LIBRATIONS

Although from the earth we can see only 50% of the moon's surface at any one time, during each lunar month we are able to see 9% more of its surface because of slight changes in its orientation toward the earth. These changes, called *librations*, are *latitudinal*, *longitudinal*, and *diurnal*.

Latitudinal libration occurs because the moon's equator is inclined about 6.5° to the plane of its orbit, permitting the observer to see a few degrees beyond the moon's north pole when the moon is on one side of the earth and a few degrees beyond the south pole when the moon has moved in its orbit to the other side of the earth.

Since the moon's rotational motion is uniform, whereas its orbital motion is variable, the moon is displaced slightly eastward and then slightly westward during each lunar month. This variation, which is called longitudinal libration, permits the observer to see a few degrees beyond in longitude at each edge of the moon.

Diurnal libration results from the earth's rotational motion and from the fact that the moon is observed from the earth's surface rather than from its center. The rotation of the earth permits the moon to be observed from two widely separated positions during a 12-hour period and enables the observer to see about 1° around both edges of the moon—western edge at moonrise and eastern edge at moonset.

7.6 ATMOSPHERE

The moon shines by reflected sunlight; therefore, its spectrum is a dim replica of the sun's spectrum. If the moon had an atmosphere, it could be detected spectroscopically by the presence of absorption bands produced by sunlight passing through it's atmosphere twice before reaching the observer on the earth. Without an atmosphere, none of the weather elements that exist on the earth are found on the moon, and the daytime lunar sky appears black.

In spite of the general absence of earth-based evidence that the moon has an atmosphere, a mass spectrometer on the Apollo 15 flight found evidences that the moon has a very thin atmosphere. The instrument detected small isolated areas of argon and neon as it orbited the moon. It also detected carbon dioxide at one point on the moon's terminator, and scientists believe that it might have come from a fissure near the terminator. We know that the earth's atmosphere moderates the temperature range between day and night. On the moon, when the sun is directly overhead (lunar noon), the lunar surface temperature is about 270°F (132°C); at lunar midnight, it is about −270°F (−168°C). This range of about 540°F (300°C) may help one appreciate the importance of earth's atmosphere.

7.7 DISTANCE FROM THE EARTH

The moon's distance from the earth can be determined by several methods. *Triangulation*, involving the solution of a triangle, is the most direct. This method has been simplified by photographing the moon from two positions on the earth's surface at the same instant when it is on the observer's horizon so as to produce two similar right triangles (Fig. 7.8). When the photographs are compared, they show that the center of the moon's disk has moved in relation to the stars an angular distance of P; therefore, the two angles at the moon in the two right triangles are each equal to $P/2$. In practice, the angle $P/2$ is called the moon's *horizontal equatorial parallax* and is defined as the angle that subtends the earth's radius. The moon's parallax is not constant because the moon's distance from the earth is variable; its average parallax is nearly 1° ($57'02.62''$). The moon's distance from the earth can be calculated by solving the right triangle when the moon's parallax and the earth's radius are known. The moon's mean distance is 238,857 miles (384,393 km).

This experiment can actually be performed fairly easily in collaboration with a friend living in a city a few hundred miles away if pictures of the moon are taken simultaneously (say at 8 P.M. C.S.T.) from both locations. Comparison of the two pictures will show the moon against a slightly different set of stars as seen from two cities.

Another method for measuring the moon's distance from the earth is to use a laser. This device consists of a short ruby rod or tube of gas in which atoms are stimulated to a high-energy level by intense lamp radiation. When the laser "fires," it emits an intense beam of coherent light, that is, light of definite wavelength which travels great distances with practically no dispersion. A laser beam has been directed to three separate clusters of 300 silica reflector cubes mounted on a square frame, which the Apollo 11, 14, and 15 astronauts left on the moon. By measuring the time between sending and receiving the signal, the moon's distance can be determined to an accuracy of about six inches! This is so accurate that it will soon be possible to directly measure motions of the earth's crust due to continental drift by observing distances from the moon to a given location on the earth change with time.

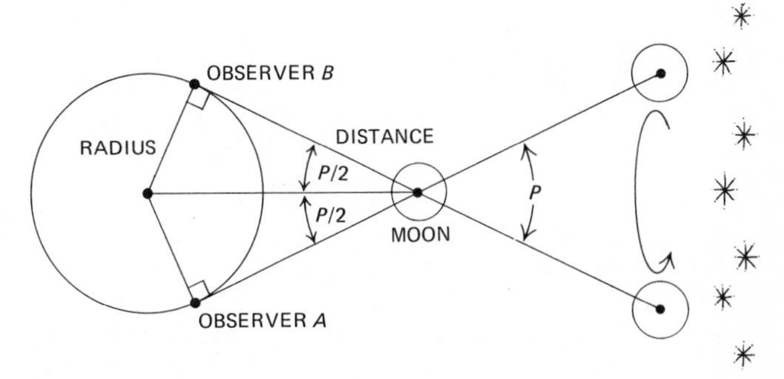

FIG. 7.8
Finding the moon's distance by the triangulation method. The solution of the right triangle involves the moon's horizontal equatorial parallax ($P/2$) and the radius of the earth.

7.8 LINEAR DIAMETER

The moon's linear diameter can be determined from the moon's angular diameter of 31.09′ (approximately ½°) at its mean distance from the earth (238,857 miles, or 384,393 km). In Fig. 7.9, a circle is drawn with the earth's center as its center and a radius equal to the moon's mean distance from the earth. The observer is assumed to be located at the earth's center. Since the angular diameter of the moon is very small, the linear diameter (d) is assumed to be equal to the segment of the circle AB. Therefore, the linear diameter is in the same ratio to the complete circle ($2\pi r$) as the angular diameter is to 360°. Solving for d in the proportion

$$d/2\pi r = \tfrac{1}{2}/360$$

gives us the moon's diameter—about 2162 miles (3479 km).

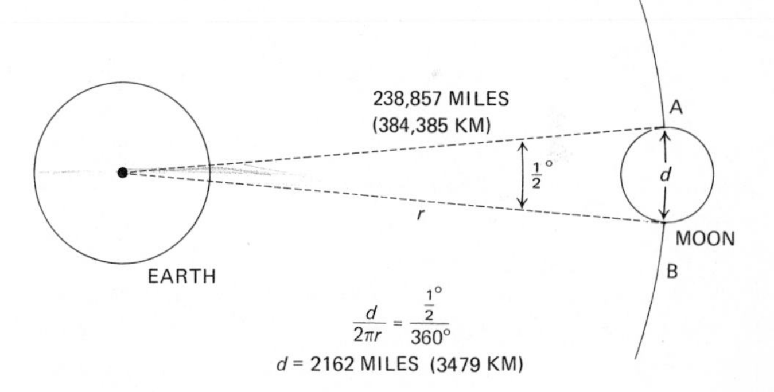

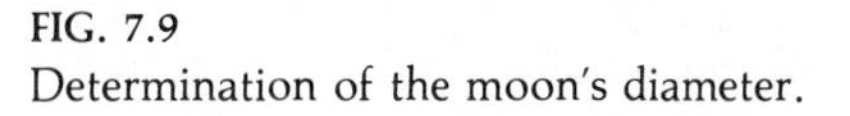

FIG. 7.9
Determination of the moon's diameter.

7.9 MASS

The moon's diameter of 2162 miles (3479 km), which is over one-fourth the size of the earth's diameter, makes it

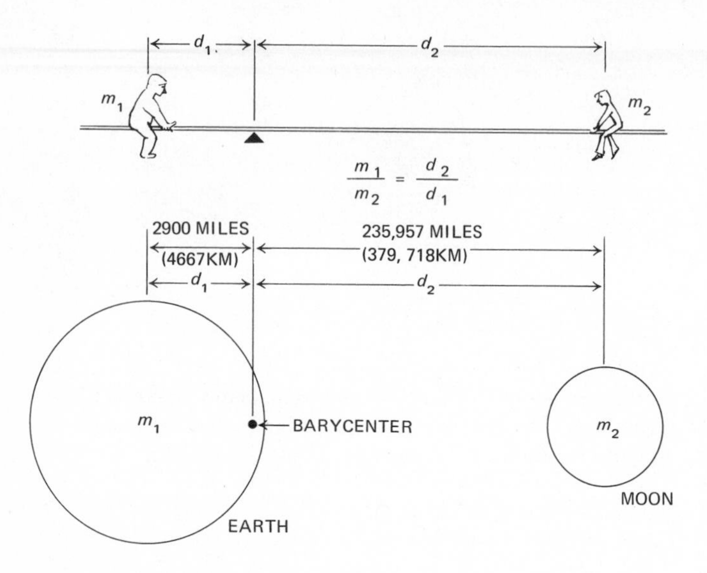

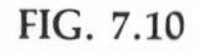
FIG. 7.10
The location of the barycenter of the earth-moon system and the determination of the moon's mass.

the largest satellite in the solar system in comparison to the size of the planet around which it revolves. The earth-moon system, sometimes called a *double planet,* is unique in the solar system not only because of the moon's relatively large size in comparison to the earth but also because of its nearness to that body (Fig. 7.10).

An astronaut in the vicinity of Venus would see the earth as a brilliant, bluish body and the moon as a beautiful, yellowish body about the same brightness as the planet Jupiter. He or she would also observe that the moon slowly oscillates from one side of the earth to the other through an angular distance of approximately ½°, which is the apparent diameter of the full moon. Therefore, the earth and the moon would appear as two planets always seen close together.

Although the moon appears to revolve around the earth as the earth revolves around the sun, the earth and moon actually revolve around their common center of mass, which is also their common center of gravity. This center of mass, called the *barycenter,* moves around the sun in an elliptical orbit. The barycenter lies on the line that joins the centers of the two bodies, and its distance from each body is inversely proportional to the mass of the body. The distance is determined by the formula

$$m_1/m_2 = d_2/d_1,$$

where m_1 is the earth's mass, d_1 is the earth's distance from the barycenter, m_2 is the moon's mass, and d_2 is the moon's distance from the barycenter. A similar situation exists when an adult and small child sit on a seesaw. If the seesaw is balanced, the adult is seated closer to the *fulcrum* (pivot) than is the child.

The location of the barycenter can be established when the sun is observed to oscillate in its motion on the ecliptic. During the first half of the lunar month, the earth in its orbit is slightly ahead of the barycenter, which causes the sun to appear on the ecliptic slightly to the east of its expected position. During the last half of the lunar month, the earth is slightly behind the barycenter, which causes the sun to appear on the ecliptic slightly to the west of its expected position. This means that during the first half of the month, the earth is farther inside the orbital path of the barycenter, and during the last half of it is farther outside. The sun's oscillation indicates that the barycenter is located about 2900 miles (4700 km) from the earth's center, which places it within the earth.

When the values for the earth's and moon's distances are substituted in the formula

$$m_1/m_2 = d_2/d_1,$$

the ratio of the earth's mass to the moon's mass (m_1/m_2) is 81. Thus, the moon's mass is 1/81 the earth's mass, or 8.1×10^{19} tons. The moon's low mass produces a force of gravity that is one-sixth that of the earth. This means that an astronaut who weighs 180 pounds on earth weighs only 30 pounds on the moon.

The moon's density, which can be found by dividing its mass by its volume, is 3.34 grams per cubic centimeter. An analysis of the lunar rocks that were returned by the Apollo astronauts (Chapter 7.14) revealed that the anorthosites have a density of 2.85 grams per cubic centimeter, and the basalts have a density of 3.3 grams per cubic centimeter. There is no close genetic relationship between these two types of rocks from the lunar seas. Since there is an amazing agreement between the chemical composition of the anorthosites and the samples analyzed from the highlands north of the crater Tycho Brahe, there is the possibility that the anorthosites recovered at Mare Tranquillitatis might have come originally from the highlands. Also, concentrations of large, dense material (Mascons) that lie beneath the moon's surface were discovered and observed in several mare basins by the Lunar Orbiters and Apollos as they circled the moon. The discovery was made when the spacecrafts unexpectedly increased their speeds as they passed over these concentrations. The increase in speed was caused by an increase in the lunar gravity due to the concentration of dense material. Thus the moon's density is not uniform; however, samples from different regions must be recovered and analyzed before this question can be resolved.

7.10 THE TELESCOPIC VIEW OF THE MOON

Telescopic views of the first- and third-quarter moons are shown in Fig. 7.11. They give a general overview of the important surface features of the moon. When the moon is seen through a telescope, its image is inverted and reversed—north is at the bottom, with the moon's western limb visible at first-quarter and its eastern limb visible at third-quarter. Note that the surface features are the sharpest at the *terminator* (the line of demarcation between the illuminated and dark portions), which is often called the *sunrise line*.

7.11 EXPLORATION OF THE MOON

For many years prior to the Apollo mission, the moon had been the subject of observations and study by both professional and amateur astronomers. During this period a great deal of knowledge about the moon had been accumulated; knowledge of the lunar structure and its physical characteristics was derived indirectly from the studies of the lunar spectra, radar penetrations of its surface, analysis of supposed meteorites and tektites from its surface, high-resolution photographs taken by the Orbiter and Ranger spacecrafts, and experiments conducted by Lunar Surveyor.

Most of the moon's surface has now been photographed by the Lunar Orbiters (Figs. 7.12, 7.13, 7.14). From the high-resolution photographs, surface features as small as eight feet across were resolved, and three general terrain characteristics were established: flat broad areas, rolling hills with slopes less than 10°, and irregular hills with slopes greater than 10°. The flat areas are the dark areas of the lunar "seas," and the rolling and irregular hills are the light areas of the highlands. Dispersed over the entire lunar surface are thousands of craters that appear to be randomly clustered, although the bright highlands are more heavily cratered than the dark seas.

The exploration of the moon took on a new dimension on July 20, 1969, when two astronauts from the Apollo 11 flight landed on the moon. The moon could now be studied at first hand, and direct confirmation of earlier observations could be obtained. By 1972, five more moon landings had been made. On November 19, 1969, the second landing took place during the Apollo 12 flight, a landing of greater scientific significance than the Apollo 11 flight because the astronauts left instruments on the moon's surface to detect the presence of an atmosphere, ionosphere, dust, and solar wind.

The Apollo 11 and 12 flights were made to establish and perfect landing techniques. Systematic exploration of

FIG. 7.11 ►
Left, third-quarter moon as seen by the unaided eye; right, first-quarter moon as seen by the unaided eye. (Courtesy of NASA.)

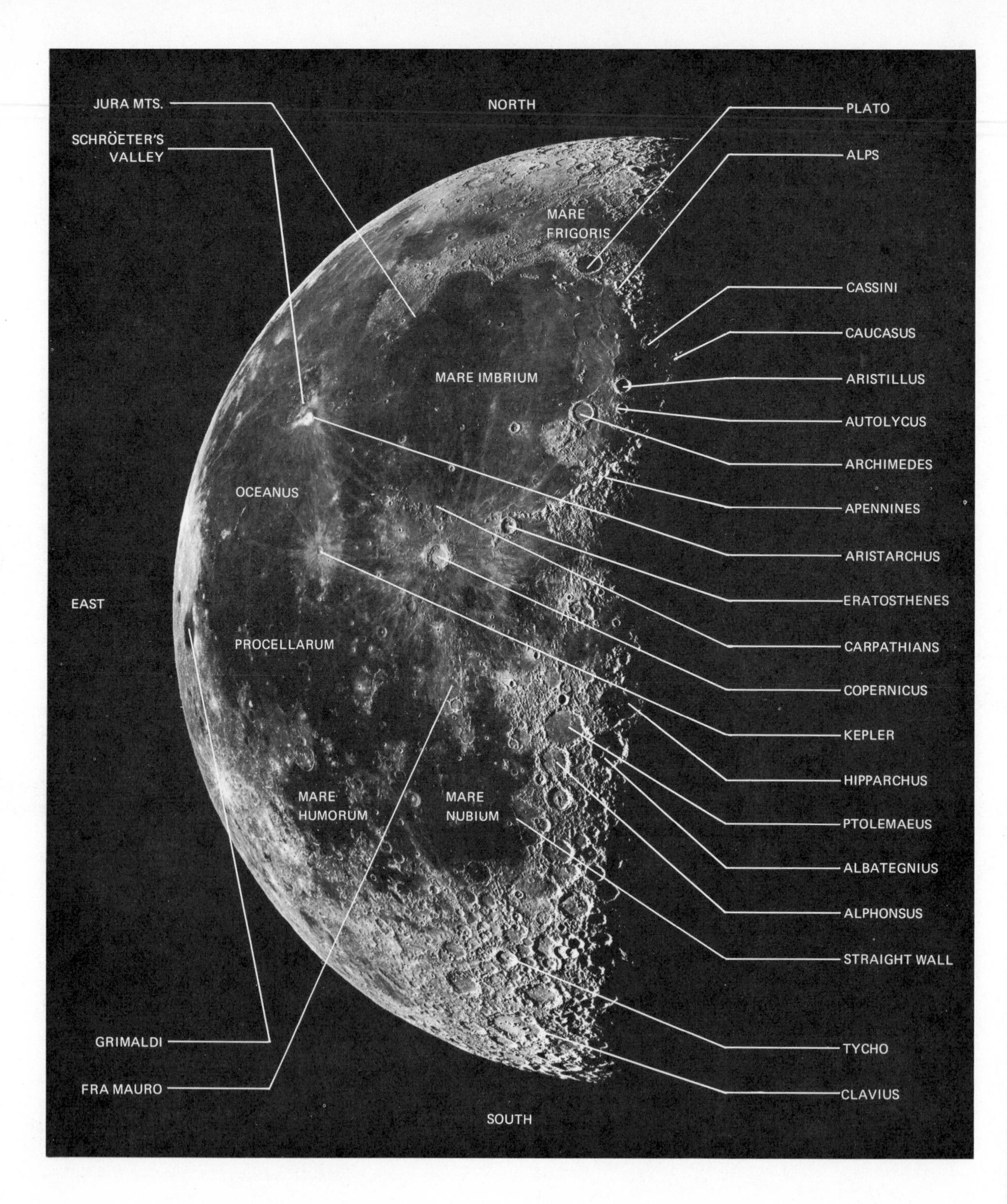
NORTH
JURA MTS.
SCHRÖETER'S VALLEY
PLATO
ALPS
MARE FRIGORIS
CASSINI
CAUCASUS
MARE IMBRIUM
ARISTILLUS
AUTOLYCUS
ARCHIMEDES
OCEANUS
APENNINES
ARISTARCHUS
EAST
ERATOSTHENES
PROCELLARUM
CARPATHIANS
COPERNICUS
KEPLER
HIPPARCHUS
MARE HUMORUM
MARE NUBIUM
PTOLEMAEUS
ALBATEGNIUS
ALPHONSUS
STRAIGHT WALL
GRIMALDI
TYCHO
FRA MAURO
CLAVIUS
SOUTH

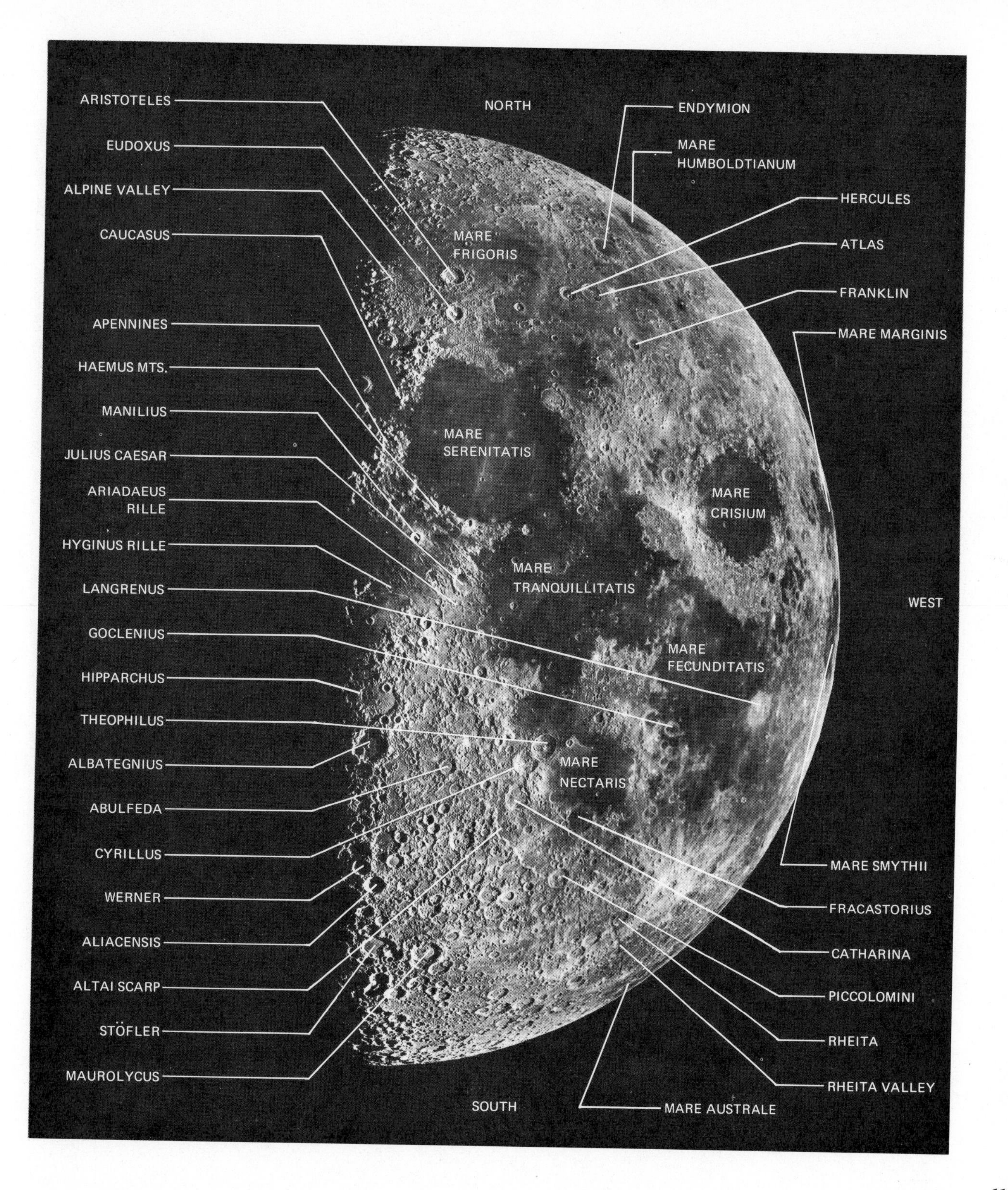
NORTH
ARISTOTELES
EUDOXUS
ALPINE VALLEY
CAUCASUS
APENNINES
HAEMUS MTS.
MANILIUS
JULIUS CAESAR
ARIADAEUS RILLE
HYGINUS RILLE
LANGRENUS
GOCLENIUS
HIPPARCHUS
THEOPHILUS
ALBATEGNIUS
ABULFEDA
CYRILLUS
WERNER
ALIACENSIS
ALTAI SCARP
STÖFLER
MAUROLYCUS
ENDYMION
MARE HUMBOLDTIANUM
HERCULES
ATLAS
FRANKLIN
MARE MARGINIS
MARE FRIGORIS
MARE SERENITATIS
MARE CRISIUM
MARE TRANQUILLITATIS
MARE FECUNDITATIS
MARE NECTARIS
WEST
MARE SMYTHII
FRACASTORIUS
CATHARINA
PICCOLOMINI
RHEITA
RHEITA VALLEY
MARE AUSTRALE
SOUTH

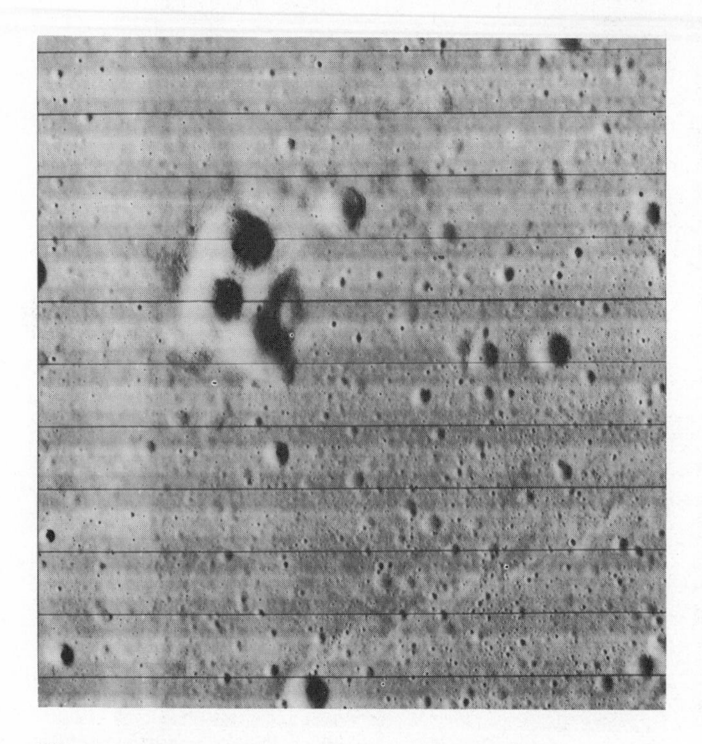

FIG. 7.12
Lunar Orbiter 2 photograph of an area covering 10 square miles in the Sea of Tranquillity. The Orbiter was 30 miles above the lunar surface when it took this photograph. (Courtesy of NASA.)

the different types of lunar terrain for clues to the origin and evolution of the moon began with the Apollo 14 flight. Apollo 14 astronauts landed near Fra Mauro Crater. They explored the North Boulder Field and the rugged terrain almost up to the rim of Cone Crater. The Apollo 15 astronauts landed at the Hadley/Apennine site and explored the Hadley rille and the slope of the Apennine mountains (Fig. 7.15). The rocks they brought back from this area are expected to reveal more scientific information than all the rocks from the three previous landings. The discoveries and knowledge gained from these four landings will be presented in the following sections.

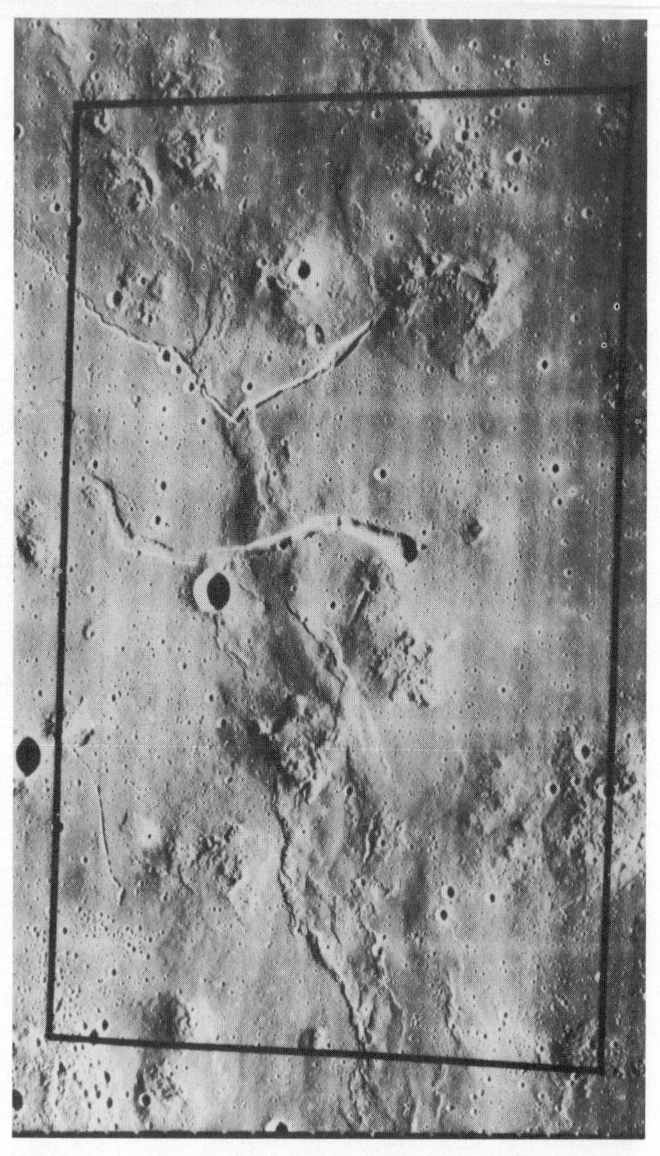

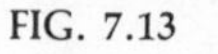

FIG. 7.13
Lunar Orbiter 5 photograph of the Marius Hills taken from 69 miles above the lunar surface. Surface features as small as eight feet in diameter are clearly visible. (Courtesy of NASA.)

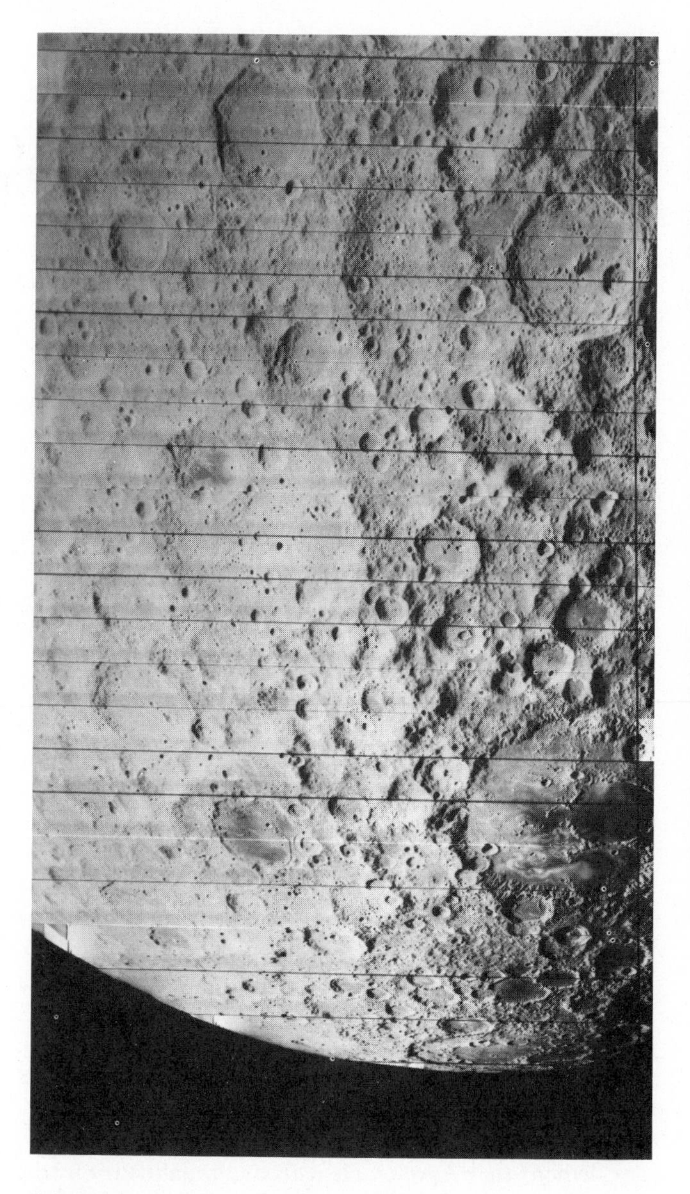

FIG. 7.14
Lunar Orbiter 2 photograph of a large, heavily cratered area of the moon. The lunar equator is at the top, and the south pole is at the bottom of the photograph. The smallest surface feature visible is about 20 miles across. (Courtesy of NASA.)

FIG. 7.15
Apollo 15 astronaut David R. Scott on the moon in August 1971, with close-up view of the lunar module "Falcon," lunar soil, and bootprints. Looking southeast, one can see the Apennine Front in the left background at a distance of about 3 miles and the Hadley Delta in the right background about 2.5 miles above the plain. (Courtesy of NASA.)

FIG. 7.16
The northern portion of Oceanus Procellarum is at the top of this Lunar Orbiter 3 photograph. The largest crater in the background is Galilei—about 10 miles in diameter, 1 mile deep, and with rims about 1000 feet above the outside terrain. The Cavalerius Hills are in the foreground. (Courtesy of NASA.)

7.12 LUNAR MARIA

The important surface features of the moon are the *maria, mountains, craters, rays,* and *rilles.* Of these, the most conspicuous are the maria—the dark, grayish areas that cover about one-half of the moon's visible disk and less than one-third of the moon's total surface area, but stand out conspicuously from the brighter and more rugged highlands. The maria have been given fanciful names such as Sea of Crisis (Mare Crisium) or Ocean of Storms (Oceanus Procellarum). These dark areas are also what gives the "man in the moon" his features. Most of the maria are located in the northern hemisphere of the visible surface from the earth and are interconnected.

The maria are nearly circular, although when seen near the moon's limb, they appear quite elliptical. They have relatively smooth, undulating surfaces that are pockmarked with craters, mountains, hills, ridges, depressions, rays, and rilles. The largest of the dark areas is Oceanus Procellarum with an area of nearly 900,000 square miles, or 2,300,000 square kilometers (Fig. 7.16). The next in size, Mare Imbrium (about 700 miles, or 1127 km, across and an area of over 300,000 square miles, or 778,000 square kilometers) is one of the most interesting because it is almost completely encircled by a series of great mountain ranges: Carpathians, Apennines, Caucasus, and the Alps. It also contains examples of each of the important lunar surface features.

Several theories have been proposed to explain the origin of the maria. Most astronomers believe that they were created by the impact of gigantic meteorites and a subsequent great lava outpouring from the moon's molten interior. Mare Imbrium is believed to have been created by the impact of a tremendous meteorite, and it's ejected material is believed to have covered the Fra Mauro Crater region.

When the Apollo 11 astronauts landed on the southern part of Mare Tranquillitatis, they found the texture of the lunar surface to be very porous, sticky, and compressible (Fig. 7.17). The material adhered to the soles and sides of their boots in layers, like graphite. An analysis of the soil indicated that it consists of a layer of fragmented material, a physical mixture of loose, unconsolidated basaltic rocks, breccias, glasses, and iron mete-

FIG. 7.17
This photograph shows the texture of the surface at the southern part of Mare Tranquillitatis and Apollo 11 astronaut Neil Armstrong's clear and well-defined footprint in the lunar soil. (Courtesy of NASA.)

oric fragments that vary in size from extremely fine particles to objects about three feet in length.

The Apollo 12 astronauts found the surface of Oceanus Procellarum somewhat different than that of Mare Tranquillitatis—much finer and dustier. The Apollo 11 astronauts' footprints were depressed fairly uniformly to about an eighth of an inch in Mare Tranquillitatis, whereas those of the Apollo 12 astronauts were at times depressed by as much as three inches. At Oceanus Procellarum, the very fine lunar dust seemed to adhere to everything as though it was electrostatically charged.

The Apollo 12 astronauts observed that the color of the lunar surface changed with the angle of the sun from very dark brown, similar to a freshly plowed field, to dull gray. The soil in some areas was so soft and fine that walking was very difficult, whereas in other areas it was so coarse and firm that walking was quite easy.

When the Apollo 14 astronauts landed in the Fra Mauro Crater region, located about 630 miles (1014 km) south of the crater Copernicus, they described the region as rolling highland, pockmarked with small shallow craters, ridges, and rocks. The terrain is so rolling that not a single level area was visible, and even when standing next to a large-size crater, the astronauts had difficulty recognizing it. The *albedo* (reflecting power) of the highland is greater than the dark areas on which Apollo 11 and 12 astronauts landed. The ridges were observed to be radially aligned to the center of Mare Imbrium.

7.13 LUNAR MOUNTAINS

Most of the moon's great mountain ranges are located in its northern hemisphere. An interesting series of mountain ranges, which almost completely encircle Mare Imbrium, is shown in Fig. 7.18. The Carpathians are located on the southern border, the Apennines and the Caucasus on the western border, and the Alps on the northern border. The eastern border of Mare Imbrium opens into Oceanus Procellarum. Its present appearance might have been the result of a gigantic meteoric impact, which obliterated the mountain range in the east and increased the height of the ranges to the west. These four

◄ FIG. 7.18
In this telescopic view of Mare Imbrium, north is at the bottom. At the upper right are the Carpathians; upper left the Apennines; lower center, the Alps; and lower right, the Jura Mountains. The large crater near the top is Copernicus; that in the Apennines is Eratosthenes; the large one in Mare Imbrium is Archimedes; and the dark crater in the Alps is Plato. (Courtesy of the Hale Observatories.)

FIG. 7.19 ►
The Apennine Mountains are in the background, with Mount Hadley to the left. The layering of Mount Hadley is clearly visible. Note the boulder in the foreground. (Courtesy of NASA.)

mountain ranges appear to slope abruptly on the side facing Mare Imbrium and to slope gradually away from Mare Imbrium on the opposite side.

Lunar mountains appear more rugged than those on the earth because of the absence of water and the weather elements which produce the drainage and erosion features that are typical of terrestrial mountains. The Apollo 15 astronauts reported that the Apennines appeared to be smooth and rounded, with many rough cratered places. The highest peak in the Apennines is about 18,000 feet (5490 meters) above its base, but the highest peak on the moon (located on the southern limb) rises to almost 30,000 feet (9150 meters) above its base (Fig. 7.19).

7.14 LUNAR CRATERS

A lunar crater is a roughly circular depression on the lunar surface ringed by a wall of rock and debris. Large craters often possess central peaks (Fig. 7.20). Craters appear on all parts of the lunar surface with the greatest

◄ FIG. 7.20
The crater Tycho Brahe is believed to have been created by the impact of a giant meteor. (Courtesy of NASA.)

FIG. 7.21
This Lunar Orbiter 2 photograph of the crater Copernicus was taken about 28 miles above the lunar surface and 91.5 miles south of Copernicus. The crater is 60 miles in diameter and 2 miles deep. The floor of the crater is shown in fine detail. The mountains clearly visible on the floor rise about 2400 feet and are covered with rubble. The rounded knolls in the foreground were formed by the debris from Copernicus when it was created after a tremendous explosion from the impact of a giant meteorite. Note that distance is very hard to judge because there is no atmosphere to soften distant ◄ details. (Courtesy of NASA.)

number appearing in the southern hemisphere around the south pole and on the far side of the moon.

Lunar craters range in size from small pitholes several feet in diameter to great craters such as Clavius with a diameter of about 146 miles (235 km). As Apollo 15 was orbiting the moon, astronaut Worden estimated the walls of the Gagarin Crater, located on the far side of the moon, to be about four miles high, making it the deepest crater that has been discovered. Generally, the inside walls of the craters are steeper than the outside, and their floors are level. Near the center of the floor, there is frequently an isolated mountain mass or a group of mountians, and smaller craters are sprinkled on the floor, inner and outer slopes, and on the rim.

In 1967 Lunar Orbiter 5 took a remarkable photograph of the crater Tycho Brahe from about 135 miles (217 km) above the moon's surface (Fig. 7.20). This crater, located near the south pole of the moon in the rugged highlands, is nearly circular and has a diameter of

FIG. 7.22
The Apollo 12 Lunar Module, seen in the background, landed in the Ocean of Storms within 600 feet of Surveyor 3, which is in the foreground. Dimpled craters are visible on the floor of the Ocean of Storms. (Courtesy of NASA.)

about 54 miles (87 km). The crater appears to be relatively young, because the material that was ejected when the crater was formed lies on the surface of the outside terrain, and the texture of the ejected material is very similar to that on the crater's floor. Old craters, such as Copernicus (Fig. 7.21), are recognized by the presence of landslides, rockfalls, and terraces. The terraces are formed by the material in the rim and wall of the crater slumping to the floor.

Photographs taken by Rangers 7 and 8 revealed a new lunar surface feature, found in abundance on the maria and on the floors of the craters Alphonsus and Ptolemaeus. They were appropriately named *dimple craters* because they appear as shallow depressions in the general terrain. It is believed that they were caused by the collapse of the surface material from internal forces. The dimple craters are clearly visible on the floor of the Ocean of Storms around the spacecraft Surveyor 3, which rests on the slope of a small crater (Fig. 7.22).

Most of the lunar craters are believed to be the result of meteoric impacts, although some of the smaller craters appear to be of volcanic origin. A few of the older and larger craters, which are called *walled plains,* appear to have been caused by the collapse of portions of the moon's surface. Astronaut Worden observed that a large area of the west wall of the large, young crater Tsiolkovsky is covered with material from a tremendous rockslide. He also observed that the crater's central peak is layered to a considerable length.

7.15 LUNAR RAYS

The surface features that are unique to the moon and Mercury are the bright rays that appear to radiate from only a few of the large craters such as Tycho Brahe, Copernicus, Kepler, and Aristarchus. These rays, clearly visible when the moon is full, range from 5 to 10 miles (8 to 16 km) in width and extend up to about 1500 miles

(2400 km) in length, without interruption or deflection across mountains, valleys, maria, and craters. The rays from Tycho appear to radiate as spokes from the hub of a wheel, whereas those from Copernicus appear to intertwine with one another. The origin of these rays is probably secondary craters, dust particles, and small piece of debris which have been ejected at great speeds from primary craters when they were formed by meteoric impacts.

7.16 LUNAR RILLES

Lunar rilles, which at this time are an enigma, appear as narrow cracks or valleys about one-half mile (.8 km) in width and up to about 300 miles (480 km) in length. They appear to be of roughly uniform width and are continuous. Many of them are nearly straight, but some are quite sinuous and strongly resemble meandering terrestrial rivers. They start in the highlands around craters and meander down to the plains of the maria. One such sinuous rille is Schröeter's Valley, shown in Fig. 7.23, which appears like a great river meandering down and to the right. Its apparent origin is near a crater located in the highland to the right and near the beautifully bright crater of Aristarchus. Rilles were probably formed by the flow of subsurface lava, which left a long hollow "tube" underground. Subsequent collapse of the tube's top produced the open rille.

While the Apollo 8 astronauts were orbiting the moon in 1968, they photographed the crater Goclenius (41 miles, or 66 km in width), which is located on the edge of Mare Fecunditatis. This photograph (Fig. 7.24) clearly shows the system of rilles that criss-cross the floor of Goclenius. One rather broad, shallow rille appears to traverse the entire length of the crater's floor, cross over the crater's west wall, and then move out into the plain.

When the Apollo 15 astronauts explored the mile-wide, 1200-foot deep Hadley rille canyon, they reported evidence of the layering of rock on one side of the canyon (Fig. 7.25). This formation suggests that the rille may have been the path of several hot lava flows occurring at different times. The material that the astronauts brought back from the rille is thought to be bedrock.

FIG. 7.23
This excellent photograph of Schröeter's Valley, a sinuous rille, was taken by the Apollo 15 astronauts. The head of the valley, which starts to the right of the bright crater Aristarchus, is called the Cobra Head. This rille meanders downward to the Ocean of Storms. (Courtesy of NASA.)

FIG. 7.24 ►
A beautiful pattern of rilles is clearly visible on the floor of the large crater Goclenius, which lies on the edge of Mare Fecunditatis. One long, straight, wide, and shallow rille appears to stretch across the floor and over the crater's west wall. From left to right the three craters in linear alignment are: Magelhaens A, Magelhaens, and Guttenberg D, all of which show evidence of meteoric impacts. (Courtesy of NASA.)

FIG. 7.25
Astronaut Scott and the Lunar Rover are at the edge of Hadley Rille on Hadley Delta. (Courtesy of NASA.)

The placement of seismographs to detect moonquakes was one of the most important aspects of the Apollo program. It was shown in Chapter 5 how earthquakes allow scientists to infer the interior structure of the earth. Likewise, moonquakes give data about the lunar interior. The small size of the moon seems to have prevented the accumulation of heat by radioactive decay which may have caused the interior of the earth to melt. The lunar interior, however, appears differentiated like the earth's. The outermost layer of the moon is a zone of rubble and smashed rock resulting from intense meteoritic bombardment. This layer is perhaps a few kilometers thick. Below the rubble layer is a mantle of solid rock roughly thousands of kilometers thick. Below the mantle is a core that may be molten. In its early days the mantle may also have been molten (heated by meteoritic impact) and the moon may have briefly possessed only a thin crust. The impact of giant meteoroids would have been able to break through the crust, allowing the molten mantle material to flood out and form the maria.

7.17 THE FAR SIDE OF THE MOON

On October 4, 1959, the U.S.S.R. Luna 3, an automatic planetary station, took the first photographs of the far side of the moon from 34,500 feet (10,516 meters) above the moon's center. Although these photographs are not of high resolution, the larger lunar surface features are recognizable. Since then, the U.S.S.R. and the U.S. have photographed the far side of the moon with a resolution equal to the side facing the earth. Both hemispheres of the moon have similar kinds of surface features; but the far side appears to be more rugged and to contain fewer and smaller maria (Plate 8).

7.18 THE LUNAR ROCK SAMPLES

On July 24, 1969, the Apollo 11 astronauts brought back from Mare Tranquillitatis the first lunar rock and soil samples; for the first time in history scientists could make direct studies of the structure and physical characteristics of the moon. The moon samples analyzed were classified into four important categories: *basaltic crystalline rocks,*

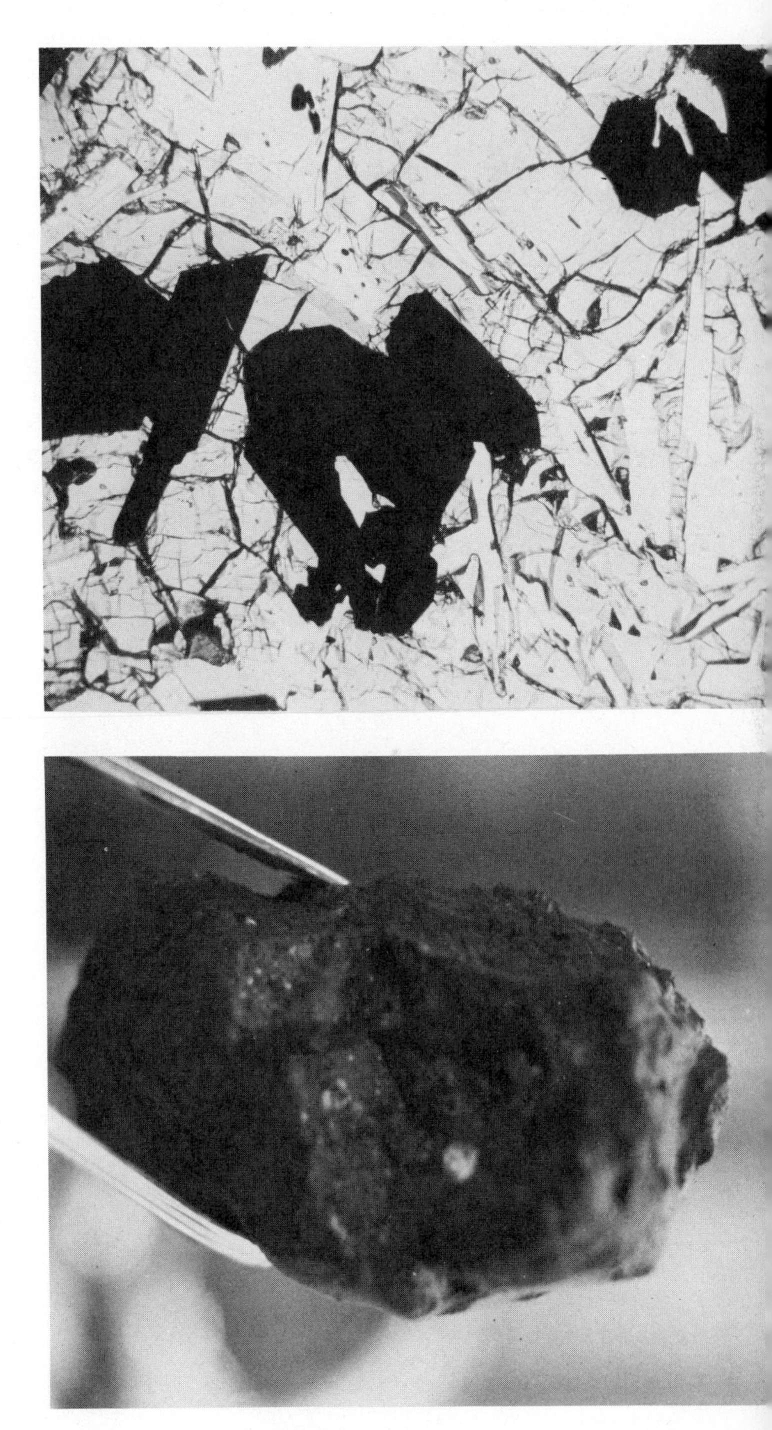

FIG. 7.26
This thin section of lunar basalt shows a continuous grading from very fine to coarse crystals, which may indicate that the basalts were cooled from lava flows. The black crystals are ilmenite, the white ones are calcic plagioclase feldspar, and the gray ones are pyroxene. (Courtesy John A. Wood, Smithsonian Astrophysical Observatory.)

soil breccias, glasses, and *anorthosites*. The basaltic crystalline rocks appear to be of igneous origin, range from fine to coarse-grain, and are unusually rich in titanium, scandium, zirconium, and hafnium, which are rare in terrestrial igneous rocks (Fig. 7.26). They also have free metallic iron, which is rarely present in terrestrial rocks, and have densities of about 3.3 grams per cubic centimeter. Their very dark color is due to the presence of very fine grains of ilmenite (Plate 9). The average age of the basalts is 3.7 billion years; however, the age of one of the highland rocks found on a subsequent Apollo mission is about 4.4 billion years, which indicates that the age of different areas of the moon's surface varies.

The soil breccias (Fig. 7.27) are similar to the igneous rock, yet distinctive for their richness in nickel, zinc, copper, silver, and gold. Large amounts of rare gases have been found in the breccias, indicating that they were probably part of the solar wind before they were eventually trapped in the lunar soil.

The presence of glasses in the soil and rocks from Mare Tranquillitatis came as a complete surprise to scientists (Plate 10). The glasses are very small and show a

FIG. 7.27
A close-up view of one of the breccia brought back by the Apollo 12 astronauts. These rocks were common in the Apollo 11 samples, but rare in the Apollo 12 collection. (Courtesy of NASA.)

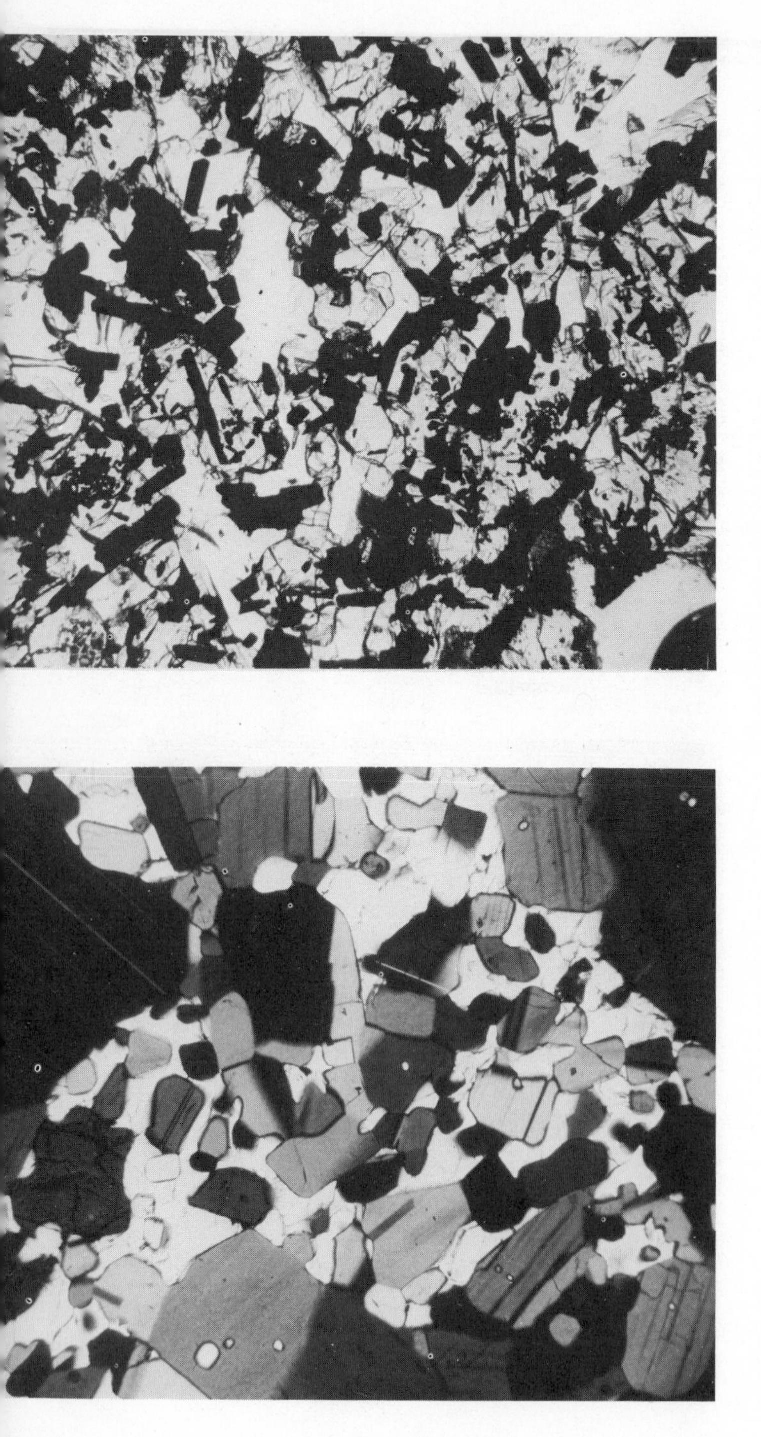

wide variety of forms, colors, and chemical properties. They are spherical and angular in shape, with highly lustrous brown, yellow, and clear colors. The glasses are believed to have been produced when meteorites hitting the moon's surface vaporized the surface rocks on contact and created gases, which condensed and solidified into the glasses. The glasses are the most abundant constituent (about 50%) of the *lunar fines* (dust) on Mare Tranquillitatis. On later Apollo flights glassy dust has been found to exist over the entire lunar surface.

The color, composition, and density of the anorthosites (Plate 11) are very different from those of the basalts and the breccias, for they are either light gray or white, granular in texture, rich in aluminum and calcium, and low in titanium (Fig. 7.28). Their most abundant mineral is calcic plagioclase. The anorthosites show evidences of having suffered severe shock—their composition is similar to the material ejected from the crater Tycho, which was analyzed by Surveyor 7, suggesting that the anorthosites are out of place on Mare Tranquillitatis, having been thrown to the mare by a meteoric impact occurring in the highlands.

The Apollo 12 astronauts found many rocks on the surface of Oceanus Procellarum that were so soft that they crumbled very easily when picked up. They also saw rocks of an odd glassy-green color, one of which was covered with many ⅜-inch pits that were glass-lined. When the Apollo 14 astronauts approached the Cone crater, they saw a large boulder field with boulders up to five feet in diameter. The largest boulders had smooth

FIG. 7.28
The crystalline structure of anorthositic gabbro shows anorthositic crystals embedded in gabbroic melt with olivine crystals. (Courtesy of John A. Wood, Smithsonian Astrophysical Observatory.)

surfaces with angular edges where pieces of the boulder had broken off (Fig. 7.29).

The age of the moon appears to be equal to that of the earth. This conclusion was reached from an analysis of a rock (83 grams) returned by the Apollo 12 astronauts from Oceanus Procellarum. The rock contained a higher percentage of uranium, thorium, and potassium, which made it about twenty times more radioactive than any other lunar sample previously studied. It has been dated at 4.6 billion years, thus making it the oldest rock recovered from the lunar surface. Its age was determined by measuring the amount of strontium 87 and rubidium 87 present in the rock. Strontium 87 is produced from the radioactive decay of rubidium 87, and the abundance ratio of these two isotopes is an indication of the time that has elapsed since the rock was formed. It is believed that the rock must have come from either the highlands, which are older than the maria, or deep below the surface of the lunar maria. Scientists believe that the oldest rock is not an isolated phenomenon but indicates that at least a part of the lunar surface is as old as the earth.

Other rocks from the Apollo 11 and 12 flights have ages that range from 3.3 to 3.7 billion years; dust samples have ages up to 4.4 billion years. The analysis of these younger rocks indicates that they were crystallized from igneous liquid. Since the solar system is approximately 4.6 billion years old, the oldest lunar rock dates to almost the same time as the solar system.

Most of the Apollo 14 samples from the Fra Mauro region were found to be fragmental, with evidences of pronounced shock effects. Their composition is definitely different from that of the basaltic rocks returned from the maria. This indicates that the rocks are probably ejected material. The moon rocks returned by the Apollo 15 astronauts contained a most interesting sample, called *Genesis rock*, which was picked up at Spur Crater. It has been described as a small, crystalline rock whose age is believed to be about 4.6 billion years.

Moon rocks returned by the astronauts show no sign of life. The soil and rock material injected into mice produced no toxic effects and no bacteriological growth. None of the complex carbon compounds that are necessary for life on the earth have been identified in any of the

FIG. 7.29
A close-up view of an interesting boulder near the rim of the Cone crater, with an apparent contrast in color and structure between the top and bottom of the boulder. (Courtesy of NASA.)

rock samples; however, all the elements necessary for life have been identified, including the most important element, carbon.

7.19 ORIGIN OF THE MOON

Although many theories have been proposed to explain the origin of the moon, the basic question of how it originated still remains unanswered. Did the combination of the earth's rapid rotation and the sun's gravitational force cause the plastic earth to extend into the shape of an unsymmetrical dumbbell and to eventually separate into two unequal spheres, with the smaller one becoming the moon? Was the moon created at the same time and from the same condensate as the sun and the earth? Or, during a close encounter was the moon captured by the earth to become its satellite?

Any acceptable theory for the origin of the moon must explain both the relative size of the moon compared to the earth and the great difference in composition of the two bodies. Most satellites in the solar system are much smaller relative to their planets than the moon is relative to the earth. For example, the ratio of the mass of Jupiter's largest moon to that of Jupiter is roughly 1/12,000. For Saturn's largest moon the ratio is about 1/4,000. On the other hand, for the earth and moon the ratio of masses is only 1/81. The relative closeness in size suggests a different kind of origin for the moon than for other satellites in the solar system.

If the moon was spun off the earth, one might expect that the moon's composition would be similar to that of the earth's crust. Likewise, if the moon formed from the same condensate that the earth did, then again the two bodies should have similar composition. If the moon was captured by the earth, it is very easy to understand the compositional differences, because the two bodies might have formed in totally different regions of the solar system. However, it is difficult to understand how so large a body as the moon could have been slowed down enough to end up in a permanently bound orbit around the earth.

One theory that may avoid some of these problems suggests that the moon may have been created by the impact of some enormous preplanetary condensation

By permission of Johnny Hart and Field Enterprises, Inc.

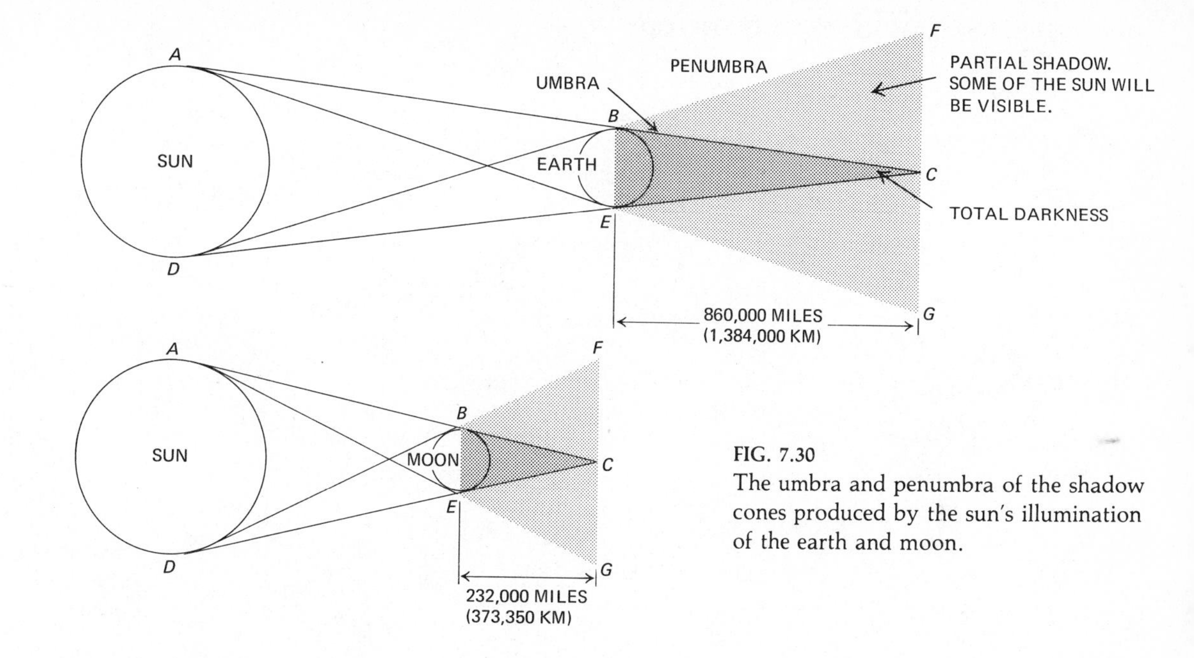

FIG. 7.30
The umbra and penumbra of the shadow cones produced by the sun's illumination of the earth and moon.

with the earth. A collision between the earth and a large body shortly after the earth formed might have "splashed" out a huge cloud of debris from the crust of the earth. The debris might have gone into orbit around the earth and gradually clumped together to form the moon. The heating associated with the original impact would evaporate many chemical elements, thereby removing them from the material that was ultimately to become the moon. Also, if the impact occurred millions of years after the earth's formation, much of the iron in the earth might have settled out of the crust into the earth's core. The ejected matter after the collision would thus be deficient in iron in accord with the low iron abundance of the moon rocks.

The rock samples from the highlands returned by the Apollo astronauts may ultimately provide a record of the early history of the moon, which may in turn give us the clue to its origin and possibly to the origin of the earth and the solar system as well.

7.20 ECLIPSES

Eclipses are one of the most beautiful and awe-inspiring spectacles of nature. Primitive people regarded them as evil omens, and their appearance always struck terror in their minds. The earliest recorded eclipse was seen by the Chinese in 2137 B.C.

Since the earth and the moon are opaque bodies and are illuminated by the sun, they cast conical shadows, which can be illustrated (Fig. 7.30) by drawing external and internal tangents to the sun and the body. The external tangents *ABC* and *DEC* produce the *shadow cone BCE* in which the sunlight is completely cut off. This region is called the *umbra.* The internal tangents *AEG* and *DBF* produce the region *FBCEG* around the umbra in which the sunlight is partially cut off. This region is called the *penumbra.* The length of the earth's umbra is about 860,000 miles (1,384,000 km), and that of the moon's umbra is about 232,000 miles (373,000 km).

Eclipses occur when the sun, the earth, and the new or full moon are approximately in line and lie in the same plane. Since the moon's orbital plane is inclined to the earth's orbital plane, the new or full moon usually lies above or below the earth's orbital plane, and an eclipse cannot occur (Fig. 7.30). The moon's orbit moves eastward in the moon's orbital plane and completes one rotation every 8.85 years. Also, the line of nodes (the intersection of the earth's and the moon's orbital planes)

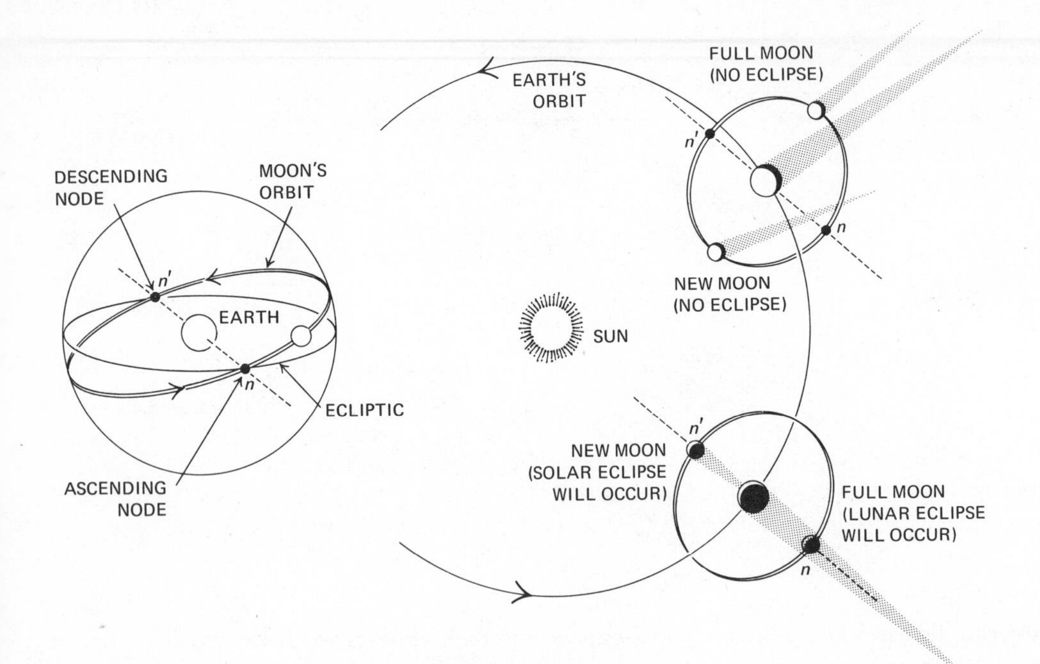

FIG. 7.31
Solar and lunar eclipses occur when the sun is at or near the ascending or descending node and the line of nodes points toward the sun.

moves westward in the plane of the ecliptic and completes one rotation every 18.6 years. The line of nodes is the straight line that connects the ascending node (point where the moon crosses the ecliptic from south to north) to the descending node (point where the moon crosses the ecliptic from north to south). These motions are caused by the moon's perturbations, which are produced by the sun's unequal attraction for the earth and moon and by the earth's equatorial bulge. In Fig. 7.31, n represents the ascending node and n' the descending node. When the moon is at or near either the ascending or descending node and the line of nodes points toward the sun, a solar eclipse will occur when the moon is new; a lunar eclipse when the moon is full.

7.21 MECHANICS OF A LUNAR ECLIPSE

On the mean cross-sectional (end-on) view of the earth's shadow cone (Fig. 7.32), the dark area is the umbra and the light area the penumbra. When the full moon passes through the earth's shadow cone and completely misses

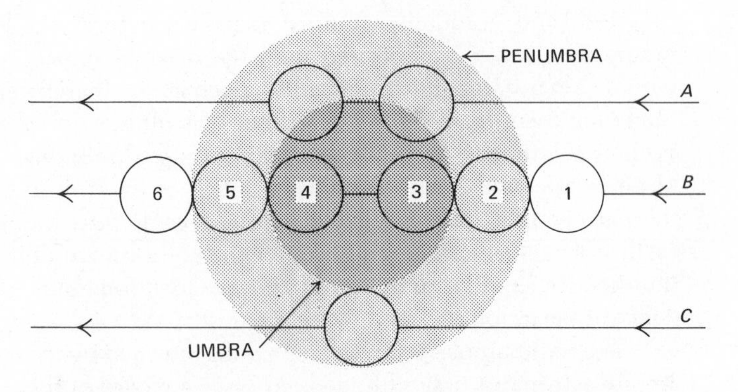

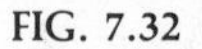

FIG. 7.32
Cross-sectional view of the earth's shadow cone (umbra and penumbra), showing the moon's paths for a partial eclipse (A), central total eclipse (B), and penumbral eclipse (C).

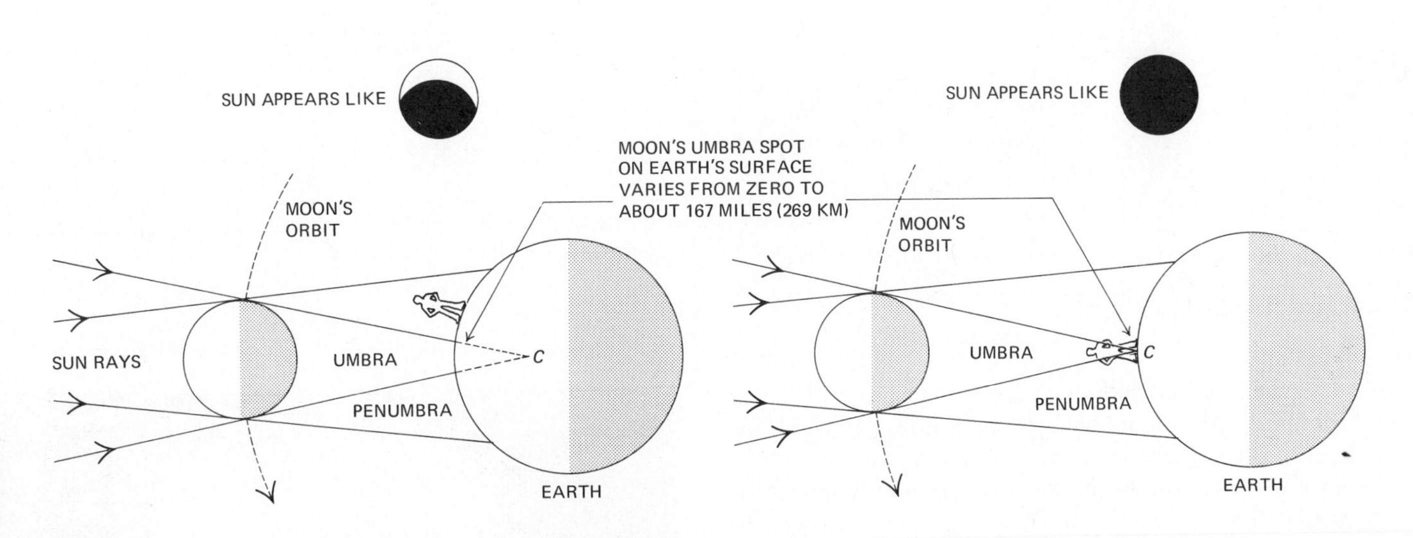

FIG. 7.33
To observe a total solar eclipse, the observer must be in the moon's umbra spot on the earth's surface, which varies from zero to about 167 miles in diameter. A partial solar eclipse will be observed when the observer is in the penumbral circle.

the umbra, the moon appears to darken and redden slightly—a phenomenon called an *appulse*; when only part of the moon passes through the umbra, it is called a *partial eclipse*; and when the entire moon passes through the umbra, it is called a *total eclipse*. The longest total lunar eclipse occurs when the moon passes through the center of the umbra.

During the penumbral phase, the full moon appears to darken and redden slightly, and during the total phase, which lasts for about two hours, it is not completely obscured, but appears as a dull-reddish disk. The red color is produced by sunlight that is refracted around the earth by its atmosphere. In passing through the atmosphere, most of the blue light is removed. Seeing a totally eclipsed moon rising blood red over the eastern horizon is enough to instill a sense of foreboding in anyone.

7.22 MECHANICS OF A SOLAR ECLIPSE

Since the moon's distance from the earth varies from about 221,000 to 252,000 miles (356,000 to 406,000 km) and the length of the moon's umbra is about 232,000 miles (373,000 km), the moon's umbra spot on the earth's surface varies from zero to a maximum of about 167 miles (269 km) in diameter. It is maximum when the moon is at perigee and the earth is at aphelion (Fig. 7.33). Since the moon revolves around the earth from west to east, its shadow cone appears to move in the same direction across the earth's surface. When in the path of the moon's umbra, an observer will see a total eclipse, and when in the moon's penumbra, a partial eclipse.

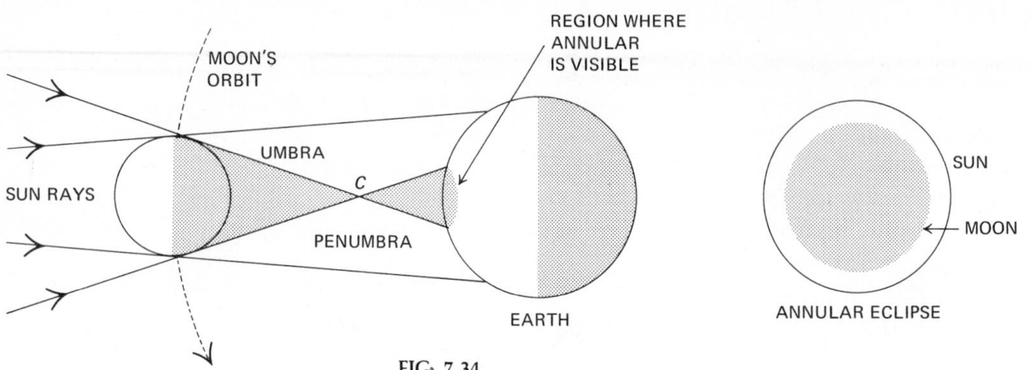

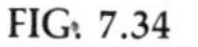

FIG. 7.34
When the moon's umbra does not reach the earth's surface, an annular eclipse is produced, because the moon's apparent diameter is smaller than the sun's.

When the sun and moon are at their mean distances from the earth, the moon's umbra spot fails to reach the earth's surface, and an annular eclipse occurs (Fig. 7.34). Since in this position the moon's apparent diameter is smaller than the sun's, an observer located within the extension of the moon's umbra will observe the annular eclipse as a dark disk (new moon) surrounded by a brilliant annular ring. The ring is the light that comes from the edge of the sun and appears less white than the sun's disk. An observer within the region of the penumbra will observe a partial eclipse. The annular ring is maximum when the moon is at apogee and the earth is at perihelion.

7.23 TOTAL SOLAR ECLIPSE

The spectacular phenomenon of a total solar eclipse commences when the moon appears to make contact with the west edge of the sun. As the moon moves slowly across the face of the sun, the sun's visible disk diminishes, and its diffused light becomes less intense. During a total solar eclipse, unusual reddish tones appear on the earth's landscape because the light comes from the sun's limb, where it is less blue.

Before totality, the sun appears as a very thin crescent of light, and as the light filters through the foliage of the trees and plants, beautiful crescent-shaped images appear on the ground. Also, peculiar shadows appear to

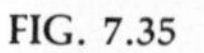

FIG. 7.35
The solar corona near sunspot maximum, photographed during a total eclipse. (Courtesy of the Hale Observatories.)

move over the landscape in ripples and wave-like motions, and animals and plants behave as they do at sunset. The crescent of light soon becomes a series of irregular bright spots of light around the moon's limb, "Baily's Beads," with the beautiful "diamond ring" effect (Plate 12). They are caused by the sunlight's passing through the irregularities of the terrain on the moon's limb. At totality, as shown in Fig. 7.35, the beads disappear, and the corona, the sun's outer atmosphere, appears. During totality there is a noticeable darkening effect on the earth, a drop of several degrees in the earth's temperature, and the appearance of the planets and brightest stars. As the moon continues to move eastward, it slowly uncovers the sun.

The average duration of totality is about three minutes; however, when the moon is at perigee, the earth at aphelion, and the observer on the equator at sea level, the maximum duration of totality is slightly over seven minutes. Such an eclipse occurred across north central Africa on June 30, 1973, and the duration of totality was seven minutes and twelve seconds.

A few of the important studies made during a total solar eclipse are the determination of the exact positions of the sun and the moon, the deviation of light as it passes close to the sun (a test of Einstein's relativity), the search for small bodies that might be inside Mercury's orbit, and the study of the sun's atmosphere and surface features.

7.24 ECLIPSE LIMITS AND SEASONS

When the new moon is near a node, the line of nodes points toward the sun, and the moon appears to be tangent to the sun; this point, the *solar eclipse limit*, marks the extent from either side of the node in which a solar eclipse is possible. This angular distance is variable, because the apparent diameter of the moon changes with its distance from the earth. When the maximum and minimum apparent diameters of the sun and moon are used, the solar eclipse limits range from 15°21′ to 18°31′.

When the full moon is near a node, the line of nodes points toward the sun, and the moon appears to be tangent to the earth's umbra; this point, the *lunar eclipse limit*, marks the extent from either side of the node in which a lunar eclipse is possible. When the maximum and minimum apparent diameters of the moon and the umbra of the earth's shadow are used, the lunar eclipse limits range from 9°30′ to 12°15′.

The time during which the sun or the earth's umbra is within the eclipse limit is called the *eclipse season*. Since the lunar eclipse limit is about 24 days and the moon-phase interval is about 29.5 days, it is possible for a year to pass without a single lunar eclipse because the moon will not be full during the eclipse season. Since the solar eclipse limit is about 31 days and two eclipse seasons occur every year, at least two solar eclipses can occur in any one year. A maximum of seven eclipses can occur in any one year—five solar and two lunar or four solar and three lunar.

7.25 PREDICTION OF SOLAR ECLIPSES

It is always a source of great amazement to the average person that the occurrence of a solar eclipse can be predicted to within seconds and its path determined to within a quarter of a mile. This requires a thorough knowledge of the motions of the sun and moon and their positions with respect to the earth's center. With these factors, the time and place that an eclipse will occur can be determined.

In predicting an eclipse, the *cycle of the saros*, an interval of 18 years 11⅓ days, or 6585⅓ days, is used. The saros, which was well known to the ancient Chaldeans, represents an interval of time that is very nearly evenly divisible by the lunar synodic month of 29.5 days and by the eclipse year of 346.6 days. This means that the saros contains about 223 synodic months and about 19 eclipse years; when an eclipse occurs, another one will follow in 6585⅓ days because in 19 eclipse years the sun will return to the same node, and in 223 synodic months the moon will again be in the new phase.

7.26 TIDES

To people living near the coast, the daily rise and fall of the ocean is one of the fundamental rhythms of life. For thousands of different animal and plant species, the ebb

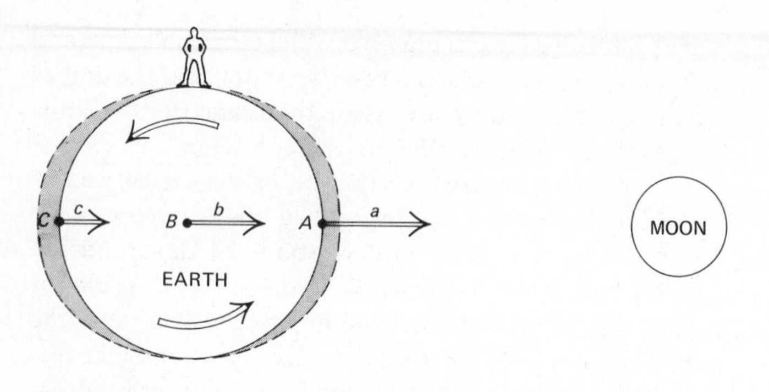

FIG. 7.36
The moon's gravitational force at any point on the earth's surface depends on its distance from the moon. The vectors (a, b, and c) indicate the relative magnitude and direction of the force.

and flow of water in response to tidal forces is essential to their existence. Like night and day, the tidal motion of the seas arises from celestial phenomena. Tides are caused by the gravitational attraction of the moon and, to a lesser extent, that of the sun. The moon's gravity produces two tidal bulges in the waters of the earth: one on the side of the earth toward the moon and pointing roughly at the moon, and one on the opposite side of the earth (Fig. 7.36). As the earth turns on its axis, a point on the earth will be swept through the two tidal bulges, thus experiencing, in general, two high tides and two low tides each day. The tidal bulges can be understood in a simplistic way as arising because the moon lifts up the water on the side of the earth nearest the moon, and pulls the earth out from under the water on the far side of the earth. A more accurate explanation will follow.

The moon's gravitational force is inversely proportional to the square of its distance from the earth. This means that the moon's gravitational force is greater for those parts of the earth that are closer to the moon.

For simplicity in determining the tide-raising force at any point on the earth's surface, we will assume that the sun's influence is negligible, the moon is stationary, and the earth is covered with a uniform layer of water. In Fig. 7.36, the moon's gravitational force at any point depends on that point's distance from the moon. It is greatest at point A, less at point B, and least at point C. These conditions are represented in a relative manner by vectors a, b, and c. If we accept the earth's center as the reference point, the tide-raising force at any point on the earth's surface can be defined as the difference between the forces exerted at the surface point and at the earth's center. Therefore in Fig. 7.37 the tide-raising force at point A is $a-b$; at point D, $d-b$; at point E, $e-b$; and at point C, $c-b$. Subtracting one vector quantity from another is equivalent to adding the negative of the vector (direction reversed) (Fig. 7.37).

If the moon is stationary and the rotating earth is completely covered with water, a point on its surface would experience two high and two low tides approximately every 24 hours. At any particular time, the earth experiences two high tides—one at the point nearest the moon (A) and the other on the opposite side of the earth (C)—and two low tides at right angles to the highs (F and G). Point A will experience the higher of the two tides because of its nearness to the moon.

In our discussion of tides, we ignored the sun's effect. Even though it is considerably less than the moon's, it does produce noticeable effects in the size of the tides. When the moon is in line with the sun and the earth (syzygy), the tide-raising forces of the sun and moon combine to produce tides with greater range than any other tide produced during that month (Fig. 7.38). These are called *spring tides* and occur twice each month when the moon is in the new and full phases. When the moon forms a right angle with the sun and the earth, the tide-raising force of the sun reduces the moon's effect to produce tides with a minimal range less than any tide produced during the month (Fig. 7.38). These are called *neap tides* and occur twice each month when the moon is in the first- and third-quarter phases.

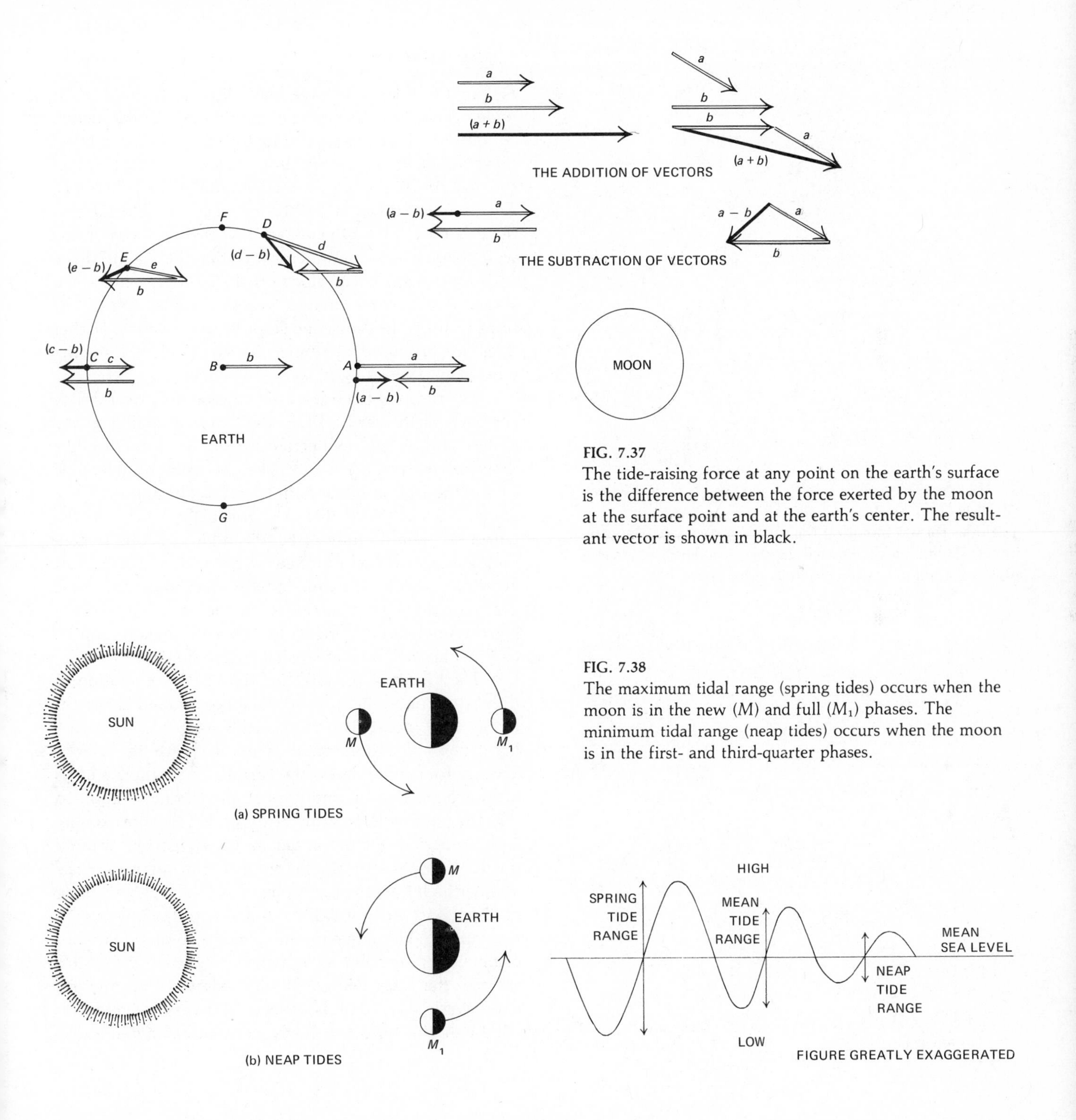

FIG. 7.37
The tide-raising force at any point on the earth's surface is the difference between the force exerted by the moon at the surface point and at the earth's center. The resultant vector is shown in black.

FIG. 7.38
The maximum tidal range (spring tides) occurs when the moon is in the new (M) and full (M_1) phases. The minimum tidal range (neap tides) occurs when the moon is in the first- and third-quarter phases.

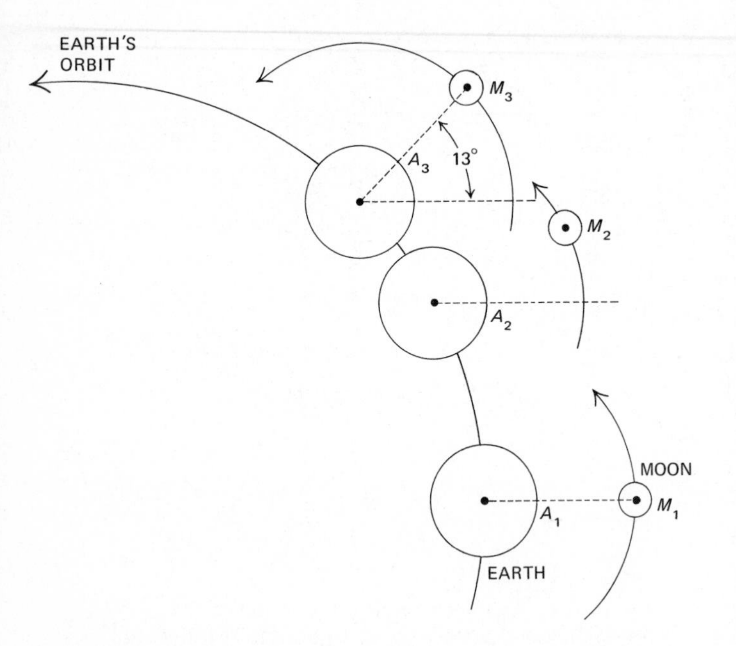

FIG. 7.39
The tides for any point on the earth's surface occur approximately 50 minutes later each day because during a 24-hour period the moon has moved in its orbit about 13°.

7.27 ACTUAL TIDES

So far, the tides that have been discussed would occur under ideal and theoretical conditions. Under actual conditions, there are several factors that affect the tides. Since the earth's surface is about 71% water, the first factor that affects the tides is the location of the land areas. Tides move in a westerly direction; therefore, the tides on the east coast of continents are higher than those on the west coast. The second factor affecting tides is the shape and slope of the ocean floor off the land areas. Gradual slopes and narrow ravines produce higher tides. The third factor is the shape and slope of the coastline. Higher tides are produced in funnel-shaped, gradually sloping bays than along straight, steep shore lines.

The range of the tide varies considerably over different parts of the earth. Tides are almost negligible at the eastern end of the Mediterranean Sea. They average less than one inch in the Great Lakes and slightly over one foot in the Gulf of Mexico. In New York harbor, they are about 2.5 feet (0.76 meters). The highest tides in the world occur in the Bay of Fundy in Nova Scotia, where they average 65 feet (20 meters) every day. The Bay of Fundy is ideally situated for the occurrence of these exceptionally high tides; it is on the east coast of the North American continent, its bay is V-shaped, and its continental shelf is channel-shaped and sloping.

The fourth factor affecting tides is the revolution of the earth around the sun and the moon around the earth. In Fig. 7.39, point A_1 on the earth's surface is experiencing a high tide because it is in line with the moon. Twenty-four hours later, the earth is in position 2 in its orbit around the sun and has completed one rotation so that the point is in position A_2. During this approximate 24-hour period, the moon has moved in its orbit around the earth to position M_2. Before the point can experience another high tide, the earth must rotate through an angle of about 13°, when it will be in line with the moon as in position A_3. This angle is equivalent to about 50 minutes of time; therefore, for any particular point on the earth's surface, the tides will occur approximately 50 minutes later every day, that is, every 24 hours 50 minutes. Although the tides are fairly constant at a particular place, they quite often differ considerably from those at an adjacent place. The times and special effects for the tides at any particular place, known as the "establishment of the port," are tabulated in the tide tables published by the *United States Coast and Geodetic Survey*.

7.28 TIDAL BRAKING

Tides are believed to play an important role in slowing down the rotation of the earth. The full details of the braking mechanism have still not been completely worked out, but it is believed that roughly 4.6 billion years ago the earth rotated once every five hours. At that time the moon was probably much closer to the earth than it is now, and its tidal forces were correspondingly larger. Because the earth rotates about its axis more rapidly than the moon moves in its orbit around the earth, there is a tendency for the earth's rotation to drag the tidal bulge produced by the moon around the earth. Thus, instead of being aligned toward the moon, the tidal bulge slightly "leads" the moon in its orbit (Fig. 7.40). This results in the moon's experiencing a slightly greater gravitational force in the direction that it is moving in its orbit. This force accelerates the moon, thereby slowly increasing its distance from the earth. In turn, frictional forces between the ocean waters and the rotating earth slow down the earth's rotational speed.

Over the course of billions of years this has lengthened the day from approximately 5 hours to the current value of 24 hours. Ultimately the earth will rotate so that it keeps pace with the moon's orbital motion. However, the moon will be much further away from the earth than it is now and the month will thus be longer. Calculations suggest that the synchronization of the earth's rotation and the moon's orbital motion will not happen for about 50 billion years and at that distant time the month will be about 47 days long. Tidal forces, though much weaker than now, will still dissipate energy so that the earth's rotation will continue to slow. The moon will then decelerate in its orbit and slowly creep in toward the earth until it gets so near that the earth's gravity tears it apart and scatters it as a ring around the earth.

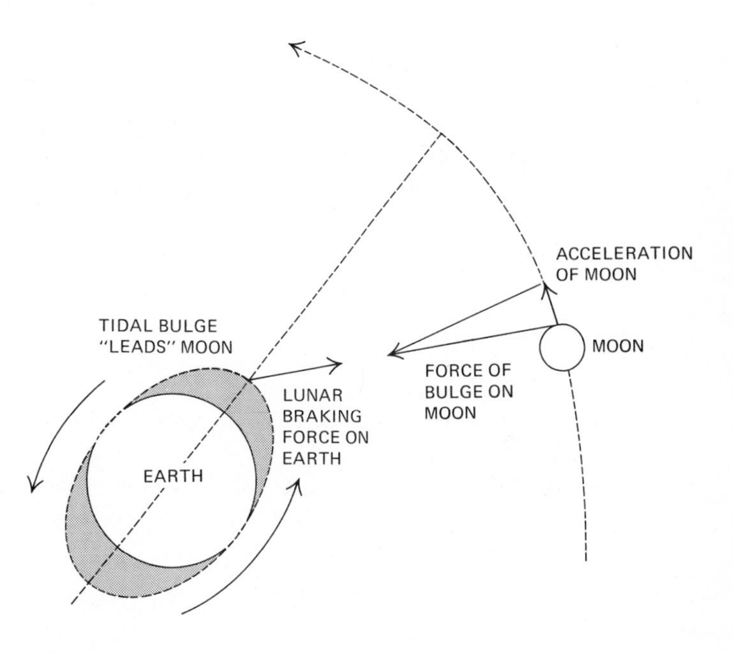

FIG. 7.40
Tidal braking and the acceleration of the moon.

SUMMARY

The moon is our nearest neighbor in space. Its landscape is austere and inhospitable in comparison to the earth's: there is no air, no water, and a much less diversified topography. The lunar landscape can be divided into two main types: low, flat, relatively smooth plains (mare) and rugged, heavily cratered highlands. In addition to craters, the lunar surface has rilles (valleys), and domes (rounded bulges). The moon appears to be as old as the earth (about 4.5 billion years), but has undergone less surface modification. It seems to have been subjected in its remote past to an intense surface bombardment by meteors, which produced most of its craters. The impact from extremely large fragments may have formed the mare.

The moon produces two astronomical effects of interest: tides and eclipses. Tides are produced by the moon's gravity tugging on the oceans. Solar eclipses occur when the moon comes between the earth and the sun, and lunar eclipses occur when the earth prevents sunlight from reaching the moon.

REVIEW QUESTIONS

1. Explain why the moon always keeps the same side toward the earth. Does the moon rotate with respect to the sun?
2. What is the direction of the moon's rotational motion? In what direction does the moon revolve around the earth? What apparent motion of the moon in the sky is produced by each of these two motions? Explain.
3. The moon is first visible at about 3:00 P.M. Where is it located in the sky and what is its approximate phase?
4. When it is midnight local time, where in the sky will the full moon appear?
5. Is it possible to see a crescent moon at sunset with the horns pointing downward toward the horizon? Why?
6. What is meant by a harvest moon? When is it visible? What causes it?
7. What is meant by lunar librations?
8. How can the moon's mass be determined? Can you suggest a method using Kepler's third law?
9. Describe the lunar mare. How might they have been formed? Referring to Fig. 7.10(b), find Mare Crisium. Does its shape suggest anything about how it was formed?
10. What are sinuous rilles? How might they have been formed?
11. Describe one theory of the moon's origin and discuss the pros and cons of this theory.
12. Under what conditions will a lunar eclipse occur? a solar eclipse?
13. If there is a solar eclipse on June 10th you might expect a lunar eclipse on June 24th, approximately. Explain.
14. Explain why eclipses tend to occur at either two-week or six-month intervals. How would this be changed if the moon's orbit were *not* inclined?
15. Describe what an observer would see during a total solar eclipse.
16. Why typically are there two high tides each day?
17. What are spring and neap tides? When do they occur?
18. Explain why a total lunar eclipse lasts longer than a total solar eclipse. What are the durations of totality?
19. In what direction does the shadow of the earth move across the moon's face during a lunar eclipse?
20. Explain what the "saros" is and how it is used in predicting eclipses.
21. At one time it was suggested that all of the lunar craters were volcanic in origin. Is this still the case? Why?
22. Compare the interior structure of the earth and moon (refer to Fig. 5.3).
23. People sometimes argue that when the moon is full, tides ought to be extremely low because the sun and moon are working against each other in raising tides. Is this correct? Why?
24. Why does the height of the tide vary so much along a coastline?
25. What kinds of information about the moon became available as a result of the Apollo program? How has the information helped our understanding of the moon's history?

CHAPTER 8
THE SOLAR SYSTEM

The solar system consists of the sun, nine planets, thirty-five natural satellites, thousands of asteroids and comets, meteoroids, and interplanetary material. (See Fig. 8.1.)

8.1 THE SUN

The sun, a large gaseous sphere 864,400 miles (1,391,000 km) in diameter, is about one-third of a million times more massive than the earth and contains over 99% of the total mass of the solar system. By virtue of its enormous mass, it exerts a gravitational force that keeps the members of its family together. The sun is a typical star—typical because it is average in size and brightness, and a star because it shines by its own light. It derives its energy from the nuclear reactions that take place in the core, producing temperatures of about 15,000,000°K and surface temperatures of about 5770°K. (Chapter 10 deals specifically with the sun.)

8.2 THE PLANETS

There are nine known planets in the solar system. Six of them—Mercury, Venus, Earth, Mars, Jupiter, and Saturn—were known to the ancients. The three most distant planets—Uranus, Neptune, and Pluto—were discovered after the invention of the telescope, which occurred in 1609. The planets are relatively cold and are visible only by reflected sunlight. Their diameters range from about 3000 miles (4900 km) for Mercury to about 89,000 miles (140,000 km) for Jupiter. All planets except Venus rotate from west to east. Jupiter has the shortest period of rotation (9 hours 55 minutes) and Venus the longest (243 days).

a. Classification

Planets may be classified according to the relationship between their respective orbital positions and that of the earth. If their orbits lie inside the earth's, as do those of Mercury and Venus, they are called *inferior planets*, and if they lie outside the earth's, as do those of Mars, Jupiter, Saturn, Uranus, Neptune, and Pluto, they are called *superior planets*.

A more meaningful classification scheme is based upon the physical properties of the planets. The family of nine planets divides very naturally into two groups, one

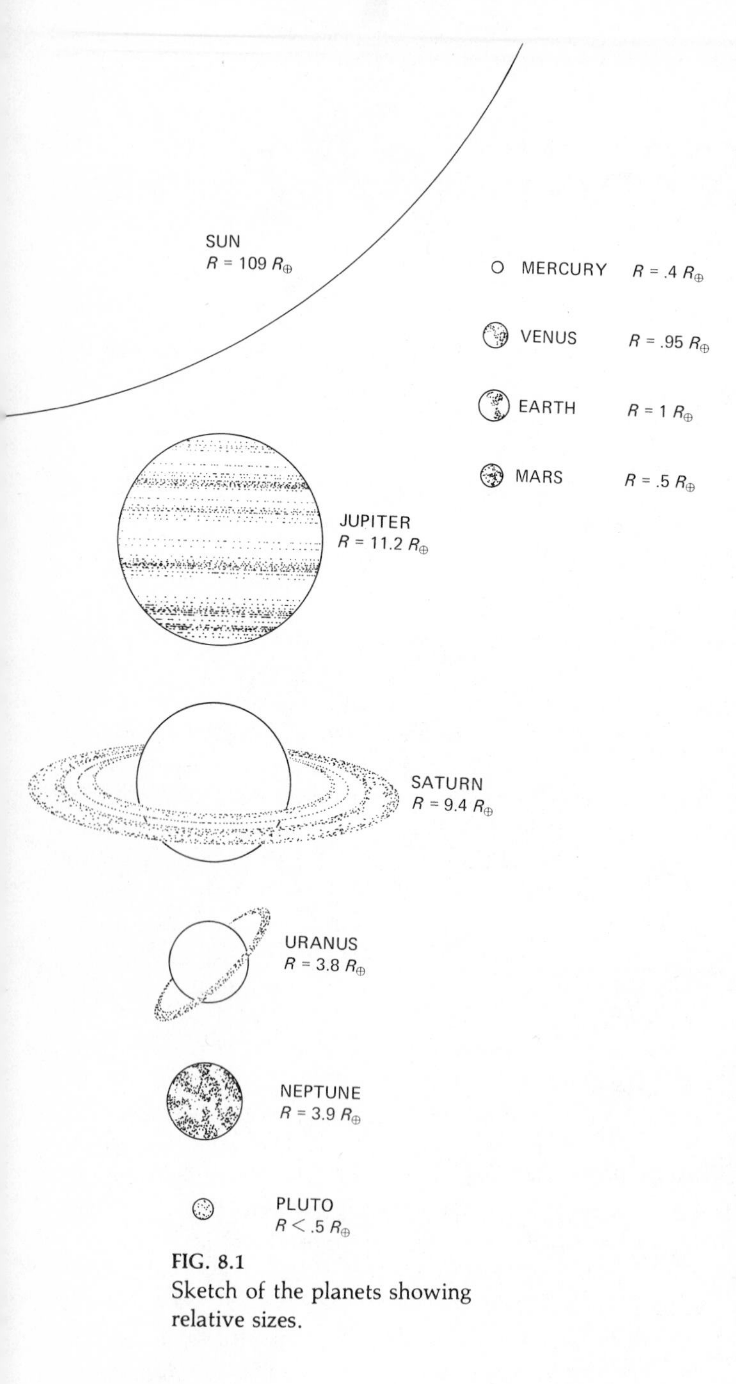

FIG. 8.1
Sketch of the planets showing relative sizes.

containing planets very much like the earth, called *terrestrial planets*, and the other containing planets similar to Jupiter, called *Jovian planets*. The terrestrial planets are Mercury, Venus, Earth, and Mars. The Jovian planets are Jupiter, Saturn, Uranus, and Neptune. Too little is known about Pluto to assign it to either group.

The terrestrial planets are all smaller in size and less massive than the earth. Their radii are all roughly similar to the earth's ($R_\oplus$), although Mercury's radius is only 0.4 $R_\oplus$. Their masses are also comparable to the earth's ($M_\oplus$), with Mercury again the smallest with a mass of about $0.05M_\oplus$. Apart from the similarity in sizes, the terrestrial planets possess few, if any, satellites and all rotate with periods of a day or longer. The terrestrial planets are all basically spheres of rock with iron cores. With the exception of Venus, their atmospheres are much less dense than those of the Jovian planets. Finally, the terrestrial planets are all relatively close to the sun. Mars, the most distant, is only half again as far from the sun as is the earth.

The Jovian planets differ dramatically from the terrestrial ones. They are all considerably larger than the earth in both radius and mass. The smallest has nearly four times the earth's radius and fourteen times its mass. Most possess well-developed families of satellites. Jupiter and Saturn, in fact, are nearly like miniature solar systems in terms of the number and extent of their satellite systems.

Despite these differences from the terrestrial planets, the most striking contrast is in the structure and atmospheres of the Jovian planets. Jupiter and Saturn appear to possess no well-defined surfaces. All the Jovian planets have deep, dense atmospheres rich in hydrogen, helium, and gaseous hydrogen compounds. The atmospheres are thought to merge smoothly into first a liquid and then a solid interior, with little evidence of the rocky matter found in the terrestrial planets. As we will see in Chapter 9.25, the striking differences between the two classes of planets are believed to be a natural outcome of the manner in which the solar system originated.

b. Orbits

Looking down from the north pole of the earth as a point of orientation, we can say that the planets revolve

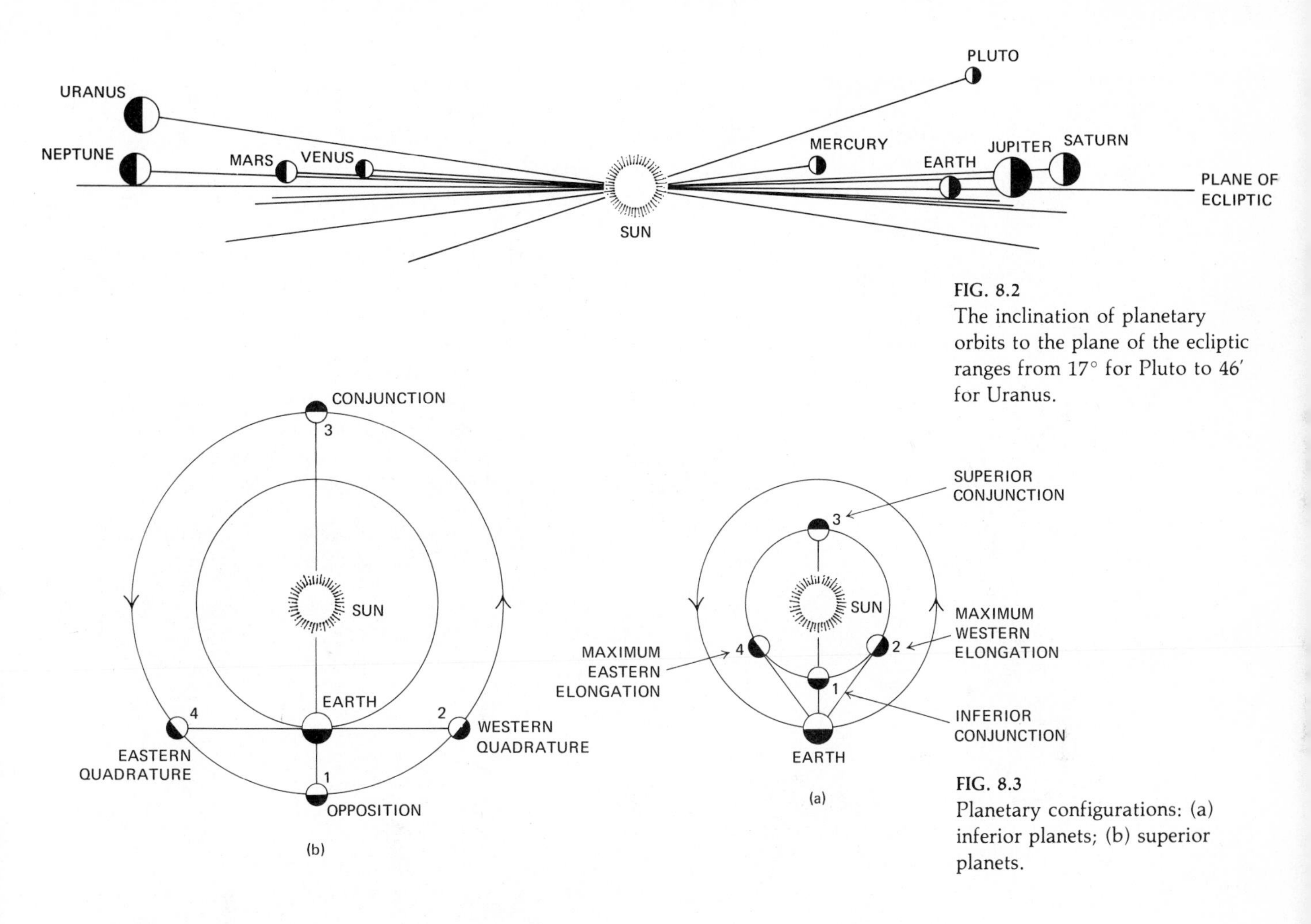

FIG. 8.2
The inclination of planetary orbits to the plane of the ecliptic ranges from 17° for Pluto to 46′ for Uranus.

FIG. 8.3
Planetary configurations: (a) inferior planets; (b) superior planets.

around the sun in a counterclockwise direction—west to east. Their mean distances from the sun range from about 36 million miles (58 million km) for Mercury to nearly 4 billion miles (6 billion km) for Pluto. The orbital velocities of the planets range from about 30 miles (48 km) per second for Mercury to 3 miles (5 km) per second for Pluto.

The orbits of all the planets lie very close to the plane of the ecliptic (Fig. 8.2). Pluto's orbit, at 17°09′, has the greatest angle of inclination to the ecliptic. Although all planetary orbits are elliptical, most of them are almost circular. Their eccentricities range from 0.007 for Venus, which is the most nearly circular, to 0.249 for Pluto, which is the most eccentric, and all are less than 0.1, except for Mercury and Pluto. Mercury has the shortest period of revolution (about 88 days), and Pluto has the longest (about 248 years).

c. Configurations

To locate the positions of a planet in its orbit with respect to the earth and the sun, special positions, called *configurations*, have been established (Fig. 8.3). For an inferior planet, the following configurations occur: (1) *inferior conjunction,* when the planet passes between the sun and the earth; (2) *maximum western elongation,* when the planet is leading the sun (west) so that its elongation (angle between the sun, earth, and planet) is maximum;

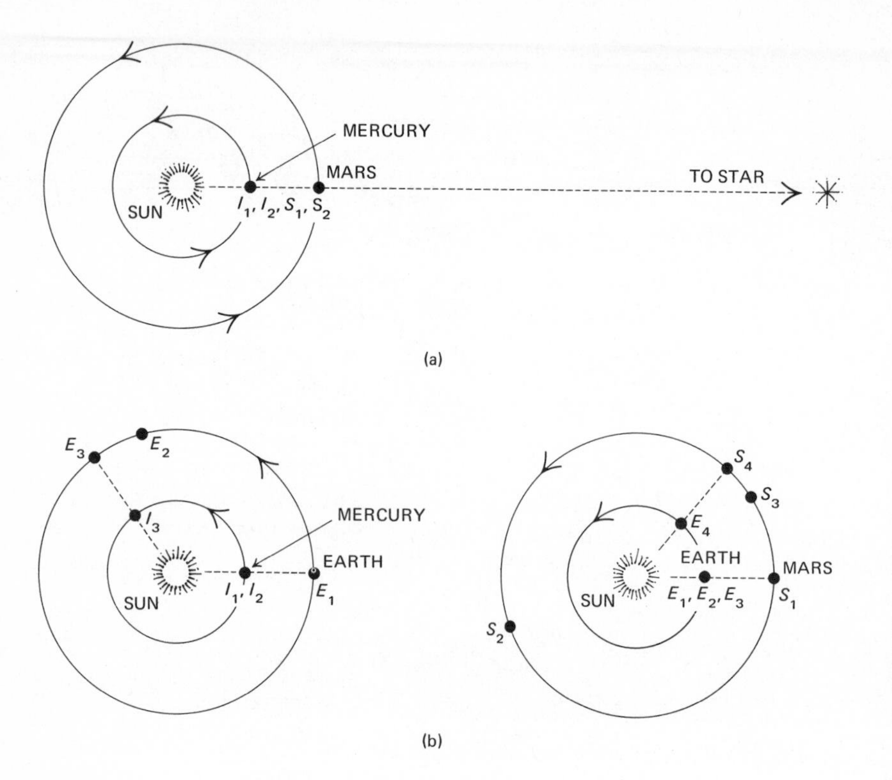

FIG. 8.4
The (a) sidereal and (b) synodic periods of planets. The sidereal period of a superior (S_1 to S_2) and an inferior (I_1 to I_2) planet is the interval of time for the planet to make one complete revolution around the sun with respect to a star. The synodic period of an inferior planet (Mercury) is the interval of time for both Earth and Mercury to move from position 1 to position 3. The synodic period of a superior planet (Mars) is the interval of time for both Earth and Mars to move from position 1 to position 4.

(3) *superior conjunction*, when the planet is on the opposite side of the sun from the earth; and (4) *maximum eastern elongation*, when the planet is following the sun (east) and the elongation is maximum. For a superior planet, the following configurations occur: (1) *opposition*, when the planet is on the opposite side of the earth from the sun; (2) *western quadrature*, when the planet is to the west of the sun and forms a 90° angle with the earth and sun; (3) *conjunction*, when the planet is on the opposite side of the sun from the earth; and (4) *eastern quadrature*, when the planet is to the east of the sun and forms a 90° angle with the earth and sun.

d. Sidereal and Synodic Periods

There is a difference between a planet's true and apparent periods of revolution around the sun. The true, or sidereal, period is the interval of time the planet takes to complete one revolution with respect to a star as seen by a hypothetical observer on the sun. The apparent, or synodic, period is the interval of time the planet takes to move between two successive identical configurations with respect to the earth. In Fig. 8.4, the sidereal period of an inferior planet is the time that it takes the body to travel from position I_1 to I_2, and for a superior planet, from position S_1 to S_2. The synodic period of an inferior planet is the time that it takes the body to travel from position I_1, through I_2, to I_3, and for a superior planet, from position S_1 to S_4. Another way of looking at the synodic period is to determine the time required for the faster-moving planet to gain one lap on the slower-moving planet.

The relationship between the sidereal and synodic periods of a planet can be expressed by the formulas:

$$1/S=1+1/P$$

for an inferior planet, and

$$1/S = 1 - 1/P$$

for a superior planet, where S is the planet's sidereal period and P is its synodic period, both expressed in years. When one of the planet's periods is known, the other can be determined from these formulas. For example, if the sidereal period of a superior planet is four years, its synodic period is determined as follows:

$$1/S = 1 - 1/P$$
$$1/P = 1 - 1/S$$
$$= 1 - \tfrac{1}{4} = \tfrac{3}{4}$$
$$\therefore P = \tfrac{4}{3}, \text{ or } 1\tfrac{1}{3} \text{ years.}$$

e. Masses

The mass of a planet measures the amount of matter in the planet and, if the dimensions of the planet are known, also gives information about the planet's composition. In Chapter 5 it was mentioned that the earth is not a sphere because its rotation makes an equatorial bulge. Many of the other planets are likewise not spheres. The difference between the planet's equatorial and polar radii when divided by the equatorial radius gives the *oblateness* of a planet. In general, the more rapidly rotating Jovian planets are more oblate than the more slowly turning terrestrial ones. The difference between the polar and equatorial diameters for the earth is about 27 miles (43.45 km) in 8000 miles (13,000 km), and about 5500 miles (8800 km) in 88,000 miles (142,000 km) for Jupiter. It is believed that since a marked difference exists in the physical characteristics of the terrestial and Jovian planets, a marked difference in their internal structure must also exist. The mean densities of Saturn and Jupiter, 0.71 and 1.4 grams per cubic centimeter, respectively, are so low in comparison to the earth's mean density of 5.5 grams per cubic centimeter that a large portion of their material must be either solidified hydrogen, the least dense solid, or solidified helium. If the material were solidified helium, the planet's densities would be considerably higher; therefore, it has been concluded that a large portion of the material in Jupiter and Saturn must be solidified hydrogen.

The most accurate method for determining the mass of a planet is to observe its gravitational influence on one of its satellites or a passing spacecraft. This information is applied to Kepler's third law of planetary motion, which Newton expressed as

$$(M+m)P^2 = ka^3,$$

where $M + m$ represents the total mass of the planet and its satellite. As we saw in Chapter 2.8, the value of k may be set equal to 1 if the units for P, a, and the masses are chosen cleverly. For a satellite in orbit about a planet, it is natural to use the period of the moon around the earth (a month), the earth-moon distance, and the earth's mass ($M_\oplus$) as the units. For example, determine the mass of Mars, given that its satellite Deimos has a sidereal period of 1.262 days and a mean distance of 14,600 miles (23,500 km) from Mars' center. *Solution*: Deimos' period in sidereal months is 1.262/27.3, or 0.0463. Deimos' distance from Mars in terms of the moon's distance from the earth is 14,600/238,000, or 0.0613.

$$(M + m) = a^3/P^2$$
$$= (0.0613)^3/(0.0463)^2$$
$$= 0.108 \text{ earth masses.}$$

This means that Mars has a mass 0.108 times that of the earth. The mass of Deimos is so small that it can be ignored. A similar calculation for Jupiter shows that its mass is $318M_\oplus$, making it not only the most massive planet in the solar system, but more massive than all the planets put together. The least massive planet is Pluto, with a mass of $0.002M_\oplus$, also measured using Kepler's third law.

f. Temperatures

A planet's temperature depends primarily on the amount of solar energy that falls on a square unit of its surface, the fraction of that energy the planet absorbs, and the blanketing properties of the planet's atmosphere, if it has one. The amount of incident energy from the sun is determined by the planet's distance from the sun. The amount of energy absorbed by the planet depends on the surface properties, being low for a surface that is highly reflective, like snow or sand. The *reflectivity of a surface* is measured by the *albedo*, defined as the fraction of incident energy reflected back to space. A perfect reflector has an albedo of 1. Dark rock, such as occurs on the surface of Mercury, has an albedo of about 0.06, meaning it reflects only 6% of the incident light, and absorbs 94%. The surface temperature of the planets ranges from about 900°F (480°C) for Venus to −400°F (−240°C) for Pluto.

g. Atmospheres

An atmosphere around a planet may be detected by a variety of means. For objects with no atmosphere, such as the moon, the dividing line (the *terminator*) between the sunlit and dark sides is sharp. An atmosphere will allow small amounts of light to be scattered or reflected into the dark side, producing a twilight zone such as we have on the earth. A high albedo may indicate an atmosphere. Most bare rock surfaces have low albedos while a layer of atmosphere, especially if it contains clouds, may be highly reflective. The dense atmosphere of Venus, for example, reflects about 74% of the sunlight it receives. Both the *gradual occultation* of a star by a planet, that is, the gradual obscuration of a star when the planet passes between it and the earth, and the appearance of *absorption lines* in the planet's spectrum further indicate the presence of an atmosphere.

Whether or not a planet is able to retain an atmosphere depends on its escape velocity and the velocity of the molecules in its atmosphere. A planet's *escape velocity*, the velocity required for a particle such as a molecule to escape the planet's gravitational force, depends on the planet's size and mass. Mercury, whose size and mass are small, has a weak gravitational force and a low escape velocity; Jupiter, whose size and mass are great, has a strong gravitational force and a high escape velocity.

The velocity of the molecules depends on their temperatures. Since the radiation a planet receives from the sun is inversely proportional to the square of its distance from the sun, the temperature of the molecules in the atmospheres of planets closer to the sun is much higher than for those farther away. Therefore, molecules with velocities greater than the planet's escape velocity will escape from the gravitational force of the planet so that eventually the planet will lose its atmosphere.

Since the velocity of the molecules in the atmosphere of a Jovian planet is so much lower than the escape velocity, these planets still have dense atmospheres. The terrestial planets, which are smaller, have lower escape velocities and their atmospheric molecules have higher velocities due to their warm upper atmospheres. Under such conditions, the terrestial planets (except Mercury) have been able to hold the heavier elements in their atmospheres, whereas the lighter elements have been able to escape into space. Mercury, because of its small size and mass and high surface temperature, has lost its atmosphere completely.

8.3 SATELLITES

The 35 known satellites in the solar system are bodies that revolve around the planets. Most of them revolve in the same direction as their planets, in orbital planes that lie very close to the planets' equatorial planes. Two of the planets—Mercury and Venus—have no known satellites. Jupiter has the largest number, Saturn has 10 (its latest, Janus, was discovered in 1967); Uranus has 5; Neptune and Mars have 2 each; and Earth and Pluto have 1 each. Six of the satellites are as large as or larger than the earth's moon. The two largest are Saturn's Titan and Jupiter's Ganymede—with diameters of about 3200 miles (5200 km) and 3500 miles (5800 km), respectively. These two satellites therefore have larger radii than the planet Mercury. The smallest are Mars' Deimos with a diameter of about 8 miles (13 km) and several of Jupiter's outer moons. Titan, Saturn's largest satellite, is one of the few known to possess an atmosphere.

8.4 ASTEROIDS

The thousands of small, irregular, slightly elongated, barren rocks that revolve around the sun in elliptical orbits are called asteroids. Their orbital eccentricities and inclinations are quite varied. Most of their orbits lie between those of Mars and Jupiter. Some extend beyond Jupiter's orbit, whereas others come within the orbit of

Venus. The largest asteroid (Ceres), with a diameter of about 660 miles (1000 km), is not large enough to be resolved into a visible disk by a telescope. Vesta, with a diameter of about 340 miles (540 km), is more reflective and more visible to the unaided eye. The total mass of the asteroids has been estimated at less than 1% of the earth's mass. Asteroids are discussed further in Chapter 9.

8.5 METEOROIDS

In addition to the asteroids, there are a great many extremely small bodies (meteoroids), too small to be observed with a telescope, which revolve around the sun. When these bodies leave their orbits and pass through the earth's atmosphere, they are heated by the friction which they encounter as they collide with the molecules in the atmosphere. As the heat increases, they partially vaporize, may become luminous, and may be visible from the earth. When a meteoroid becomes visible, it is called a *meteor,* or a shooting star. It has been estimated that over 200 million meteoroids enter the atmosphere during a 24-hour period over the entire earth. On a clear, moonless night, an observer can see about six meteors every hour. (For further discussion, see Chapter 9.)

8.6 COMETS

A comet consists of a swarm of meteoritic material embedded in ice with such frozen gases as ammonia and methane. Comets revolve around the sun in very eccentric orbits so that they are visible for periods ranging only from several days to several months. Many of them are visible only through a telescope or appear as hazy spots of light to the unaided eye. When they are near the sun, comets consist of two parts—a head and tail. The head comprises (1) the *nucleus,* consisting of one or more solid chunks perhaps a few miles across; and (2) the *coma,* consisting of gases evaporated from the nucleus and surrounding it. A typical comet head has a diameter of about 50,000 miles (90,000 km). The *tail* consists of a material swept out of the coma by the combined effect of solar radiation and matter flowing out from the sun, and may extend for a distance of 100 million miles (160 million km).

8.7 MERCURY

As a bright object in the evening or morning sky, Mercury was well known to many of the ancient people. Because of its rapid motion with respect to the stars, the ancient Greeks associated this celestial body with the swift messenger of the gods, Mercury. Since Mercury is the planet nearest the sun, it appears to oscillate about the sun to a maximum angular distance of about 28°. When Mercury is to the west of the sun, it appears in the morning before sunrise; when it is to the east of the sun, it is seen in the evening and sets after the sun.

Mercury's nearness to the sun hampers observation with the unaided eye when the planet is high in the sky; therefore, the best time to view it is just before sunrise or just after sunset. Since the planet is very bright, it can be viewed telescopically when it is high above the horizon, where atmospheric interference to viewing is greatly reduced. Even under these conditions, only a few of the many ground-based photographs taken of Mercury show any of its surface markings.

a. Orbit

Mercury revolves around the sun in an elliptical orbit at a mean distance of 36 million miles (58 million km). Its orbit is inclined about 7° to the plane of the ecliptic. At perihelion it is about 29 million miles (47 million km), and at aphelion it is about 44 million miles (71 million km). With this difference of 15 million miles (24 million km), its eccentricity is about 0.206. As the fastest planet in the solar system, it has a mean orbital velocity of about 30 miles (48 km) per second. It sidereal period of about 88 days and its synodic period of about 116 days are the shortest of any planet. Since its orbit is inside the earth's, its phases are similar to those of the moon, although its apparent size changes considerably more than does the moon's. At inferior conjunction its apparent diameter is about three times greater than at superior conjunction.

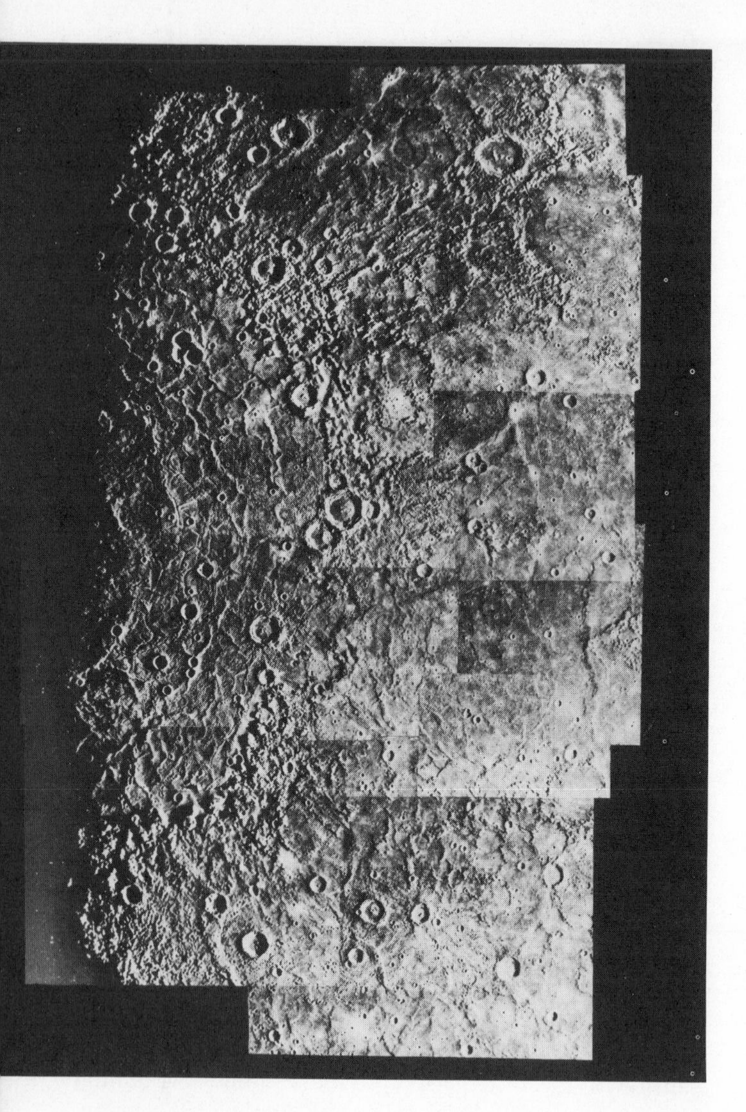

FIG. 8.5
Caloris Basin on Mercury. This giant impact feature is similar to lunar mare and is about 800 miles (1300 km) across. (NASA/JPL photograph.)

When Mercury is at inferior conjunction and at the same time at either of the nodes, it transits the sun's disk. This is a telescopic phenomenon that is observed as a small black spot, about $\frac{1}{160}$ the sun's diameter, moving across the sun's face for about eight hours. The next transit will occur on the morning of November 13, 1986. The astronomical importance of these transits is that the event can be timed most accurately, and from these times our knowledge of Mercury's orbit and motions can be improved.

b. Physical Properties

With a diameter of 3032 miles (4878 km), which is about 0.38 that of the earth, Mercury is the smallest planet in the solar system. Since Mercury has no known satellites and produces very small perturbations on the orbit of Venus, its mass has been determined by observing the perturbations it produces on the orbit of the asteroid Eros as the latter passes close to the planet.

More recently an improved value has been determined from the change in orbit of Mariner 10 as it passed the planet. The best value is about 5% of the earth's mass. From this value, its density (mass per unit volume) is found to be 5.4 grams per cubic centimeter, which is slightly less than that of the earth. The high density of the earth is due in large measure to the compression of material in its core by the overlying layers of rock. Mercury's high density must be due mainly to a higher iron content in its interior, as its smaller mass is unable to compress the interior material.

c. Rotation

For many years astronomers believed that Mercury's period of rotation was synchronous with its period of revolution (88 days). However, while using the 1000-foot radar telescope at Arecibo, Puerto Rico, in 1965, Gordon H. Pettengil and Robert B. Dyce found that Mercury's period of rotation is about 59 days. Subsequent radar measurements indicated that the planet's rotational period is 58.65 days, which to within the measurement errors is exactly two-thirds of the planet's orbital period of 87.97 days. This means that the planet rotates three times in two revolutions around the sun. The planet

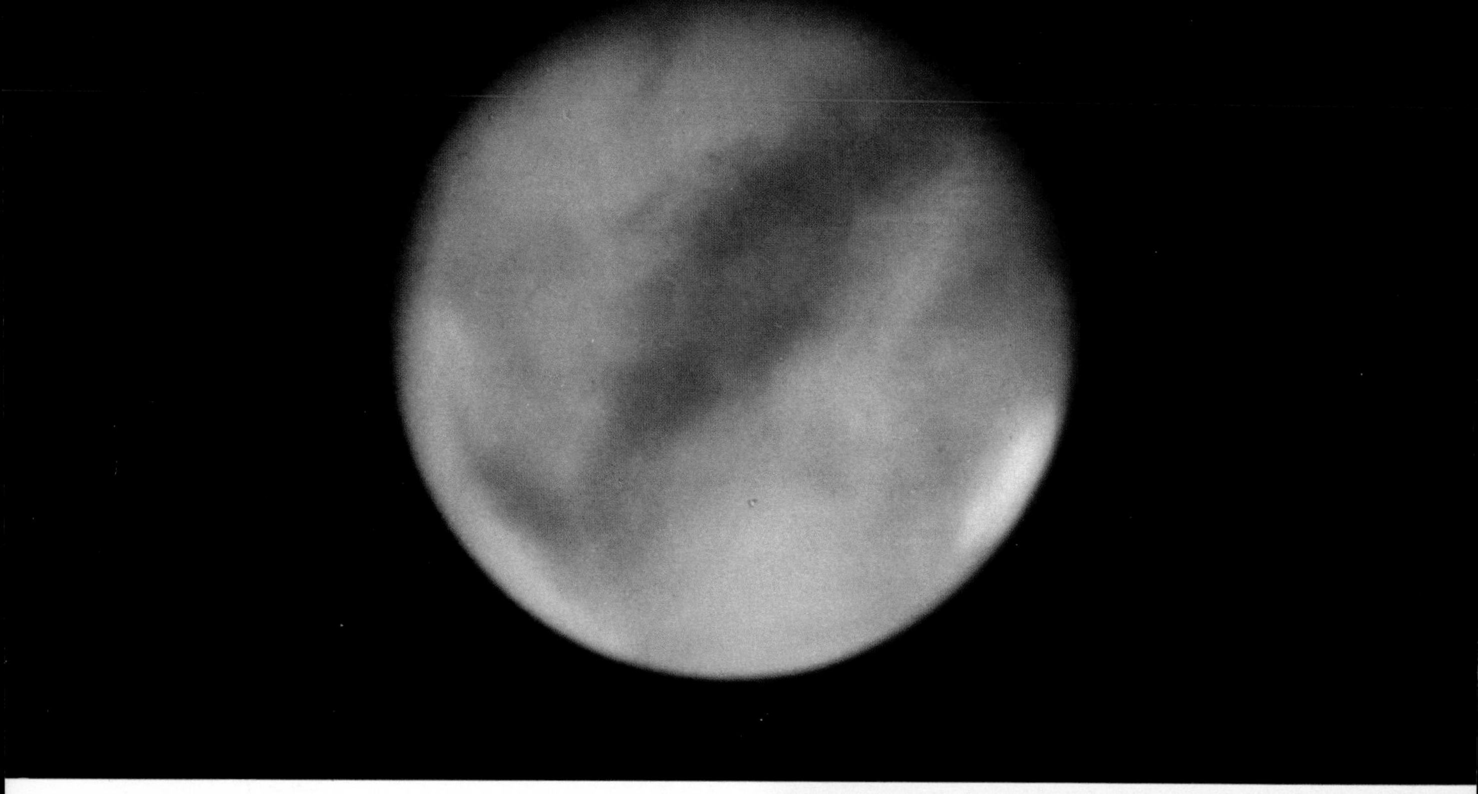

PLATE 14 (preceding page) Mars as seen from the approaching Viking spacecraft. Note the frost on the volcanic peaks (white spots). (NASA/JPL photograph)

PLATE 15 (above) Sixty-inch photograph of the planet Mars, Earth's neighbor and the solar system's most intriguing member. (Photograph from the Hale Observatories)

PLATE 16 (right) The Great Red Spot of Jupiter, photographed by a Pioneer spacecraft as it passed by the planet. The spot is roughly 30,000 miles (45,000 km) across and may be a vortex in Jupiter's atmosphere similar to a terrestrial hurricane. (NASA photograph)

PLATE 17 (above) Saturn and its ring system, photographed with the 60-inch reflector. The belt markings and the Cassini division (which separates the inner and outer rings) are clearly visible. (Photograph from the Hale Observatories)

PLATE 18 (next page, left) The Crab Nebula in Taurus, an expanding remnant of a supernova explosion in A.D. 1054. Most of the white light comes from high-speed electrons spiraling in the nebula's magnetic field. Most of the red light comes from glowing hydrogen. Photographed in red light with the 200-inch telescope. (Photograph from the Hale Observatories)

PLATE 19 (next page, right) The Ring Nebula in Lyra, a planetary nebula containing a hot, blue star surrounded by a thick spherical shell of rarefied gas. Photographed in red light with the 200-inch telescope. (Photograph from the Hale Observatories)

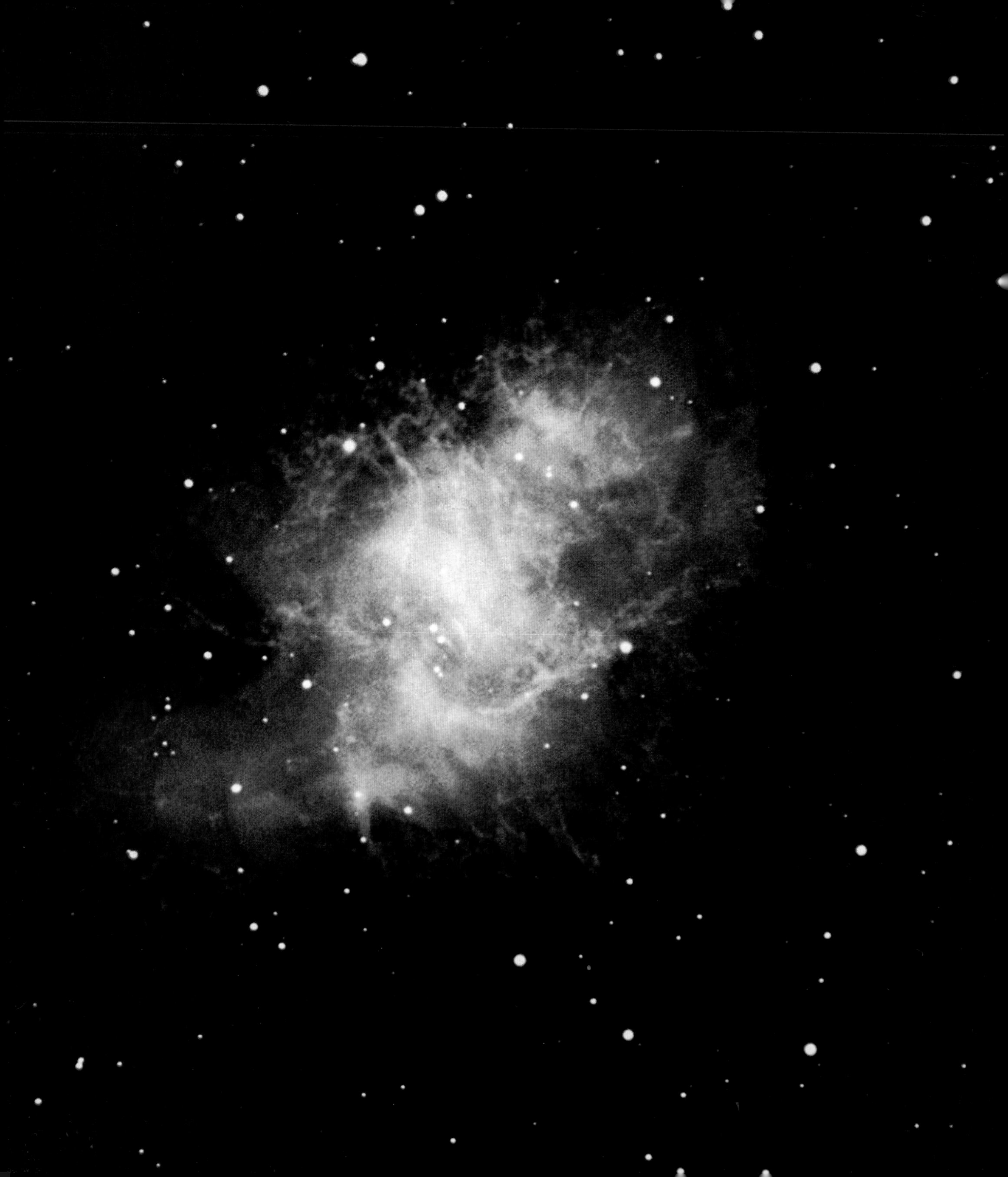

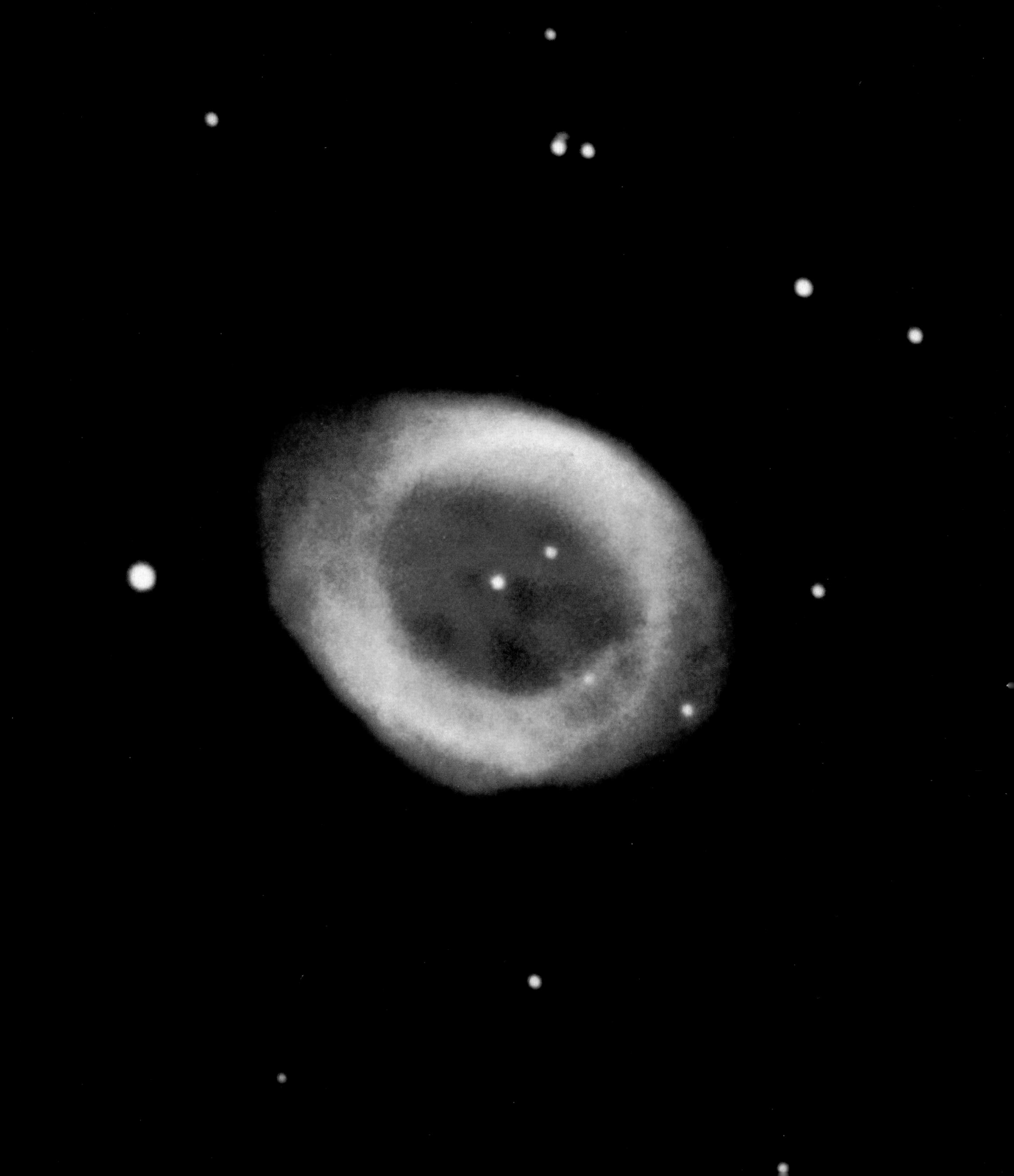

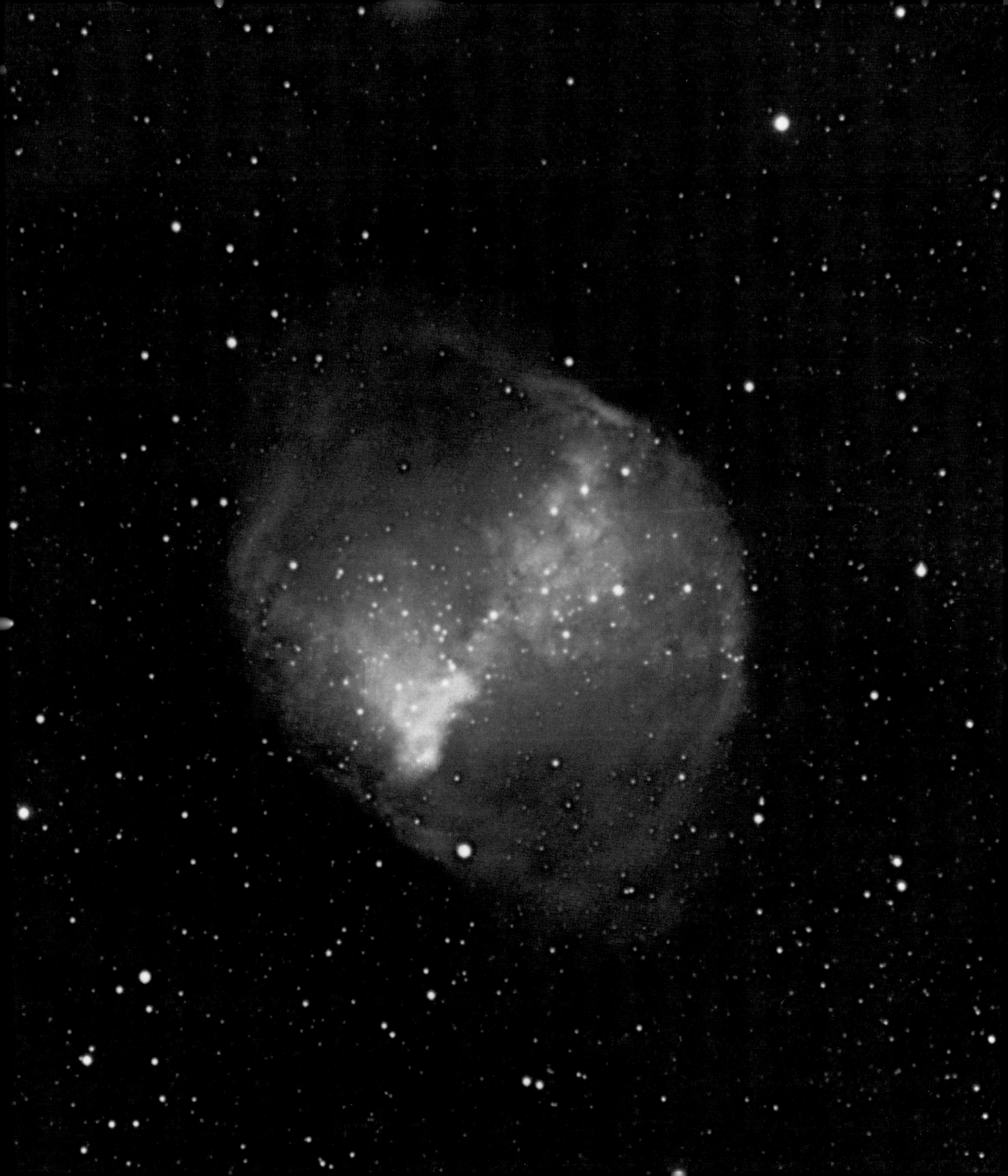

PLATE 20 (preceding page, left) Planetary Nebula in Aquarius, NGC 7293, photographed with the 200-inch telescope. (Photograph from the Hale Observatories)

PLATE 21 (preceding page, right) The Dumbbell Nebula in Vulpecula, a planetary nebula with a small, hot, blue star surrounded by a sphere of tenuous gas. Photographed with the 200-inch telescope. (Photograph from the Hale Observatories)

PLATE 22 (above) The central region of the Great Nebula of Orion. The ultraviolet energy from the hot stars embedded in the nebula makes the great cloud of interstellar dust and gas fluoresce. (Photograph from the Hale Observatories) rotates in a counterclockwise direction, with its axis of rotation nearly perpendicular to its orbital plane.

d. Surface Features and Atmosphere

While it had long been suspected that Mercury's surface was cratered like the moon's, direct observations of the planet's pitted, scarred surface were not obtained until 1974, when Mariner 10 sent back the first TV pictures as the spacecraft passed within 500 miles (750 km) of the surface. The photos reveal the planet's surface to be heavily pockmarked and to have huge flat plains (such as the Caloris basin in Fig. 8.5) similar to lunar mare. The craters, often surrounded by secondary craters produced by ejecta, presumably were formed by meteoritic impact as described in our discussion of the moon. The light streaks radiating from some craters and wrinkles on the smooth basin floors are also similar to lunar features. Mercury's crust seems to have cracked, producing a set of *scarps,* where the surface is shifted vertically by amounts ranging from a few hundred meters to as much as 3 km. These huge cliffs run for hundreds of kilometers across the surface. It is believed the crustal cracking was produced by a slight shrinking of the interior of the planet, perhaps as the planet's iron core cooled.

The surface temperature varies widely over the planet because of its slow rotation. Noontime equatorial temperatures are as high as 800°F (430°C) and nighttime values are as low as −280°F (−170°C). Because of the high surface temperature and low gravity, it is not surprising that Mercury has no atmosphere (see Chapter 8.2g).

One of the many surprises of the Mariner 10 mission was the discovery of a magnetic field on Mercury. It had previously been thought that only rapidly rotating planets could have a magnetic field.

8.8 VENUS

The planet Venus, which was named after the Roman goddess of love and beauty, is one of the most beautiful objects in the sky. It is third only to the sun and moon in brightness, and when its position is known, it is clearly visible in the daytime sky.

The Babylonians first recorded the appearance of Venus nearly 4000 years ago. As an inferior planet, it appears to oscillate about the sun and is never seen more than 48° on either side of it. As happened with Mercury, this confused the ancients, who believed that they were two separate stars—a *morning star* rising just before sunrise and an *evening star* setting just after sunset. The ancient Greeks called the morning star Phosphorus and the evening star Hesperus. About 500 B.C., the Greek scientist/philosopher Pythagoras recognized that the two "stars" were actually the same body.

a. Telescopic View

Galileo was the first to observe Venus telescopically. When he noted that it displayed phases similar to those of the moon and that it changed its apparent size over a continuously recurring cycle (Fig. 8.6), he became convinced that Venus revolves around the sun. Thus, he realized that the sun is the center of revolution for all the planets and that the Copernican model of the universe is correct.

Even under the most favorable viewing conditions, the telescopic view of Venus is almost featureless and unimpressive. The few observable features consist of small differences in color contrast as shown in the unusual photograph taken at the New Mexico State University (Fig. 8.7). Quite often the cusps appear brighter than the rest of the crescent, and a narrow strip along its entire limb appears brighter when the planet is in the crescent phase. There is also a darkening effect of the illuminated edge of the terminator (the line of demarcation between the illuminated and dark portions of the face of Venus).

b. Orbit

At a mean distance of 67 million miles (108 million km), Venus revolves around the sun in an elliptical orbit that is the most nearly circular of any planet's. The difference between its aphelion and perihelion distances is only 910 thousand miles (1.5 million km), which gives its orbit an eccentricity of 0.007, the lowest of any planet's. As shown in Fig. 8.8, Venus is the closest planet to the earth when at inferior conjunction, a distance of 26 million miles (42 million km). At superior conjunction, it is

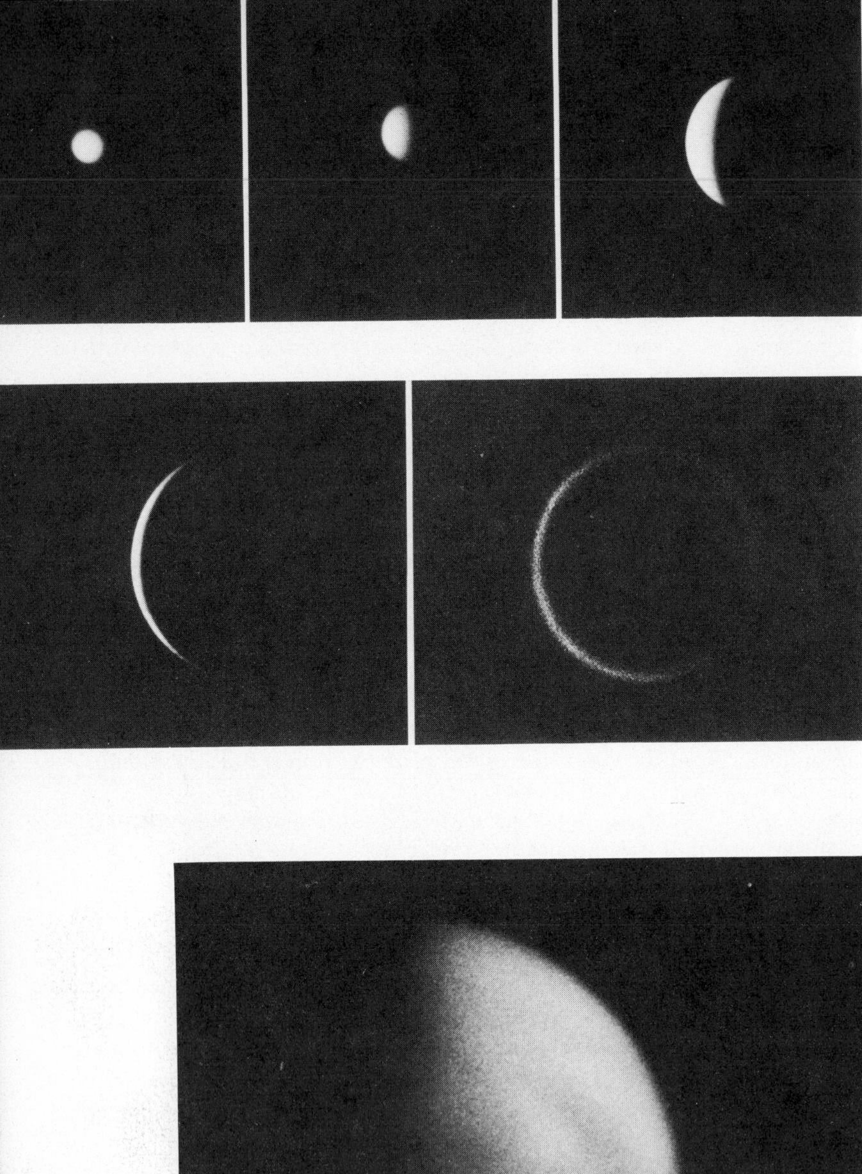

FIG. 8.6
These five photographs of Venus are in the same scale and show that in the full phase Venus appears smaller because it is farther from the earth; in the crescent phase it appears larger because it is nearer the earth. (Courtesy of the Lowell Observatory.)

FIG. 8.7
Cloud structure of Venus in ultraviolet light, 24 May 1967, 0132 U.T. (Courtesy of the New Mexico State University Observatory.)

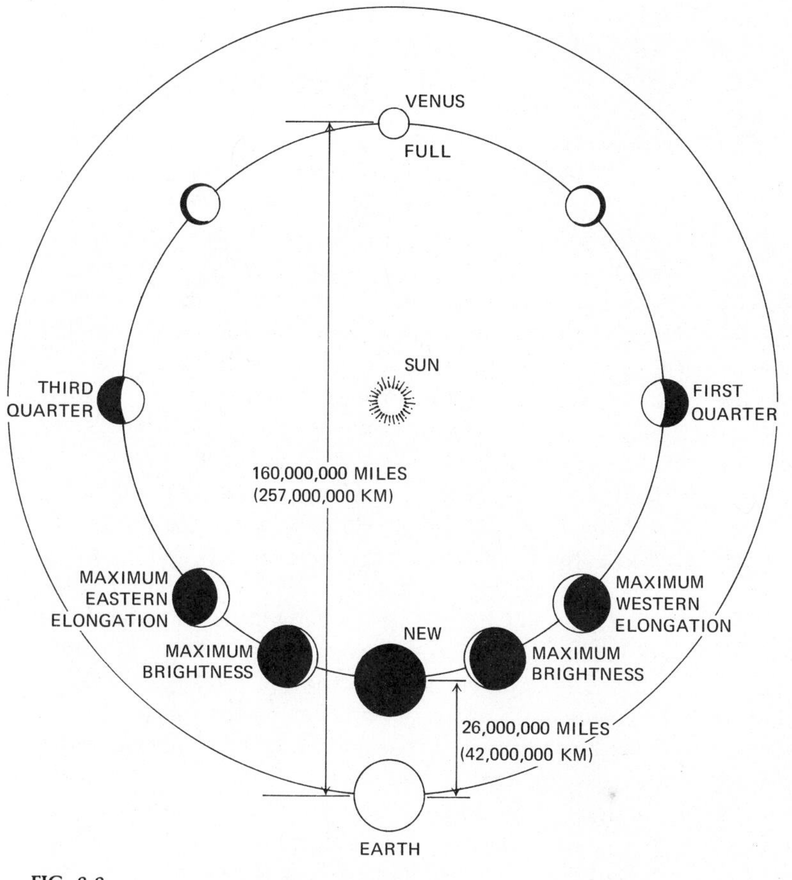

FIG. 8.8
The phases of Venus, its average maximum and minimum distance from the earth, and its relative apparent size as seen from the earth.

FIG. 8.9
Venus, photographed with the 200-inch telescope near inferior conjunction in the crescent phase, appears over six times greater in size than when at superior conjunction in the full phase. (Courtesy of the Hale Observatories.)

nearly 160 million miles (257 million km) from the earth. This great difference in its distance from the earth produces a large variation in its apparent size. At superior conjunction, its apparent diameter is about 10″ and at inferior conjunction, about 64″ (Fig. 8.9).

The orbital plane of Venus is inclined 3°24′ to the plane of the ecliptic. At an orbital speed of about 22 miles (35 km) per second, its sidereal and synodic periods are about 225 and 584 days, respectively.

The transit of the sun's disk by Venus is a rare phenomenon and is visible to the unaided eye. It occurs when Venus is at inferior conjunction and close to one of the nodes. The last two transits occurred in 1874 and 1882, and the next two will occur on June 8, 2004 and on June 6, 2012.

c. Physical Properties

Venus is often called the earth's twin because of the similarity in the general physical properties of the two planets. The exact size of Venus is uncertain because dense clouds completely shroud the planet. A diameter of about 7600 miles (12,200 km) was obtained from data that were based on visual, photographic, and photoelectric observations of the occultation of the star Regulus by Venus on July 7, 1959. This diameter is about 0.96 that of the earth. Venus has an escape velocity of about 6.5 miles (10.46 km) per second as compared to about 7 for the earth.

d. The Venusian Atmosphere and Surface

The basic features of the Venusian surface and atmosphere have eluded visual observation, again because of the dense cloud cover of the planet. It has thus proved important to utilize information gathered from parts of the electromagnetic spectrum that can penetrate the clouds. Most of our knowledge of Venus has been garnered from radar, radio, and infrared astronomical techniques. In addition, at least four probes from the U.S.S.R. have descended to the Venusian surface to radio back data about both atmospheric and surface conditions.

Direct sampling of the atmosphere has confirmed the high abundance (~95%) of carbon dioxide (CO_2) inferred by spectra taken with terrestrial telescopes. In addition, precise measures have now been made of the abundance of oxygen (O_2), carbon monoxide (CO), water (H_2O), and nitrogen (N_2). Other substances detected from ground-based observations, such as hydrochloric, hydrofluoric, and sulphuric acids (HCl, HF, H_2SO_4), were also confirmed. In fact, the Venusian atmosphere seems more like a chemistry laboratory than a canopy of "air."

Not only has the atmosphere proved to be far more complex than previously thought, but the influence of the atmosphere on the planet's temperature has been clarified. Early radio observations suggested what seemed an implausibly high surface temperature of nearly 900°F, making Venus the hottest planet in the solar system. The observations were in fact correct, as has now been amply verified by both the Soviet Venera landers and the United States Mariner flyby spacecraft. The reason for the high temperature lies in what is called the "greenhouse effect," which occurs as follows. The temperature of any object is set by a balance between the energy it receives and the energy it radiates. A brick put out in the sun will slowly absorb solar energy and warm up until it radiates heat as rapidly as it receives it. Planets are heated the same way. Solar energy flows down through the atmosphere and reaches the surfaces, where it is absorbed. The ground then reradiates the energy. The solar energy illuminating a planet typically has a relatively short wavelength of several thousand angstroms. The planet, being cooler, typically radiates at much longer wavelengths (recall Wien's law in Chapter 3.8), usually in the infrared. Most atmospheric gases are relatively transparent to radiation in the visible part of the spectrum, but many, especially CO_2 and H_2O, are very "opaque" in the infrared. Thus the short-wavelength energy heating the planet can reach down to the surface and warm it, while the long-wavelength energy emitted by the planet cannot escape through the planet's own atmosphere. The atmosphere therefore acts like a blanket, making the surface of the planet warmer than if the planet had no atmosphere.

It was at one time thought that the glass in a greenhouse acted the same way, allowing the warming rays of the sun to penetrate and then trapping the heat inside. A similar effect occurs in an automobile parked in the sun. It is now known that greenhouses are hot simply because the warm air cannot flow out. Any enclosure transparent to sunlight will get hot whether or not it allows infrared radiation to escape. Nevertheless, the phrase "greenhouse effect" still lingers. Regardless of the name, the earth is about 15°F warmer due to the trapping of heat in its atmosphere, and Venus is hundreds of degrees hotter due to its extremely dense atmosphere of CO_2, which is highly opaque to infrared radiation.

In addition to having a high temperature, the lower atmosphere of Venus is under tremendous pressure. The

▲
FIG. 8.10
A view of the surface of Venus transmitted to earth by a Soviet Venera lander. Rocks are visible as well as the horizon on the right. (Courtesy of Novosti Photo.)

surface atmospheric pressure is about 90 times that of the earth. The surface humidity is extremely low: about 0.01% water vapor, as compared to a typical value of 2% for the earth.

The structure of the Venusian atmosphere is somewhat puzzling. The exact nature of the clouds is not yet certain. Currently, it appears that droplets of sulphuric acid match best the reflected color and spectra of the clouds. Although the high pressure and temperature combined with the dense clouds were at one time thought to produce a murky light at the surface, with distant objects including the horizon grossly distorted by mirage effects, the Russian Venera landers sent back pictures which revealed a rather prosaic landscape of rubbly rocks casting distinct shadows (Fig. 8.10).

The manner in which the atmosphere circulates is also unclear. Observations from the Mariner 10 show that in the ultraviolet (UV) distinct cloud patterns exist (Fig. 8.11). This is in agreement with observations of dark markings in the UV as made from the earth. The clouds seem to be of two types. Because of the planet's

FIG. 8.11
The clouds of Venus seen in ultraviolet light, photographed from Mariner 10. Note how the clouds seem to wind around the planet in a spiral pattern. (Courtesy of NASA/JPL.)

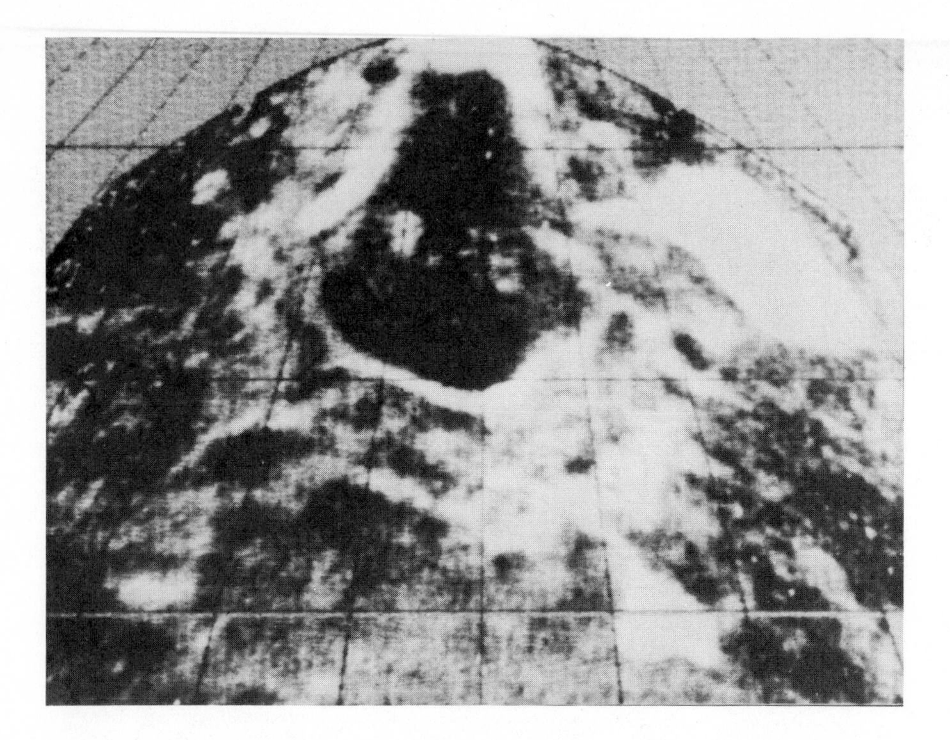

FIG. 8.12
A radar map of a portion of Venus. Note the large dark basin with bright rim. Mapping was done at Arecibo Observatory by D. B. Campbell and R. B. Dyce (both from the National Astronomy and Ionosphere Center) and G. H. Pettengell (MIT).

slow rotation, the area immediately below the sun is intensely heated. Here, a bubbly, convective cloud cover showing small, roughly circular features is visible. Away from the subsolar point, long cloud streamers appear to spiral out and wind around the planet like stripes on a barber pole. Observations of the clouds suggest that the atmosphere rotates once in four days. Since the planet itself rotates once in 243 days, winds of hundreds of miles per hour must blow in the upper atmosphere. Surface winds measured by Venera landers were low.

Despite the landings of four Venera craft on the Venusian surface, little is known about the general features of the planet. The thick clouds have prevented mapping of the surface by the means so successful in the case of Mercury and Mars. However, a crude idea of the nature of the surface has been obtained by radar observations from the earth. A radar pulse that travels across space is sent out, penetrates the planet's atmosphere, strikes the surface, and is reflected back to a receiver on the earth. By carefully timing the signal and measuring its strength, it has been possible to map the large-scale structure of the Venusian surface. Venus seems to have craters and relatively smooth plains like the moon (Fig. 8.12).

Radar pulses have also been used to determine the planet's rotation rate. A radar pulse striking the planet's turning surface is reflected off one edge moving toward us and the other edge moving away from us. This difference in motion slightly changes the wavelength (by the Doppler shift) of the radar reflected from the two edges of the planet and allows a measure to be made of the length of the Venusian day. Venus rotates more slowly than any other planet and, more peculiarly, rotates *clockwise* as seen from above its north pole. "Backward" rotation such as this is referred to as *retrograde*. Venus is the only planet rotating in a retrograde sense. Even more peculiarly, Venus rotates at such a speed that when it is closest to the earth it always presents the same side to the earth, rather like the moon. It has thus been suggested that gravitational forces produced by the earth on Venus have slowed down and reversed its spin to the present low retrograde value of 243 days.

e. Interior and Magnetic Fields

From the mass and radius determined by space probes the mean density of Venus is found to be 5.2 grams per cubic centimeter, slightly less than that of the earth. The close similarity in density, mass, and radius suggests to many that Venus and the earth should be nearly identical geophysically. However, observations of Venus by radar suggest that its surface may not have undergone the volcanism and crustal motion that have reshaped the earth's surface. Whether the absence of tectonic activity is somehow related to the planet's slow rotation or to a lower abundance of the radioactive elements that are believed to be responsible for heating the earth is still unknown. The slow rotation rate *is* believed to be responsible for the extreme weakness of Venus' magnetic field, which is at least ten thousand times smaller than the earth's. One final riddle is why there is so much CO_2 and so little water on Venus. Perhaps the high surface temperature has prevented CO_2 from being assimilated into the Venusian rocks the way it has on the earth. (Limestone, for example is rich in CO_2.) Such assimilation is further enhanced by water, whose rarity on Venus may also be due to the high temperature. Existing only as a gas, water may diffuse slowly into the upper atmosphere of Venus, where sunlight breaks it down into hydrogen and oxygen, and the light hydrogen escapes.

8.9 MARS

Mars, with its colorful history, is one of the most fascinating planets in the solar system (Plate 15). Since antiquity, its bright, reddish orange disk has inspired people to associate disaster, war, and death with the planet. The Greeks called it "Ares," which means disaster; the Chaldeans called it "Nergal," the god of the dead; and the Romans called it "Mars," their god of war.

The planet has played a very important role in the development of astronomical thought. Kepler deduced the three laws of planetary motion from the voluminous visual observations of Mars recorded by Tycho Brahe and from the planet's eccentric orbit. In 1877 the American astronomer Asaph Hall discovered the planet's two small satellites, and the Italian astronomer Giovanni Schiaparelli discovered the controversial Martian canals, seeing them as dark lines across the face of the planet. The theory presented by Percival Lowell that they were constructed by a race of intelligent beings started the great dispute as to the nature of the canals and the speculation of the possibility of life existing on its surface. While now recognized as resulting mainly from the tendency of observers to link random features into straight lines, the "canals" have encouraged study of and created considerable interest in the red planet.

Science fiction writings about Mars have had a tremendous hold on our imagination, as witnessed by Orson Welles' dramatic radio presentation of the "War of the Worlds" in 1938. In the form of a special news broadcast, this imaginary invasion of the United States by Martians was so realistic that many people panicked when they tuned in on the program.

With the Mariner 4, 6, 7, and 9 and Viking orbiters and landers, our knowledge and understanding of the planet's physical characteristics and its biological conditions have increased significantly. As with so many scientific studies, there are still many unanswered questions about the planet; the more we learn, the more we recognize how much remains to be understood.

a. Telescopic View

Telescopically, Mars appears as a bright, reddish orange disk, brighter than any planet except Venus. Even under the most favorable atmospheric conditions for viewing, the 200-inch Hale telescope cannot obtain high resolution photographs of Mars, yet a small telescope shows the basic features of the planet that are visible in Plate 16: reddish orange desert areas, dark markings, and polar caps.

b. Physical Properties

Mars is, in its general physical properties, similar to but somewhat smaller than the earth. Mars is an oblate spheroid with an average equatorial diameter of 4220 miles (6792 km) and a polar diameter of 4197 miles (6751 km). These values were measured by Mariner 9. The planet's mass is 0.11 that of the earth, and its mean density of 3.92 grams per cubic centimeter is about 0.71 that of the earth. These values give Mars a surface gravity that is 0.38 that of the earth, which means that a 100-pound girl would weigh only 38 pounds (17.25 kg) on Mars. Mars' escape velocity of 3.1 miles (4.99 km) per second is about half that of the earth; its albedo is 0.15.

The Martian atmosphere is extremely thin, exerting a pressure about 0.6% that of the earth; it is comparable to what would be found around the earth at a height of about 20 miles (30 km). The thin atmosphere permits observation of the Martian surface features, which have enabled astronomers to measure the planet's period of rotation and the angle of inclination of its axis of rotation. Its rotational period is 24 hours 37 minutes 23 seconds, and its equator is inclined 24°48′ to its orbital plane. The north pole of its axis points toward the star Deneb in the constellation of Cygnus the Swan. The planet's axis of rotation precesses like the earth's, but at a much slower rate. Mars' precessional period is about 183,000 years.

The surface temperature of Mars varies considerably. When the planet is at perihelion, the highest noon temperature at the equator is 80°F (27°C); the lowest midnight temperature is −90°F (−68°C). The average temperature at the poles is 50°F (10°C) in the summer and −130°F (−90°C) in the winter. The temperature of the south polar cap in the winter reaches about −150°F (−101°C). The average temperature of the illuminated surface of Mars has been determined from infrared measurements to be −30°F (−34°C).

c. Orbit

Mars travels around the sun in a counterclockwise direction at a mean distance of about 141,690,000 miles (228,020,000 km). It has a rather eccentric elliptical orbit whose eccentricity is 0.093. Its perihelion distance is 128,400,000 miles (206,600,000 km) and its aphelion distance is 154,900,000 miles (249,300,000 km)—a difference of about 26 million miles (42 million km). Its orbital path is inclined 1°51′ to the plane of the ecliptic. With a mean orbital speed of about 15 miles (24 km) per second, its sidereal period is about 687 days, and its synodic period is about 780 days, the longest of any planet. During its synodic period, Mars moves eastward with respect to the stars (direct motion) for 710 days and then moves westward (retrograde motion) for 70 days. It retrogrades as it approaches the earth at opposition, and at the same time it appears to increase its brilliance.

The most favorable time for viewing Mars occurs when the planet is close to opposition. Because of its orbital eccentricity, the opposition distance varies from the most favorable, at about 35 million miles (56 million km)—which occurred in 1971—to the least favorable, at 62.5 million miles (100.6 million km). The dates and distances of opposition are shown in Fig. 8.13. The most favorable oppositions occur in August or September, when the planet is near perihelion; the least favorable oppositions occur in February or March, when the planet is at aphelion. Unfortunately for observers in the northern hemisphere, the most favorable oppositions occur when Mars is not well positioned for viewing in the sky. At these times, the planet is almost 30° below the celestial equator, which means that the planet is seen low in the sky.

d. Satellites

Mars has two known satellites, which were discovered in 1877 (when Mars was in a most favorable opposition) by Asaph Hall, director of the U.S. Naval Observatory. He named them Phobos (fear) and Deimos (panic). Phobos (about 13 miles wide and 16 miles long) is the inner satellite, and Deimos (about 7.5 miles wide and 8.5 miles long) is the outer satellite (Fig. 8.14). The orbits of the satellites are almost circular and lie very close to the planet's equatorial plane. Phobos revolves around the planet at a distance of 5800 miles (9300 km) from the planet's center in 7 hours 39 minutes, and Deimos at a distance of 14,600 miles (23,500 km) in 30 hours 18

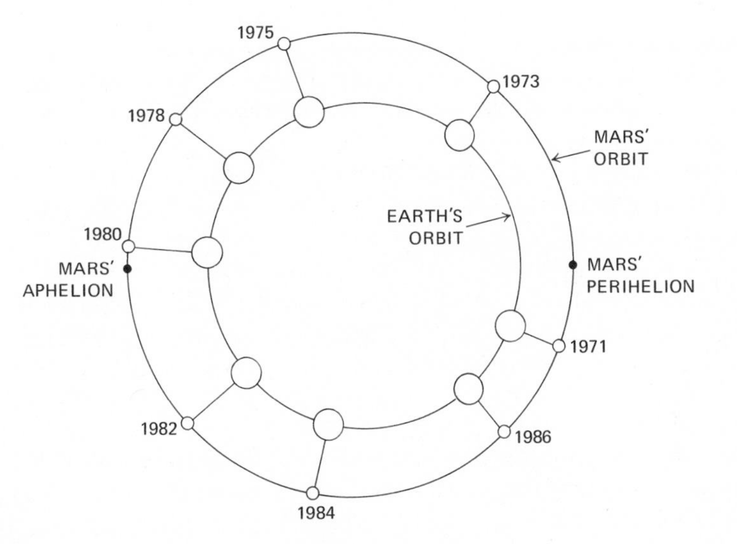

FIG. 8.13
The oppositions of Mars occur at intervals of about 780 days, Mars' synodic period. Favorable oppositions usually occur in either August or September at intervals of about 15 years, when Mars is at or near perihelion.

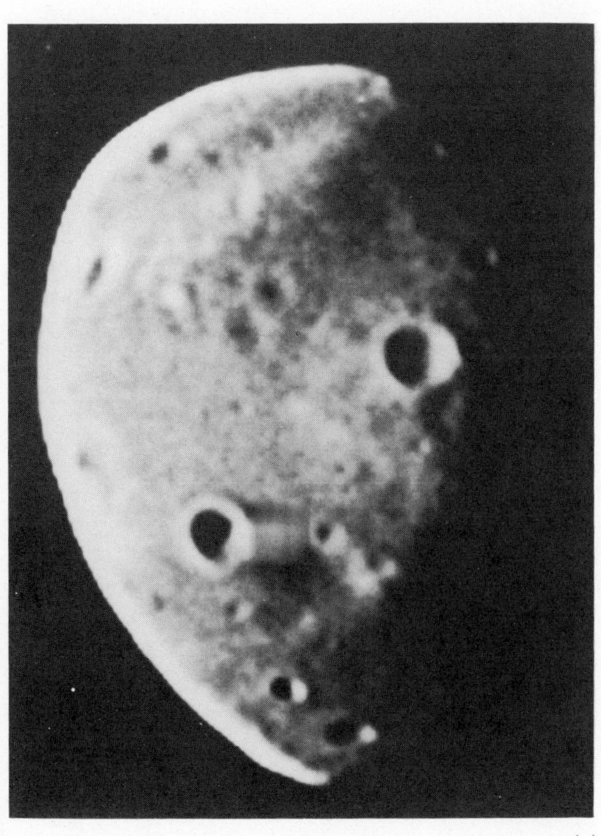

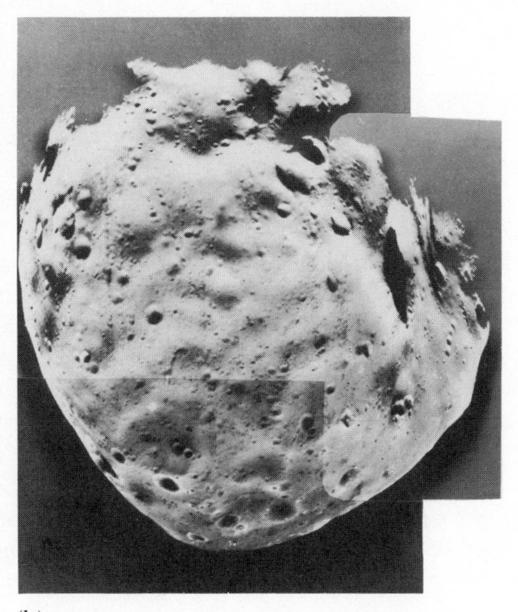

(a) (b)

FIG. 8.14
Mars' satellites. (a) Deimos, Mars' outermost satellite, photographed by Viking Orbiter I from 3300 km (2050 miles). The two large craters are roughly 1.3 km (.8 miles) in diameter. (b) Photograph of Phobos taken from a distance of 300 miles (400 km) by Viking Orbiter I. Note the striations and numerous craters. The disk of Phobos is not quite fully illuminated. (Courtesy of NASA/JPL.)

minutes. Phobos' period of revolution is unique in the solar system because it revolves around the planet in less time than it takes the planet to rotate, completing about three revolutions every day.

Although their presence was not known prior to their discovery in 1877, Kepler, in 1600, wrote a letter to Galileo in which he speculated on the possibility that Mars has two small moons revolving around it. In his imaginative *Micromegas*, Voltaire mentions the presence of these satellites. Also, Dean Jonathan Swift, a contemporary of Newton, wrote in his *Gulliver's Travels* (1726) that the Laputan astronomers had discovered two satellites revolving around the planet Mars. Their sizes and distances from the planet as given by Swift were remarkably close to their actual values.

8.10 THE MARTIAN SURFACE

On July 14–15, 1965, Mariner 4 became the first spacecraft to successfully fly by Mars and complete its mission. It came within 6118 miles (9846 km) of the planet's surface and radioed back to earth 21 photographs of Mars' surface. The resolution of these photographs produced the amazing discovery that the planet has a moon-like surface of densely packed impact craters.

Mariner 4 photographs produced no evidence of the controversial Martian "canals" that were observed and reported by Giovanni Schiaparelli in 1877, although some of the dark lines he discovered seem to lie along what at high resolution are seen to be a series of chains and clusters of small, secondary craters.

On July 31, 1969, Mariner 6 came within 2100 miles (3380 km) of the planet's surface along an equatorial path, and on August 5, 1969, Mariner 7 came within 2000 miles (3219 km) of the planet's surface along an angular path over the south pole. The photographs from both probes have a resolution of about 900 feet (275 meters) and clearly show that except for the polar regions and areas where clouds appear temporarily, the planet's surface is usually visible.

When Mariner 9 went into orbit around Mars in November 1971, it marked the first time that human beings had placed a spacecraft in orbit around another planet. All previous missions were flybys, that is, the spacecrafts were in the vicinity of the planet for only a brief period of time. Mariner 9 is expected to remain in orbit for several decades.

Although a massive, severe dust storm enveloped the planet during the spacecraft's first two months in orbit, Mariner 9 still further increased our knowledge of Mars. Mariner 4 photographs had shown the surface of Mars to be a cratered landscape like that of the moon, but Mariner 9 revealed a planet whose surface displays a richness and diversity found on no planet other than the earth. As it happened, the first glimpses of the surface from the early flights had been of very unrepresentative areas.

The surface of Mars appears to be divided into about half a dozen basic "geological" provinces (see Fig. 8.15). Each pole is covered with a cap of terraced ice. Layers of ice about 30 ft (10 meters) thick appear stacked up on each other, building up a total ice cover of a few hundred feet. The permanent ice cap is overlain with a frost of CO_2, which appears to sublime in the summer and then to recondense to frost in the winter months. During the winter nearly one-fourth of the Martian atmospheric CO_2 may be locked up in the polar frosts.

Immediately surrounding the poles are relatively smooth areas showing highly weathered craters. The craters are shallow, although often large—up to 80 miles (100 km) across. Interestingly, the northern subpolar areas appear less cratered than the southern areas.

Closer to the equator, Mars shows signs of volcanic activity. Like the other terrestrial planets and our moon, Mars exhibits a planetary asymmetry. One side of Mars contains enormous lava-covered areas (1000 miles across) punctuated with volcanic peaks rising to 70,000 feet (20 km) above the surrounding plains. The other side of the planet displays almost no evidence of volcanic activity. The volcanoes appear to be millions of years old and are almost certainly no longer active, given the presence of a small number of craters on their slopes.

Still closer to the equator, a giant series of canyons is found on that side of Mars which shows the signs of vulcanism. Small canyons with the appearance of old rivers (winding channels) merge with features looking

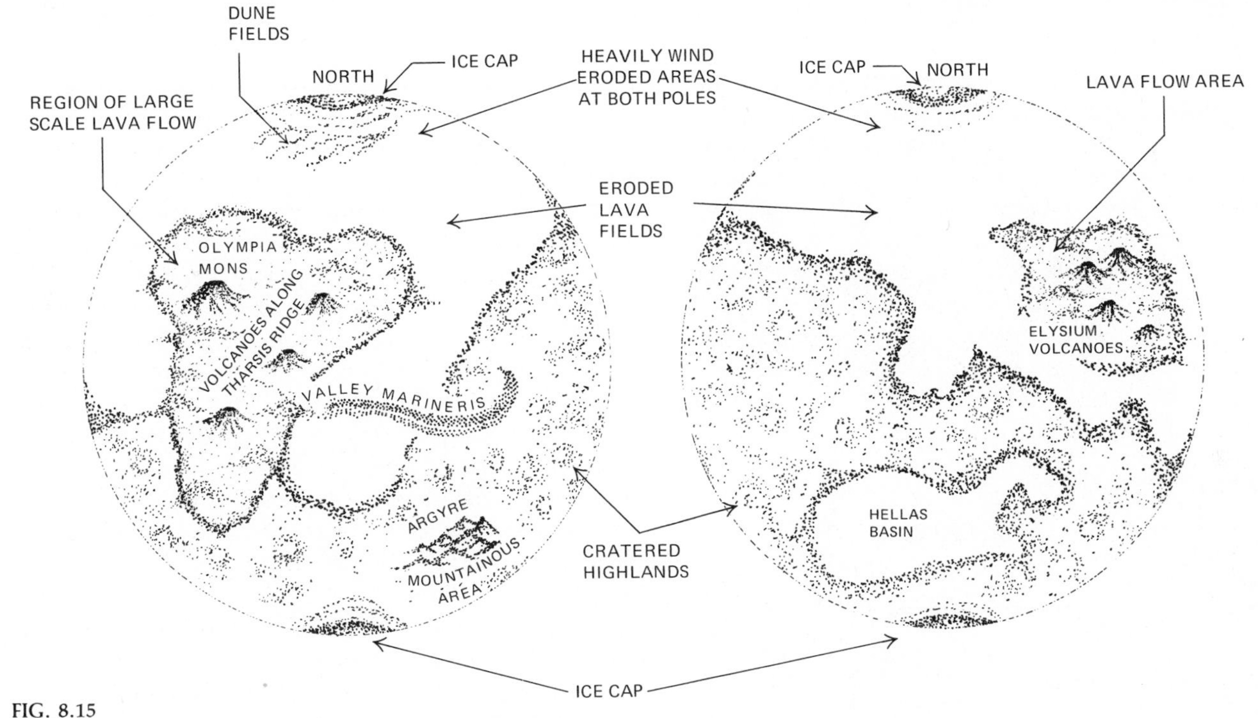

FIG. 8.15
Map of Mars. (Adapted from a map by James Pollack.)

nearly identical to the rugged canyon lands and arroyos of the American west. Large canyons, in turn, gradually empty into a monster rift spreading nearly one-sixth of the way around Mars just south of the equator. These features suggest very strongly that massive volumes of water flowed on the surface of Mars in the distant past. Island-like features complete with sandbars line the floor of the larger canyons, but all is now dry.

Many other features familiar to students of land forms on the earth are found on Mars. Extensive fields of sand dunes are found near the polar regions and elsewhere. Huge landslides have caved in segments of steep canyon walls. "Hummocky" and "chaotic" terrain in which it appears that the ground has slumped into ragged, irregular pits is found bordering some of the canyon areas. These areas suggest withdrawal of subsurface material—perhaps by the melting of a permafrost—and a subsequent settling of the ground as water drained out from under the surface.

8.11 THE MARTIAN ATMOSPHERE

It has long been known from earth-based observations that the atmosphere of Mars, like that of Venus, is composed largely of carbon dioxide. Oxygen and water are present in trace amounts. Water vapor in the Martian atmosphere was detected spectroscopically in 1963 by Munch, Kaplan, and Spinrad when they observed the water-vapor lines in the spectra taken with the 100-inch Mount Wilson telescope. This was accomplished by obtaining the spectrum when Mars was receding from the earth at a high velocity so that the Doppler effect caused the Martian water-vapor lines to be separated from the stronger ones of the earth's atmosphere.

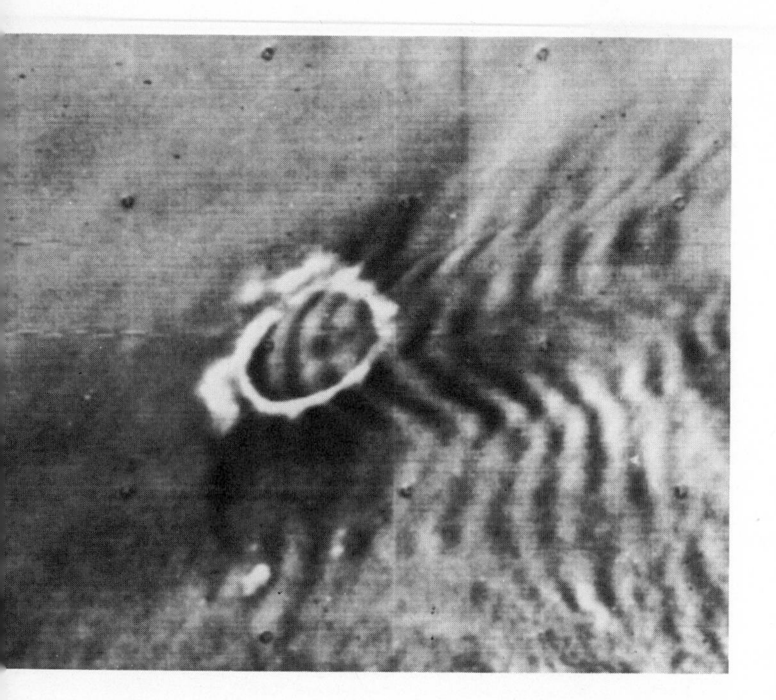

FIG. 8.16
Rows of clouds produced by wind blowing over a crater. Mountains on earth often produce similar cloud patterns. (Courtesy of NASA.)

The Mariner flybys and orbiters and, more recently, the Viking lander have now allowed a precise measure of the composition and properties of the atmosphere of the red planet. Carbon dioxide makes up 95%; nitrogen, 3%; argon, 2%; oxygen, a few tenths of a percent; and water, 0.01% to 0.1%. The low water abundance, although disappointing to those looking for life on the planet, is not unexpected in view of the cold, which tends to freeze water out and precipitate it as snow. The atmosphere is very thin, and because of Mars' distance from the sun, very cold. At the site of the Viking lander the atmospheric pressure was 0.007, or roughly 1% of that of the earth. The upper atmosphere seems somewhat similar to the higher layers of the earth's atmosphere. Sunlight produces an ionosphere and an ozone layer.

Mars has a variety of clouds. Earth-based observations revealed that areas of the planet were often obscured by white, blue, and yellow clouds. It was early recognized that the yellow clouds were gigantic dust storms, and in fact the photo reconnaissance of Mariner 9 was delayed for more than a week by a dust storm which enveloped nearly the entire planet. The blue and white clouds appear to be composed of either frozen carbon dioxide or water, depending on their height and location. Clouds seem to form as the Martian wind blows over crater rims and peaks, just as clouds form on the earth (Fig. 8.16). Often canyons and crater bottoms are obscured by cloud layers, much as terrestrial valleys fill up with fog. The Martian "ground fog" dissipates during the day, as do low terrestrial clouds. The large amount of dust in the air has the strange consequence that the sky on Mars is pink.

Mars is often a windy planet. From the persistence of dust storms it was inferred that Martian winds may blow at hundreds of miles per hour. In fact the scouring effect of these strong winds is now believed to explain many of the seasonal changes in the appearance of the light and dark areas of the planet. Windborne dust settling on a dark lava field makes it look light, and when the wind removes the dust, a "darkening" occurs.

Evidence for water erosion on the surface is overwhelming. Did water come from rainfall or from the melting of subsurface ice? In either case, the Martian atmosphere could not always have been the same as it is now. Under current conditions liquid water would rapidly evaporate. It is therefore believed that at some period in the past, possibly following an epoch of intense volcanic activity associated with the formation of Olympia Mons and the Tharsus region, the Martian atmosphere was much denser than it is now, and that subsequently the atmosphere has leaked away.

If volcanic activity and possibly crustal rifting (Coprates region) have occurred on Mars, then the Martian interior must have been hot in the past. It is too early yet to make definite predictions about the present state of the core of Mars, but it seems probable that like the earth it possesses a core-mantle structure, heated by radioactive decay. The planet is so much smaller than the earth

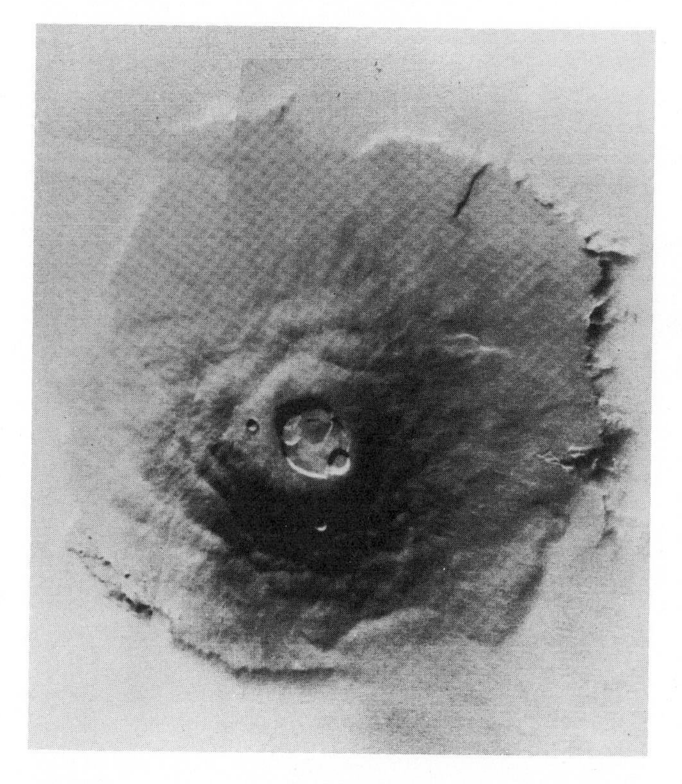

FIG. 8.17
Olympia Mons, the largest known volcano. It stands 15 miles high and is 300 miles across at its base. Note the cliff running around the edge of the main peak. This cliff is itself 10,000 ft high in some places. (Courtesy of NASA/JPL.)

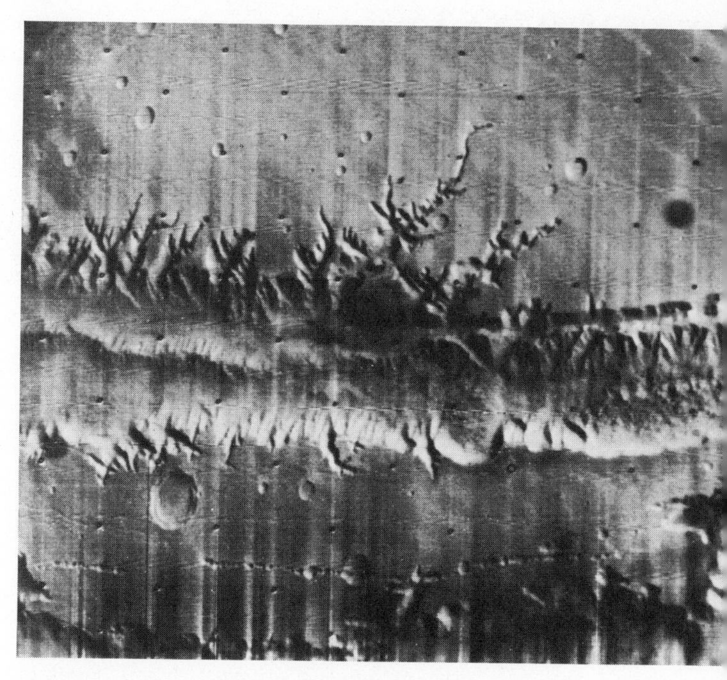

FIG. 8.18
Mariner 9 photograph of a unique Martian landform consisting of treelike canyons located in Tithonius region, 300 miles south of the equator. (Courtesy of NASA/JPL.)

that tectonic activity may be less intense (a smaller bulk means less heating). That the core must differ from the earth's in some details is suggested by the weakness of the Martian magnetic field.

There are two Martian features that are worthy of special mention. The first of these is the Olympia Mons volcano (Fig. 8.17). Olympia Mons is larger than any volcano known on the earth. Its height is well over 15 miles (20 km) and its base spreads across 300 miles (500 km). It has apparently been built up by gradual eruptions over millenia. Lava flows are clearly visible on its slopes. The summit has collapsed with the subsidence of lava to form a caldera or crater 40 miles (65 km) across. The shield of volcanic rock has been eroded away along its base to form a set of cliffs running around the volcano for hundreds of miles. The cliffs themselves are nearly a mile high.

The second feature of special interest is Tithonia Chasma, part of the huge canyon system referred to collectively as Valles Marineris. More popularly it is

FIG. 8.19
The surface of Mars as seen from the Viking Lander I site. Note the sand dunes and numerous rough boulders strewn everywhere. (Courtesy of NASA/JPL.)

called the Grand Canyon of Mars (Fig. 8.18). This feature runs for about 2000 miles (3000 km) across the surface of Mars just south of the equator. It is typically 60 miles (100 km) wide and 5 miles (7 km) deep. Many canyons feed into it and much of its bottom has the scoured appearance of a stream bed. While almost certainly modified by water flow, the feature may well represent an incipient rifting of the Martian crust. Many areas of the earth's surface (see Chapter 5) have been torn asunder by the subsurface motion that gives rise to continental drift. Perhaps we see here on Mars the initiation of a similar phenomenon. If interplanetary parks are ever established, Olympia Mons and Tithonia Chasma deserve to be included.

8.12 MARS AS VIEWED BY VIKING

In the summer of 1976, two American spacecraft landed on the Martian surface. The first, Viking I, landed in the Chryse channel, a desert region at latitude 22°N. Viking II, after a careful analysis of a variety of possible landing sites, descended in the Utopia region at a latitude of 48°N. Although it was Martian summer, temperatures at both sites, even the one near the equator, were frigid by terrestrial standards. Daily highs were −20°F (−29°C), with the temperature falling to lows of −120°F (−84°C) immediately before dawn.

The Viking landers were programmed to perform a variety of experiments, as well as to transmit pictures of the area back to earth (Fig. 8.19). Measurements of Martian "weather" phenomena—e.g., temperature, pressure, wind speed and direction—were made and samplings of atmospheric composition were performed. In addition, a seismometer to record Marsquakes was landed successfully at the Utopia region. But perhaps the most exciting experiments were those designed to help answer the question of whether life has ever existed on Mars. Specifically, the experiments were designed to (1) detect evidence of photosynthesis, (2) test whether any kind of organism was assimilating nutrients from a culture medium, and (3) identify any organic compounds in the soil. No organic molecules have yet been detected,

FIG. 8.20
Jupiter, 23 October 1964, 0858 U.T., Blue Light. (Courtesy of the New Mexico State University Observatory.)

but the response of the Martian soil to the photosynthesis and nutrient experiments was peculiar. The soil appeared to release oxygen and in many ways mimicked what one might expect from living cells. The absence of organic molecules, however, suggests that the Martian soil may simply have chemical properties very different from those of the earth, and that the results by chance are similar to what living organisms would cause. The question whether there is life on Mars remains unanswered.

8.13 JUPITER

a. Physical Properties

The fifth planet from the sun has been appropriately named after the ruler of the Greek gods, Jupiter. The planet deserves this title because, except for the sun, it is the most prominent body in the solar system (Fig. 8.20). It is the largest planet (88,640 miles, or 142,650 km, in diameter, which is about 11 times larger than the earth) and the most massive (318 times greater than the earth and 2.5 times greater than all the other planets combined). Its mean density is 1.33 grams per cubic centimeter, which is about one-fourth that of the earth and slightly less than that of the sun. Its surface gravity is 2.64, which means that a 100-pound girl would weight 264 pounds (119.75 kg) on Jupiter. Its velocity of escape is 37 miles (59.54 km) per second, which is the greatest of any planet.

b. Orbit

Jupiter revolves around the sun in an elliptical orbit whose eccentricity is quite small (0.05) at a mean distance of 484 million miles (779 million km). The orbit is inclined 1°18′ to the plane of the ecliptic, and the planet's equator is inclined 3°07′ to its orbital plane. At an orbital speed of 8.1 miles (13.04 km) per second, Jupiter's sidereal period is 11.86 years, and its synodic period is 398.9 days.

c. Telescopic View

To the unaided eye, Jupiter appears as a bright yellowish object. When Jupiter is faintest, it is slightly fainter than the star Sirius; when it is at its brightest, it is almost twice as bright as Sirius. After Venus and Mars, it is the brightest planet.

When Jupiter is seen through a telescope, it appears as a bright disk with prominent bands, or belt markings, that run nearly parallel to the planet's equator. These markings have a wide variety of colors. The belts range from gray to bluish; the zones between them tend to be yellowish or white. The belts are the result of currents in

Jupiter's atmosphere similar to the wind systems found on the earth. The belts nearest the poles appear to be more stable in position, arrangement, and size than those nearest the equator.

One of Jupiter's most interesting atmospheric features appeared suddenly in August 1878 in the planet's south tropical zone at about 20° south of the equator. It was "The Great Red Spot," 30,000 miles (48,000 km) long and 8000 miles (13,000 km) wide, with its major axis parallel to the planet's equator (Plate 16). Giovanni Cassini, the first director of the Paris Observatory, first saw the spot and from its motion determined the planet's rotational period to be 9 hours 55 minutes. This was the first time that the rotational period of any planet had been determined. The planet's visible surface, however, does not rotate as a solid body. The equator rotates in 9 hours 50.5 minutes, while the temperate zones take about five minutes longer.

The spot has undergone many changes in color, brightness, and size. It appears to brighten, fade, then brighten again in irregular periods. It remains prominently bright for several years, then fades slowly until it is barely visible. The shape of the spot always appears elliptical, and its width remains fairly constant; however, its length has varied considerably over the years. At present it is augen-shaped (eye-shaped), the interior is lighter than its edges, and dark areas appear at both ends. It also appears to drift in the planet's atmosphere at a variable rate—sometimes faster and at other times slower than the planet's rotational rate. Elmer Reece and H. Gordon Solberg, Jr., at the New Mexico State University Observatory, have found from accurate measurements of the spot's position on photographic plates that its longitude changes along a complex sinusoidal oscillation curve during a period of about 88 days.

Recently, a new, bright circular spot (about 4000 miles in diameter) in the northern hemisphere of Jupiter was discovered photographically by Elmer Reece. Its rotation period of 9 hours 47 minutes 5 seconds is the shortest of any of the planet's markings. Because of its brightness when photographed in ultraviolet light, this new spot is believed to be located in the planet's upper atmosphere. This is the sixth disturbance that has occurred in this region—the first was in 1880.

d. Atmosphere

In 1932 Rupert Wildt made a spectroscopic discovery of ammonia (NH_3) and methane (CH_4) in Jupiter's atmosphere, with methane in more abundance. In 1958 W. Demarcus concluded from theoretical studies that Jupiter's atmosphere contains about 78% hydrogen by weight. His model of Jupiter's structure consisted of a solidified hydrogen core surrounded by an atmosphere of hydrogen. Subsequently, spectroscopic studies have revealed the presence of hydrogen as well as helium and water.

One of the prespace-age techniques for studying the Jovian atmosphere was that of *stellar occultations*. A star is occulted when some object such as a planet passes between the earth and the star, creating a condition similar to an eclipse. If the planet has an atmosphere, then starlight passing through the atmosphere will be refracted and partially absorbed. An analysis of the changes produced in the star's light as it passes through successively deeper layers gives information about the structure and composition of the planet's atmosphere. The occultation technique, first used to study Jupiter, has recently been applied to several of the outer planets and their satellites.

In 1955 astronomers detected that Jupiter was emitting occasional bursts of radio noise at frequencies near 20 megacycles per second. Individual bursts are about one second in duration and occur in groups that last for about one hour. The bursts appear to be related to the position of the planet's satellite Io, to the planet's rotation, and to a particular alignment of Jupiter's magnetic field with the earth.

e. Satellites

Jupiter has 14 known satellites. Four are about as large as or larger than the moon, whereas the others are less than 100 miles (160 km) in diameter. The four largest—Io, Europa, Ganymede, and Callisto—were discovered in

1610 by Galileo and can easily be observed with an ordinary pair of binoculars. The mean values for their important physical characteristics are as follows.

	IO	EUROPA	GANYMEDE	CALLISTO
Magnitude (apparent)	5	5	5	6
Diameter (miles)	2260	1900	3280	3050
Diameter (km)	3640	3050	5270	4900
Mass (moon's mass=1)	1	0.6	2	0.6
Density (grams/cm^3)	3.5	3.3	2	1.6
Sidereal period (days)	1.77	3.55	7.16	16.69

It is interesting to note that if either Ganymede or Callisto orbited the sun instead of Jupiter, it would be classified as a planet because both have diameters larger than Mercury's. With densities near that of the moon, Io and Europa could be similar to the moon in composition, and Callisto, with a density of 1.6, could be similar to Jupiter.

E. E. Barnard discovered the fifth satellite in 1892. It is about 150 miles (240 km) in diameter and is located at a distance of about 112,500 miles (181,000 km) from the planet's center. At an orbital speed of 17 miles per second, the fastest of any satellite in the solar system, its sidereal period is about 12 hours. The last nine satellites, with diameters considerably less than 100 miles (160 km), were discovered photographically, and only five of them have been observed visually.

Jupiter's satellites may be classified according to the position of their orbits into the inner, middle, and outer groups. The orbits of the five satellites in the inner group are small, circular, and lie close to the planet's equator. The three satellites in the middle group have orbits that are eccentric and inclined about 30° to the planet's equator. The six satellites in the outer group have orbits with large eccentricities and inclinations, and all six move around Jupiter in a retrograde (clockwise) motion.

Interest in the Jovian satellites has increased recently because they form a system around Jupiter, rather similar to the system of the planets around the sun. New information about them has been gained from the Jupiter fly-by missions (Pioneers 10 and 11) and from observations of stellar occultations. This latter technique, described in the previous section, has been applied now to several of the Jovian satellites. Also, more sensitive spectroscopic studies have improved our understanding of the surfaces of these satellites. Io, for example, has been discovered to possess a tenuous atmosphere of sodium and hydrogen and may have a surface covered with a layer of ordinary table salt (sodium chloride). Europa and Ganymede appear to be covered with a shell of ordinary ice and perhaps to have a zone of liquid water in their interiors. The sodium atmosphere on Io has the interesting property that it should appear faintly yellow at night.

f. Pioneer 10 and 11

On December 4, 1973, the first close-up pictures of Jupiter were sent back to earth from Pioneer 10. Launched one year and nine months earlier, Pioneer 10 crossed the asteroid belt, swung up and over Jupiter's north pole, and was accelerated by the giant planet's gravity onto a new trajectory which will carry it out of the solar system. The spacecraft came within about 130,000 km of Jupiter and made a series of measurements on the planet's magnetic field, atmospheric composition, and gravitational field. Almost exactly a year later, on December 3, 1974, Pioneer 11 repeated the trip, but after passing Jupiter it was put into an orbit to carry it past Saturn.

Our knowledge of Jupiter has been increased enormously by these two space vehicles. The minute deviations in Jupiter's gravitational field from that expected of a perfectly spherical planet have made it possible to infer something about the planet's interior structure. It is believed that as one penetrates down through the atmosphere one encounters in succession thin layers

FIG. 8.21
Great Red Spot and cloud belts of Jupiter. The red spot is approximately four times the size of the earth and appears to be an atmospheric vortex perhaps similar to terrestrial hurricanes. (Courtesy of NASA.)

of gaseous hydrogen, ammonia-ice crystals, ammonium hydrosulfate crystals, ordinary water-ice crystals, and water droplets. Below, the atmosphere thickens into a layer of liquid hydrogen which perhaps changes to metallic hydrogen in the planet's deepest interior.

Although the cloud-top temperature is a frigid −220°F (−140°C), infrared observations suggest that the temperature rises steadily as one penetrates deeper into the atmosphere. From earth-based observations it has been known since 1968 that Jupiter radiates about twice as much energy as it receives from the sun. Its interior must therefore be very hot. Radiometric observations from the Pioneer probes suggest an internal temperature of 54,000°F (30,000°C), roughly five times the sun's surface temperature. The energy to produce this high temperature probably is produced by gravity compressing the gases out of which the planet developed.

The heat is believed to be carried to the surface by convection currents similar to those found just beneath the solar surface (which are responsible for granulation—see Chapter 10). On Jupiter, the currents are deflected by the planet's rotation and on reaching the surface give rise to the cloud bands that are observed. The red spot was found by Pioneer 10 to be a huge atmospheric vortex, similar to terrestrial hurricanes (Fig. 8.21).

8.14 ROEMER'S EXPERIMENT IN THE SPEED OF LIGHT

In 1675 the Danish astronomer Olaus Roemer successfully conducted the first experiment to determine the speed of light by measuring the orbital periods of Jupiter's satellites. Roemer observed that the period was minimum when Jupiter as at eastern quadrature and maximum when at western quadrature. The period changes accumulate and result in a difference between the predicted and observed time of eclipse. The eclipse occurs earliest with respect to its predicted time when Jupiter is at opposition. The eclipse occurs latest compared to its predicted time when Jupiter is at superior conjunction (Fig. 8.22). Since the orbital period of a satellite is constant, Roemer concluded that the difference between the two periods is due to the time it takes light to travel the diameter of the earth's orbit. From this, he was able to

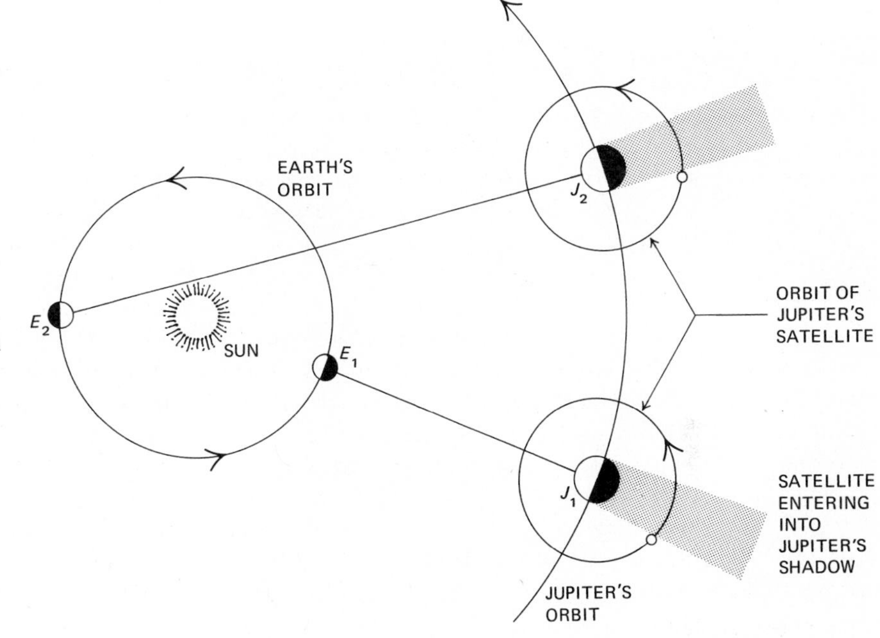

FIG. 8.22
Roemer's experiment in the speed of light. He observed that the eclipse of the satellite occurs earlier than the predicted time when the earth is in position E_1 and later when in position E_2.

show that the speed of light is finite. His value for the speed of light was slightly inaccurate because he used an inaccurate value for the diameter of the earth's orbit. With new techniques, determination of the speed of light has been improved so that the presently accepted value of light in a vacuum is about 186,282 miles (299,783 km) per second.

8.15 SATURN

a. Physical Properties

Saturn, with an equatorial diameter of about 75,100 miles (121,000 km) and a polar diameter of about 67,800 miles (109,000 km), is the most oblate planet in the solar system. Its mass is 95 times that of the earth, and its mean density of 0.7 grams per cubic centimeter (which is also 0.7 that of water) is the lowest of any planet. With such a density, the planet could easily float in water. If the composition of Saturn is similar to Jupiter's (mostly hydrogen and helium), its smaller size would produce a smaller gravitational force; consequently, the gases would not be as compressed as they are in Jupiter. This would explain the planet's low density. Its surface gravity is 1.2 times that of the earth, and its escape velocity is 22 miles (35.4 km) per second.

Although white spots appear occasionally in the equatorial region of the planet, they are usually too faint and short-lived to be used to determine the planet's rotational period. Spots are rarely seen near the polar regions. One was observed in 1960, in latitude 58°N, that gives the planet a rotational period at this latitude of 10 hours 40 minutes. In 1969 the first measurable spot was observed in the south polar region in latitude 57°S. It was about 4000 miles (6400 km) long and 5000 miles (8000 km) wide and gives the planet a rotational period of 10 hours 36.5 minutes.

Infrared measurements indicate that the planet's temperature is −290°F (−180°C). The atmosphere of Saturn is similar to that of Jupiter, except that its spectrum shows stronger methane and weaker ammonia bands. Since Saturn's temperature is lower than Jupiter's, the ammonia freezes out of the atmosphere, settles to a lower level, and thus permits the viewing of the methane to a greater depth. No trace of hydrogen has been found in Saturn's atmosphere. This is not surprising because although hydrogen is believed to be abundant, it would be very difficult to detect.

b. Orbit

Saturn, the outermost planet known in antiquity, revolves around the sun in an elliptical orbit whose eccentricity is 0.056 and at a mean distance of about 887 million miles (1.4 billion km). Its orbit is inclined 2.5° to the plane of the ecliptic, and the planet's equator is inclined 26°45′ to the plane of its orbit. In January 1974 the planet was at perihelion, and when it is in opposition to the earth, it will be at its shortest distance to the earth. With an orbital speed of 6 miles (9.66 km) per second, the planet's sidereal period is 29.5 years, and its synodic period is 378 days.

c. The Ring System

Viewed with a telescope, Saturn is one of the most impressive objects in the sky, appearing as a large, yellowish disk with faintly defined markings that are parallel to the planet's equator. Surrounding the disk and concentric to it is a unique ring system that lies parallel to the planet's equator. In 1610, one year after the invention of the telescope, Galileo observed Saturn through his crude, homemade telescope. Since the instrument could not resolve the rings, he announced that he had observed two satellites revolving around the planet at a very short distance from its surface. In 1655 Christian Huygens gave the first correct explanation for the rings when he described them as thin, flat, concentric rings completely separate from the planet (Plate 17).

The system consists of three concentric rings (known as A, B, and C, counting from the outside), whose outside diameter is about 171,000 miles (275,000 km) and whose thickness is less than 10 miles (16 km). The inner edge of the C-ring, which is also known as the *inner* or *crepe ring,* is about 7000 miles (11,000 km) from the top of the planet's atmosphere and has a width of about 11,000 miles (18,000 km). The B-ring, which is also known as the *middle* or *bright ring,* is about 16,000 miles (26,000 km) in width and is in juxtaposition with the C-ring. Between the B-ring and the A-ring is a 2500-mile (4000-km) gap, which is called the *Cassini division*. The A-ring, also known as the *outer ring,* is about 10,500 miles (16,900 km) in width. In 1970 the French astronomer Pierre Guerin discovered a fourth ring, the D-ring, next to the C-ring, which makes it the ring closest to the planet. Although it has been observed visually by other astronomers, as yet there are no satisfactory photographs of it. There is also some evidence for an extremely faint ring lying outside the A-ring and tentatively called the E-ring.

The ring system lies in the plane of the equator, which is inclined about 27° to the plane of the ecliptic. The planet and its ring system maintain a fixed orientation in space as they revolve around the sun; therefore, when seen from the earth, the shape of the ring system changes in a recurring cycle. As shown in Fig. 8.23 when the planet is in position 2 or 4, the ring system and the earth lie in the same plane; therefore, the rings are observed edgewise and appear as very thin, dark lines across the planet and as short, faint, thin lines on each side of the planet. When the planet is in position 3, the top of the ring system is visible; when in position 1, the bottom is visible. Since the sidereal period of the planet is 29.5 years, the ring system will pass through this cycle during this period of time.

The rings are not solid; their spectrum clearly shows large Doppler shifts, with the inner parts of the rings moving more rapidly than the outer parts, as shown in Fig. 8.24. If the rings were solid, the opposite effect would be observed. The rings were originally believed to be composed of small pieces of water, ice, and ice-coated rocks that revolved around the planet in the same direction in which the planet rotated, but at different speeds. Observations of the infrared spectrum of Saturn's rings, made in 1969 by Kuiper, Cruikshank, and Fink at the University of Arizona, indicate that the rings are composed of ice. It is also believed that the particles are small and very cold; otherwise, their evaporation rate in space would be quite rapid.

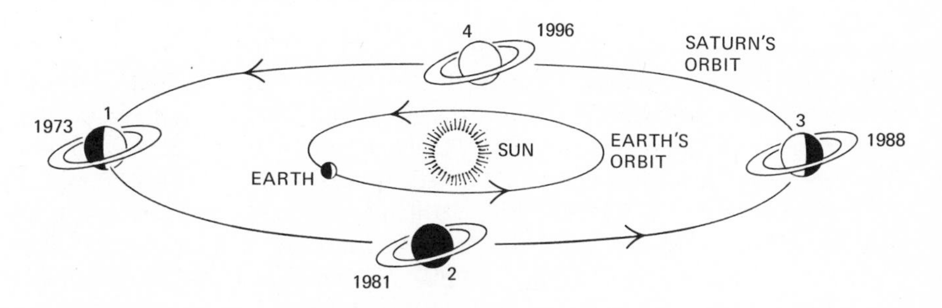

FIG. 8.23
The aspects of Saturn's rings. In 1973 the southern side of the rings was visible. In 1981 the rings will be seen edgewise. In 1988 the northern side of the rings will be visible. In 1996 the rings will again be seen edgewise. These changes are caused by Saturn's revolving around the sun once every 29½ years and by the plane of its rings being inclined 27° to the planet's orbital plane.

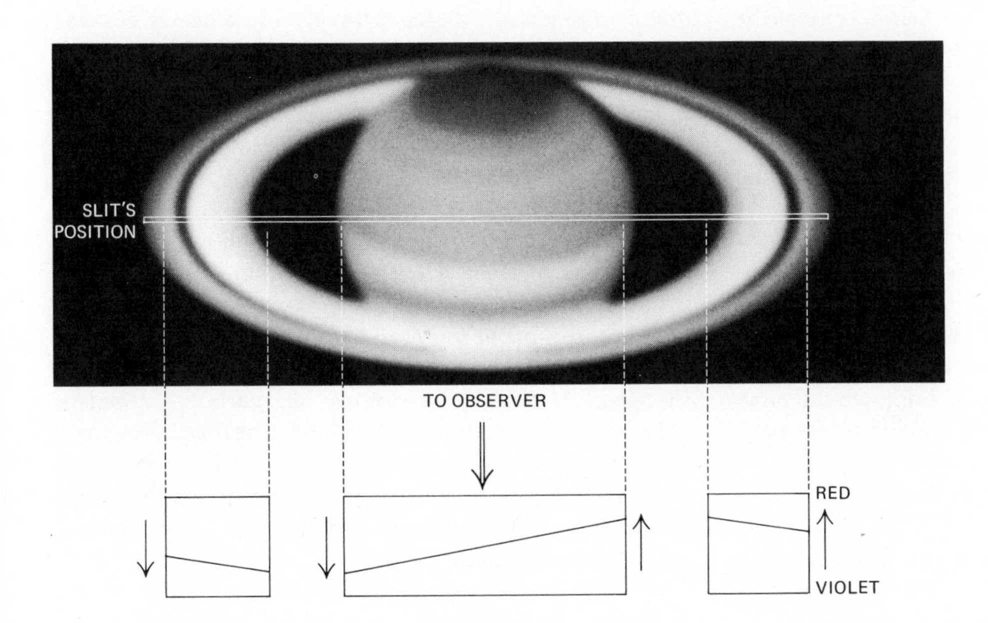

FIG. 8.24
Placing the slit of a spectrograph across Saturn's sphere and rings provides proof that both are rotating and that the rings consist of small particles, because the spectral lines shift to the red when the bodies are receding and to the violet when they are approaching the observer. (Courtesy of the Hale Observatories.)

A controversy still exists about the origin of the rings. Some believe that they represent the debris of a satellite that approached the planet too closely and was broken up by its gravitational force. Others believe that they represent the original material that failed to coalesce either with Saturn or into a satellite of the planet.

In 1850 E. A. Roche stated that the disruptive tidal force of a planet will exceed the cohesive gravitational force of a satellite of the same density as that of the planet within a distance of about 2.44 times the planet's radius. This disruption can be easily understood as follows: since the force of gravity depends inversely on the square of the distance, the side of a satellite that is nearer a planet is pulled on more strongly by the planet than the side of the satellite that is farther way. There is thus a different (or differential) gravitational attraction exerted by the planet across the satellite, tending to pull it apart. The satellite, of course, has its own gravitational force holding it together. However, if the differential gravitational attraction produced by the planet exceeds the self-gravity of the satellite, the satellite will be disrupted. The differential pull tending to disrupt the satellite depends upon its proximity to the planet, getting stronger as the satellite nears the planet's surface. The distance at which a satellite held together by gravity is disrupted is called the *Roche limit*.

The outer edge of Saturn's ring system is 2.3 times the planet's radius, which is within the Roche limit; therefore, it is possible that a satellite approached the planet too closely and disintegrated. Alternatively, the rings may represent matter which is left over from Saturn's formation. Matter inside the Roche limit would have been unable to condense into a satellite due to the tidal forces of Saturn.

d. Satellites

Saturn has ten known satellites, which in many ways are similar to, but smaller than, those of Jupiter. Saturn's nearest satellite, located just outside its outer ring, is Janus, which was discovered on December 15, 1966 by Audouin Dollfus at the Meudon Observatory in France. Janus follows a circular path at a distance of about 98,000 miles (158,000 km) from the planet's center, and has an orbital period of about 18 hours. The most distant satellite is Phoebe, 180 miles (290 km) in diameter, which revolves in 550 days at a distance of 7,700,000 miles (12,390,000 km) from the planet's center. Phoebe is Saturn's only satellite with a retrograde motion. The largest and brightest satellite is Titan, 3600 miles (5800 km) in diameter. Titan is one of the few satellites in the solar system with an atmosphere; in 1944 Kuiper found methane in its spectrum.

8.16 URANUS

The discovery of the first planet since antiquity was made by William Herschel on March 13, 1781, while he was making a routine observation of the faint stars in the vicinity of Gemini the Twins with his 7-inch reflecting telescope. In his report to the Royal Society, he announced that he had discovered a comet. About five months later, A. J. Lexell announced that Herschel's comet was actually a planet traveling in a nearly circular orbit around the sun. Herschel proposed the name Georgium Sidus for the new planet, in honor of King George the Third, during whose reign the planet was discovered. This broke with tradition in the naming of planets and it was not received favorably. At the suggestion of J. E. Bode, tradition was followed by naming the planet Uranus, after the Greek deity who sprang from Chaos and became Heaven. The planet was appropriately named, because Uranus was the father of Saturn, who in turn was the father of Jupiter.

The visual brightness of Uranus is such that under the most favorable conditions on a clear, moonless evening it can be seen with the unaided eye as a very faint star. Nearly 100 years before its discovery, Uranus was seen at least 23 times by several observers and was recorded on charts as a star. The first such recording was made by Flamstead in 1690.

a. Telescopic View

Through a telescope the planet appears as a small, pale greenish disk because of the abundance of methane in its atmosphere (Fig. 8.25). The high-resolution photographs of Uranus taken in 1970 by the Princeton University

FIG. 8.25
Uranus with three of its satellites, photographed with the 120-inch telescope. (Courtesy of the Lick Observatory.)

Observatory Stratoscope II unmanned balloon show the planet to be slightly oblate with a darkened limb. Although observers have reported the presence of very faint equatorial belt markings on the planet similar to those of Jupiter and Saturn, the Stratoscope II failed to reveal any of them. It is possible that they do exist but that they are difficult to observe because their contrast might be even lower than Saturn's.

By observing the Doppler shift across the disk of the planet, a rotation period for Uranus has been determined. (Recall the discussion of Saturn's rings.) The period has been determined to be about 24 hours.

b. Orbit

Uranus revolves around the sun at a mean distance of about 1,783,000,000 miles (2,869,000,000 km) in an elliptical orbit whose eccentricity is 0.047. The discovery of this planet almost doubled the known size of the solar system, since Uranus' distance is nearly twice that of Saturn. The plane of the planet's orbit is inclined 0°46′ to the plane of the ecliptic, which is the smallest angle of inclination of any planet. Its orbital speed is about 4.5 miles (7.25 km) per second; its sidereal period is 84 years, and its synodic period is 369 days.

A remarkable feature of Uranus is the extreme tilt of its rotation axis. If the north pole of Uranus is chosen to be the one about which the planet turns counterclockwise (as is the case for the earth's north pole), then Uranus is tipped over 98° with respect to its orbit. Uranus lies nearly on its side, and its axis of rotation is presently pointed almost in the direction of the sun (Fig. 8.26); therefore, its seasons would be 42 years of summer and 42 years of winter. An observer at the poles would also observe 42 years of daylight and 42 years of darkness.

c. Physical Properties

Uranus has a mean diameter of about 32,000 miles (52,000 km) and a mass of 14.5 times that of the earth. Its mean density of 1.2 grams per cubic centimeter is close to Jupiter's. The planet has a surface gravity 0.88 times that of the earth and an escape velocity of about 13 miles (21 km) per second. The spectrum of Uranus shows very strong bands of methane and absorption lines of molecular hydrogen. Ammonia has not been detected in the spectrum of Uranus, although astronomers believe that the gas is present in its atmosphere. Uranus has a radiometric surface temperature of about −360°F (−218°C), which would cause any ammonia to freeze and sink to a

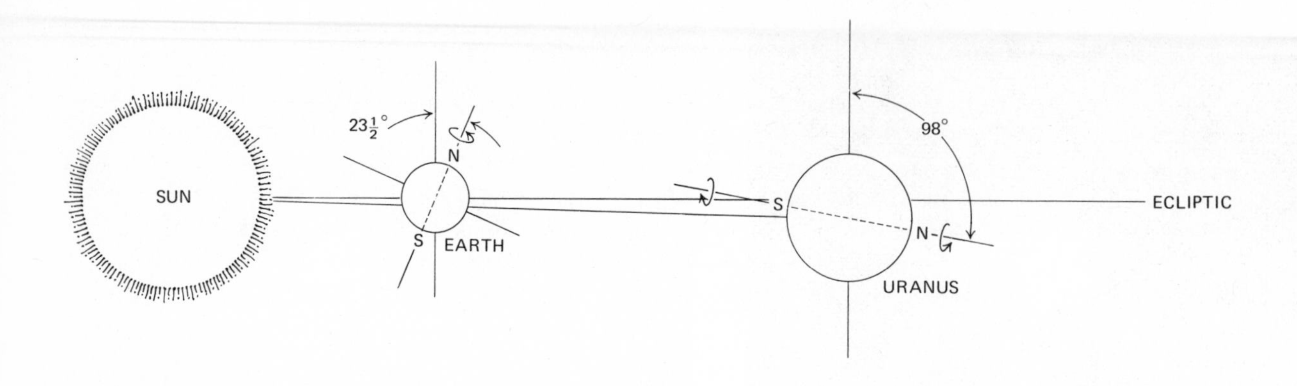

FIG. 8.26
Uranus' axis of rotation. The orbit of Uranus is tipped less than 1° from the earth's orbital plane and is thus too small to show in the diagram.

lower level in the atmosphere. Spectral observations of Uranus are probably limited to the top of the planet's atmosphere, where the ammonia would not be visible.

d. Satellites

Uranus has five known satellites—Ariel, Miranda, Oberon, Titania, and Umbriel. Their diameters range from about 150 miles, or 240 km (Miranda) to 600 miles, or 966 km (Titania). Their distances from the center of the planet range from 77,000 miles (124,000 km) for Miranda to 364,000 miles (586,000 km) for Oberon. Miranda has the shortest peiod of revolution, less than 1.5 days, while Oberon has the longest, with 13.5 days. All the satellites revolve around the planet in circular orbits that coincide with the equatorial plane of the planet. Since the equatorial plane of the planet is nearly perpendicular to its orbital plane, the satellites appear to move in an almost north and south direction.

e. The Rings of Uranus

In March 1977 a group of American astronomers discovered what appears to be a set of rings around Uranus. The rings were discovered accidentally during an attempt to observe the occultation of a faint star by Uranus. The original goal of the experiment was to study the Uranian atmosphere by observing the dimming of the star's light as the planet moved between earth and the star (see Chapter 8.11e). About half an hour prior to the expected time of occultation, the light from the star dimmed unexpectedly. Four other dimmings occurred and were observed when the star reappeared after the occultation. The dips in the star's light were observed by two groups of astronomers, and so are not attributable to a malfunction in the equipment. Like the rings of Saturn, the Uranian rings lie inside the Roche limit for the planet. The Uranian rings seem to be too faint to allow direct photographs at this time. However, in the next decade a spacecraft may fly close enough to Uranus to allow pictures to be made.

8.17 NEPTUNE

The discovery of the planet Neptune was one of the most interesting achievements of mathematical astronomy. It was a tremendous triumph for the validity of Newton's laws of gravitation, on which the calculations for the discovery were based. The existence of Neptune was detected by its gravitational effect on Uranus before Neptune itself was actually discovered. When finally discovered, it was within 52 minutes of arc of its calculated predicted position. The details of the story are as follows.

Twenty years after the discovery of the planet Uranus, a difference between its observational and calculated positions appeared. Some astronomers believed that Newton's laws of gravitation did not apply to bodies at great distances from the sun and that therefore these laws could not be used to determine the elements of their orbits. Others believed that the perturbations of Uranus were caused by the gravitational force exerted by a body farther out in space.

John Couch Adams, a brilliant young mathematician from Cambridge University, decided to solve this unprecedented problem. His treatment of the problem was most systematic. First, he evaluated all the available observations pertaining to the problem; then he calculated the perturbations of Uranus that would be produced by the gravitational forces of Jupiter and Saturn and found them insufficient to account for the observed perturbations. After he had discussed with others the possibility that Newton's laws did not apply to distant bodies, he concluded that they did apply and that the perturbations of Uranus were caused by the presence of a body farther out in space. By 1845 Adams had calculated the position of the unknown body and sent the information to Sir George Airy (the Astronomer Royal of England), requesting that a search be made to locate the body. Unfortunately, Airy believed that Newton's laws of gravitation had no bearing on the perturbations of Uranus. Having very little faith in Adams' theoretical calculations, he devoted his time and effort to trying to discover a way to refine Newton's laws of gravitation.

In 1845, while Adams was making his unsuccessful attempts to get Airy to search for the unknown body, the French astronomer and mathematician Urbain Leverrier began to work on the problem, unaware that Adams was also working on it. In 1846 Leverrier completed his calculations and sent the position of the unknown body to Johann Galle, the director of the Berlin Observatory. The same night that he received the information from Leverrier, Galle was able to locate and identify the planet.

FIG. 8.27
Neptune and the larger and brighter of its two satellites, Triton, photographed with the 120-inch telescope. (Courtesy of the Lick Observatory.)

Credit for the discovery of Neptune is shared by both Adams and Leverrier. Tradition was followed when the planet was named for Neptune, the ancient Greek god of the sea.

a. Telescopic Appearance and Orbit

When Neptune is seen through a telescope, it appears as a small, greenish disk with no conspicuous surface markings (Fig. 8.27). It has a small oblateness of about $\frac{1}{40}$. It revolves around the sun at a mean distance of 2.8 billion miles (4.5 billion km) in a very nearly circular orbit whose eccentricity is 0.009. Its orbit is inclined $1°47'$ to the plane of the ecliptic, and its equator is inclined $29°$ to the plane of its orbit. At an orbital speed of 3.3 miles (5.3 km) per second, the planet's sidereal period is 164.8 years, and its synodic period is 367.5 days. Its period of rotation, about 22 hours, was determined recently from the Doppler shift in its spectrum.

b. Physical Properties

Since their size and other physical characteristics are similar, Neptune and Uranus are considered to be twins. Neptune's mean diameter is about 31,000 miles, or 50,000 km (Uranus' is 32,000 miles, or 52,000 km) and its mass is 17.3 times that of the earth (Uranus' is 14.5 times that of the earth). It has a density of 1.7 grams per cubic centimeter (as compared to Uranus' 1.2 grams per cubic centimeter), a surface gravity 1.14 times that of the earth (Uranus' is 0.88 times that of the earth), and an escape velocity of about 15 miles (24 km) per second (that of Uranus is 13 miles, or 21 km per second).

The planet's high albedo of 0.55 suggests that it is surrounded by a dense atmosphere. Its spectrum shows strong methane bands and molecular hydrogen absorption lines. It has a surface temperature of about −380°F (−228°C), and its internal structure is probably the same as that of the other giant planets.

c. Satellites

Neptune has two known satellites—Triton and Nereid. Triton is larger, brighter, and nearer than Nereid. Triton has a diameter that is probably slightly larger than that of the moon. It revolves around Neptune in a retrograde motion at a distance of 220,000 miles (354,000 km) from the planet's center in a nearly circular orbit inclined about 160° to the plane of the planet's orbit. Its orbital period is 5 days 21 hours.

Nereid, which was discovered by Kuiper in 1949, has a diameter of about 200 miles (320 km), revolves around the planet in a direct motion, and has the most eccentric orbit of any satellite (0.75). At its nearest approach to the planet, Nereid is 730,000 miles (1,175,000 km) away; at its farthest, it is 7 million miles (11 million km) distant. Its orbital period is 359.9 days.

8.18 PLUTO

Two American astronomers, Percival Lowell and W. H. Pickering, predicted the existence of a planet beyond Neptune's orbit. Their predictions were based on the slight residual deviations between the predicted and observed motion of Uranus. Because Neptune had traveled only a short distance in its orbit since discovery, there was little information available on its orbital perturbations. Independently, each astronomer calculated that in order to produce Uranus' observed perturbations, the unknown planet would have to be in the constellation of Gemini the Twins. The mass of the unknown planet was calculated by Lowell to be nearly 6⅔ times that of the earth; Pickering calculated it as slightly less than that of the earth. Both undertook an intensive search of the sky to locate the unknown planet. It was a monumental task, because Gemini lies near the Milky Way, in which thousands of stars appear in a single photographic field of view. Both men were denied success; Lowell died in 1916. In 1919 Pickering photographed the area in which the planet was located, but failed to recognize it because of a flaw in the photographic plate and the appearance of a very bright star near the planet.

After Lowell's death, his associates at the Lowell Observatory continued the search with the new 13-inch astrographic telescope given to the observatory by Lowell's brother. On March 13, 1930, Clyde W. Tombaugh announced the discovery of the planet. He used the new blink comparator, a device which projects in rapid sequence on one screen two photographs taken several days apart of the same field of view. If the star images on both photographs are the same, an observer sees a series of flickering, identical photographs. If a body such as a planet appears in the photographs, its position in the two photographs will be different because of its apparent motion in relation to the stars; therefore, the observer sees a series of flickering photographs in which the planet appears to jump back and forth as the two photographs are alternated by the blink comparator. The unknown planet was located within 6° of Lowell's predicted position. It was named Pluto after the Greek god of the lower world, the world of the dead (Fig. 8.28). Ironically, it now appears that the predicted position was in error and the planet's discovery was a lucky coincidence.

a. Orbit

Pluto, which is the remotest of the known planets, revolves around the sun at a mean distance of

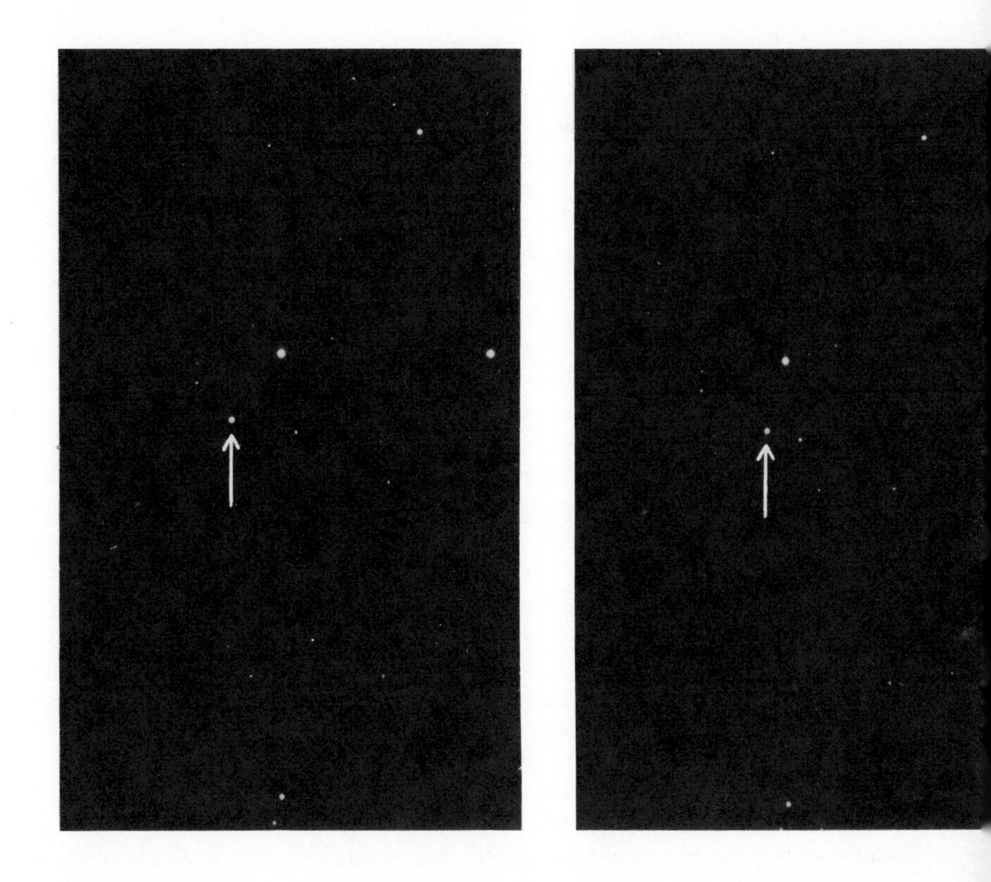

FIG. 8.28
Two photographs showing Pluto's motion in a period of 24 hours as recorded by the 200-inch telescope. (Courtesy of the Hale Observatories.)

3,675,000,000 miles (5,914,000,000 km) in an elliptical orbit whose eccentricity of 0.249 is the largest of any planet. Its orbit is so eccentric that it perihelion distance of 2.8 billion miles (4.5 billion km) is within Neptune's orbit. Although one point of Pluto's orbit lies nearly 35 million miles within Neptune's orbit, there is no chance of a collision, because Pluto's orbit has an inclination of 17°09′, which is the highest of any planet. At an orbital speed of 2.9 miles (4.7 km) per second, the planet's sidereal period is 248.4 years, and its synodic period is 367 days.

b. Physical Properties

Our knowledge of Pluto is incomplete and inconclusive because of the planet's great distance from the earth. Soon after its discovery, a very disturbing problem developed. Although Pluto was found to be moving in an orbit very close to the one calculated for it, its observed apparent size and brightness did not agree with Lowell's calculated values. The planet appeared much smaller and dimmer. The problem became more confused when Pluto's mass was estimated at $\frac{1}{10}$ that of the earth—an estimate based on its observed size and dimness. Later, two prediscovery observations of Neptune made by M. J. Lalande in 1795 were used to determine Pluto's gravitational effect on Neptune's orbit. From these calculations, Pluto's mass was estimated to be equal to that of the earth. In 1950 Kuiper was able to observe Pluto in the 200-inch Hale telescope as a tiny disk with an angular diameter of about one-quarter arc-second. This value corresponds to a linear diameter of about 3700 miles (5954 km), which is about half that of the earth, and with a volume that is one-tenth that of the earth.

Kuiper's value for Pluto's diameter has been supported by evidence obtained when Pluto passed close to a star on April 28-29, 1965, during a predicted possible occultation. Although the occultation failed to occur, several major North American observatories took a series of photographs showing Pluto passing south of the star. Ian Halliday and his colleagues at the Dominion Observatory in Ottawa, Canada, established the minimum separation between Pluto and the star at 0.125 seconds of arc, making it possible for them to set the upper limit for Pluto's diameter at about 3600 miles (5800 km).

On the basis of recurring light variations measured with a photoelectric photometer, Pluto's rotational period is 6.39 days, longer than that of any of the giant planets. These light variations indicate that there might be two different types of surfaces on Pluto. The planet's surface temperature is probably near −380°F (−230°C). No gases have been observed in its spectrum, and there is no evidence that the planet has an atmosphere.

Recently, astronomers at the University of Arizona and the University of Hawaii have found evidence that Pluto is covered with a frost of methane. Pluto appears to be out of place in its position along with the giant planets. It is considerably smaller, less massive, and rotates more slowly. Its orbit has the greatest inclination and is the most eccentric of any planet's. (Because of its high orbital eccentricity, in 1979 Pluto will have moved to a location in its orbit that places it closer to the sun than Neptune is.) All these facts have raised the possibility that Pluto was once a third satellite of Neptune. This suggestion is further supported by the unusual orbit of Triton, Neptune's largest moon. The theory is that Neptune originally possessed three satellites. At some juncture in the past, their orbits brought them so close together that one—now known as the planet Pluto—was ejected into an orbit around the sun. The near-collision threw Nereid into a very eccentric orbit. Triton was left in a nearly circular but highly tilted and retrograde orbit. Triton's orbit is slowly changing and it appears that the satellite will eventually move inside the Roche limit for Neptune. If this occurs Neptune may end up with a ring system like Saturn and Uranus.

In 1978, J. Christy at the U.S. Naval Observatory detected what may be a satellite of Pluto, moving in an orbit 20,000 km from the planet. On the basis of its period of about 6 hours, he determined a mass for Pluto of approximately .002 earth masses. This, if confirmed by later observations, would make Pluto the smallest planet and comparable in size and mass to the moons of Uranus.

SUMMARY

The solar system contains nine planets orbiting the sun, thirty-four satellites orbiting their respective planets, and many smaller bodies including comets, meteoroids, and asteroids.

The planets can be divided into two main families: those resembling the earth (terrestrial planets) and those resembling Jupiter (Jovian planets). The terrestrial planets tend to be small in both mass and diameter; they are rocky spheres with thin atmospheres, and they have few or no satellites. The Jovian planets have atmospheres rich in hydrogen or hydrogen compounds such as methane and ammonia; they have many satellites, and they are very large compared to the earth, both in mass and in diameter. The terrestrial planets are nearest the sun; the Jovian planets are the distant ones. The size of the solar system is best grasped in terms of astronomical units (one astronomical unit is roughly the earth's distance from the sun). Mercury, at .4 a.u., is the planet closest to the sun, and Pluto, at 40 a.u., is the most distant.

A few of the properties of each planet are listed below.

Mercury — Moonlike; no atmosphere; cratered.

Venus — About the same size as the earth; dense carbon dioxide atmosphere which traps solar energy; surface temperature highest of any planet in the solar system.

Mars	Variety of surface features: craters, volcanoes, canyons, old river beds, deserts, ice, and frozen CO_2 polar caps; thin CO_2 atmosphere.
Jupiter	Larger than all the other planets put together; massive hydrogen, helium, methane, and ammonia atmosphere; rotates rapidly—possesses cloud bands; has an internal heat supply and strong magnetic field.
Saturn	Similar to Jupiter but smaller; ring system a remnant of a disrupted satellite or one that failed to form.
Uranus/ Neptune	Similar in their sizes; Uranus has a small ring system; atmosphere rich in hydrogen compounds.
Pluto	Little known; size and mass smaller than earth's; perhaps an escaped satellite of Neptune.

All the planets move around the sun in the same direction in a flattened system, and most (Venus and Uranus are exceptions) rotate in that same sense, which is counterclockwise as seen from above the earth's north pole. The satellites, with a few exceptions, also exhibit counterclockwise motion.

REVIEW QUESTIONS

1. Name two ways in which planets may be classified. List the planets in both classifications.
2. What are the characteristics of planetary orbits?
3. What are the criteria for membership in the solar system?
4. What is meant by planetary configurations? Assuming that an observer is looking down on the north pole of the earth, list the configurations in proper order for an (a) inferior planet and (b) superior planet.
5. Explain the retrograde motion of planets. Do superior planets retrograde? Near what configuration do inferior planets retrograde?
6. Distinguish between a planet's sidereal and synodic periods.
7. Write the formula for the relationship between the sidereal and synodic periods for a superior planet. Determine the synodic period of Jupiter given its sidereal period is about 12 years.
8. Explain how the mass of a planet can be determined.
9. What factors determine whether a planet is able to retain its atmosphere?
10. Briefly state the differences between satellites, asteroids, and meteoroids.
11. Which one of the following planets can never appear at opposition: Venus, Mars, Jupiter, Saturn, and Neptune? Why?
12. Explain why Mercury and Venus show phases like those of the moon.
13. When is the best time for viewing Mercury? Why?
14. What is Mercury's rotational period? How was this determined? Is there anything about it that is unique?
15. Explain why the apparent diameter and brightness of Venus change. Who first observed this? Was it an important astronomical discovery? Why?
16. Explain the role that Mars has played in the development of astronomical thought.
17. When is the best time for viewing Mars? Why? How often does Mars appear in this favorable position?
18. Describe the different kinds of surface features found on Mercury, Mars, and the earth.
19. Describe and compare the atmosphere of the terrestrial planets.
20. How is it known that Venus has craters?
21. Why are people hopeful that Mars may now or in the past have had life? What are the observations that encourage or discourage this view?
22. What is unique about the satellites of Mars? What has Mariner 9 revealed about Phobos and Deimos?
23. Discuss the controversial canals on Mars.
24. Describe Jupiter's atmosphere. How does it compare with those of the other giant planets?
25. How can Jupiter's period of rotation be determined? How does it compare with that of the other planets?

26. Is it likely that we will ever land on the planet Jupiter? Why might Ganymede be a better choice?

27. Compare the color of the sky on the earth, Mars, and Io.

28. Although Jupiter's four large satellites appear as stars (point source of light) through binoculars, what observational evidence can you use to show someone that they are really satellites?

29. Explain how Roemer was able to show that the speed of light is finite.

30. How many rings does Saturn have? How are they oriented? Describe their structure and composition. Discuss their origin.

31. Why might Uranus and Neptune be called twin planets?

32. How were the rings of Uranus discovered?

33. What were the events, beliefs, and reasoning that led to the discovery of Neptune?

34. What made Lowell institute a search for a planet beyond Neptune? When and how was Pluto discovered? How was it verified that it is a planet and not a star? Give reasons for the suggestion that Pluto might once have been a satellite of Neptune.

CHAPTER 9
ASTEROIDS, COMETS, AND METEOROIDS

Although the sun, the moon, and the planets are the most conspicuous bodies in the solar system, there are many thousands of minor bodies, called *asteroids, comets,* and *meteoroids,* which exist between the planets. Most of the asteroids (small planet-like objects) are orbiting the sun between Mars and Jupiter and appear even in large telescopes only as points of light. Comets (accretions of frozen gases and rock) appear as hazy, luminous clouds as they approach the sun in their orbit. Meteoroids (small particles of interplanetary matter orbiting the sun) appear as streaks of light as they pass through the earth's atmosphere.

9.1 THE DISCOVERY OF ASTEROIDS

In 1766 Johannes Titius, professor of mathematics at Wittenberg, developed an interesting and curious mathematical relationship that expressed rather closely the distances of the planets from the sun. This relationship, known as the Bode-Titius rule, was published in 1772 by Johann Bode, director of the Berlin Observatory. Developed from the series 0, 1, 2, 4, 8, 16, 32, 64, 128, 256, each term was multiplied by 3 to produce the new series 0, 3, 6, 12, 24, 48, 96, 192, 384, 768. Four was added to each term, then each sum was divided by 10, producing the final series 0.4, 0.7, 1.0, 1.6, 2.8, 5.2, 10.0, 19.6, 38.8, 77.2. Titius used this rather convoluted procedure so that the third number in the series, 1, represented earth's distance from the sun. The series of numbers represents the distances of the planets from the sun in astronomical units (1 a.u. equals about 93 million miles, the earth's distance from the sun). When these distances were compared with the actual distances (Table 9.1), their closeness made the Bode-Titius relationship most interesting.

When William Herschel discovered the planet Uranus in 1781 and found its actual distance to be very close to that predicted by the Bode-Titius rule, many astronomers were convinced that the rule was valid and that a planet existed at a distance of 2.80 astronomical units.* To expedite its discovery, the zodiac was divided

* The Bode-Titius rule is no longer seen as a law because of the discrepancies between Neptune's and Pluto's actual distances and those predicted by the rule. It is recognized today as simply a convenient, empirical rule-of-thumb.

TABLE 9.1
Planetary distances
by Bode-Titius rule and actual measurements

PLANET	BODE-TITIUS RULE (A.U.)	ACTUAL DISTANCE (A.U.)
Mercury	0.4	0.39
Venus	0.7	0.72
Earth	1.0	1.00
Mars	1.6	1.52
Missing planet	2.8	2.80
Jupiter	5.2	5.20
Saturn	10.0	9.54
Uranus	19.6	19.19
Neptune	38.8	30.08
Pluto	77.2	39.46

into 24 regions, and each region was assigned to an astronomer. One of these was the Italian Father Giuseppe Piazzi. On January 1, 1801, before he had received his assignment, he observed during a routine survey of the sky a star-like object in the constellation of Taurus, which he believed was a new comet. After sufficient data had been obtained and its orbit calculated, astronomers believed that the missing planet had been found. Piazzi named the body Ceres, after the protecting goddess of Sicily.

In 1802, slightly over one year after the discovery of Ceres, another star-like object, similar in appearance and close to Ceres, was discovered by Heinrich Olbers. He named the body Pallas. In 1804 a third object, named Juno, was discovered, and in 1807 a fourth object was discovered, which was named Vesta. With these four discoveries, it became apparent that several small objects rather than one large planet exist at the mean distance of 2.80 astronomical units from the sun. Since these objects are small and have orbital paths around the sun, they are called *planetoids* (minor planets), and since they appear as star-like objects in a telescope, they are called asteroids (little stars). As a result of these discoveries, the "planet" Ceres was demoted to the status of another asteroid.

During the nineteenth century, over 300 asteroids were discovered. In 1891 the discovery rate increased tremendously when Max Wolf at Heidelberg introduced the camera as part of the discovery technique. The simple and efficient photographic method involves a long photographic exposure of an area in the sky. On the photographic plate, the stars appear as points of light, the asteroids as a streak. Several hundred asteroids have been discovered by this method.

A newly discovered asteroid is assigned a number only after it has been observed in at least two oppositions and its orbital elements have been computed. In some instances, the confirming sightings are extremely difficult because the asteroid's orbit has been greatly altered by the gravitational effects of the earth, moon, and the near planets.†

9.2 PHYSICAL CHARACTERISTICS

Even though the observed asteroids have diameters that range from about 1 mile (1.6 km) to 600 miles (1000 km), it is believed that there are many more smaller ones with diameters the size of pebbles. The four largest and brightest asteroids are Ceres (600 miles, or 1000 km), Pallas (380 miles, or 600 km), Vesta (340 miles, or 540 km), and Hygeia (280 miles, or 450 km). Their albedoes range from 0.02 to 0.38.

† One fictional astronomer had an even more difficult time getting his asteroid authenticated. His asteroid was the one on which Saint-Exupéry's little prince lived: "This asteroid has only once been seen through the telescope. That was by a Turkish astronomer, in 1909. On making his discovery, the astronomer had presented it to the International Astronomical Congress, in a great demonstration. But he was in Turkish costume, and so nobody would believe what he said. Grown-ups are like that . . . Fortunately, however, for the reputation of Asteroid B-612, a Turkish dictator made a law that his subjects, under pain of death, should change to European costume. So in 1920, the astronomer gave his demonstration all over again, dressed with impressive style and elegance. And this time everybody accepted his report." Antoine de Saint-Exupéry, *The Little Prince*, tr. Katherine Woods. New York: Harcourt, Brace, and World, 1943, p. 15.

Reflective studies of Vesta indicate that its surface is covered with some kind of dust, probably formed by extreme temperature changes, meteoric impacts, or high radiation exposure. The brightness of the larger asteroids appears to be steady, whereas the smaller ones show a continuous variation. This indicates that the larger ones may be almost spherical and that the smaller ones may be block-shaped with angular faces. When the asteroid Eros came within 14 million miles (23 million km) of the earth in 1931, observations revealed that it is an elongated object about 17 miles (27 km) in length and 4 miles (6 km) in width, and that it rotates about its short diameter in 5¼ hours (the periods of rotation of the asteroids range from about 2 to 18 hours). Eros also exhibited variations in brightness (2 maxima and 2 minima), indicating that as it rotates, it presents different surface reflectivities to the earth.

There is little definite information on the exact mass of any of the asteroids. Ceres, Pallas, and Vesta have masses of 1.2×10^{24}, 2.3×10^{23}, and 2.4×10^{23} grams, respectively, and are believed to contain more than half the total mass of all the asteroids. This is much smaller than the mass of the moon.

FIG. 9.1
The asteroid Icarus, photographed in 1949 with the 48-inch Schmidt telescope, appears as a streak of light moving at about 20 miles per second. (Courtesy of the Hale Observatories.)

9.3 ORBITAL CHARACTERISTICS

The orbits of most of the asteroids lie between the orbits of Mars and Jupiter. A few have their perihelion within Mars' orbit, and still fewer have it within the earth's orbit. The orbital motion of all the asteroids is direct, that is, they move from west to east in the same direction as the planets. Most of their orbits, which lie close to the plane of the ecliptic, have a mean angle of inclination of 9.5° (some 30 of them have angles greater than 25°), a mean value of eccentricity of 0.15, and a mean orbital period of five years. In comparison to the planetary orbits, the asteroid orbits are more highly inclined to the ecliptic and slightly more eccentric.

9.4 UNUSUAL ASTEROIDS

Asteroid 1566 Icarus, discovered in 1949 by Walter Baade at Mount Palomar Observatory, has the smallest orbit (1.08 a.u.) (Fig. 9.1). It is the only planetary body, other than meteors and comets, that passes within the orbit of Mercury; at perihelion it is inside Mercury's orbit, and at aphelion it is outside of Mars' orbit. In 1968 it passed within 4 million miles (6 million km) of the earth. Icarus is nearly spherical, less than one mile in diameter, and has a surface that is quite jagged. From its periodic light variations, its rotational period is about 2¼ hours. Hidalgo, which has the largest orbit (5.79 a.u.), is close to the orbit of Mars when at perihelion and close to the orbit of Saturn when at aphelion. When Hermes was discovered in 1937, it became the nearest planetary body to the earth, with an approach of less than 1 million miles (1,609,000 km).

Recently a peculiar "mini-planet" with an orbit lying between those of Saturn and Uranus has been discovered by C. Kowal of the Hale Observatories at Mt. Palomar. It is approximately 300 miles (500 km) in diameter, making it large for an asteroid but small for a planet. It has tentatively been given the name "Chiron" after one of the centaurs noted for its wisdom and kindness. It will be interesting to see if further searches will lead to the discovery of other such objects beyond Saturn.

9.5 DISTRIBUTION OF THE ASTEROIDS

The asteroids are not uniformly distributed between the orbits of Mars and Jupiter. Gaps exist in regions whose periods are simple fractions of Jupiter's orbital period of 12 years, such as one-third, two-fifths, one-half, and so on. An explanation for these gaps was presented in 1866 when Daniel Kirkwood deduced that when an asteroid drifts close to these regions, Jupiter's gravitational force perturbs and prevents the asteroid from moving into the region.

It is believed that the asteroids are slowly "grinding" themselves up in collisions. A number of asteroids move along orbits that if traced back intersect in a rather small area of the asteroid belt. There are several such "families" of asteroids, each with its own location for the orbit intersections. It was suggested by the Japanese astronomer Hirayama that the families represent fragments of larger asteroids that were broken into smaller pieces following a collision.

9.6 THE TROJAN ASTEROIDS

In 1772 the French mathematican Joseph Lagrange proved that the position in Jupiter's orbit either 60° ahead or behind the planet is gravitationally stable. A body that occupies either position thus forms an equilateral triangle with Jupiter and the sun. It always remains in this position as it revolves around the sun, because the gravitational effects of the sun and Jupiter on it are in equilibrium. It came as no surprise, therefore, when an asteroid discovered in 1906 was found to revolve around the sun in the same orbital path and 60° ahead of the planet Jupiter. By 1959, 14 asteroids had been discovered—9 in a group 60° ahead of Jupiter and 5 in a group 60° behind Jupiter (Fig. 9.2). These asteroids are called the Trojans after the Homeric heroes; the largest, Hector, named after the greatest warrior of Troy, has a diameter of about 50 miles. Although only 14 Trojans have been discovered, there may be many more that are too small and faint to be detected.

9.7 THE ORIGIN OF THE ASTEROIDS

Two basic hypotheses have been presented to explain the origin of the asteroids. The first is that asteroids are fragments of a large planet or several small planets that disintegrated from either a collision with one another or excessive rotational speeds. According to the second theory, asteroids are bits of the original material from which the solar system was formed that for some reason failed to condense into a satellite, coalesce with either or both of the planets Mars and Jupiter, or form into a separate planet.

9.8 COMETS

Of all the astronomical bodies that are visible to the unaided eye, comets are probably the most awesome because of their unexpected appearances, spectacular changes in brightness and shape, and in some cases long tails that fan out behind them as they approach the sun. According to Webster, the word "comet" is derived from the Greeks, who called these bodies *kometes*, which

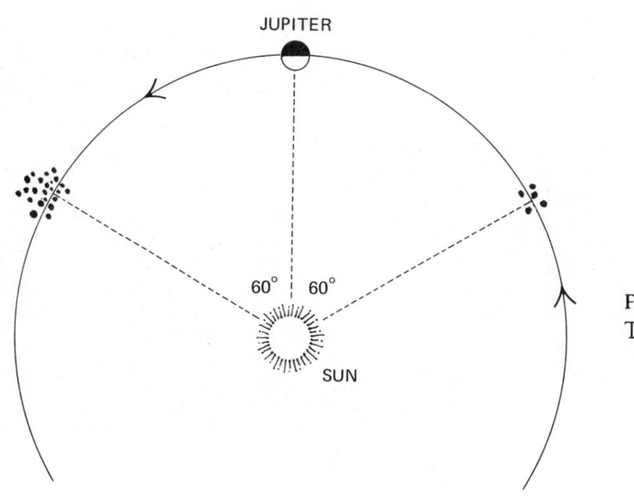

FIG. 9.2
The Trojan asteroids.

means "long-haired." Most comets, however, do not have tails.

Since time immemorial, comets have been regarded as the most puzzling and mysterious of the astronomical bodies. Their appearance caused fear among the people, who regarded them as omens of some great calamity. When Halley's comet appeared in 1066, many people assumed it to have been directly responsible for the death of King Harold at the Battle of Hastings and the conquest of England by the French.

The fear of comets that accompanied the belief that they are a part of the earth's atmosphere was lessened when Brahe showed that the 1577 comet was a body some three times as distant as the moon and proposed that the comet was probably revolving around the sun. This concept was firmly established when Edmund Halley predicted the reappearance of the comet that bears his name. After noting the similarity of the orbits of the comets that appeared in 1456, 1531, 1607, and 1682, he concluded that it was a single comet traveling around the sun in an elliptical orbit once every 75 or 76 years. He predicted its reappearance in 1758. The earliest record of its appearance is 467 B.C. Its last appearance was in 1910, when it appeared to the unaided eye as a faint, hazy object on April 19, 1910, at the U.S. Naval Observatory. One month later (Fig. 9.3) it became a magnificent spectacle—a spot with a beautiful tail that was as bright as the Milky Way. Halley's comet will appear again in 1986 (Fig. 9.4).

9.9 THE DISCOVERY OF COMETS

Many of the visual discoveries of comets (such as the Comet Ikeya-Seki, which was discovered in 1965, Fig. 9.5) are made by amateurs using wide-field, low-power telescopes. When an astronomer discovers a comet, it usually appears on a photographic plate that was taken for other purposes.

The temporary designation of newly discovered comets is made up of the year of their discovery, followed

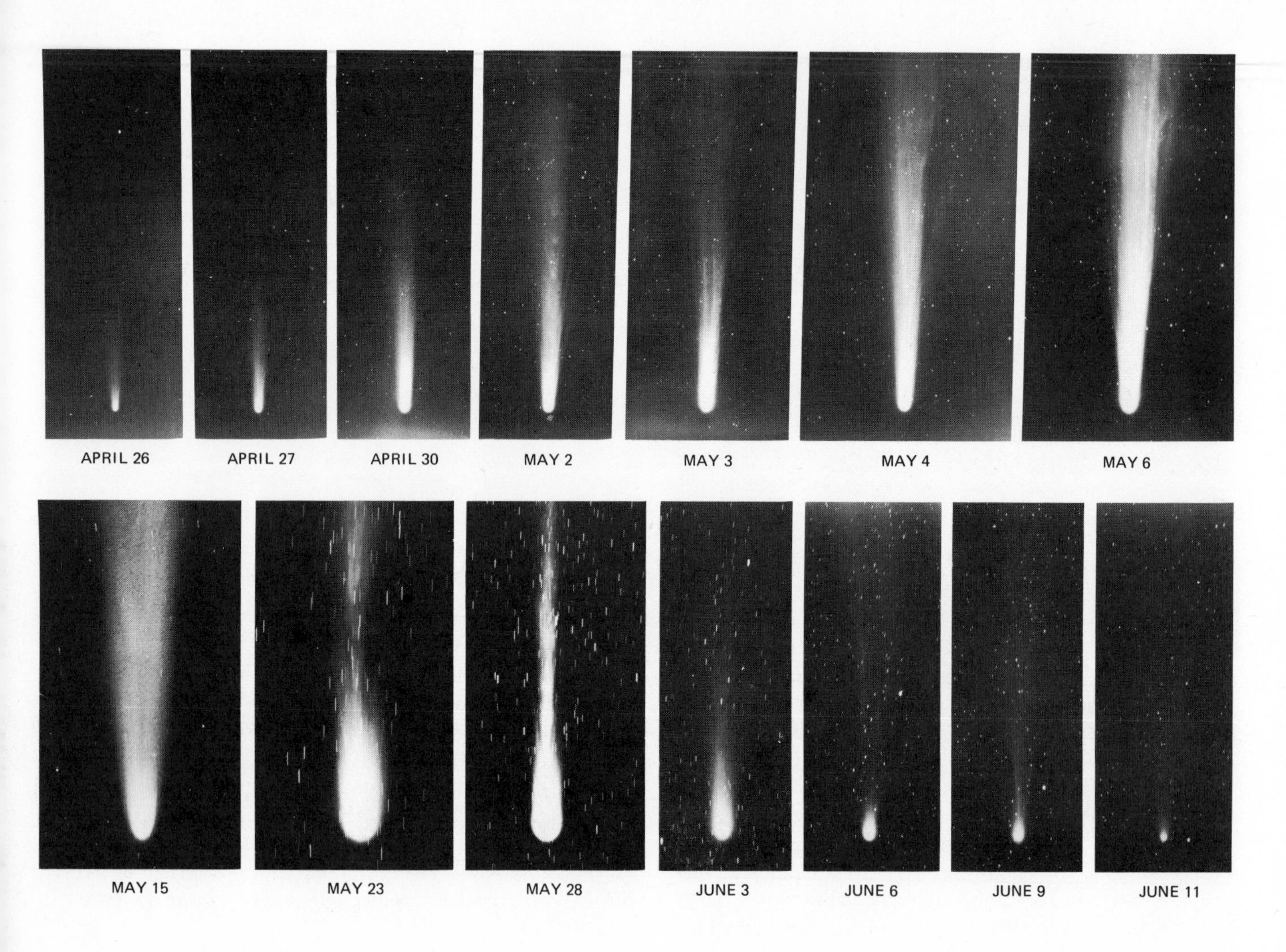

FIG. 9.3
Halley's comet in 1910. Note the development of its tail as the comet approached the sun and its decline as the comet receded from the sun. (Courtesy of the Hale Observatories.)

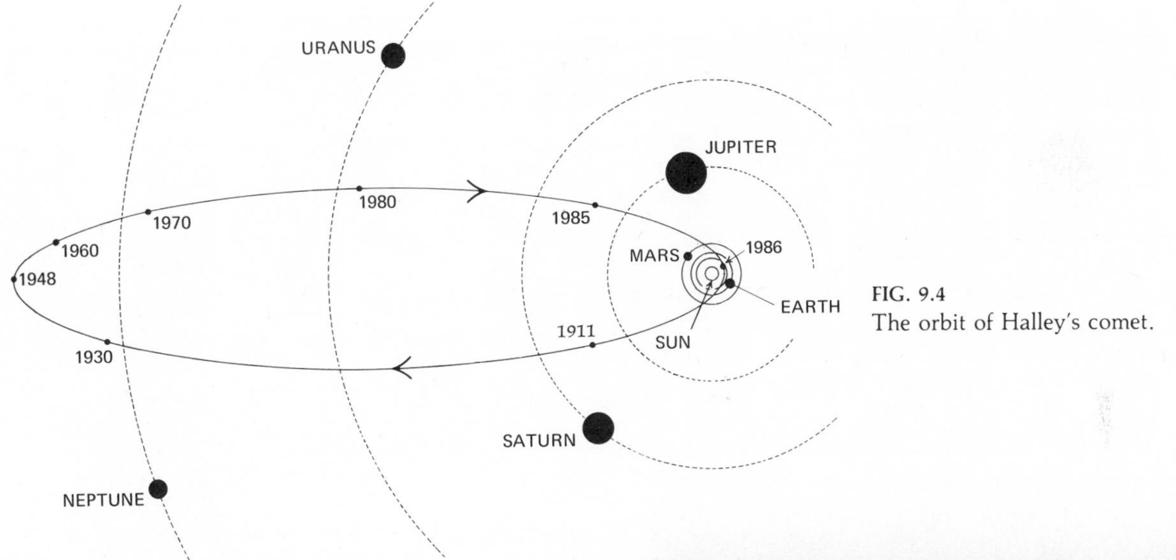

FIG. 9.4
The orbit of Halley's comet.

FIG. 9.5
Comet Ikeya-Seki, November 1965. (Courtesy of NASA.)

by a letter which indicates the order of their discovery, e.g., Comet 1973b designates the second comet discovered in 1973. The permanent designation is made by the order of their passage of perihelion. Comet 1973 II designates the second comet to pass perihelion in 1973.

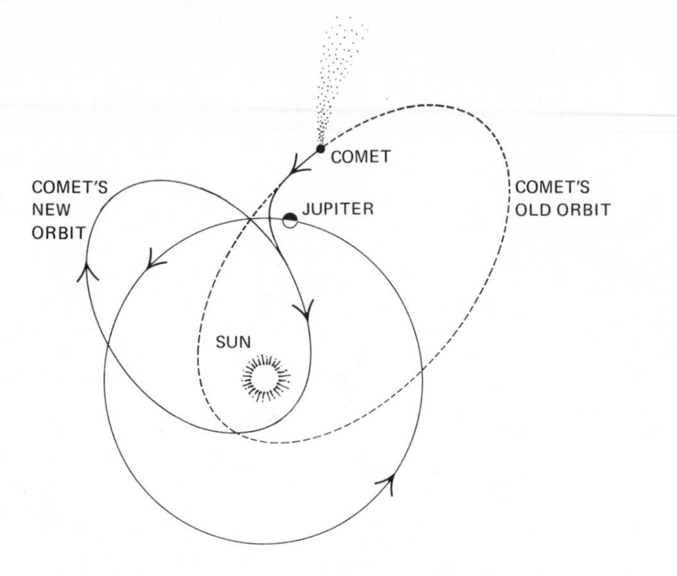

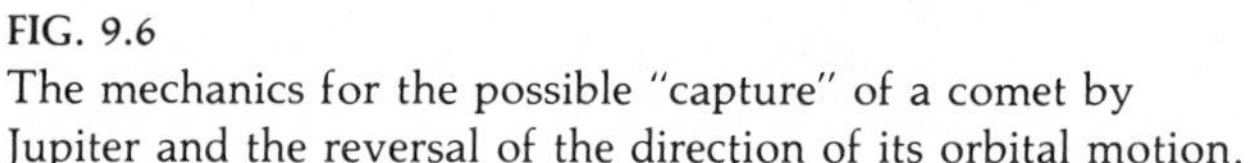

FIG. 9.6
The mechanics for the possible "capture" of a comet by Jupiter and the reversal of the direction of its orbital motion.

9.10 ORBITS

Since about 75% of the observed comets for which there is sufficient and reliable information have orbits that are nearly parabolic and the remaining have less eccentric elliptical orbits, the available information supports the theory that comets originate within the solar system. An analysis of those with parabolic orbits reveals that they initially moved in highly eccentric elliptical orbits, which were altered when the comets made a close approach to either Jupiter or Saturn. There is also the possibility that if more time were available to observe the individual comets and to calculate their orbital elements more accurately, it would reveal that they follow an elongated elliptical rather than a parabolic or hyperbolic orbit.

From a statistical study of the known cometary orbits and their distribution in space, the Dutch astronomer Jan Oort in 1950 proposed a hypothesis for the source of comets. His work indicates that the sun is capable of holding a cloud of orbiting comets at a distance of about 50,000 to 150,000 a.u. The gravitational attraction of a passing star perturbs these comets and alters some of their orbits so that they make a close approach to the sun in an elliptical, parabolic, or hyperbolic orbit.

Of the nearly 200 comets that have been observed with elliptical orbits, about 85 have periods of less than 200 years, and about 40 have been observed to pass perihelion at least twice. These are called *periodic comets*. Their orbital characteristics are different from those of the planets in that generally they are more eccentric and have inclinations that are less than 30° to the plane of the ecliptic. Most of them move in the same direction as the planets, although a few of them retrograde.

Comet Oterma, a periodic comet with a most unusual orbit, was discovered by Schwassmann and Wachmann at the Hamburg Observatory in 1927, two years after it had passed perihelion. Its orbit is more like that of a planet—nearly circular with a small inclination to the plane of the ecliptic (9.5°). Another unusual feature about its orbit is that it lies completely outside of Jupiter's orbit, between the orbits of Jupiter and Saturn. The comet also shows remarkable fluctuations in brightness. In one day its brightness was observed to increase over 100 times, and during a two-month period in 1945, it increased over 2500 times from causes that are still not known. Comet Oterma has an eight-year period in the most nearly circular orbit known.

A few comets have orbital periods that are less than Jupiter's 12-year period. These are believed to have been "captured" by Jupiter and are referred to as Jupiter's family of comets. When a comet makes a close approach to Jupiter, the planet's gravitational attraction is sufficient to alter the comet's orbit by retarding it, thereby placing it in a smaller orbit with a shorter period (Fig. 9.6). There are about 45 comets that have periods between 5 and 10 years. They all move in a direct motion in orbits that are inclined less than 45° to the plane of the ecliptic. Their orbits are oriented in space so that their aphelion and either of their nodes are close to Jupiter's

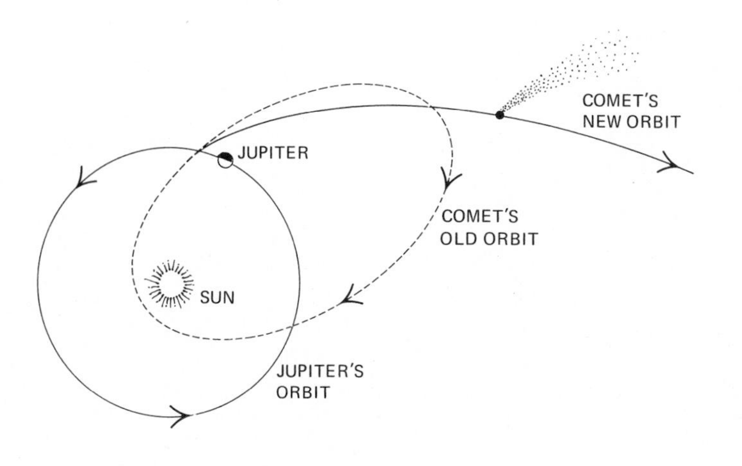

FIG. 9.7
The mechanics for the possible "expulsion" of a comet by Jupiter.

orbit. The one notable exception is Encke's comet, which has the shortest period of any known comet, 3.3 years. Its perihelion distance of 32 million miles (51 million km) places it within the orbit of Mercury.

Jupiter is also capable of "expelling" a comet from the solar system. At a close approach, Jupiter's gravitational attraction can accelerate the comet, thereby placing it in a larger orbit with a longer period (Fig. 9.7). Successive similar encounters could change the comet's orbit from elliptical to parabolic or hyperbolic, thus causing the comet to leave the solar system. Space scientists have taken advantage of Jupiter's ability to alter orbits by using the giant planet to fling spacecraft into larger orbits that will carry them to the outer planets or beyond.

9.11 THE STRUCTURE OF COMETS

A complete comet has a head and a tail. The head consists of a star-like nucleus (frozen particles) surrounded by a hazy, luminous, spherical coma (gas cloud). The tail is a faint streak of luminous material that stretches away from the head and away from the sun. The nucleus, from which the coma and the tail are derived, is extremely small—less than 10 miles (16 km) in diameter. The head diameters range from about 10,000 miles (16,000 km) to nearly 1.5 million miles (2.4 million km). The average head diameter is 80,000 miles (129,000 km), which is approximately the diameter of Jupiter. Tail lengths range from a few million miles to more than 100 million miles (160 million km). With such values, a comet is easily the largest object in the solar system.

The extremely small mass of a typical comet has been estimated to be less than that of a very small asteroid. Comets have been observed to pass within 1.5 million miles (2.4 million km) of the earth and very close to planetary satellites without the slightest effect on their orbits. From these observations, the upper limit for the mass of a typical comet has been established at less than one trillionth that of the earth. With a very low mass and a very large volume, the density of a typical comet is extremely low—so low that the comet head appears transparent. When Halley's comet appeared in 1910, the stars were clearly visible through its 50,000-mile (80,000-km) head, and when it passed between the sun and the earth, no trace of its tiny nucleus was visible against the background of the solar disk. Also, the outer portions of the comet's tail swept across the earth without any noticeable effects.

9.12 THE SPECTRA OF COMETS

The average comet becomes visible at about 3 a.u., when the frozen gases in the nucleus (ammonia, methane, and

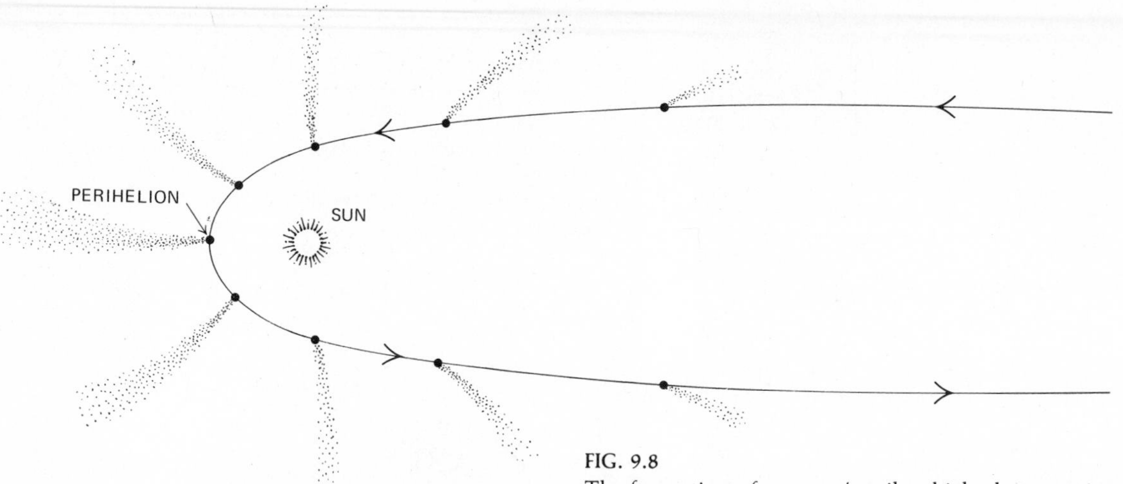

FIG. 9.8
The formation of a comet's tail, which always points away from the sun.

water) begin to evaporate and reflect sunlight. This is indicated by its spectrum, which before that is a faint replica of the solar spectrum. As the comet approaches the sun the bright bands of carbon (C_2 and C_3), cyanogen (CN), imino (NH), amide (NH_2), and hydroxyl (OH) appear in the comet's spectrum. All are formed from the four abundant elements: carbon, hydrogen, oxygen, and nitrogen. When the comet is near perihelion, the bright emission lines of the metals chromium, iron, nickel, and sodium become visible in the spectrum. The spectrum of a comet's tail reveals the presence of positively charged molecules of carbon monoxide (CO^+) and nitrogen (N_2^+).

9.13 A COMET'S APPEARANCE

Faint comets are the general rule, bright comets are very few, and extremely bright comets are rarities. A typical comet is visible for only a few days, some are visible for several weeks, and relatively few are visible for several months. Usually a comet is first seen telescopically as a faint, hazy spot of light. As it approaches the sun, its brightness and size appear to increase, and quite often a bright, star-like nucleus is visible within the coma. Although many comets never produce a tail, those that do so appear to form the tail when the comet is about 2 a.u. from the sun (Fig. 9.8). The tail reaches its maximum length and brightness soon after the comet has passed perihelion; then the tail gradually decreases as the comet recedes from the sun. The tail always extends away from the sun, so that when the comet approaches the sun, its tail follows the head. When the comet moves away from the sun, its tail precedes the head.

Two forces act on the particles in the comet's nucleus to cause the formation of the comet's tail—the *solar wind* and the *solar-radiation pressure*. The solar wind is a hot, low-density gas of charged atomic particles (a hydrogen plasma of protons and electrons) that continuously streams outward from the sun. The interaction of the solar wind with the particles in the nucleus produces the tail; however, it is not fully understood how this is accomplished. When the solar-radiation pressure exceeds the solar gravitational force, the particles in the comet's nucleus are forced out away from the sun, forming the comet's tail. This occurs when the ratio of surface area to mass is relatively large, which is the case when considering small particles. The radiation pressure acts on a particle in the same way that the wind acts on a sail.

The tails of comets may be classified into two important types: ion and dust. The ion and dust tails are clearly

FIG. 9.9
Comet Mrkos, 27 August 1957. (Courtesy of the Hale Observatories.)

FIG. 9.10
Comet Arend-Roland, 27 April 1957, with its unusual, sharp spike pointing toward the sun. (Courtesy of the Hale Observatories.)

visible in the remarkable Comet Mrkos, which was discovered in 1957 (Fig. 9.9). The ion tail consists of ionized particles projected in a stream away from the sun. Excitation by solar radiation makes the ions luminous. The dust tail consists of small dust particles that curve backward in the plane of the comet's orbit. The dust particles are luminous because of reflected and scattered sunlight.

Photographs of the April 1957 Comet Arend-Roland (Fig. 9.10) displayed not only a normal tail streaming away from the sun, but also a long, straight, slender, bright tail that appeared to point directly toward the sun

FIG. 9.11
This time-exposure photograph shows the zodiacal light beyond the observatory dome and the star trails moving down in the western sky. (Courtesy of the Yerkes Observatory.)

This unusual tail looked like a spike because it consisted of dust particles that were diffused from the comet's head into a thin sheet that was lying completely in the comet's orbital plane. Actually, the "spike" was pointing away from the sun, but its position in space made it appear to be pointing towarJ the sun.

9.14 LIFE HISTORY OF A COMET

It has been stated that comets are believed to originate in the *Oort comet cloud* somewhere near the outer portion of the solar system. Here, far from the sun, the temperature is so low that all ordinary gases and liquids are frozen into solids. A comet originates as a nucleus, perhaps 15 miles (25 kilometers) in diameter, which is disturbed from its orginal orbit and begins to move toward the sun on a path that will take it into the inner solar system. The change in the orbit of the nucleus may be due to a passing star. As the frozen mass of gas moves in toward the sun it warms up. Upon reaching the orbit of Saturn some of the solids begin to turn into a gaseous state. A weak coma forms around the nucleus. As the comet approaches still nearer to the sun, heating drives more frozen gases out; the coma expands and begins to respond to the force of the solar wind and radiation. A tail may form. Matter swept out into the tail is of course lost from the comet, as is most of the material in the coma. If a comet moves along an orbit that takes it close to the sun on many trips, each passage erodes more of the nucleus until all of the frozen gases are gone. Only solid matter such as dust particles and aggregates of dust remain. Even these are vaporized if the comet passes close enough to the sun, as occasionally happens.

9.15 THE ZODIACAL LIGHT

On a clear, moonless spring evening at the end of twilight, the zodiacal light can be seen extending above the western horizon. Just before twilight on an autumn morning, it can be seen extending above the eastern horizon. It appears as a faint, diffuse, conical-shaped patch of light. The zodiacal light's base lies on the horizon, its axis coincides approximately with the ecliptic, and its hazy tip extends to a great height (Fig. 9.11). In the evening sky

FIG. 9.12
Meteor trail among the stars in the constellation of Orion. (Courtesy of the Yerkes Observatory.)

this means that in spring, the zodiacal light appears to make a large angle with the horizon; in autumn, the angle is small. The above situation would be reversed in the morning sky.

The zodiacal light appears to come from an edge view of a lens-shaped cloud of small particles centered around the sun and lying close to the plane of the ecliptic. It is believed that the particles are probably the material ejected from comets and the debris from colliding asteroids, which eventually spiral toward the sun because of their size. Still smaller particles, however, would be blown away from the sun due to radiation pressure. The light appears bright along its axis; it fades gradually outward and becomes so diffused that it is difficult to establish its edges. Its spectrum is similar to the solar spectrum, indicating that it is either reflected sunlight or light scattered by small particles. Since the zodiacal light is slightly polarized and is not red, small particles such as atoms and gas molecules have been eliminated as its source. To produce the observed brightness, particles with diameters of about one micron (0.00004 inch) and albedoes of about 7% would be required.

A smaller, extremely faint, nearly circular patch of light called the *gegenschein* (counterglow) appears in the night sky opposite the sun. Although its true nature is not known, it is believed to be the end-on view of the earth's tail, which is similar to that of a comet, is produced by the solar wind, and is made visible by reflected sunlight. Instrumentation aboard the Pioneer 10 probe to Jupiter observed the *gegenshein* but failed to confirm the above.

9.16 METEORS

Almost everyone has seen star-like objects shoot across the evening sky. They are called meteors and are popularly known as shooting stars (Fig. 9.12). The term

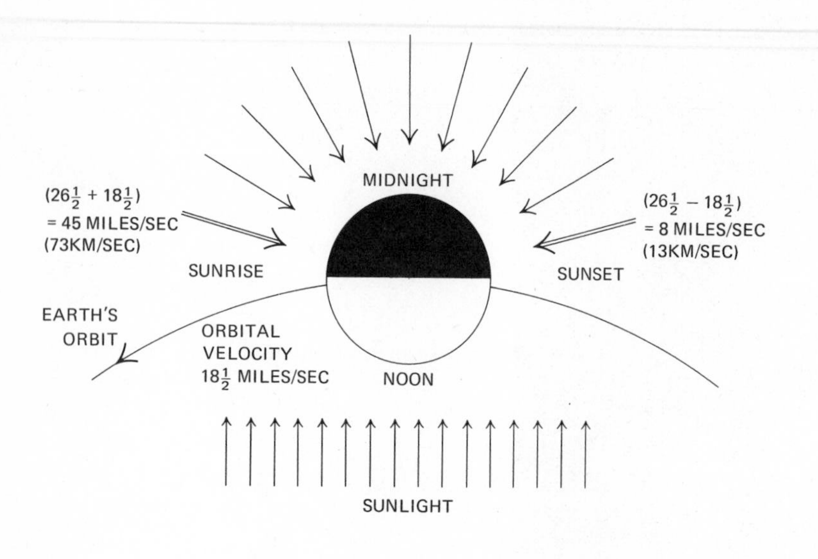

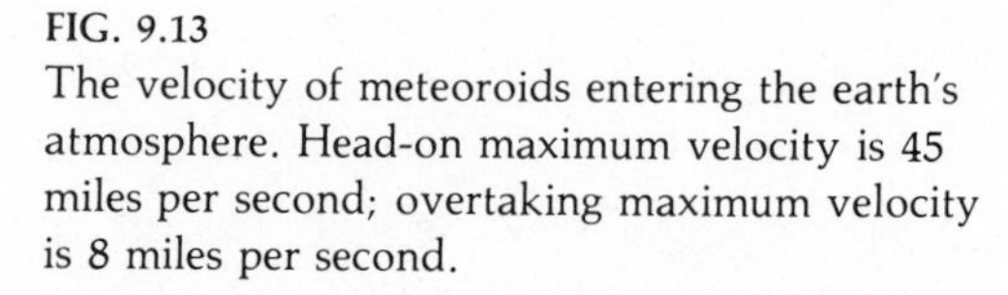
FIG. 9.13
The velocity of meteoroids entering the earth's atmosphere. Head-on maximum velocity is 45 miles per second; overtaking maximum velocity is 8 miles per second.

"meteor" does not refer to the actual body, but rather to the luminosity that is produced as it passes through the earth's atmosphere. The body that produces the meteor is called a *meteoroid*. Occasionally, when a meteor is bright enough to be visible in the daytime or to cast shadows at night, it is called a *fireball*; and more rarely, when it explodes in the atmosphere and produces thunder-like sounds, it is called a *bolide*. A meteoroid that is large enough to survive its flight through the earth's atmosphere and fall on the earth's surface is called a *meteorite*. When a meteorite seen to fall is recovered, it is called a *fall*; when an unknown meteorite is discovered, it is called a *find*.

Many people confuse meteors with comets. Meteors are very small bodies, usually about the size of pebbles, which become visible for only a few, fiery seconds as they pass through the earth's atmosphere. Comets are the largest bodies in the solar system and are visible for several weeks as hazy spots of light. Although comets move rapidly through space, their apparent motion relative to the stars is very slow.

There are two possible sources for meteors—either inside or outside the solar system. If meteors were produced by bodies within the solar system, they would travel at speeds ranging from about 8 to 45 miles (13 to 72 km) per second; if produced by bodies outside of the solar system, they would travel faster. Since the velocity of escape from the sun (velocity in a parabolic orbit) at a distance of 93 million miles (150 million km) is about 26.5 miles (42.7 km) per second and since the earth's orbital velocity is 18.5 miles (29.8 km) per second, a meteoroid's maximum velocity as it enters the earth's atmosphere head-on (Fig. 9.13) is 45 miles (72.4 km) per second (26.5+18.5). If the meteoroid overtakes the earth, its velocity is 8 miles (12.9 km) per second (26.5−18.5). No meteor has been observed with a velocity greater than 45 miles (72 km) per second, which indicates that meteors come from within the solar system rather than from interstellar space.

9.17 OBSERVATIONS

An observer on a clear, dark evening may be able to see from six to ten meteors per hour. More are visible after midnight than before, and the number increases to a maximum just before dawn. This is understandable when

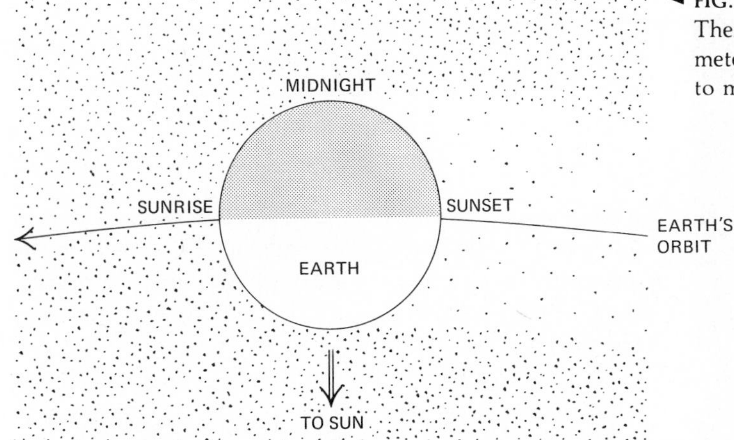

◄ FIG. 9.14
There are fewer meteors in the earth's wake; therefore, more meteors are visible from midnight to sunrise than from sunset to midnight.

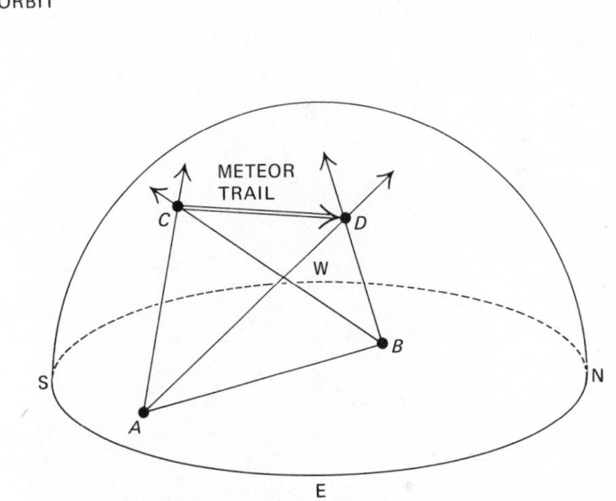

FIG. 9.15 ►
Simultaneous photographs of a meteor taken from two stations (*A* and *B*) can establish the initial point (*C*) and the final point (*D*) of its tail.

the orbital and rotational motions of the earth and the motion of the meteors are considered. Between sunset and midnight (Fig 9.14), the observer is on the side of the earth that is moving away from the meteors; therefore, only those meteors that overtake the earth will be visible. After midnight, the observer is on the side of the earth that is moving head-on into the meteors; therefore, he or she will see more meteors, and they will appear brighter because they are entering the earth's atmosphere with higher velocities.

The average meteor becomes visible when it is about 60 miles (100 km) above the earth's surface; it has been completely consumed by the time it reaches a height of about 50 miles (80 km). However, if the meteor is sufficiently large, it will remain visible to a height of about 30 miles (50 km). It has been estimated that nearly 200 million meteors become visible to the unaided eye over the entire earth's surface during a 24-hour period. There are also many more that are visible telescopically. Their average total mass has been estimated at about 100 tons (91,000 kg). Only about one ton is deposited on the earth; the rest is consumed in the earth's atmosphere.

Two important techniques, photography and radar, are used to determine the direction and velocity of a meteor. In the photographic technique (Fig. 9.15), simultaneous photographs of the meteor trail (*CD*) are taken with wide-angle telescopic cameras from two stations (A and B) that are about 25 miles (40 km) apart. When the meteor is first observed, the simultaneous photographs show the initial point of the meteor trail with respect to the stars from each station. From the photographs the spatial position for the beginning of the meteor trail (*C*) is

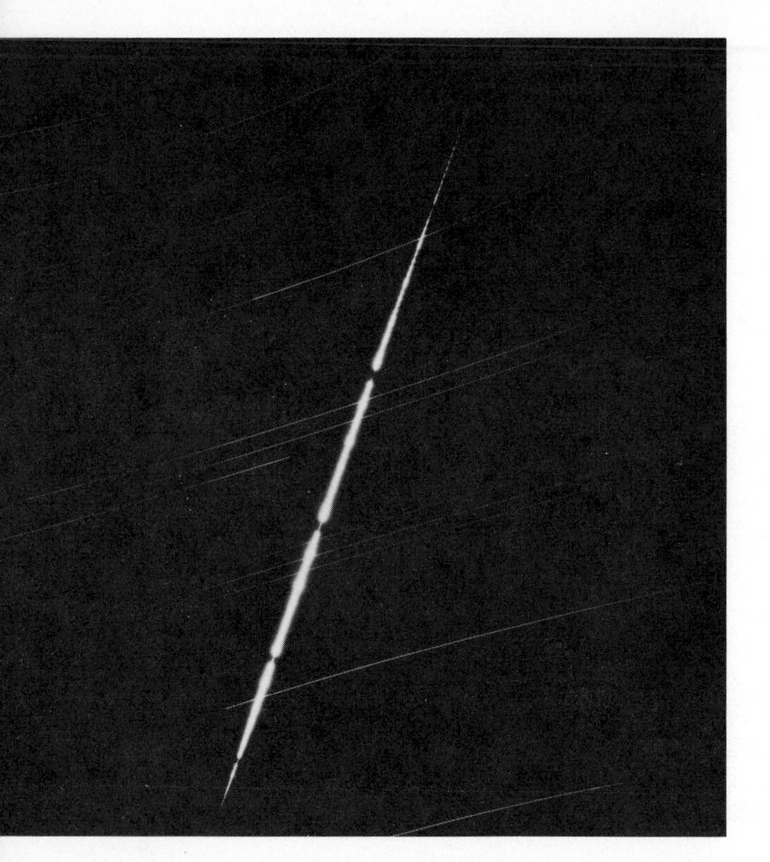

FIG. 9.16
This is the bright, interrupted meteor tail of the Lost City meteorite fall, 7 January 1970. (Courtesy of the Smithsonian Astrophysical Observatory.)

established by triangulation. The end of the meteor trail (*D*) is established in the same manner. The velocity of the meteor is determined by a shutter placed in front of each camera, which rotates at a uniform rate and interrupts the meteor trail at 20 breaks per second. Figure 9.16 shows the segments of the interrupted image. When the segments are measured, the angular and meteor velocities can be determined.

In the radar technique, developed after World War II, wavelengths of about 3 meters that are easily reflected by the ionized gases in the meteor trail are sent out and received by the radar transmitter. With this technique the range, direction, and velocity of the meteor can be measured. The results, however, are less accurate than those derived from the photographic technique. The most important advantage of the radar technique is its ability to detect faint meteors when they enter the earth's atmosphere in the daytime.

9.18 METEOR SHOWERS

The meteors that have been discussed are called *sporadic* because they have no common point of origin, that is, they appear at any time and at any place in the sky. When meteors appear to radiate from a common point in the sky, they are called *meteor showers*. The common point (*radiant*) is one of perspective. Since the meteors in a shower move along parallel paths, they appear to diverge from the radiant (Fig. 9.17), in the same manner

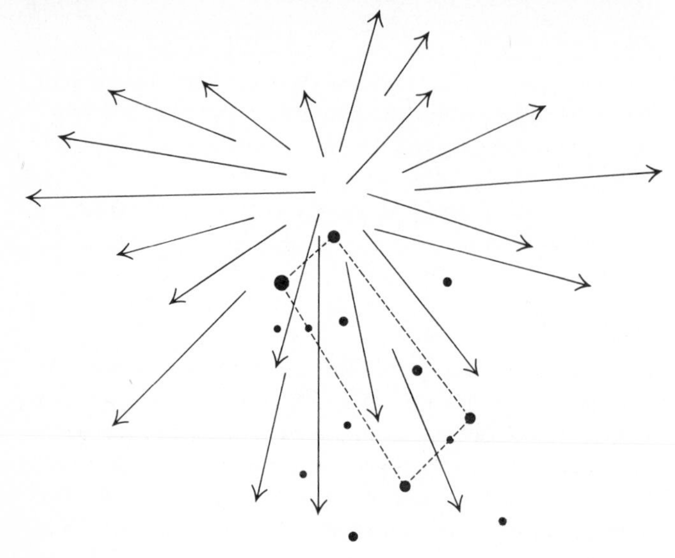

FIG. 9.17
The Geminids, a beautiful meteor shower that reaches its maximum rate of 55 meteors per hour about December 14. The meteors appear to radiate from a common point (the radiant) in the constellation of Gemini the Twins.

TABLE 9.2
Maximum visual display of meteor showers

SHOWER	DATE	RADIANT R.A. DEC.	HOURLY RATE	ASSOCIATED COMET
Quadrantids	Jan 3	$15^h20^m48°$	30	——
Lyrids	Apr 21	$18^h00^m33°$	5	1861 I
Aquarids	May 4	$22^h24^m0°$	5	Halley
Perseids	Aug 12	$3^h04^m58°$	35	1862 III
Draconids	Oct 10	$16^h56^m54°$	(periodic)	Giacobini-Zinner
Orionids	Oct 22	$6^h16^m16°$	15	Halley
Taurids	Nov 1	$3^h28^m21°$	5	Encke
Leonids	Nov 17	$10^h08^m22°$	5	Temple
Geminids	Dec 14	$7^h32^m32°$	55	——
Ursids (Ursa Major)	Dec 22	$13^h44^m80°$	15	Tuttle

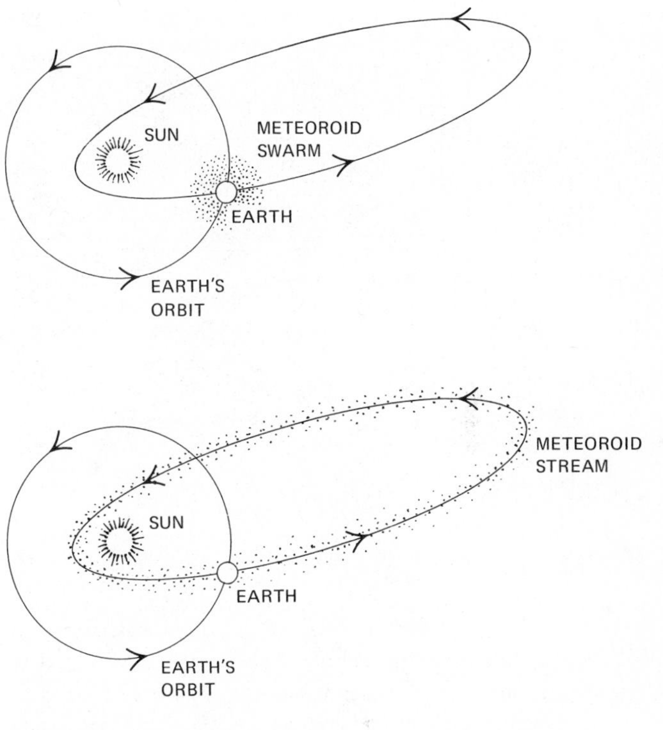

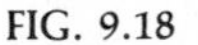

FIG. 9.18
An annual meteor shower occurs when the earth is at the intersection of its orbit and the orbit of the meteoroid stream. A periodic shower occurs when the earth and the meteroid swarm are both at the intersection.

as a road appears to diverge from a common point. Meteor showers are identified by the constellation in which the radiant appears to be located (Table 9.2).

Many of the meteor showers are associated with comets. The particles that have been ejected from the nucleus by solar radiation pressure move in an orbit around the sun that is very close to the cometary orbit. As shown in Fig. 9.18, these particles may be either concentrated in a pile (*meteoroid swarm*) behind the nucleus or distributed over the entire orbit of the comet (*meteoroid stream*). A meteor shower occurs every year when the earth crosses the meteoroid stream. Periodic showers occur when the earth passes through a meteoroid swarm.

9.19 METEORITES

In the Bible, Joshua 10:11, we find ". . . and it came to pass, as they fled from before Israel, and were in the going down to Beth-horen, that the Lord cast down great stones from heaven upon them unto Azekah, and they died." The philosopher Anaxagoras and the Chinese scholars have also eloquently described stones falling

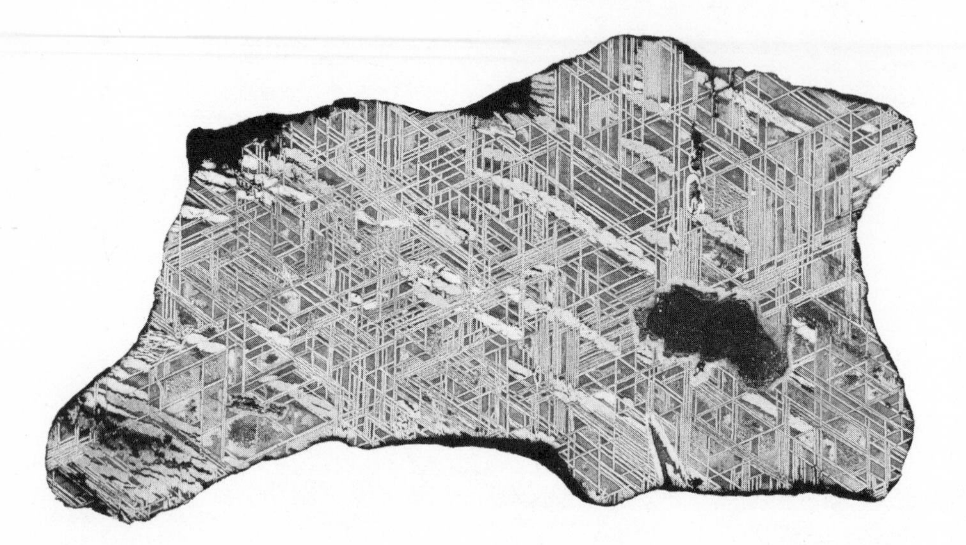

FIG. 9.19
Widmanstätten figures. The polished, etched surface of a section of the Edmonton, Kentucky iron meteorite clearly shows the low nickel crystalline structure of kamacite as white bands. (Courtesy of the Smithsonian Institution Museum of Natural History.)

from the sky. Even though this phenomenon was observed, for centuries scholars chose to believe that it was an impossibility and that it had to be an earthly phenomenon. Only at the beginning of the nineteenth century did people first begin to put together the pieces that convinced them they were truly stones from the sky.

9.20 COMPOSITION

Meteorites are classified into three major categories: iron, stones, and stony-irons. It is interesting to note that from the study of the known meteorites, about 95% of the falls are stones and 65% of the finds are irons. Despite the fact that the stones are more abundant in space, more of the finds are irons; the stones are more difficult to differentiate from terrestrial rocks, and they decompose more rapidly.

The average composition by weight of the important elements in iron meteorites is 90% iron and 8.5% nickel; in stone meteorites it is 36% oxygen, 24% iron, 18% silicon, and 14% magnesium. The stony-iron meteorites are composed of iron and stone in equal amounts by volume, with the stony material imbedded in a sponge-like mass of iron. The irons are usually heavy, nearly three times heavier than a terrestrial rock of the same size. The stone meteorites are similar to terrestrial rocks and are composed of silicate materials, nickel, and aluminum. Most of the iron meteorites can be identified by their unusual and distinctive internal structure, which was discovered by Alois de Widmanstätten and bears his name. When a small surface area of an iron meteorite is ground, polished, and etched, the beautiful *Widmanstätten figures* are usually revealed. They consist of large, parallel, intersecting crystalline bands of low nickel kamacite formed by slow cooling (Fig. 9.19).

The stones are further classified as *chondrites* and *achondrites* (Fig. 9.20). The chondrites consist mostly of silicate materials, similar to terrestrial rocks, and contain tiny, almost spherical drops of glass (magnesium and iron silicates) called *chondrules*. Their composition is variable, indicating that they must have evolved from different materials, but under similar conditions. The two dominant minerals in chondrites are olivine and orthopyroxene. The abundance of iron in the chondrites is also quite variable. The achondrites, which are also similar to terrestrial rocks, do not contain chondrules or nickel-iron.

(a) (b)

FIG. 9.20
(a) Plainview, Texas chondrite is six inches long and clearly shows regmaglyphs (thumbprint-like depressions); (b) Pasamonte, New Mexico achondrite is two inches long and shows thread lines. (From the collections of Ronald A. Oriti, Griffith Observatory; photographs by James E. Klein.)

9.21 APPEARANCE

Most people are surprised to learn that a moderate-sized meteorite is not flaming hot when it strikes the earth's surface; rather, it is cool enough to touch, does not burn for days, and does not make a glowing crater. Beyond the earth's atmosphere, meteoroids are very cold. When they pass through the atmosphere, friction heats them to incandescence; however, their flight through the atmosphere takes but a few seconds so that only their surfaces become hot enough to melt. Also, much of the melted material is swept away into the earth's atmosphere so that when the meteoroid strikes the earth's surface, it is relatively cool. A meteorite that has just fallen is usually covered with a very thin, dark crust, the molten material that has solidified rapidly toward the end of its flight through the earth's atmosphere. Many meteorites show flow-lines (threads or small ridges) where the molten material has flowed because of the meteorite's forward motion. Weathering quickly changes the color of the crust to a dull gray or brown.

9.22 SPECTACULAR METEORITES

The largest known meteorite fall ever observed in the United States was the 2000-pound stone meteorite that fell in Furnas County, Nebraska, on February 18, 1948. Recovered six months later, it is now on display at the University of New Mexico (Fig. 9.21). The second largest was the 800-pound stone meteorite which fell near Paragould, Arkansas, on February 14, 1930. It was recovered one month later and placed on display at the Chicago Natural History Museum.

The largest meteorite find in the United States was the Willamette meteorite (15 tons), which was discovered in 1902 near the Willamette River, Portland, Oregon, and is on exhibit at the Hayden Planetarium, New York. It is conical-shaped, probably because it did not tumble as it passed through the earth's atmosphere, and it has several large cavities on one of its surfaces.

In 1894 the arctic explorer Commodore R. E. Peary discovered three iron meteorities near Cape York, Greenland: "The Ahnighito" (34 tons) (Fig. 9.22), "The

FIG. 9.21
The Furnas County, Nebraska stone meteorite. (Courtesy of the Institute of Meteoritics, Department of Geology, University of New Mexico.)

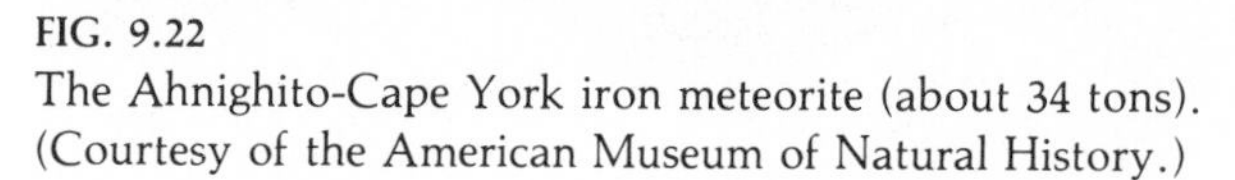

FIG. 9.22
The Ahnighito-Cape York iron meteorite (about 34 tons). (Courtesy of the American Museum of Natural History.)

Woman" (2500 pounds), and "The Dog" (1000 pounds). They were brought to the United States in 1897 and are on exhibit at the Hayden Planetarium.

In this century, Soviet territory has been hit twice by great meteorites—first on June 30, 1908 in central Siberia, and second on February 12, 1947 north of Vladivostok. The great Siberian meteor of 1908 fell in a forest near the Tunguska River and was seen by many people over an area of several thousand square miles. It first appeared above the southern horizon at 7:00 A.M. and moved northward along a meridian line at great speed. One observer described it as "like a piece broken off the sun"; another said that "the ground suddenly rose and fell like a wave." Seismographs and microbarographs throughout Europe recorded the tremor and the pressure waves. After the Russian Revolution, an investigation of the site revealed more than 100 depressions, which indicated that a cluster of meteorites rather than a single one had hit the earth. In spite of all this evidence, no one has ever been able to find a single meteorite in the entire area. Soviet scientists have concluded that a small comet struck the earth, because the comet's nucleus, which contains frozen water, ammonia, methane, and carbon dioxide, would have completely evaporated on contact with the earth's surface.

One of the finest meteorite craters is the Barringer, located near Winslow, Arizona. It is a nearly circular depression with a diameter of 4200 feet (1280 meters) and a depth of 600 feet (180 meters) from its rim. The rim rises about 130 feet (40 meters) above the surrounding plain and appears somewhat upturned, similar to the appearance of lunar craters (Fig. 9.23). Over 30 tons of iron meteoric material have been found on the floor of the crater and for several miles around it. The largest

FIG. 9.23
An aerial view of the Barringer meteorite crater in Arizona. (Courtesy of the Yerkes Observatory.)

single fragment weighs over 1400 pounds (635 kg). It is believed that the crater was formed about 50,000 years ago by a meteorite whose mass has been estimated to range from 63,000 tons, or 57,000 metric tons (by E. M. Shoemaker) to 2,600,000 tons, or 2,400,000 metric tons (by E. J. Öpik). These estimates are based on theoretical reasoning and calculation.

9.23 MICROMETEORITES

A large number of very tiny particles, or micrometeorites (about 0.0001 inch), enter and pass through the earth's atmosphere, especially during a meteor shower. Because

FIG. 9.24
Tektites from Thailand.

of their small size, they pass through the earth's atmosphere at low speeds, thereby reaching the earth without vaporizing.

9.24 TEKTITES

Small glass-like stones (tektites) have been found in large numbers in a few scattered places in Indo-China, Java, the Philippines, Australia, the United States (Texas and Georgia), Czechoslovakia, and the Ivory Coast of Africa. A chemical analysis has revealed that they are rich in silicates and are dissimilar in composition to any of the rocks in the area in which they were found. When they were first observed on the ground, they appeared as dirty brownish or greyish rocks; however, when they were scrubbed and cleaned, they were black and shiny. Their shapes fall into three general groups: *roughly spherical*, *disk*, and *tear-drop*, which indicates that they were all involved in a flight through the earth's atmosphere (Fig. 9.24).

Although the ages of the tektites over the world vary considerably, the ages of those in any one area are identical. The oldest tektites (34 million years) are found in the United States. Others range in age from 15 million years (Czechoslovakia) to 700,000 years (Australia). Their distribution on the earth suggests that they probably are particles splashed out by large meteoric impacts.

9.25 THE ORIGIN OF THE SOLAR SYSTEM

We have seen in our earlier discussion that the solar system possesses many regularities. The planets all move about the sun in the same direction and in basically the same plane. Most of the planets and their satellites rotate in this same direction. The planets divide naturally into two groups, the terrestrial and the Jovian planets, whose properties differ dramatically. The terrestrial planets are separated from the Jovian planets by the asteroid belt, an aggregate of relatively small objects similar to the terrestrial planets in that they are rocky, while the Jovian planets are separated from the surrounding interstellar space by the comet cloud, whose particles are like miniature Jovian planets.

These regularities have been recognized for decades, even centuries, and have fascinated scientists and natural philosophers alike. There have been basically two schools of thought about the solar system's origin. One school argues that the system formed as a natural consequence of the condensation of the sun. According to this point of view, one could expect to find planets orbiting around most stars. The other school argues that the solar system formed as the result of special circumstances. One of the most common variants of this school is the *encounter theory*, in which the planets were supposed to have formed as the result of a near-collision between the sun and a passing star. The vagrant star lifted tides on the sun, causing a cloud of hot gases to be torn off the solar surface. Pulled out into space, the gases cooled, condensed, and formed planets. It is relatively easy to show that such a "close encounter" between stars is very rare. If the planets were formed in this fashion, there are probably only a few more planetary systems in our whole galaxy. While the unlikelihood of this event is not reason enough to reject the encounter theory, there are other objections to it as well. The earth, for instance, contains some elements that are found only in minute traces in the sun (lithium is the most striking). More importantly, however, calculations show that if hot gas is pulled off the sun it will disperse into space and not condense into planets. For these and other reasons, the majority of astronomers now believe the solar system originated when the planets and sun formed by *condensation from an interstellar cloud*. How this might have happened is discussed below. Formation of the solar system by condensation will be seen to offer elegant explanations of many of the solar system's properties.

It is generally believed that stars form from the collapse and condensation of gas and dust clouds in space. A typical interstellar cloud (Chapter 15.4) is composed of slightly less than 75% hydrogen and 25% helium, with perhaps 1% dust and other gases. The dust particles are believed to contain aluminum, iron, and magnesium silicates. The gases in the cloud are initially very cold and very rarefied. At the low temperature thought to prevail in the cloud, the motion of the particles is not sufficient to offset their mutual gravita-

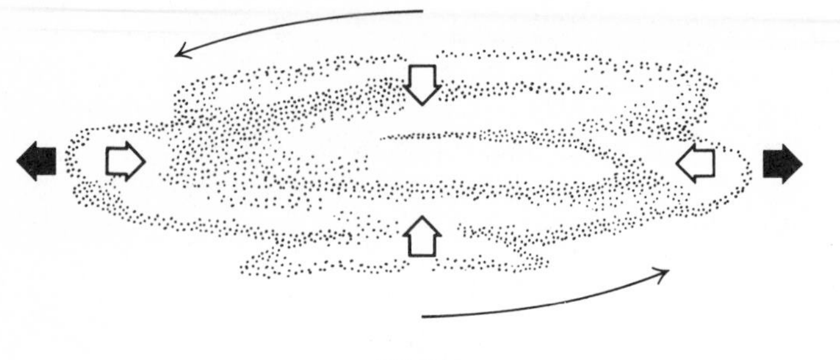

FIG. 9.25
Rotation causes the early solar system to flatten. Solid arrows represent rotational forces, open arrows gravity.

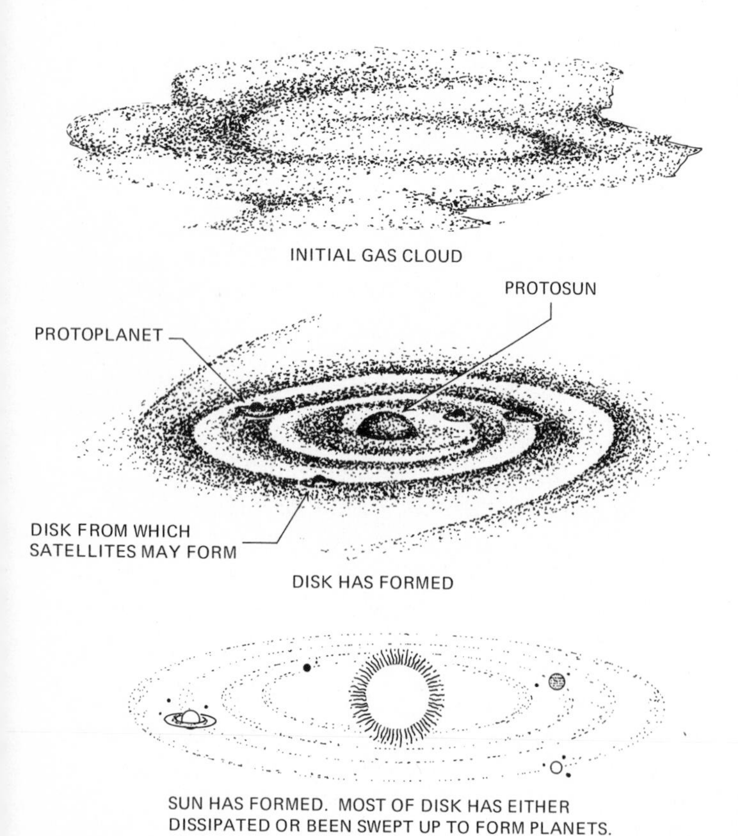

tional attraction. Gravity thus slowly draws them together. As they move close together, any initial rotation the cloud has (even if it is only a few inches per minute) is increased. The gases in the cloud become more compressed and begin spinning still faster. Eventually rotation becomes so fast that it is able to offset the force of gravity perpendicular to the rotation axis (Fig. 9.25). The cloud ceases to shrink in toward the axis, but can still collapse parallel to the axis. The cloud therefore flattens into a pancake-like shape. A dense central core develops, which will soon become the sun. This protosun finds itself surrounded by a whirling, flattened disk of gas and dust. Gravity compresses the central core and it eventually becomes hot enough to begin emitting light and heat. The gases in the inner part of the disk are warmed, but those further out remain cold. The protosun begins to shine like a normal star and its heat begins to drive out the gases in the inner disk, leaving behind only the dust particles. The dust particles collide with one another; small irregularities form in the disk and act as new centers for condensations (Fig. 9.26). Gravity pulls

FIG. 9.26
The formation of the solar system.

these condensations into smaller and smaller aggregates. Those aggregates in the outer part of the cloud are cold enough to be able to accumulate both dust and gas. In the inner part of the system only dust is swept up by the condensations. Turbulence and eddying motions in the disk cause the condensations (protoplanets) to begin rotating, and where it is cold enough they develop small disks of their own. The process repeats and each outer protoplanet spawns a family of satellites. All of these processes have taken millions of years. The sun is now shining as a star. It is surrounded by a family of condensations (now planets), some of which have satellites. Having formed from the circulating matter in the original disk, all the planets move in the same plane and direction. The inner planets, formed of the dust, are rich in the heavy elements and devoid of hydrogen and helium, which was driven away by the sun's heat. The outer planets, further from the sun and thus colder, have retained their gases. Thus the sun has two families of planets.

The terrestrial planets must pass through at least two more phases. The accumulation of dust probably proceeds in stages whereby the initially microscopic dust particles grow first to BB size, then to pebble size, then into boulders, and finally into objects the size of asteroids. Some of these will have already undergone collisions to form the bulk of the terrestrial planets, but for perhaps nearly a billion more years the inner solar system will contain debris not yet accumulated into planets. The collisions between asteroid-size chunks of matter and a protoplanet will generate enormous heat, melting the outer layers of the planets. Eventually fewer and fewer asteroids remain in the inner solar system, and the surfaces of the planets congeal and finally solidify. The remaining debris now plummets down onto a hard crust and blasts craters in the planets, leaving them heavily pockmarked by the final stage of their accretion. The largest of the terrestrial planets now begin to heat up by radioactive decay of elements in their interiors, causing their cores to melt. For the largest of the inner planets geological activity begins, including volcanoes which pour forth lava and eject gas. Volcanic gases are rich in carbon dioxide, water, and nitrogen, and an atmosphere accumulates around the largest planets.

Mercury is believed to be too small to have the extensive radioactive heating of the interior found in a larger planet like the earth. Therefore, volcanic activity and mountain building, such as are seen on earth, are unlikely to have broken through Mercury's crust, and it remains devoid of atmospheric gases. Mars is so small most of the gases ejected from volcanoes have escaped. Only Venus and the earth retain the bulk of the ejected gases and perhaps a small amount of the original atmosphere, which they might have accumulated at their formation. It is unclear why the earth and Venus are so different. Some astronomers theorize that Venus is just enough closer to the sun that surface waters on the planet were hotter and thus better able to escape the gravity of the planet. In any case, on the earth oceans formed, and in the oceans, it is theorized, life was born.

SUMMARY

Just as the solar system includes two families of planets, it includes two families of minor objects: asteroids and comets. The asteroids are small rocky chunks, most of which orbit the sun between Mars and Jupiter. The largest known asteroid, Ceres, is about 300 miles (500 km) in radius. The comets are thought to be mixtures of frozen gases and liquids which are in turn mixed with dust and gravel-size solid material, all contained in the cometary nucleus. Normally comets move in orbits in the extreme outer reaches of the solar system (50,000 a.u. away from the sun) in what is believed to be a "comet cloud." When a comet moves in close to the sun, it is heated and the frozen material melts and vaporizes. Radiation pressure and the solar wind (see below) then push the thawed matter out into a tail. The total mass in a comet is not known but is believed to be very much smaller than the moon's mass. Asteroids and comets probably originated in the cloud of dust and gas from which the solar system itself formed.

There is a general outflow of gas (mostly hydrogen and helium) from the sun called the solar wind. This extremely tenuous material streams outward through the inner solar system in approximately radial fashion. The solar wind interacts with the earth's magnetic field and as mentioned above, is important in producing the tail of comets.

Solid material occasionally enters the earth's atmosphere, where it is heated by friction with the atmospheric gases. The heated material streaking across the sky is called a meteor, or shooting star. Typical meteors are only the size of grains of rice, and most are totally burned up before they can reach the surface. If a piece of material survives passage through the atmosphere and reaches the ground it is called a meteorite. Both comets and asteroids produce meteors. Asteroidal material tends to arrive sporadically and produces stoney or iron meteorites; cometary material often arrives in "showers" at predictable times of year.

The solar system is believed to have originated from a gas cloud which contracted as a result of its gravitational forces. Rotation of the contracting cloud made it flatter. The sun formed at the center and the planets formed from smaller condensations in the flattening disk.

REVIEW QUESTIONS

1. What is the Bode-Titius rule? Is it valid? How are the asteroids related to this rule?
2. Discuss the location, distribution, orbits, size, and mass of the asteroids.
3. Do any asteroids have orbits not between Mars and Jupiter?
4. What are the Kirkwood gaps? How are they produced?
5. What is the nature of a comet?
6. List the characteristics of cometary orbits.
7. How do we know that comets do not contain large amounts of material?
8. What is Jupiter's family of comets? Describe how they might have been "captured" by Jupiter. Describe and explain the changes that may occur in the appearance of a comet as it approaches and leaves perihelion.
9. Explain how the tail of a comet is formed and describe its shape and size. Do all comets have tails? Why? Can a comet develop more than one tail? If so, how is this accomplished? Give an example.
10. What is meant when people refer to the Tunguska event?
11. What evidence from comet orbits suggests they are not from interstellar space?
12. Explain the nature of the zodiacal light and the counterglow. When are the most favorable times for viewing the zodiacal light? Explain.
13. Halley's comet will appear about 1986. At the present time: (a) near what planetary orbit is it the closest; (b) how far is it from the earth; (c) is it visible with a telescope? Why?
14. What source has been proposed for comets? Explain what three basic types of paths may be followed by a comet.
15. What bodies not native to the earth, other than lunar rocks, are available for study and analysis? What can be learned from such bodies?
16. Cite one evidence that meteors come from within the solar system.
17. Explain why a meteor appears as a bright streak of light.
18. What is a meteor shower? What is meant by the radiant of a meteor shower? Name three of the most prominent annual meteor showers and state when their maximum is visible.
19. What is the composition of meteorites?
20. Why do astronomers believe the encounter theory is an unsatisfactory explanation for the origin of the solar system?
21. Describe the history of the earth from its earliest beginnings to the development of its atmosphere.

CHAPTER 10
THE SUN: EARTH'S NEAREST STAR

From its prominent position in the sky, the sun has exerted a hypnotic influence on human consciousness. Early in history, people recognized that the yellow disk seen moving across the sky every day was the key to existence on earth. It provided the heat that allowed the crops to grow; it dispelled the foreboding darkness of the night and the fears that lurked in the shadows. The sun had a tremendous impact on the psyche of humanity—the human imagination began to associate the concepts of creator, heaven, and life with this body.

Until the emergence of science, the sun was worshipped as a god, a powerful deity that humans could little afford to ignore. The Egyptians worshipped the sun in many forms; one such representation depicted it as an egg that was laid each morning by the sky-goose. The Greeks, who called the sun god Helios, believed that he rose from the swamps of Ethiopia every day, traveled across the sky in a chariot, and set in the land of the Hesperides. The Aztecs sacrified captives to their sun god, Tezcatlipoca, in the hope that they would keep the sun from collapsing.

Although we have eliminated myths from our ideas of the sun, we recognize that it contains over 99% of the solar system's mass and dominates everything that comes within its influence. The very presence of the solar system would be impossible without the binding grip of the sun's gravitational force. Life on earth is dependent on the sun—eliminate it and life would cease to exist in a very short period of time. From a physical perspective, worship of the sun by the ancients was not as naive as one might believe.

10.1 PHYSICAL PROPERTIES

Let us take a closer look at the dominant body of the solar system. At 92,960,000 miles, the sun is the earth's nearest star. The next nearest star (Proxima Centauri) is over one-quarter million times farther away. These distances can be better understood when one realizes that the light from the sun takes only 8 minutes to reach the earth, while the light from Proxima Centauri reaches the earth in about 4.3 years. The sun is the only star close enough to the earth to allow its features to be observed and studied in detail (Fig. 10.1). The knowledge of its structure and behavior enables us to better understand the stars in general.

By permission of Johnny Hart and Field Enterprises, Inc.

The sun is a gaseous sphere whose visible surface has a linear diameter of about 864,400 miles (1,391,100 km) and an angular diameter of just under 32 minutes. Its shape, as far as we can tell, is spherical. The sun's mass, which is 2.2×10^{27} tons (1.99×10^{33} gm) is about 300,000 times greater than the earth's. Its volume is over one million times greater than the earth's; therefore, its average density of 1.41 grams per cubic centimeter is about one-fourth that of the earth.

FIG. 10.1
The world's largest solar telescope (60-inch diameter) is located at the Kitt Peak National Observatory near Tucson, Arizona. The heliostat is mounted at the top of the vertical tower. The optical tunnel is slanted to permit the instrument to be oriented to the north celestial pole in order to facilitate the daily tracking of the sun. The major portion of the instrument is located in the subterranean chamber under the telescope. (Courtesy of Kitt Peak National Observatory.)

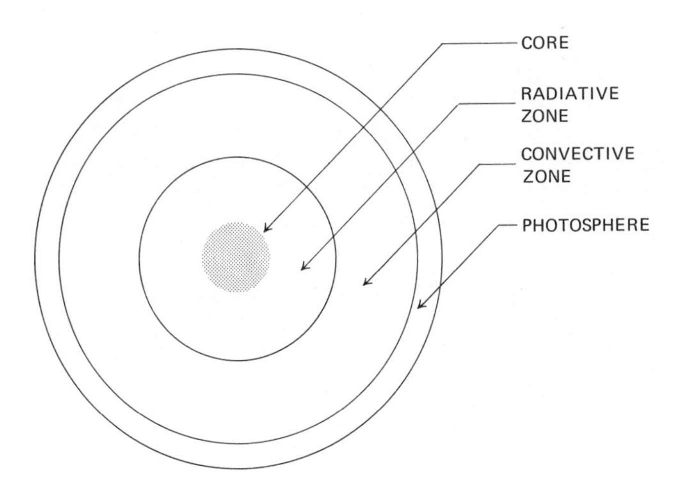

FIG. 10.2
The sun's interior. The different layers are not shown to scale. The sun is gaseous throughout its interior.

The movement of the solar surface features reveals that the sun rotates from west to east, in the same direction as the earth's rotation, and that its equator is inclined about 7° to the plane of the ecliptic. A careful observation of these features reveals that the sun does not rotate as a solid body. The sunspots near the equator move faster than those in the higher latitudes. Since sunspots do not normally appear in latitudes greater than about 45° or near the equator, the rotational period for these latitudes is determined from the Doppler shift in the sun's spectrum. The spectra of the approaching and receding edges are taken simultaneously. The spectral lines of the approaching edge are shifted toward the violet and those of the receding edge toward the red. The apparent rotational velocity at any latitude is proportional to one-half the difference between the spectral shifts at that latitude. The actual rotational period is determined by dividing the circumference of the sun at that particular latitude by the apparent rotational velocity. The sun's rotational period is about 25 days at the equator, 27 days at 35° latitude, 33 days at 75° latitude, and about 35 days near the poles.

The energy that the sun emits comes from thermonuclear reactions (Chapter 13.5) that occur in its *core* (Fig. 10.2). Energy flows outward to the sun's surface within the *radiative zone* by radiation and within the *convective zone*, which is close to the sun's surface, by convection. The earth intercepts an extremely small amount of the total energy emitted by the sun. On one square meter of surface that is perpendicular to the solar beam, the earth receives about 1.36 kilowatts. This is called the *solar constant*.

By Wien's law the sun's surface temperature is about 6164°K; by Stefan's law, about 6050°K. These values differ because they are based on different characteristics of the energy distribution curves. If the sun were a perfect radiator, the effective temperature of its surface, which is the average temperature of the entire disk, would be 5750°K. As one moves inward, the temperature of the sun's interior rises rapidly to an estimated value of 15,000,000°K at the center.

10.2 COMPOSITION

Solar radiation produces an absorption spectrum characterized by dark lines superimposed on a continuous spectrum. The presence of absorption lines shows that the temperature rises steadily with increasing depth below the solar surface. The hotter and denser gases in the interior produce the continuous spectrum. The

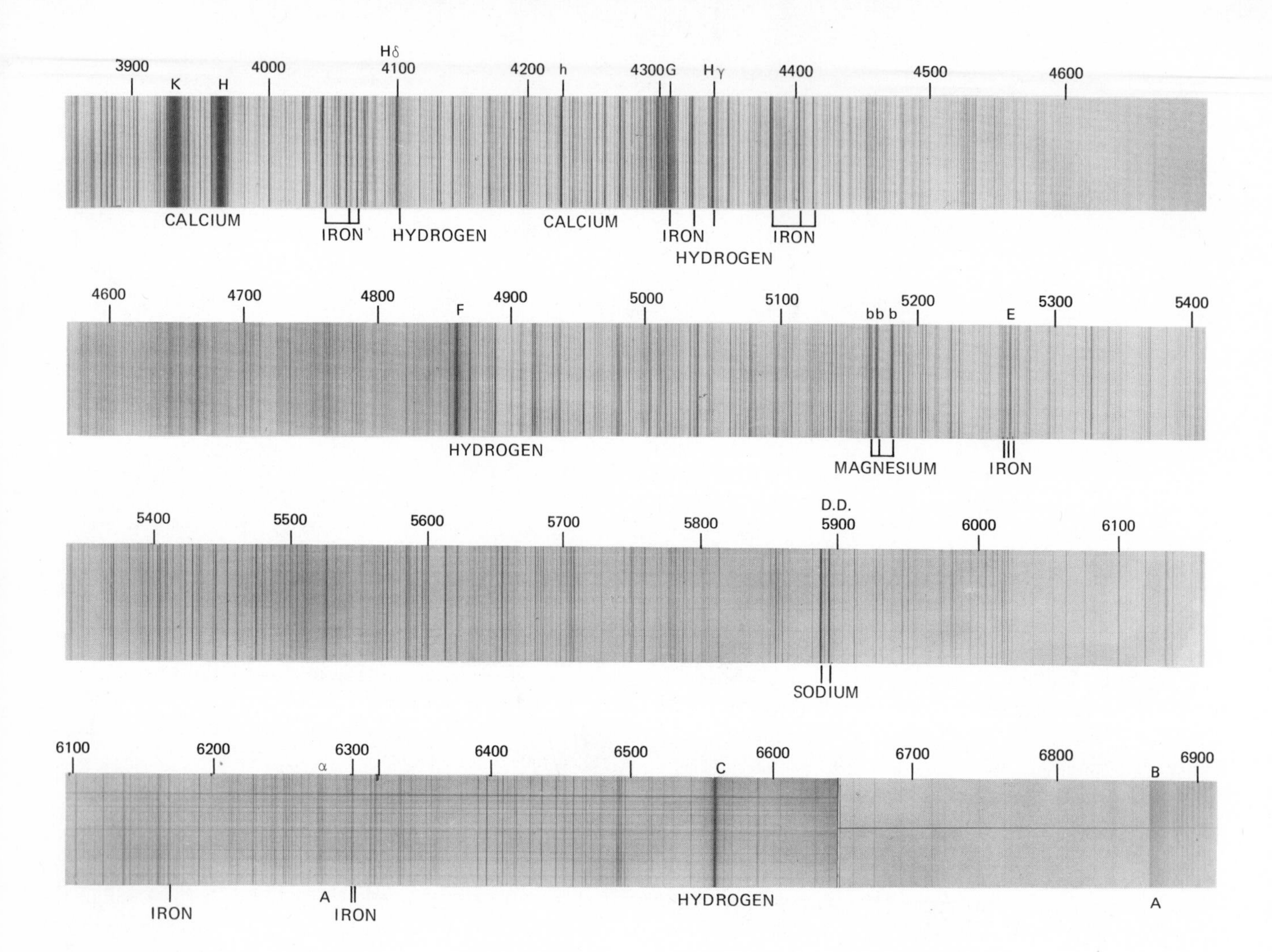

gradual decrease in temperature toward the surface allows absorption lines to be produced in the cooler outer layers (Chapter 3.10).

The composition of the sun is determined by comparing its spectrum with the spectra of the known elements (Fig. 10.3). Although more than 60 elements have been identified in the solar spectrum, it is believed that all the known elements are present, since there are many absorption lines that have not as yet been identified, and the quantities of some elements are insufficient to produce absorption lines. The two dominant elements

FIG. 10.3
The solar spectrum from 3900 Å to 6900 Å, taken with the 13-foot spectroheliograph. Only a few of the lines are identified. (Courtesy of the Hale Observatories.)

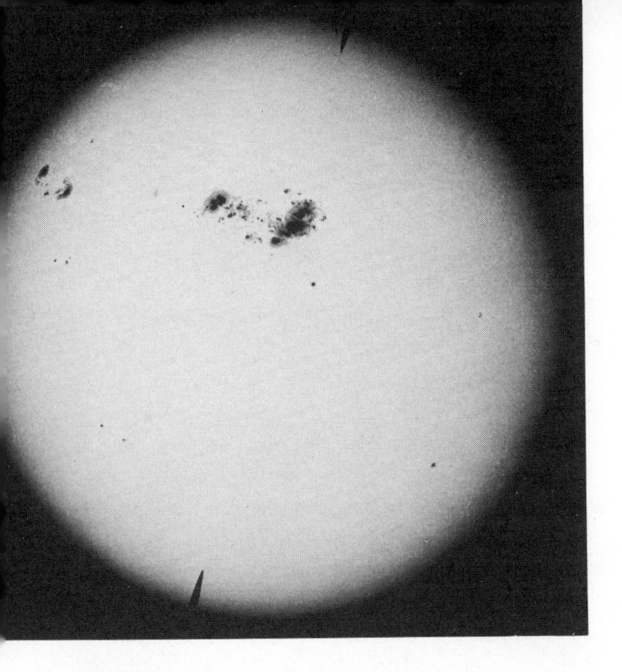

FIG. 10.4
The darkening of the sun's limb. Light at the sun's limb comes from higher levels, where it is cooler; light at the sun's center comes from a much greater depth, where it is considerably hotter. (Courtesy of the Hale Observatories.)

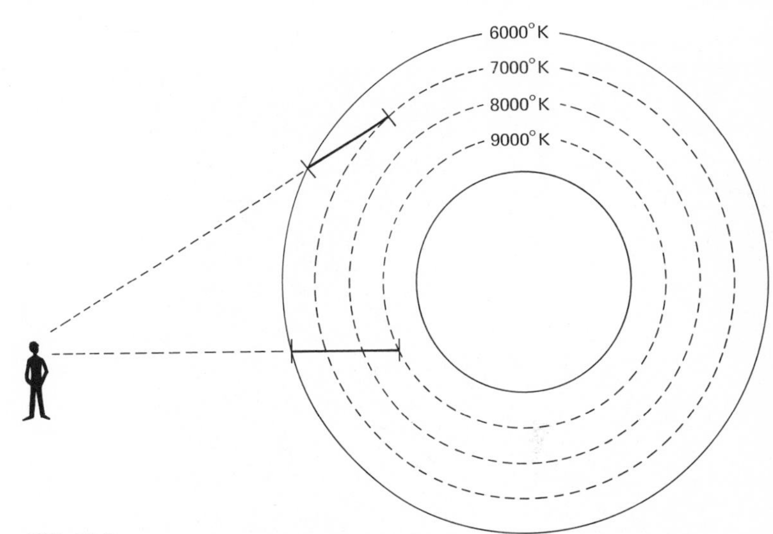

FIG. 10.5
An observer looking in through the hot gases to a distance of 10,000 km (for example) sees to a depth where the temperature is 7000°K at the limb, but to a depth where the temperature is 9000°K near the center. The scale is exaggerated.

in the sun's interior and atmosphere are hydrogen and helium, which are also the dominant elements in the universe. The sun is sufficiently hot that most chemical compounds are broken down. However, a few types of molecules such as titanium oxide and compounds of calcium and magnesium have been identified in the sun's atmosphere.

10.3 THE PHOTOSPHERE

Since the sun's effective temperature is 5750°K, all of its matter is in the gaseous state; therefore, it cannot have a distinct surface. The *photosphere* is the visible surface of the sun. The term "surface" as applied to the sun means the highest layer of its gases visible to the human eye. The photosphere is an envelope of glowing gases about 300 miles (480 km) in depth. It is from this layer that the continuous spectrum is emitted. The opacity of the photosphere is produced mainly by the presence of negative hydrogen ions (hydrogen atoms that have acquired an extra electron). These ions absorb the wavelengths of visible light so effectively that one can see into the sun's atmosphere only to about a depth of 500 miles. Above the photosphere is the sun's upper atmosphere (chromosphere and corona), in which the gases, because of their low density, emit and absorb so little radiation by comparison that the region is transparent.

The photosphere is not uniformly bright. The edge of the sun's disk (the limb) appears darker than the center. This appearance, called the *limb-darkening effect,* occurs because at the limb we see only a short distance below the surface, whereas near the center of the disk we see much deeper to hotter and therefore brighter material (see Figs. 10.4 and 10.5).

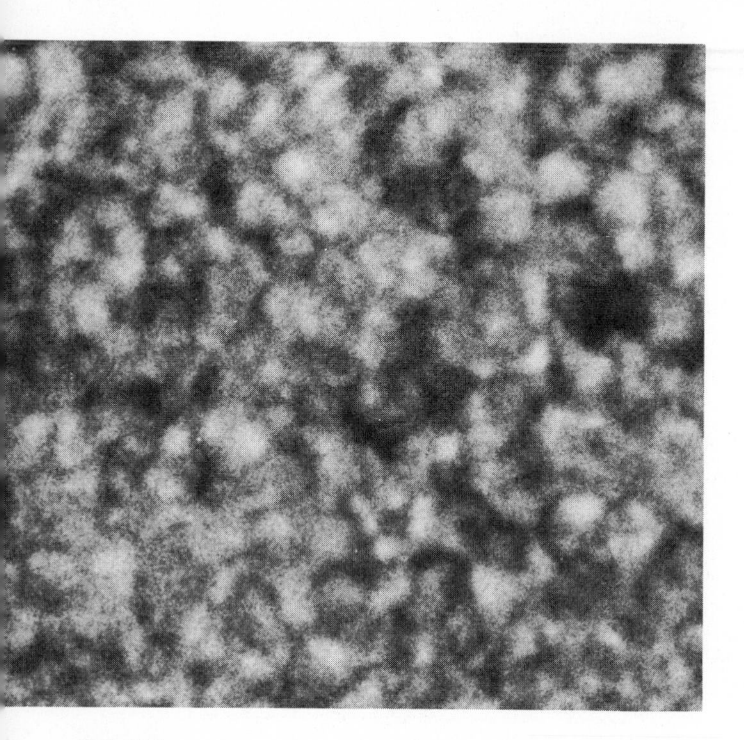

FIG. 10.6
A highly magnified section of the sun's surface, showing the granulation. The size of a small light area (hot rising matter) is about 1000 km across. The small black region at the center may turn into a sunspot. (Photograph from Sacramento Peak Observatory, Association of Universities for Research in Astronomy, Inc.)

When seen through a telescope, the photosphere appears mottled (granulated). (See Fig. 10.6.) Do not attempt to view the sun through a telescope without a special filter. You may burn your eye seriously. The bright, irregular granules, whose diameters range from about 150 to 800 miles (240 to 1300 km), resemble rice grains (Fig. 10.6). These bright patches are 50°–100°K hotter than the photosphere and are produced by the gases that rise convectively from below. The average granule remains visible for a few minutes until it cools to the temperature of the surrounding photosphere. The process which produces the convective motions of the gases takes place in the region directly below the photosphere, in a layer called the *hydrogen convective zone.*

10.4 THE CHROMOSPHERE

The chromosphere is a layer of highly structured, nearly transparent gases whose height extends to about 8000 miles (13,000 km). The density decreases rapidly, and the temperature increases slowly outward in the lower part of the atmosphere. The opposite occurs in the upper part—the density decreases slowly, and the temperature increases very rapidly. The hydrogen in the lower part is nearly neutral, whereas in the upper part it is nearly ionized.

During a total solar eclipse, the chromosphere becomes visible for a very brief moment. When the moon completely covers the sun's photosphere, the chromosphere appears as a red, irregular fringe of light around the moon's disk. When this occurs, the absorption spectrum of the photosphere changes abruptly to a *brightline emission spectrum* (Fig. 10.7), or *flash spectrum.* Each crescent in the flash spectrum represents the radiation in one wavelength (spectral line). The thin visible segment of the sun's limb acts like the slit of a spectroscope. Normally spectroscopes have a straight slit so the spectral lines produced are straight. A curved slit produces curved lines. The brightest and strongest lines are those of hydrogen, calcium, and helium—the elements that produce these lines emit strongly at the greater heights above the photosphere. The hydrogen (Hα) produces the characteristic red color of the chromosphere.

A permanent feature of the chromosphere is the presence of many fine, hair-like, luminous threads (*spicules*) about 325 miles (523 km) in diameter and 8000 miles (13,000 km) in height which are visible in the lower part of the chromosphere. Each spicule lasts for several minutes; however, since there are so many of them, they are always visible near the sun's limb. They also appear to rise vertically above the level of the photosphere at speeds of about 12 miles (19 km) per second.

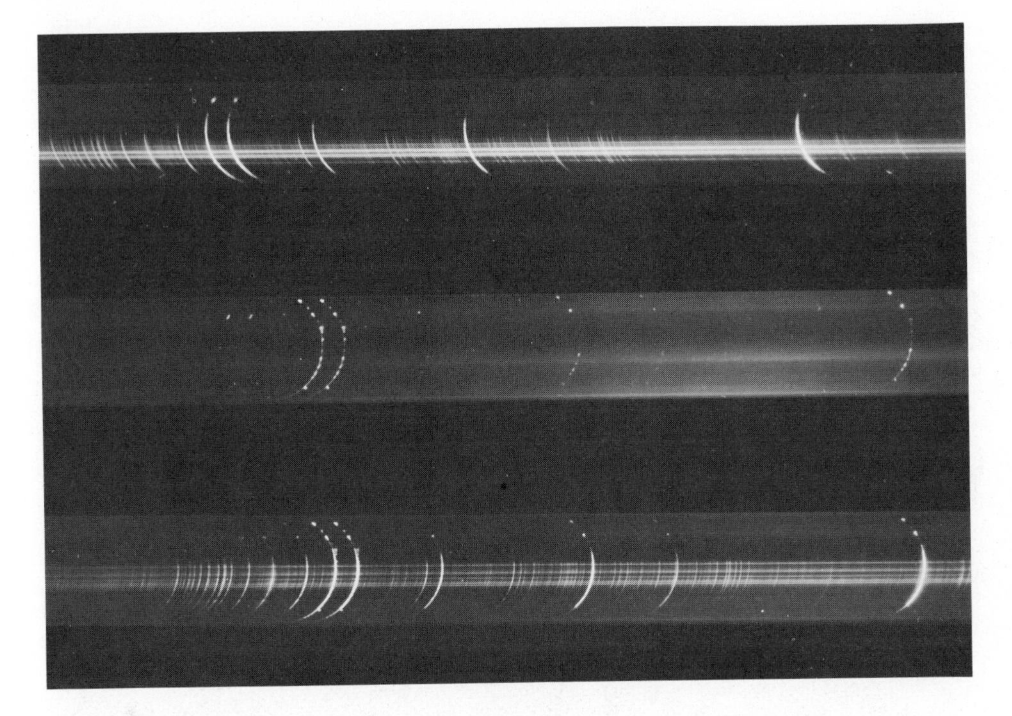

FIG. 10.7
The flash spectrum of the sun's chromosphere, taken during the total eclipse of 24 January 1925, Middletown, Connecticut. (Courtesy of the Hale Observatories.)

10.5 THE CORONA

The corona, the tenuous outer layer of the sun's atmosphere, is visible during a total solar eclipse as a faint, pearly halo whose brightness is comparable to that of the full moon. Since the invention of the *coronograph* (an instrument which produces an artificial solar eclipse), the bright inner corona can be observed and studied whenever the sun is visible and the earth's atmosphere is sufficiently transparent. Currently the best means for studying the upper solar atmosphere is from spacecraft. Experiments conducted from Skylab, for example, have dramatically improved our knowledge of the corona.

The corona is divided into inner and outer parts (Fig. 10.8). The inner corona displays a *continuous spectrum*, which results from the the scattering of the radiation from the photosphere by the free electrons in the corona. Throughout the continuous spectrum there are several bright emission lines whose wavelengths do not coincide with the Fraunhofer lines. For years, astronomers believed that they were the spectral lines of a new element, *coronium*. In 1941 the Swedish physicist B. Edlén resolved the problem when he discovered that the emission lines were produced by atoms of perfectly ordinary elements that were so highly ionized that it had not been possible to observe them in terrestial laboratories. Ionization is produced by the impact of electrons moving at high speeds.

To account for the high degree of ionization, the temperature of the corona has been estimated at 1,500,000°K. (This is a kinetic rather than a radiation temperature.) The coronal green line at 5303 Å is produced by the ionized iron atom with 13 of its electrons missing, and the coronal red line at 6374 Å is produced by

FIG. 10.8
Composite photograph of the sun's inner and outer coronas.
(Courtesy of NASA/Ames Research Center.)

the ionized iron atom with 9 of its electrons missing. Other spectral lines have been identified as having been produced by ionized atoms of argon, calcium, and nickel. The presence of the coronal emission lines indicates that the corona's density is extremely low. Even with an extremely high temperature, the corona thus contains very little heat. The coronal gases are so rarefied that, despite the enormous temperature of the outer solar atmosphere, we are not incinerated.

The outer corona displays a continuous spectrum with many absorption lines, which is similar to the solar spectrum. The visible part of the outer corona extends to a height of about 1,000,000 miles (1.6 million km) above the chromosphere.

The structure of the corona varies with sunspot activity. During sunspot maximum it is more uniform and concentric around the sun's disk, whereas during sunspot minimum it shows streamers extending to great heights above the solar equator.

The high temperature of the corona and chromosphere is thought to be produced by waves generated in the photosphere, possibly by the granulation. These waves travel outward through the ever more rarefied atmosphere, accelerating as they rise. Eventually they are traveling faster than the gas atoms and produce what are called "shock waves" (a familiar example of a shock wave is a sonic boom). Because they move faster than sound, shock waves can compress the gas through which they are traveling and heat it. It is possible that the high temperatures of the solar corona are maintained by this process.

10.6 SOLAR VIBRATIONS

We have seen in earlier chapters how the study of seismic waves traveling through the interior of a planet reveals information about the interior of the planet that is not otherwise obtainable. The same is true of the sun. While the sun does not pulsate dramatically as many stars do (Chapter 14), some astronomers believe it quivers. On the smallest scale, rising and sinking motions are clearly discernible in the granulation. On a slightly larger scale, intermittent oscillations can be detected from the Doppler shifts of photospheric spectrum lines. The atmosphere appears to rise and fall with a period of about 5 minutes. In addition there are surface flow patterns, called *supergranulation,* which last for about a day. The solar magnetic field is slightly concentrated at the boundaries between adjacent supergranulation cells. On an even larger scale, the whole sun seems to "vibrate." This motion is not easy to detect and is evident only as a slight brightening of the limb every 52 minutes. It has so far not been possible to construct a complete picture of the solar interior from these many kinds of oscillations, but they will surely be as useful to solar astronomers as earthquakes are to geophysicists.

10.7 THE SPECTROHELIOGRAPH

One of the most important instruments used to study the solar structure and atmosphere is the spectroheliograph, invented independently by George E. Hale and H. Deslandres in 1890. It is an adaptation of the spectrograph (Chapter 4) with two slits that permit the sun to be photographed in the light of a single wavelength. The sun's image is brought to a focus at the first slit by a telescope objective. The light passes through the slit and then through a prism or a grating, which produces the solar spectrum. A mirror then reflects and focuses the spectrum on a second slit, located in front of a stationary photographic plate. Rotating the mirror makes it possible to pass any single spectral line through the second slit. If the image of the sun is allowed to move across the first slit while the photographic plate is moved at the same rate across the second slit, the image of the sun is photographed in the light of the single wavelength which passes through the second slit. These photographs are called *spectroheliograms.* Hydrogen and calcium spectroheliograms are preferred because these two elements produce the strongest lines formed in the higher levels of the sun's atmosphere. Figure 10.9 shows photographs of the sun in white light, in the light of the alpha line of hydrogen, and in the light of ionized calcium. Spectroheliograms in the light of the alpha line of hydrogen and the K line of ionized calcium display regions of bright light called *plages,* and *faculae* (Fig. 10.10). They are asso-

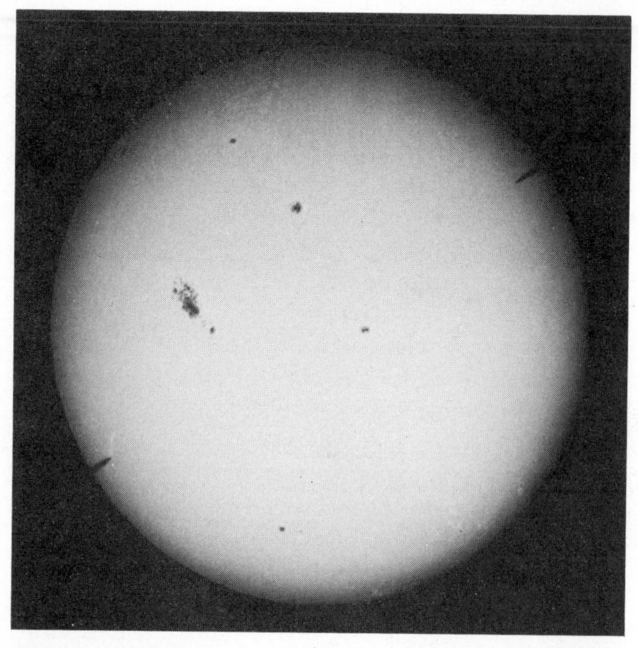
(a)

(b)

(c)

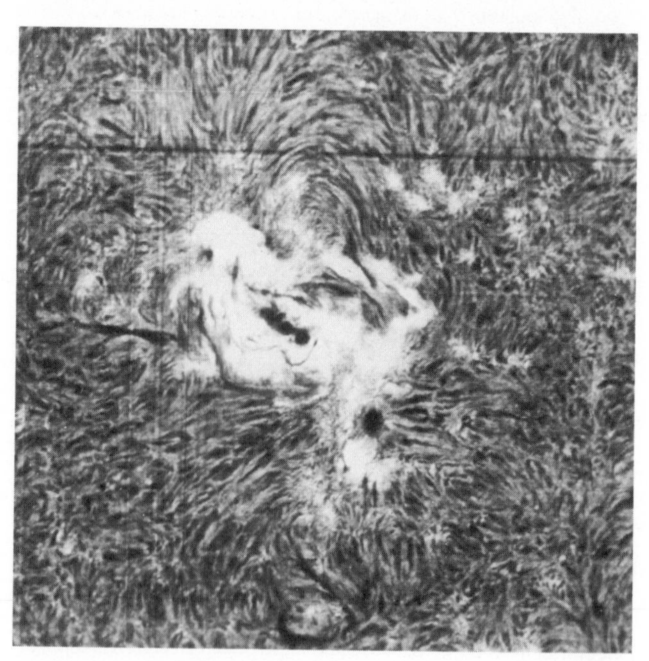
(d)

ciated with sunspots, becoming visible just before the spot appears, and remaining visible for several days after the spot has disappeared. Faculae appear to be regions that are hotter and brighter than the photosphere; when they disappear, they blend slowly and smoothly into the photosphere.

10.8 SUNSPOTS

When the sun is active, the photosphere is marred by the appearance of dark splotches called sunspots. References to naked-eye sunspots are found in ancient Chinese, Japanese, and Greek literature. The first telescopic observation of sunspots was made in 1610 by Galileo, who suggested that they might be clouds in the sun's atmosphere. Astronomers have been investigating these occurrences for over 300 years to determine what causes them and, as yet, no final answer has been found. Herschel believed that they were openings in the bright solar surface which enabled astronomers to see into the sun's dark interior. Some astronomers believed that they were openings caused by the gases' rising from the interior to the surface as the result of explosive action. Others believed that the sunspots were great, whirling fountains of gases flowing toward the center of a low-pressure area. None of these theories appears to fulfill the observed conditions. The present theory is that a spot is produced by a strong magnetic field that inhibits the flow of hot gases to the surface. Deprived of heat from below, the spot cools slightly with respect to its surroundings. Because the spot is cooler, it looks darker.

Before a sunspot becomes visible, the region shows a magnetic field when observed with a magnetograph

FIG. 10.10
Spectroheliogram. A detailed, greatly magnified section of the sun's surface taken in red light of the hydrogen alpha line, which shows the bright regions of the faculae. (Courtesy of NASA/Ames Research Center.)

◄ FIG. 10.9
Four views of the sun: (a) white light (ordinary photograph); (b) hydrogen alpha line; (c) calcium K line (calcium spectroheliogram); (d) enlarged hydrogen spectroheliogram (hydrogen alpha line) showing a sunspot group. (Courtesy of the Hale Observatories.)

(Chapter 10.11). As the magnetic field increases in area, sunspots usually emerge and grow. Doppler studies have revealed that the gases within spots move radially outward at the lower levels and radially inward at the higher levels.

When studied by either direct visual observation or photography in hydrogen light, the gases in a sunspot are seen to move rapidly along curved lines that are similar to those formed by iron filings placed along a magnet. The gases appear to flow away from the spot at the photosphere and toward the spot at the chromosphere level. This phenomenon, named after the Indian astronomer who discovered it while measuring the Doppler displacement of the spectral lines near the edge of the sunspots, is known as the *Evershed effect*.

When first observed, a sunspot appears as a small dark area (pore) with an average diameter of 1000 miles and an average life span of a few hours. It appears dark because its temperature is about 1500°K cooler than the photosphere; however, if it could be seen alone, it would appear brighter than many stars. The pores that persist increase their size and darkness and eventually develop into the umbra of the spot (inner dark core). When the umbra is fully developed, the penumbra (lighter area) appears to surround the umbra. When the sunspot begins to decay, the penumbra makes inroads into the umbra and divides it into smaller spots. This process continues until it completely obliterates the umbra. A spot has a rapid growth and a relatively slow decay. Its average life span is about two weeks. The longest life span recorded was nearly 1½ years.

10.9 SUNSPOT GROUPS

Sunspots usually appear in groups of two or more. When several spots are in a group, there are usually two large spots that dominate the group, called *principal spots*. The first spot of the pair, which one can see emerging around the limb of the sun as it rotates, is called the *leading spot* and is usually the largest spot in the group (Fig. 10.11). The other principal spot is called the *following spot*. Leading and following spot pairs are usually oriented in an east-west direction, whereas the other spots in the group are distributed randomly around the two.

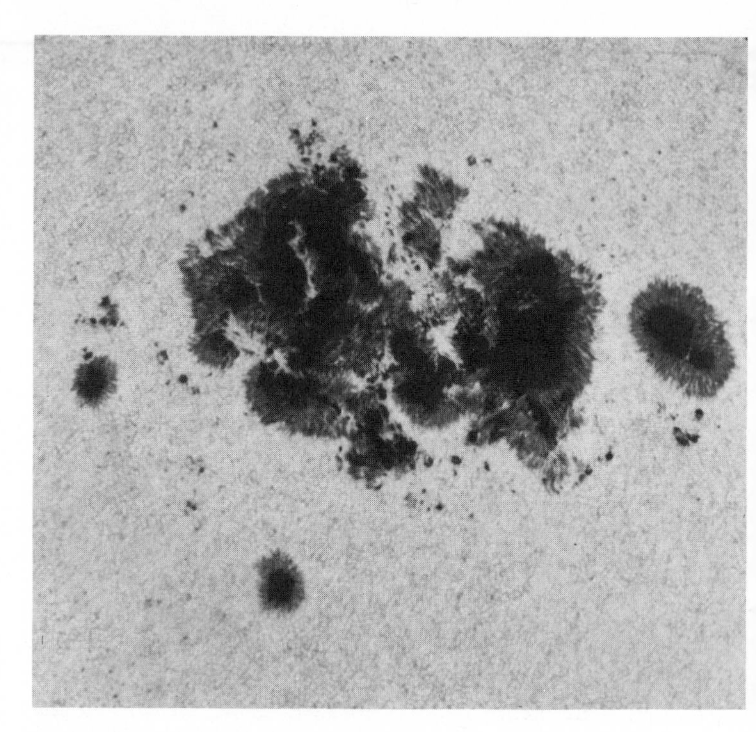

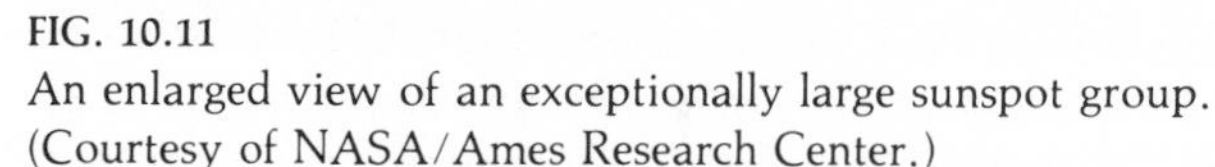

FIG. 10.11
An enlarged view of an exceptionally large sunspot group. (Courtesy of NASA/Ames Research Center.)

10.10 THE SUNSPOT CYCLE

The number of spots visible on the sun's disk varies daily and annually. After nearly 20 years of maintaining a systematic record of the number of spots appearing daily on the sun's disk when it was visible, the German apothecary Heinrich Schwabe in 1843 announced the discovery of a sunspot cycle of approximately 11 years. During the cycle, the number of visible spots increases from 0 to a maximum (more than 100 spots), then decreases to 0. From actual observations, the cycle appears to be irregular—during the past 70 years, it has ranged from about 7.5 to 16 years, with an average of 11 years.

At the beginning of a cycle, several spots appear at approximately 30° north and south latitudes. As the cycle progresses, the old spots die and are replaced by new spots. Their number increases, and their average position shifts toward the equator, reaching a maximum number at approximately 15° north and south latitudes. At sunspot maximum, the number of visible spots may range from about 50 to nearly 200. After the maximum number has been reached, the number begins to decrease, and the spots' positions shift toward the equator, reaching a minimum number or completely disappearing at approximately 8° north and south latitudes. Spots rarely appear near the equator and never between 45° latitude and the poles (Fig. 10.12).

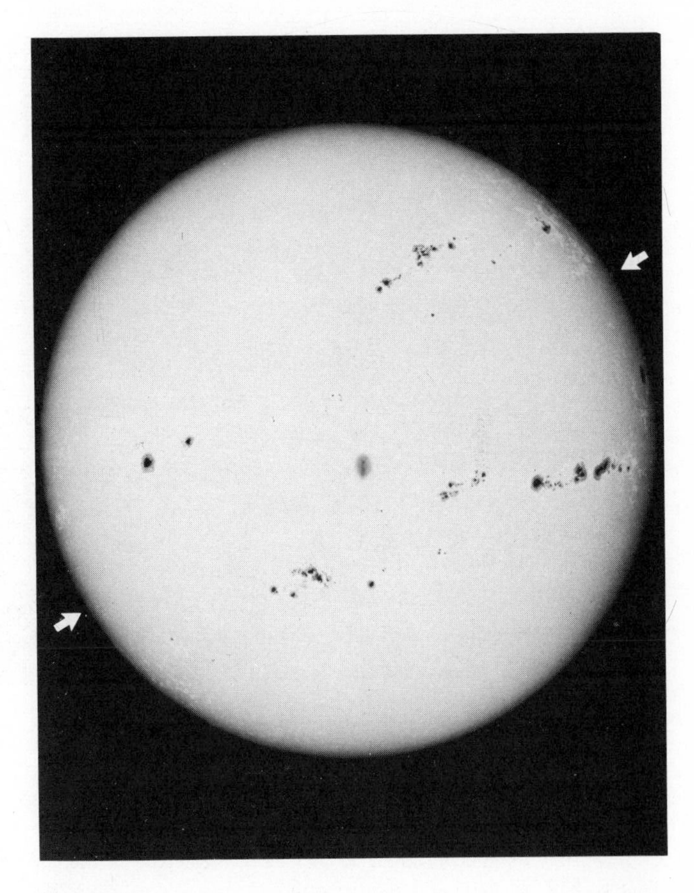

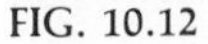

FIG. 10.12
The sun at sunspot maximum, 21 December 1957, showing sunspots, faculae, granular structure, and limb-darkening effect. Note that the spots lie in two bands roughly equidistant from the solar equator (indicated by the arrows). (Courtesy of the Hale Observatories.)

10.11 THE ZEEMAN EFFECT

In 1896 the Dutch physicist P. Zeeman discovered that when a source of light is placed in a magnetic field, its spectral lines are split into two or more components and that the degree of splitting is proportional to the magnitude of the magnetic field. In 1908 the American astronomer George E. Hale discovered the presence of magnetic fields when he observed the Zeeman effect in the spectra of spots. The strength of the magnetic fields of sunspots ranges from about 100 gauss (a unit of magnetic field intensity) for small spots to nearly 4000 gauss for large spots. (The earth's magnetic field has a strength of about 1 gauss.) The spot's magnetic field is oriented perpendicular to the sun's surface in the spot's center and becomes more tangential as the distance from the center increases until it disappears at the edge of the penumbra.

By means of a scanning process that allows the sun's image to move across the slit of a spectrograph, a *magnetogram* is produced (Fig. 10.13). This is a series of traces that show the position, intensity, and polarity of the magnetic fields over the entire solar disk. Spots or spot groups that exhibit the same polarity are called *unipolar groups.* Two spots or magnetic fields that are quite close together and exhibit opposite polarities are called *bipolar groups.* Spots that have opposite polarities irregularly distributed are called *complex groups.* Nearly 90% of the spots are in the bipolar group; less than 1% are in the irregular group.

During a sunspot cycle, the leading spots in each group in one hemisphere are of one polarity. When the

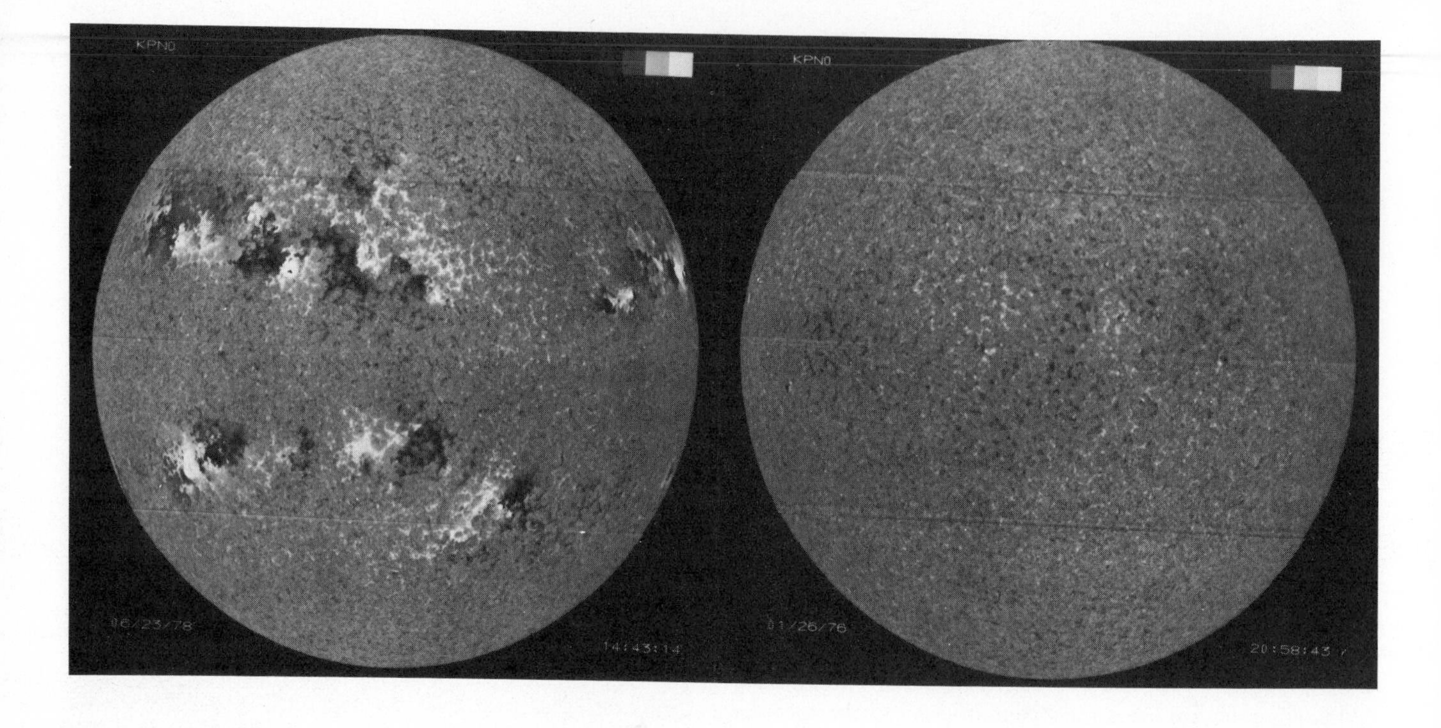

FIG. 10.13
Solar magnetic fields as measured with the Kitt Peak National Observatory magnetogram. The two views show the sun at an active phase (left) and a quiet phase (right). Light and dark areas represent opposite polarities of the field. (Courtesy of the Kitt Peak National Observatory.)

leading spots in each group in the northern hemisphere are north poles, their following spots are usually south poles. At the same time, the leading spots in the southern hemisphere are south poles, and their following spots are north poles. When a new cycle begins, a complete reversal in the polarities of the spots occurs in both hemispheres. Based on the changes in the polarity, the sunspot cycle becomes 22 years—twice the length of the cycle based on the maximum-minimum number of spots.

10.12 SOLAR PROMINENCES

Photographs taken in monochromatic light of the solar disk show long, dark, thread-like filaments around the faculae and sunspot groups. When they are seen on the solar limb, they are called *prominences* and display a wide variety of forms (streamers, loops, arches, and hedges) and activity (quiescent and eruptive).

Prominences are clouds of hot gas that are actually *cooler* than their surroundings. Recall that the corona has a temperature of about a million degrees but a very low density. For the internal and external pressures to balance, the cooler (20,000°C) prominence must be denser. Since the amount of light emitted by a hot gas increases sharply at higher densities, the prominence is brighter even though it is cooler.

The *eruptive prominences* occur around active sunspot groups during the early period of the growth of the spots (Fig. 10.14). They are characterized by hot, lumi-

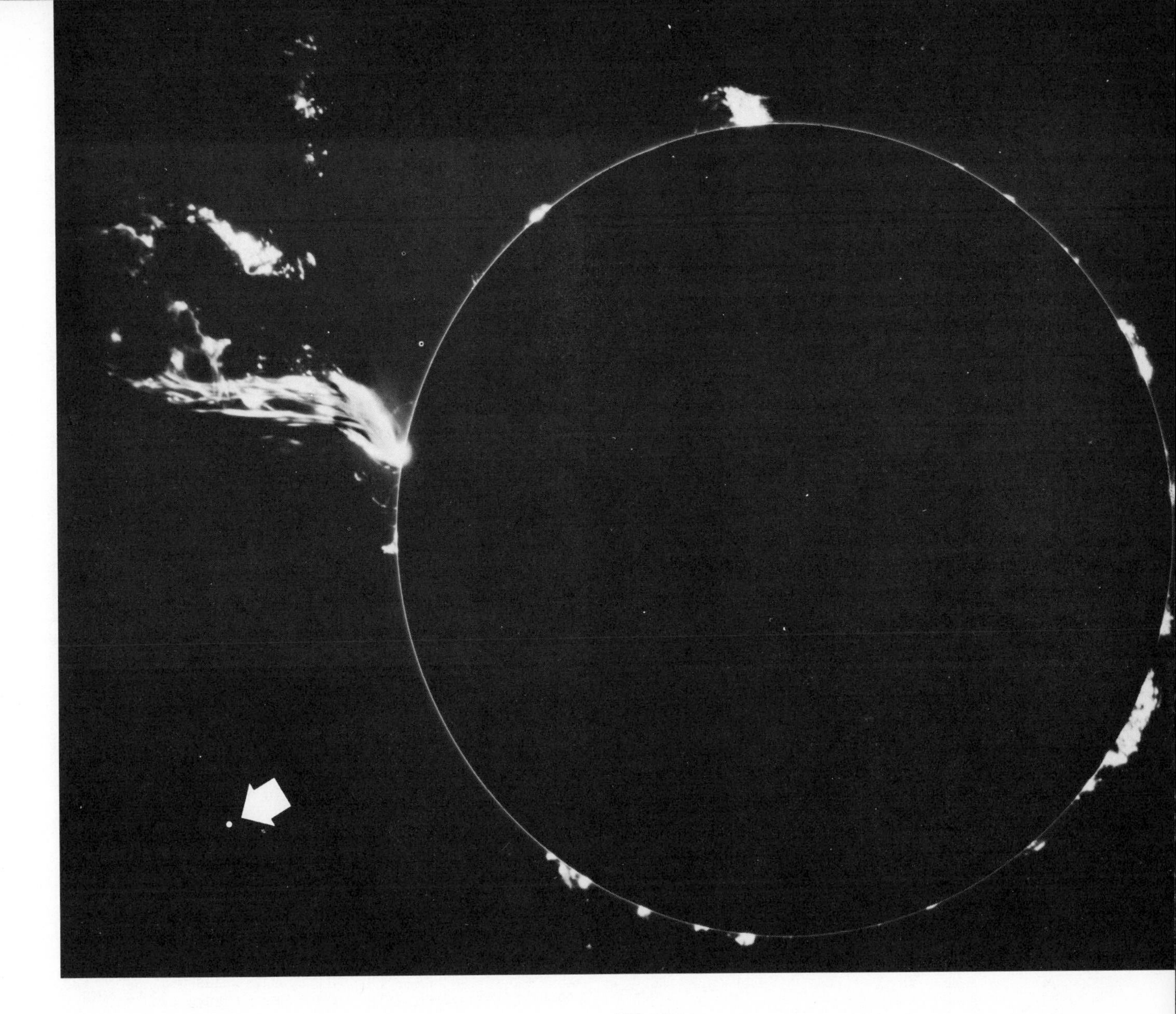

FIG. 10.14
The spectacular eruptive prominence of 1 March 1969, which appeared on the sun's western limb, was photographed from the University of Hawaii's Mount Haleakala Observatory. Its maximum visible height above the solar surface was nearly 375,000 miles. The prominence was associated with a solar flare. Many lesser prominences are also visible along the limb. The dot in lower left corner represents the size of the earth. (Courtesy of the Institute of Astronomy, University of Hawaii.)

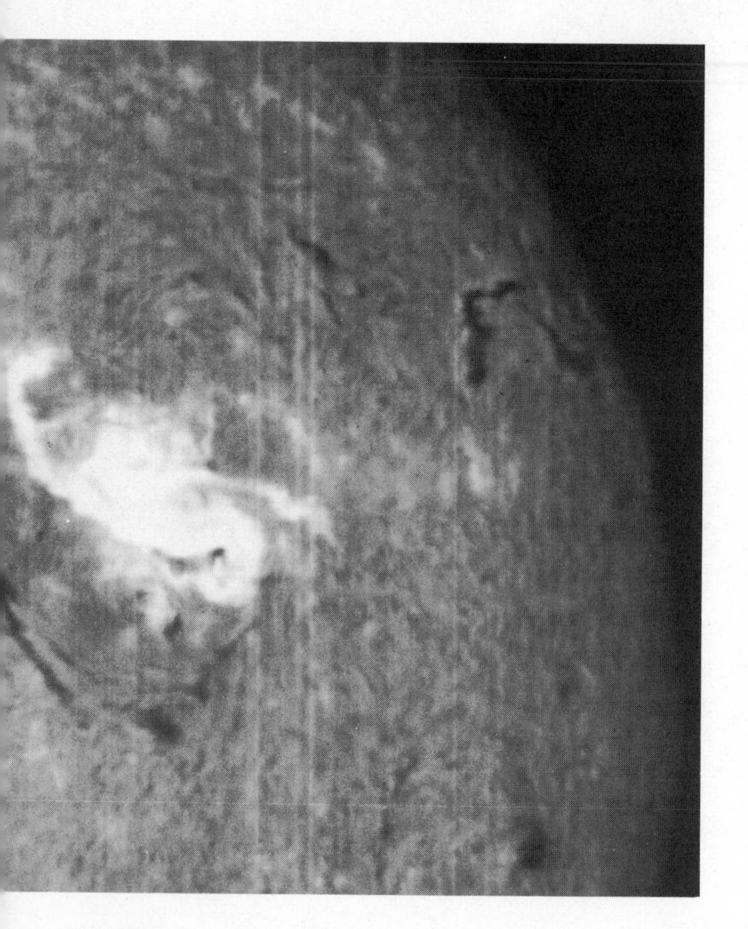

FIG. 10.15
A solar flare photographed in red light of the hydrogen alpha line, 16 July 1959. Bright faculae and dark filaments are also visible. (Courtesy of the Hale Observatories.)

nous sheets of gases that are elevated above the photosphere and the chromosphere to average heights of 50,000 miles (80,000 km). At first it was believed that the motion of the gases in the prominences was generally upward from the photosphere; however, motion picture photography has revealed that in most prominences the motion is downward.

Quiescent prominences, found around faculae and sunspot groups about one month after the first spot has appeared, are characterized by hot, luminous gases that seem to remain motionless above the solar surface, i.e., their general shapes remain the same for several hours to several months before they start to dissipate. Although they appear to be motionless, they are actually highly turbulent. Sometimes apparently placid prominences suddenly erupt with such force that part of the gas escapes from the sun (Fig. 10.14).

Coronal prominences originate high in the corona above centers of activity where the material is hotter and denser than the rest of the corona. The material appears to condense around these regions. When a sufficient amount has condensed and the magnetic field in the sun's atmosphere can no longer support it, the material starts to rain downward on the chromosphere. It is believed that the changes in the prominences result from the abrupt changes that occur in the sun's magnetic fields. The process by which this is accomplished is not yet understood.

10.13 SOLAR FLARES

Flares are outbursts of unusually high-intensity light that occur near large, active, complex sunspot groups. They are best seen in hydrogen and calcium light (Fig. 10.15); however, when the bursts are very intense, they are visible in white light. Their light curves show that they rise rapidly to maximum intensity, remain there for a short period of time, then drop slowly, fading away. The duration of the flares varies from several minutes to several hours. At sunspot maximum, the frequency of the flares reaches a maximum of about four each day.

Flares involve more energy than does any other solar activity. Since they are locally about 10 times brighter in

Hα than the average chromosphere, flares are the brightest objects visible on the solar surface. Flares eject clouds of high-energy particles into space that disturb the earth's atmosphere, produce aurora borealis, and interfere with short-wave radio transmission. Although the cause of flares has not been determined with certainty, it is believed that they are produced when the magnetic field associated with sunspots becomes highly twisted. The field seems to readjust itself like a rubber band which, when stretched too tightly, snaps. The restructuring of the magnetic field presumably causes intense local heating.

10.14 THE SOLAR WIND

The tenuous, high-temperature gases in the corona are not bound to the sun as tightly as cooler surface layers, partly because they are so hot, but also because the gravitational attraction of the sun diminishes with distance from the solar center. As a result, there is a steady escape of gas from the sun into space. This outflow of the upper solar atmosphere is called the solar wind. The solar wind steadily accelerates as it moves away from the sun and is moving at about 250 miles per second (400 km/sec) when it crosses the earth's orbit. The density of the gas in the solar wind is very low, but as we saw in Chapter 9.13, it is sufficient to affect the matter in comets. The solar wind appears to extend out to at least the orbit of Jupiter and perhaps further before it is slowed down by the interstellar gases that surround the solar system.

The solar wind seems to originate in a set of large cavities in the corona where the magnetic field opens radially into space. These coronal holes, discovered only recently by the Skylab and orbiting solar observatories, are still not completely understood. The outflow of solar matter is very small, amounting to less than a trillionth of a solar mass per year lost into space. However, the solar wind carries magnetic fields and clouds of charged particles outward through the solar system so they occasionally impinge upon the earth. Upper atmospheric disturbances result, and there is a growing body of evidence that there may be some effect on terrestrial weather. Clearly this is an area which will be very important to study in the future.

10.15 COMMUNICATION DISTURBANCES

With the appearance of a brilliant solar flare within a large sunspot group, excessive amounts of ultraviolet, x-ray, and radio radiation and streams of highly charged particles are emitted. The electromagnetic radiation travels at the speed of light and reaches the earth in about eight minutes; the charged particles travel at about 600 miles (1000 km) per second and reach the earth in about one day. The ultraviolet electromagnetic radiation disrupts the ionized layers of the earth's upper atmosphere and causes radio fadeouts that often last for several hours. The charged particles produce magnetic storms in the ionosphere and induce electrical currents within the earth, which create noise in long-range radio transmission and disrupt telegraph and telephone communications.

10.16 THE AURORAS

Streams of highly charged particles produce the aurora borealis and the aurora australis (the northern and southern lights), which are visible within a region extending about 40° from the north and south magnetic poles. Maximum auroral activity occurs within a region extending about 20° from the magnetic poles at altitudes of 60 miles (100 km). More and brighter auroras appear after sunspot maximum. The auroras appear just above the horizon in the early evening as one or more arches of soft red, green, and violet lights accompanied by streamers, which are visible to a height of about 40° above the horizon (Fig. 10.16). Their position and brightness appear to be in a constant state of flux.

The light emitted by the aurora results from the collision of incoming particles (associated perhaps with a solar flare) with the nitrogen and oxygen atoms in the earth's upper atmosphere. The collision of a solar particle with an atom in the terrestrial atmosphere lifts (excites) an electron to an upper energy level in the atom. The return of the electron to its original orbit is accompanied by the emission of light (recall Chapter 3.11). Transitions between different levels are responsible for producing the different auroral colors.

The shape of the aurora, whether it is in the form of a drapery or an arch, is determined by the magnetic field

FIG. 10.16
A view of the aurora borealis as seen from Alaska. (Courtesy of Gustav Lamprecht, College, Alaska.)

of the earth. Since the incoming particles are charged, they tend to be guided along the magnetic field. Changes in the point of injection then cause different parts of the sky to be illuminated.

10.17 THE SOLAR CYCLE

We saw in Chapter 10.10 that the number of sunspots varies from year to year, reaching a maximum roughly every 11 years. Many of the other solar phenomena that have been discussed (such as prominences, flares, and related terrestrial events like aurora) are also found to vary in their frequency of occurrence with the same 11-year periodicity. This is not surprising in view of the close relationship between sunspots and the other solar surface phenomena. All these solar events are related to the underlying magnetic field of the sun. Although the exact mechanism is not understood, the solar magnetic field undergoes changes of strength and polarity. The strength (in terms of numbers of spots) changes with a roughly 11-year cycle, while the polarity changes with a 22-year cycle. Maximum solar activity has occurred in 1947, 1958, 1969, and is due to reach maximum again in 1980. Apart from the fact that in years of maximum solar activity there is an increase in sunspot numbers and auroral activity, there is growing evidence that the solar cycle may influence terrestrial climate and weather. Preliminary analysis of drought data suggests that dry spells may recur with a 22-year cycle. There is also fairly strong evidence that during the late seventeenth century, which was a period of abnormally cold winters and wet summers in the northern temperate zone, the solar cycle seems to have turned off completely.

Two possible ways that solar activity might affect the earth's weather are (1) the solar wind carries magnetic fields and particles across space to the earth, and (2) solar flares emit large amounts of ultraviolet radiation. Perhaps these solar phenomena are able to influence the ozone layer, which in turn influences the temperature and hence the circulation patterns of the upper atmosphere. The exact explanation is still a mystery, but clearly the effects of solar activity pose a puzzle it would be advantageous for us to unravel.

10.18 THE AIRGLOW

Even when the sun is quiet—when there are no centers of activity visible—it is continuously emitting charged particles which produce a permanent "airglow" over the entire sky. The airglow provides twice as much light as the stars, although both serve as sources of night-sky illumination when the moon is not visible. The spectrum of the airglow shows that its light comes mainly from oxygen and sodium atoms and hydroxyl (OH) molecules.

SUMMARY

The sun is the star nearest to us and dominates the solar system by illuminating and warming the planets. Its structure may be divided into four zones: a core, where energy is generated; a radiative and a convective zone, which carry energy outward; and the photosphere, the visible surface. The sun's atmosphere can also be conveniently divided into a thin lower layer (the chromosphere) and an extremely hot but tenuous outer atmosphere (the corona). The latter two regions cannot be seen easily except during a total solar eclipse.

The sun is not a quiescent sphere of gas. Convective motions disturb the surface layers, producing granulation and larger-scale flows. A general outflow of matter (the solar wind) continuously causes gas to stream into space. The sun also is magnetically active. Sunspots, flares, prominences, and plages all appear to be associated with magnetic effects. Their numbers fluctuate with a period of roughly 11 years.

REVIEW QUESTIONS

1. What is the composition of the sun? Is it solid, liquid, or gas?
2. What is limb darkening? Why does it occur?
3. Draw a cross-sectional view of the sun's interior. Label the zones and indicate their thickness, temperature, and dominant method of energy transfer.
4. What is the chromosphere? the corona? Are these zones hotter or cooler than the photosphere?
5. Do all parts of the sun rotate at the same rate? Explain what this indicates.
6. Discuss the number of sunspots and the shifting of the sunspot zones during the sunspot cycle.
7. Describe the magnetism of sunspot groups in both hemispheres of the sun, from one cycle to another.
8. What are solar prominences? Describe the important types. When and where are they visible?
9. What are solar flares? What terrestrial effects are associated with the appearance of solar flares?
10. Where on the sun do spots usually appear at the start of a sunspot cycle? How do sunspot positions change as the cycle progresses?
11. What is the solar wind?
12. How can solar magnetic fields be detected?
13. How are auroras produced?
14. What is the Zeeman effect? Who discovered it? How is it detected?
15. What is a flash spectrum? What does it reveal about the sun's atmosphere?

CHAPTER 11
THE PROPERTIES OF STARS

Early people regarded stars as fiery objects located at an equal distance from the earth on the underside of a large, inverted dome that was suspended over the earth. They believed that the stars were "fixed" because they maintained their positions with respect to the other stars. Aristotle reasoned that if the earth revolved around the sun, the shifting of the nearer stars in relation to the more distant stars would be visible. Since he was not able to detect such motion, he concluded that the earth was stationary. Although he did realize that if the stars were sufficiently far away he would not be able to detect their shift, Aristotle found it easier to believe the earth was at rest than that the stars were so distant. Later, when the sun-centered Copernican system of the universe was established, efforts to observe the apparent motion of the stars were still unsuccessful; however, by that time beliefs had shifted and it was accepted that the stars were too far away from the earth for such motion to be detected.

11.1 STELLAR PARALLAX

In 1838 the German astronomer Friedrich W. Bessel observed and measured the apparent motion of the star 61 Cygni in relation to the more distant stars. His discovery made it possible to measure the distance to the nearer stars by a simple method, which was further simplified in 1903, when F. Schlesinger added the technique of photography. A star's apparent motion was measured from photographs taken of the area around the star at intervals of six months for at least two years. Although the angular value P represents the annual maximum apparent shift (parallax) in the position of the star (Fig. 11.1), for practical reasons, the parallax of a star is defined as one-half this value: the angle p subtended by 1 a.u. at the distance d of the star. When the star's parallax has been determined, its distance can be found from the proportion: the star's distance is to the radius of the earth's orbit as 206,265″ is to the star's parallax (Fig. 11.2).

The use of the parallax method is limited to the several thousand stars that are near the sun, because the more distant stars produce small displacement angles on photographic plates that are difficult to measure accurately. Some of the other methods that are available for measuring stellar distances of the more distant bodies will be presented later in the text.

By permission of Johnny Hart and Field Enterprises, Inc.

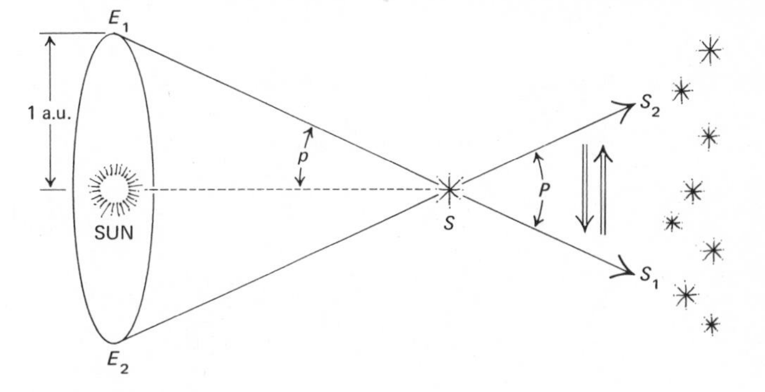

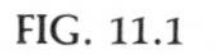

FIG. 11.1
The parallax of a star. As the earth revolves around the sun, the nearer stars appear to move in relation to more distant stars.

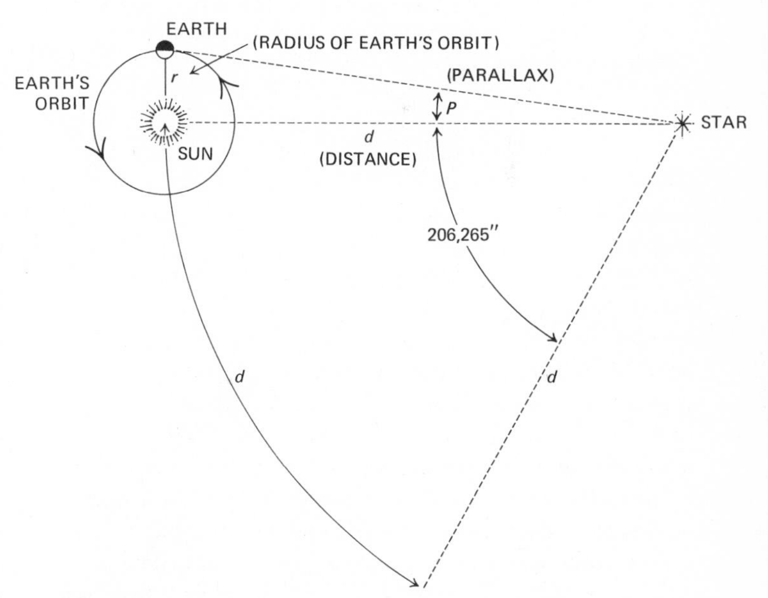

FIG. 11.2
The relationship between stellar distance (d) and stellar parallax (p). When the radius of a circle is laid along the circumference, it subtends an angle of 206,265 seconds. If the star's parallax is known, its distance can be determined from the proportion: star's distance (d) is to the radius of the earth's orbit (r) as 206,265″ is to the star's parallax (p).

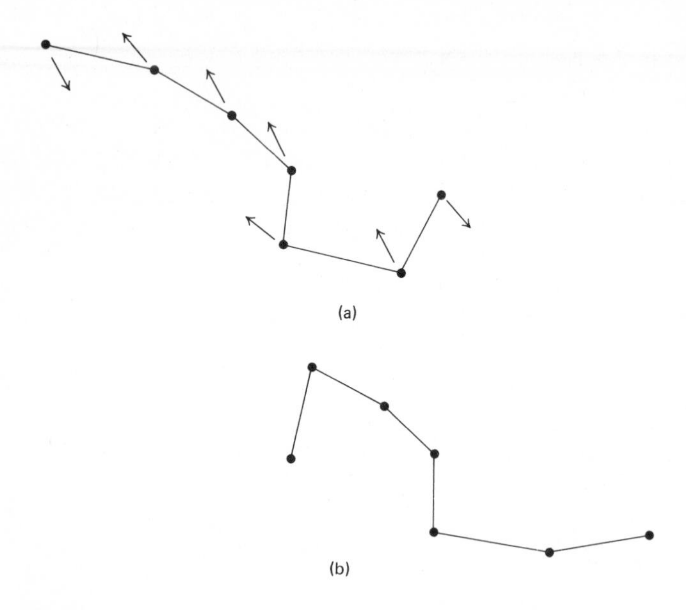

FIG. 11.3
The Big Dipper: (a) as it appears today; (b) as it will appear in 100,000 years. Arrows indicate stellar directions and magnitudes of their proper motions.

11.2 DISTANCE UNITS

Stellar distances become more meaningful when two new distance units are introduced—the *light year* and the *parsec*. The light year is the distance that light travels in one year at about 186,000 miles (300,000 km) per second. The light year is equivalent to nearly 6 million million miles (9.7 million million km). The parsec is the distance of a body when its parallax is one second. The word parsec is derived from the words "*par*allax" and "*sec*onds." Star distances vary inversely as their parallax,

$$d = 1/p.$$

Therefore, a star with a parallax of one second is at a distance of one parsec, whereas a star with a parallax of one-half second is at a distance of two parsecs. Since one parsec equals 3.26 light years, the relationship between distance in light years and parallax in seconds becomes

$$d = 3.26/p \text{ light years.}$$

The star 61 Cygni has a parallax of 0″.293 and therefore its distance from the earth is more than 11 light years. The bright, first-magnitude, double star Alpha Centauri has a parallax of 0″.75, so its distance from the earth is 4.3 light years. The faint eleventh-magnitude star Proxima Centauri has the largest observed parallax, 0″.76 at 4.3 light years, making it the nearest star to the earth after the sun.

11.3 STELLAR MOTIONS

Although stars move at great speeds and in different directions, they appear nearly stationary because of their great distances from the earth. In 1718 Edmund Halley was the first to indicate that stars are in motion when he pointed out that the star Sirius had moved about one-half degree, the apparent width of the full moon, from its position as listed by Ptolemy in his star catalog. The stars in the Big Dipper have shown no apparent motion in nearly 2000 years, but in the next 100,000 years, their motion will be quite obvious (Fig. 11.3).

The apparent position of a star can be determined by its right ascension and declination. This position is altered slightly by precession and other motions of the earth. Allowing for these changes, the change in the apparent position of the star, that is, the apparent angular motion across the observer's line of sight, expressed in seconds of arc per year, is called its *proper motion*. "Barnard's star," named after the astronomer who recognized its fast motion, has the largest known proper motion, 10″.25 per year, which is about 1° in 350 years. Proper motion varies inversely as the distance because a given motion becomes harder to detect at larger distances. Thus, a star's distance can be estimated from its proper motion. The value derived by this method is less accurate than that determined by the parallax method, however.

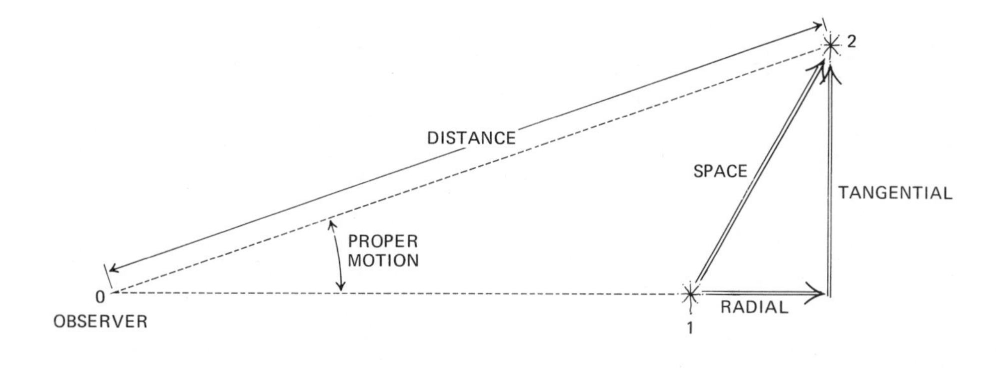

FIG. 11.4
A star's space, radial, and tangential velocities; proper motion; and distance.

A star's motion in space—its motion with respect to the sun—is called *space motion.* Space motion can be resolved into two components: *radial* and *tangential.* We see the tangential space motion as proper motion. The star's velocity with respect to the sun is called space velocity, denoted by $\mathbf{V}_s$. The space velocity can also be resolved into radial and tangential components denoted $\mathbf{V}_r$ and $\mathbf{V}_t$ respectively. In Fig. 11.4, the radial velocity ($\mathbf{V}_r$) is in the direction of the line of sight of the observer and is expressed in miles per second. Its value is obtained from the Doppler shift in the star's spectrum. A shift of the spectral lines toward the violet indicates that the star is moving toward the observer; a shift toward the red indicates that it is moving away from the observer. The amount of the displacement is directly related to and a measure of the star's velocity. The tangential velocity ($\mathbf{V}_t$) is perpendicular to the line of sight and is expressed in miles per second. Its value in astronomical units per year can be determined by dividing the star's proper motion by its parallax. When the radial and tangential velocities of a star are known, its space velocity can be determined by using the Pythagorean theorem in the solution of a right triangle,

$$\mathbf{V}_s^2 = \mathbf{V}_r^2 + \mathbf{V}_t^2.$$

11.4 STARLIGHT MEASUREMENT

The most basic quantity used in the study of the structure and behavior of stars is the light that these bodies emit. Several methods are used to measure the intensity of this light. Since many students own or use telescopes, a simple though imprecise method, based on the light-gathering power of a telescope, will be explained. This method was proposed by William Herschel and uses the principle that the light-gathering power of a telescope is proportional to the area of its objective lens or mirror. The relationship is expressed algebraically by

$$B = \pi r^2,$$

where B is the brightness of the image and r is the radius of the objective. Since the radius is one-half of the diameter ($d/2$), the relation can be expressed:

$$B = \pi r^2$$

$$B = \pi(d/2)^2 = \pi(d^2/4)$$

$$B \sim d^2.$$

Therefore, the light-gathering power of a telescope is proportional to the square of the diameter of its objective. This means that a telescope gathers more light with a large objective than with a small one. For example, let us point a 7-inch and a 5-inch telescope in the same direction. Suppose that in the field of view two stars are visible through the 7-inch—one of them fairly bright, the other just visible. Then suppose that in the field of view of the 5-inch telescope, only the brighter star appears, and that now it is just barely visible. Since the 7-inch gathers about twice as much light as the 5-inch ($7^2=49$, $5^2=25$), we can deduce that the brighter star is twice as bright as the fainter one.

Another, more precise method of judging intensity uses a photographic procedure. Star images are recorded on a photographic negative and their blackness is measured. The degree of blackness of a star's image on a negative increases as the star's brightness increases; therefore, the image of a bright star is blacker than the image of a dim star. The degree of blackness is measured by passing a light beam through the negative with the star image in its center. A bright star image will filter out the light beam more effectively than will a faint star image, and the amount of light filtered is a measure of the star's brightness.

The best and most precise method for measuring the brightness of a star is the photoelectric method, which makes use of a *photomultiplier.* Starlight is brought to a focus at the focal point of a telescope, passed through a small hole in a metal plate, and projected onto the surface of a photomultiplier. The brightness of the star is determined by the current produced in the photomultiplier.

11.5 THE LUMINOSITY AND BRIGHTNESS OF A STAR

The size and temperature of a star determine its *luminosity,* the rate at which the star radiates electromagnetic energy. If two stars are of equal size, the hotter star will radiate the greater energy. If two stars are of equal temperature, the larger one will radiate the greater energy.

The luminosity and distance of a star determine its apparent brightness. If two stars appear to be of equal brightness, the more distant star is the more luminous. The apparent brightness of a star follows the inverse square law, which states that the amount of light reaching an object varies inversely as the square of the distance from the source. A star at a given distance with a brightness of 1 will appear only one-fourth as bright when its distance is doubled, and only one-ninth as bright when its distance is tripled.

11.6 APPARENT-MAGNITUDE SCALE

As one views the evening sky, it is obvious that stars differ considerably in their apparent brightness. In the second century B.C., Hipparchus was the first astronomer to devise a system for identifying stars according to their apparent brightness. He compiled a list of over 1000 stars and assigned numbers from 1 to 6 to indicate their apparent brightness. The smaller the number on the magnitude scale, the brighter the star. First magnitude was assigned to the several brightest stars that were visible, sixth magnitude to those that were just barely visible. Second, third, fourth, and fifth magnitudes were assigned to stars between these two extremes.

In 1856 Norman Pogson proposed the present magnitude system, which is based on the work of Fechner and Herschel. The physiologist Fechner presented what is known as Fechner's law, which states that equal differences in the perception of light by the human eye correspond to equal ratios in the intensity of light. The astronomer Herschel developed the relationship that a first-magnitude star is approximately 100 times brighter than a sixth-magnitude star. Pogson proposed that the ratio of brightness between any two consecutive magnitudes be the fifth root of 100, which is equal to 2.512. This means that the difference in brightness between any two consecutive magnitudes is 2.512. In Table 11.1 the light ratios for the various magnitudes are listed. A first-magnitude star is 2.512 times brighter than a second-magnitude star; $(2.512)^2$, or 6.3, times brighter than a third-magnitude star; $(2.512)^3$, or 15.9, times brighter than a fourth-magnitude star; and $(2.512)^5$, or 100 times, brighter than a sixth-magnitude star. The magnitude of a star is only a measure of its brightness and does not take into consideration its size, temperature, or distance.

The apparent-magnitude scale has been extended at both ends, and its reference point (zero magnitude) has been established by means of several stars whose brightnesses have been determined very accurately. Zero

TABLE 11.1
Light ratios

MAGNITUDE DIFFERENCE	LIGHT RATIO
1	2.512
2	6.3
3	15.8
4	39.8
5	100.0
10	10,000.0
20	100,000,000.0
25	10,000,000,000.0

TABLE 11.2
Apparent magnitudes of familiar celestial bodies

CELESTIAL BODY	MAGNITUDE
Sun	−26.7
Moon (full)	−12.5
Venus at its brightest	− 4.2
Jupiter at its brightest	− 2.5
Mars at its brightest	− 2.0
Sirius	− 1.4
Arcturus	− 0.1
Vega, Capella	0.0
Aldebaran	+ 0.9
Polaris (north star)	+ 2.0

magnitude is 2.512 times brighter than first magnitude, and −1 magnitude is $(2.512)^2$, or 6.3 times brighter than first magnitude. Sirius, one of the original first-magnitude stars, was found to be about 10 times brighter than the average first-magnitude star, or −1.4 magnitude on the extended scale. The sun's magnitude is −26.7.

Optical aids can extend human vision from the unaided-eye limit of +6 magnitude to +23.5. A pair of binoculars brings into view stars of +10 magnitude; the 6-inch telescope used by many amateur astronomers brings into view stars of +13 magnitude; and the 200-inch Hale telescope brings into view stars of +20 magnitude (when used visually) and about +25 magnitude (when used with new high-sensitivity photographic plates). This is the present limit, because the residual night-sky light which comes from starlight, zodiacal light, and other sources is brighter than the light from stars of lesser magnitudes.

Table 11.2 lists the magnitudes of some of the most conspicuous and familiar celestial bodies.

11.7 ABSOLUTE-MAGNITUDE SCALE

Since the brightness of a star depends on its luminosity and distance, the apparent-magnitude scale cannot be used to compute the actual amount of light that stars emit. To accomplish this, the absolute-magnitude scale was introduced. The absolute magnitude of a star is the magnitude that it would have when seen from a standard distance of 10 parsecs (32.6 light years). The absolute magnitude may be either larger or smaller than the star's apparent magnitude because apparent magnitude is partly determined by the star's distance. For example, the sun has an apparent magnitude of −26.7. However, if it were moved from its distance of 1 a.u. ($=5\times10^{-6}$ parsecs) to the standard distance of 10 parsecs, it would appear very much fainter, with an apparent magnitude of 4.56. Note that by definition 4.56 is thus the sun's absolute magnitude.

On the other hand, suppose we consider an intrinsically luminous star such as Rigel, which is approximately 300 parsecs away and has an apparent magnitude of 0.0. Rigel would appear dramatically brighter if it were brought in to the standard distance of 10 parsecs. According to the inverse square law (Chapter 3.2), Rigel's brightness would be increased by a factor of 900 ($=(300/10)^2$). According to Table 11.1, Rigel would appear 7.5 magnitudes brighter at 10 parsecs than it does at 300 parsecs.

There is a mathematical relationship between apparent magnitude, m, absolute magnitude, M, and distance d (in parsecs), which is an extremely important method for measuring distances in astronomy. This relation can be expressed as

$$m=M+5\log(d/10),$$

where "log" is the logarithm to the base 10 of $(d/10)$. It is not hard to derive this relation from the inverse square law and the definition of apparent and absolute magnitude, but we introduce it here only to demonstrate that distances can be found if the apparent and absolute magnitude of a star are known. Distances can also be found without recourse to the formula above by direct application of the inverse square law. As an example, if two stars have the same luminosity (known from the fact that they have identical spectra, for example) but one is nine times fainter than the other, it must be three times further away, since tripling a star's distance makes it appear nine times fainter. Similar proportions can be worked out for other pairs of stars. As a final example, Regulus and the bright stars in the Pleiades have very similar spectra but Regulus appears three magnitudes brighter than the Pleiades stars. From Table 11.1, three magnitudes corresponds to almost a factor of 16 in brightness. A factor of 16 difference in brightness, according to the inverse square law corresponds to a factor of four times the distance for the fainter star. Thus, the Pleiades stars are about four times more distant than Regulus.

A comparison of the 10 nearest stars with the 20 brightest stars reveals that the sun is the most luminous of the 10 nearest stars, except for Sirius and Alpha Centauri. Of the 20 brightest stars, all are more luminous than the sun.

Since so many of the stars we see with our naked eye are intrinsically very luminous, we might get the impression that such bright stars are fairly common. Quite the reverse is true, however. The reason we can see so many luminous stars is that they are visible for enormous distances. Their less luminous cousins (although far more numerous) are simply too faint to be seen, unless they happen to lie very close to the sun. A measure of the relative number of faint and bright stars (which is very important in studies of the formation of stars and the history of the galaxy) is the *luminosity function* (Fig. 11.5). This measures the number of stars of each absolute magnitude in a given volume of space, typically a cubic parsec. However, it is important to be wary in inferring anything about the number of faint stars from the luminosity function. There may be stars so faint that even if they were at the edge of the solar system we could not see them.

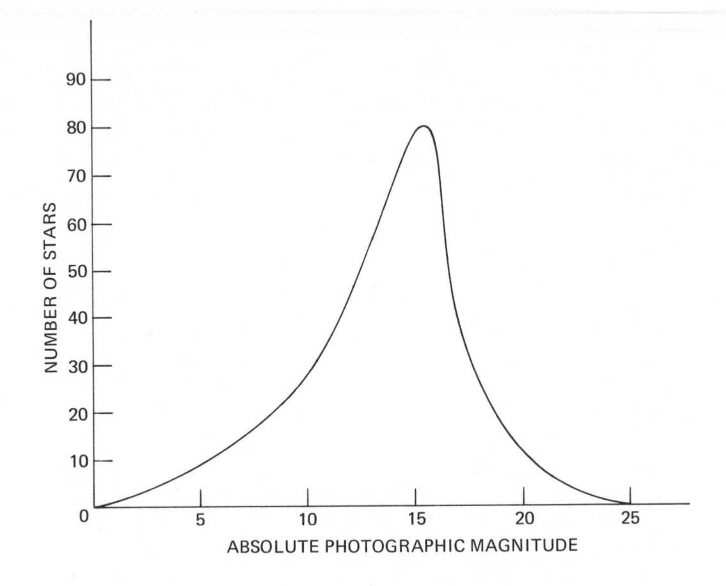

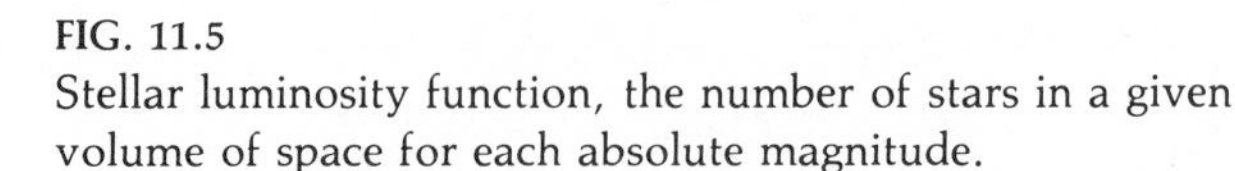

FIG. 11.5
Stellar luminosity function, the number of stars in a given volume of space for each absolute magnitude.

11.8 THE COLORS OF STARS

The brightness of a star depends on the color that is used to observe it. Stars emit energy in all wavelengths, but not in equal amounts. Because the human eye is more sensitive to yellow-green light, a star that emits its maximum energy at long red wavelengths, for example, may appear dimmer than a star that emits its maximum energy in yellow-green light.

The brightness of a star can be measured in several ways. When a star is observed visually, its brightness is called *visual magnitude.* When a photographic plate with an emulsion sensitive to blue-violet is used, the brightness of the star is called *photographic magnitude.* Since the human eye is not a reliable instrument for measuring the brightness of a star, a photographic emulsion has been developed that is sensitive to yellow-green light. When such a plate is used, the brightness of a star is called *photovisual magnitude.*

Since a blue star appears brighter photographically than visually, the difference between the two methods for measuring brightness can be used to measure the colors of the stars. The difference between the photographic and photo-visual magnitudes is the *color index* of the star. In the magnitude scale, the smaller numbers represent the brighter stars; therefore, the color index of a star is negative when the photographic (blue) magnitude is numerically smaller than the photovisual, and positive when it is larger. A very blue star thus has a negative color index, while a red star has a positive one. Since a star's color is related to its temperature, the color index is a measure of the star's surface temperature. The scale is chosen so that a color index of 0.0 corresponds to a temperature of 10,000°K. The values of the color index typically range from −0.6 for a blue star with a surface temperature of 25,000°K to 2.0 for a red star of temperature 3000°K.

All the methods that have been described for measuring the magnitude of stars use the limited range of wavelengths of visible light. When the energy emitted at all wavelengths is used, the brightness of a star is called the *bolometric magnitude.* Since this magnitude cannot be observed directly (much of a star's energy is absorbed by the earth's atmosphere), the magnitude is determined theoretically by adding a bolometric correction to a star's absolute visual magnitude.

11.9 THE SPECTRA OF STARS

The spectroscopic analysis of starlight provides us with nearly all the important information we have about stars. Spectra give information on composition, temperature, pressure, magnetic fields, motions, rotation, atmospheric motions in the star, radii, and distances.

Most stars produce absorption spectra that indicate they have hot interiors surrounded by cooler atmospheres. Their spectral lines are identical with those of terrestrial elements, indicating that matter in the stars is the same as that on the earth.

Stellar spectra were first observed in 1824 by the Bavarian optician Joseph Fraunhofer, when he discovered that the stellar absorption lines are similar to those in the solar spectrum. About 40 years later, the Italian astronomer Pietro Secchi was able to classify stellar spectra into four general groups. When photography was applied to the study of stellar spectra, the Harvard University astronomers instituted a program for classifying stellar spectra that reached its culmination with the publication of the *Henry Draper Catalogue,* the most extensive catalogue of stellar spectra, with nearly 225,000 stars catalogued. The production and development of this catalogue in its present form is largely the lifetime work of Miss Annie J. Cannon of the Harvard College Observatory.

11.10 THE HENRY DRAPER STELLAR SPECTRA CLASSIFICATION

The Henry Draper classification organizes the stars into a continuous sequence according to the appearance of their spectra. The great majority of the stars are grouped into seven principal classes designated by the letters O, B, A, F, G, K, and M.

Spectra were originally classified on the basis of complexity. A star having a spectrum showing only a few lines, regularly arranged, was denoted class A. Slightly more complex spectra were denoted B, and so on, with very complicated-looking spectra being assigned the classification M. It has since been recognized that the appearance of a spectrum is set basically by the star's surface temperature. The O stars (blue-white) have high temperatures, whereas the M stars (red) have low temperatures. When the stars are reordered into a temperature sequence, one ends up with nonalphabetical ordering O, B, etc. However, the sequence can be easily remembered by the first letters of the words in the phrase "Oh, be a fine girl, kiss me." Since each spectral class

TABLE 11.3
Familiar stars in the spectral sequence

STAR	COLOR	SPECTRAL CLASS	TEMPERATURE (°K)	COLOR INDEX	IMPORTANT FEATURES
—	Blue	09	30,000	−.32	Ionized helium, oxygen, nitrogen, and carbon
Spica	Blue	B1	20,000	−.23	Ionized oxygen, nitrogen, neutral helium, and hydrogen
Sirius	Blue	A1	10,000	0.0	Ionized calcium, hydrogen at maximum strength
Procyon	Blue to white	F5	7,000	.41	Strong lines of ionized calcium, strong lines of ionized metals
Sun	White to yellow	G2	6,000	.65	Strong lines of ionized calcium, neutral metals
Capella	White to yellow	G8	5,500	.79	Strong lines of neutral metals
Arcturus	Orange	K2	4,500	1.23	Lines of neutral metals
Betelgeuse	Red	M2	3,000	1.86	Strong lines of neutral metals, molecular bands of titanium oxide

blends with the adjacent classes, a refinement was achieved by dividing each class into 10 subclasses and numbering them from 0 to 9. Thus a star of spectral class F7 has a spectrum located seven-tenths of the way from class FO to class GO.

The differences in the stellar spectra are due primarily to the differences in the temperatures of the gases in the outer layers of the stars' atmospheres. At high temperatures atoms are often ionized, and molecules tend to be broken down into their constituent atoms. Thus in the spectra of the hottest stars, only the lines of highly ionized atoms are visible. In the F stars, located in the middle of the spectral sequence, lines of neutral metals appear because the temperatures are lower. In the M stars (with the lowest temperatures), lines of neutral metals and simple molecular compounds are most prominent. Table 11.3 lists the spectral classes and important features of some of the conspicuous and familiar stars.

In addition to the seven important spectral classes, there are several classes for unusual stars. Spectral class W has been assigned to the Wolf-Rayet stars. Since these stars are extremely hot and show strong emission lines of ionized helium, this spectral class has been placed before spectral class O. At the other end of the sequence, after spectral class M, are classes R, N, and S. The R and N stars show extensive bands in their spectra with strong molecular bands of carbon. The spectra of S stars show metallic lines and strong bands of zirconium oxide.

11.11 THE HERTZSPRUNG-RUSSELL DIAGRAM

Probably the most significant diagram drawn in astronomy was developed independently by the Danish astronomer Ejnar Hertzsprung and the American astronomer Henry N. Russell. They discovered the relationship between the luminosity of a star, its spectral class, and its color (Fig. 11.6). The Hertzsprung-Russell (or simply H-R) diagram, named in honor of the two astronomers, shows the position of several thousand stars by plotting their absolute visual magnitude against their spectral class or color index.

The stars are not distributed randomly on the H-R diagram, but appear to be concentrated in several regions. Most of the stars lie on a narrow diagonal band, called the *main sequence*, which runs from the upper left to the lower right of the diagram. The stars on this band

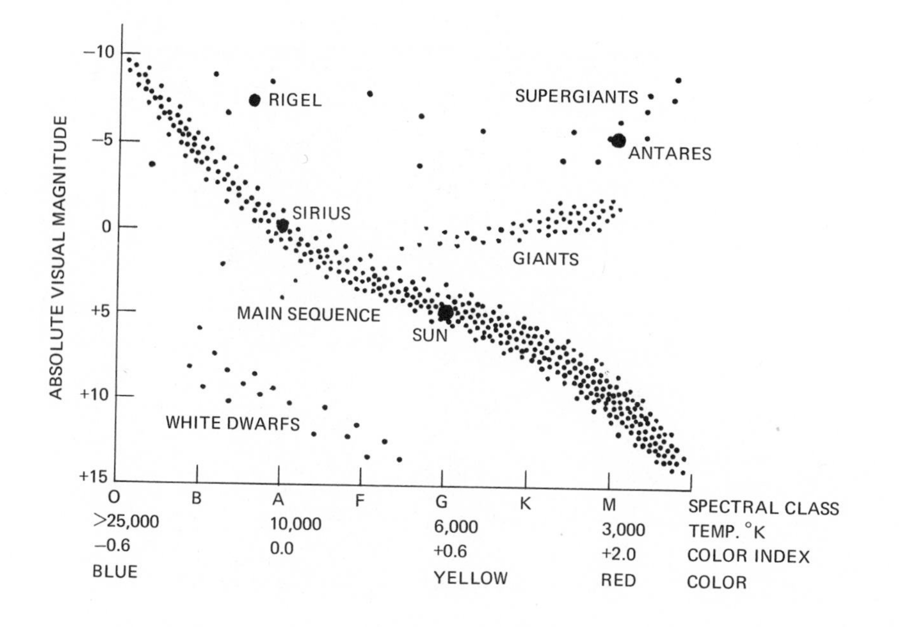

FIG. 11.6
Hertzsprung-Russell diagram, with a few of the more familiar stars shown.

are called *main-sequence stars.* Those at the upper left of the main-sequence band are hotter, larger, more luminous, and more massive than those at the lower right. Average stars, like the sun, are located in the center of the main-sequence band.

A few of the stars appear to be concentrated on a shorter band, located above and to the right of the main sequence. Since these stars have the same spectral class and temperature as the main-sequence stars, but higher luminosities, they must be considerably larger. Their red color indicates a low surface temperature, and their high luminosity, L, indicates a large radius, R, and surface area. (Recall the relation we discussed in Chapter 3.8, where we showed that $L=4\pi R^2\sigma T^4$. If two stars have the same temperature but one star is more luminous than the other, it must have a larger surface area and hence a larger radius.) Stars with large radii that are cool, and therefore red, are called *red giants.*

The very few stars that appear to be sprinkled across the top of the diagram are the *supergiants,* which represent the largest, most luminous stars.

Below the main sequence and in the lower left of the Hertzsprung-Russell diagram are the *white dwarf stars.* Their position on the diagram indicates that they are very hot with low luminosities. This implies that their total radiation, size, and surface area are extremely small—some are even smaller than the earth. Their tiny size and blue-white color leads to their name. The masses of several of the white dwarfs have been determined and are comparable to that of the sun.

11.12 THE LUMINOSITY CLASSIFICATION

Upon closer analysis of the spectra of two stars in the same spectral class, differences in the intensity of certain spectral lines are observed. When W. Morgan and P. Keenan of Yerkes Observatory recognized that certain spectral characteristics could be correlated to the star's luminosity, they established what is known as the

TABLE 11.4
The Morgan-Keenan luminosity classification

STAR	DESCRIPTION	CLASS
Rigel, Deneb	Most luminous supergiants	Ia
Antares	Least luminous supergiants	Ib
Canopus	Bright giants	II
Arcturus, Capella, Aldebaran	Normal giants	III
Procyon, Altair	Subgiants	IV
Vega, Sirius, Sun	Main sequence (dwarfs)	V

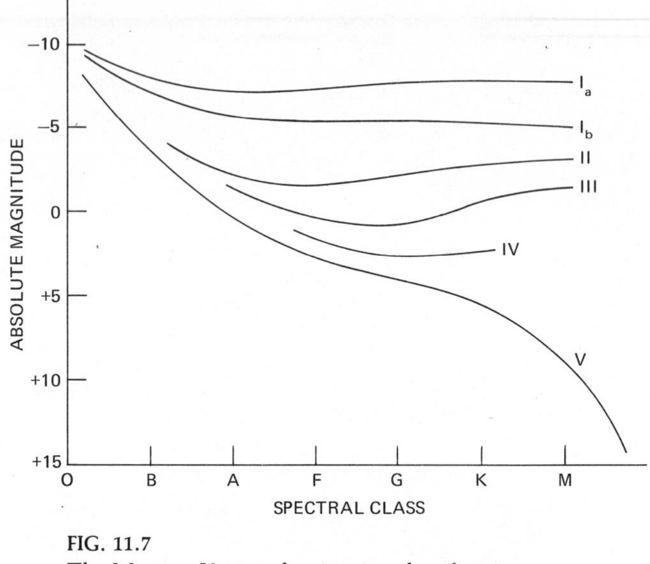

FIG. 11.7
The Morgan-Keenan luminosity classification.

Morgan-Keenan system, which divides the stars (with the exception of the white dwarfs) into five luminosity classes. This vertical refinement in the classification of stars is shown in Table 11.4 and Fig. 11.7.

11.13 STELLAR DIAMETERS

The diameter of a star can be estimated when its spectral class and absolute magnitude are known. The spectral class provides the star's surface temperature, which is used to determine the radiation that the star emits from one square centimeter of its surface every second. The absolute magnitude is used to determine the total radiation that the star emits per second. Dividing the total radiation by the radiation from one square centimeter of surface gives the total surface area of the star; from this, its diameter can be easily determined. (Recall the worked problem in Chapter 3.8.)

A star's diameter can also be determined when its angular diameter and parallax are known by the relationship

$$D = d/p,$$

where D is the star's diameter expressed in astronomical units, d is its angular diameter, and p is its parallax in seconds of arc. Even with the 200-inch Hale telescope, which theoretically should be able to show the largest stars as disks, perceptible disk images cannot be obtained because of atmospheric disturbances. However, the measurement of stellar angular diameters, achieved in 1920 by Albert A. Michelson with an interferometer (Chapter 4), produced results that are in close agreement with the estimates based on spectral class and absolute magnitude. The diameters of some of the red giants that were measured with the interferometer are shown in Table 11.5. More recently, measurements of the angular diameters of a number of stars have been made by either Speckle or Hanbury-Brown & Twiss interferometry (Chapter 4.16).

SUMMARY

This chapter explains how astronomers determine the basic properties of stars. The properties and techniques discussed in this chapter will be referred to often in later chapters, not only with reference to stars, but also in rela-

TABLE 11.5
Interferometer diameters of red giants

STAR	ANGULAR DIAMETER	DIAMETER (SUN=1)
Arcturus (Alpha Bootis)	0.020″	22
Aldebaran (Alpha Tauri)	0.020″	45
Beta Pegasi	0.021″	90
Betelgeuse (Alpha Orionis)	0.034″–0.054″	400–600
Mira (Alpha Ceti)	0.056″	460
Alpha Hercules	0.030″	500
Antares (Alpha Scorpii)	0.040″	640

tion to other astronomical objects. Many of a star's properties (e.g., color, temperature, luminosity) can be ascertained from its spectral class. The table below summarizes the methods used to determine various properties.

Property	*Method*
Distance	By observations of parallax—the shift in the apparent position of a star in relation to other stars as the earth moves around the sun.
Velocity	Along the line of sight, by the Doppler effect; across the sky, by proper motion.
Luminosity	By measuring the apparent brightness and then using distance and the inverse square law.
Color	By measuring the brightness through filters with a photometer.
Temperature	By observing color, or by analyzing a spectrum.
Radius	By interferometry.

REVIEW QUESTIONS

1. Define parallax. If a star at a distance of one parsec has a parallax of one second, and one parsec = 3.26 light years, what is the parallax of the nearest star (Proxima Centauri, at a distance of 4.3 light years)?
2. Extend one arm full length, with the thumb pointing up. Close one eye and move your thumb so that it blocks your view of some object in the room. Now open the closed eye and close the open one. Can you see the object now? Keeping in mind that there is some distance between your right and left eyes, explain the connection between this experiment and the phenomenon of stellar parallax.
3. Describe how you might measure the width of a river using a method similar to parallax.
4. Describe the principle of the parallax method for determining stellar distances.
5. To an observer located one parsec from the solar system, what would appear to be the maximum angular distance of Jupiter from the sun? (Jupiter is 5 a.u. from the sun.)
6. Why might parallax not be a good method for finding distances to stars 10,000 parsecs away? How could you use the inverse square law and knowledge of a star's luminosity to find the distance to such a remote star?

7. Distinguish among a star's proper motion, tangential velocity, and radial velocity. Explain how each is determined.
8. A star is moving straight toward the earth at 10 km/sec. Does it have a proper motion? What are its radial velocity and tangential velocity?
9. Distinguish between the brightness and the luminosity of a star. Which one can be observed directly? Explain.
10. What is the difference between the apparent and the absolute magnitude of a star?
11. How are distance, apparent brightness, and luminosity related? Think of two everyday examples of this relation.
12. How bright would the sun appear from Pluto?
13. Why do stars appear to be of different colors?
14. Rigel is a blue star. Betelgeuse is a red star. Which is hotter? Explain. Which might be spectral class B, and which spectral class M?
15. Match up the following stars with the descriptions of their spectra.

a)	Class B	1)	Ionized helium lines
b)	Class F	2)	Lines of molecules (TiO)
c)	Class M	3)	Ionized calcium lines, weak hydrogen lines
d)	Class O	4)	Strong hydrogen lines
e)	Class A	5)	Weak hydrogen lines, some helium

16. Describe the graph of the Hertzsprung-Russell diagram and show the relationship that exists between spectral class, temperature, color, and color index.
17. Referring to Appendixes 7 and 8, plot an H-R diagram for the nearest and for the brightest stars.

CHAPTER 12
MULTIPLE STAR SYSTEMS

Although most stars appear to the unaided eye as single points of light, many stars belong to multiple star systems—stars that are held together by mutual gravitational force. Such star systems range from two stars, called *binaries,* to star clusters of thousands of stars.

The discovery of multiple star systems was made by Giovanni Riccioli in 1650, when he observed that the star Mizar in the handle of the Big Dipper appeared as a double star in his telescope. By 1777 Christian Mayer had observed over 100 double stars and concluded that the fainter star in each "double" was a planet revolving around the "true" star. His conclusion caused quite a discussion among astronomers. The astronomer Father Hell in Vienna disagreed with Mayer's conclusion and explained that the fainter star was simply an optical illusion. Both were wrong.

William Herschel's work with double stars helped to establish certain characteristics about them. They differ in size and brightness, and both are in motion. Herschel also differentiated between *apparent* and *real double stars.* An apparent double (optical double) is two stars that appear to be close together but actually are not physically associated because they are separated by a large distance along the line of sight. A real double (physical binary) consists of two stars that are bound together by their mutual gravity and revolve around a common center of mass (barycenter) (recall Chapter 7.9).

Binaries have been classified according to the technique that was used in their discovery and form four classes: (1) *Visual binaries* are those whose stars are so far apart that they can be detected visually as two stars either through a telescope or on a photographic plate. (2) *Spectroscopic binaries* are those discovered spectroscopically from variations in the stars' radial velocities, indicated by the variation in the Doppler shift of their spectral lines. (3) *Astrometric binaries* are stars whose path across the sky (proper motion) shows tiny wiggles due to the presence of a companion star too faint to be seen directly. (4) *Eclipsing binaries* are those whose orbits are nearly in the line of sight of the observer (so that each eclipses the other) and are detected by a periodic variation in brightness.

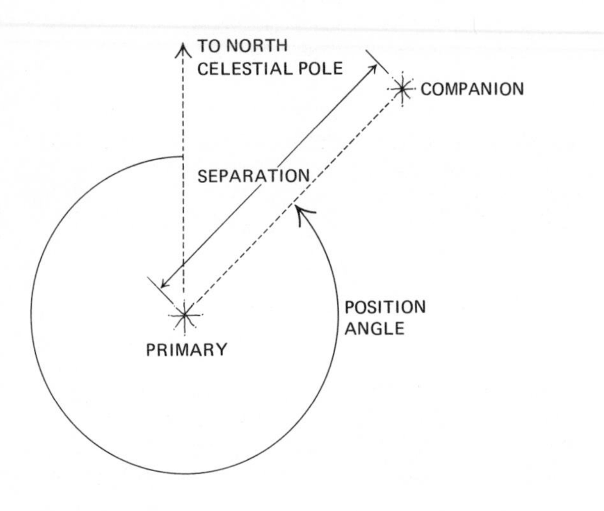

FIG. 12.1
The position angle and separation of binary stars.

12.1 VISUAL BINARIES

The position and motion of the stars in a visual binary system can be determined by observing the motion of the fainter (companion) star as it moves about the brighter (primary) star and measuring the position angle and the separation of the stars. The position angle (Fig. 12.1) is the angle measured from the north celestial pole eastward around the primary to the companion. The separation is the angular distance between the two stars. The position angle and the separation are measured with a filar micrometer or on a photographic plate. Three prominent visual binaries and their periods of revolution are Procyon (21 years), Sirius (50 years), and Castor (420 years).

12.2 SPECTROSCOPIC AND ASTROMETRIC BINARIES

Many binaries appear even in the largest telescopes as single stars; however, their binary status is clearly apparent from the periodic variation in their spectra. These stars are called spectroscopic binaries. The first to be discovered was the brightest star of the double star Mizar, which was itself shown to be a binary in 1889 by E. Pickering. Among the brightest of the spectroscopic binaries are Capella in Auriga the Charioteer and Spica in Virgo.

When the two stars of a spectroscopic binary system are about equal in brightness, they can be detected by their Doppler displacements. As shown in Fig. 12.2, when both stars are moving at right angles to the observer's line of sight, the spectral lines are superimposed. When one star is receding and the other is approaching the observer, the spectral lines appear double, that is, the spectral lines of the receding star are displaced toward the red and those of the approaching star toward the violet (Fig. 12.3). Many of the spectroscopic binaries show a single spectrum, which indicates that the primary is much brighter than its companion, consequently drowning out the spectrum of the dimmer star. The periodic variation in the Doppler shift of the spectrum indicates the presence of two stars revolving around their barycenter.

When the companion is too close to the primary and very faint, its presence can be detected by its gravita-

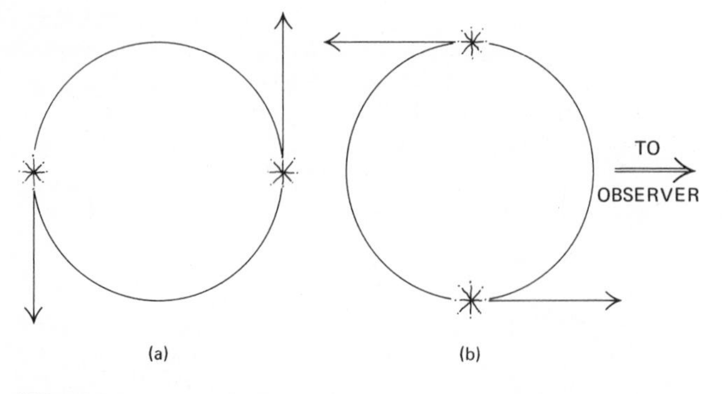

FIG. 12.2
Spectroscopic binary star system: (a) stars are moving at right angles to the observer (spectral lines are superimposed); (b) stars are approaching and receding from the observer (spectral lines appear double).

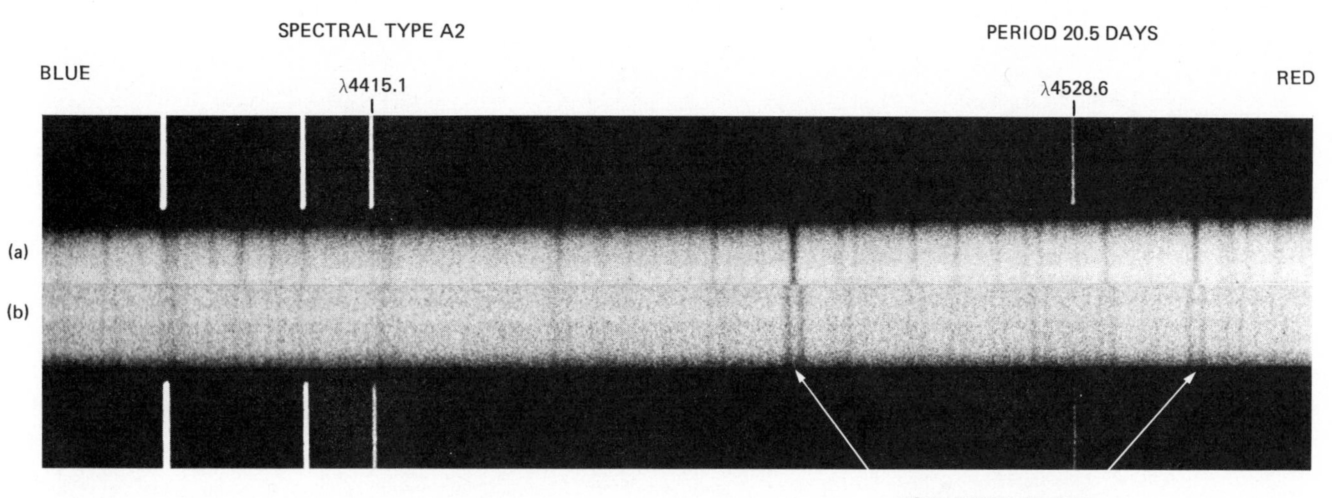

FIG. 12.3
Spectrum of a spectroscopic binary star, Zeta Ursa Majoris (Mizar): (a) lines of the two components are superimposed on 11 June 1927; (b) lines of the two components are separated by a difference in orbital velocity of 87 miles per second on 13 June 1927. (Courtesy of the Hale Observatories.)

tional effect on the motion of the primary. It causes the primary to move along a wavy path (Chapter 12.5). This was first detected by Bessel when he observed that the line of Sirius' path is slightly wavy rather than straight. Such star systems are called astrometric binaries.

12.3 ECLIPSING BINARIES

The orbital plane of an eclipsing binary lies very close to the observer's line of sight, so that each star appears periodically to eclipse the other. The first to be discovered, and probably one of the finest examples of an eclipsing binary, is the star Algol, the "blinking demon" located in the constellation of Perseus. The companion has a diameter that is about 20% greater and a brightness that is three magnitudes less than the primary. The two stars are about 7 million miles (11 million km) apart, and their orbital plane is inclined about 8° to the observer's line of sight. The bright star revolves around the dim star in about 2 days and 21 hours. During each revolution the companion passes between the observer's line of sight and the primary star, partially eclipsing it for ten hours, and at maximum eclipse, reducing the light of the binary system to one-third its normal brightness.

12.4 LIGHT CURVES OF ECLIPSING BINARIES

By plotting the brightness of an eclipsing binary against time, one can obtain a light curve of the binary. The light curve of Algol is shown in Fig. 12.4. The maximum and minimum points on the curve represent maximum and minimum brightnesses; the difference between the two brightnesses is called the *amplitude*; the interval between two consecutive maximum or minimum points is called the *period*.

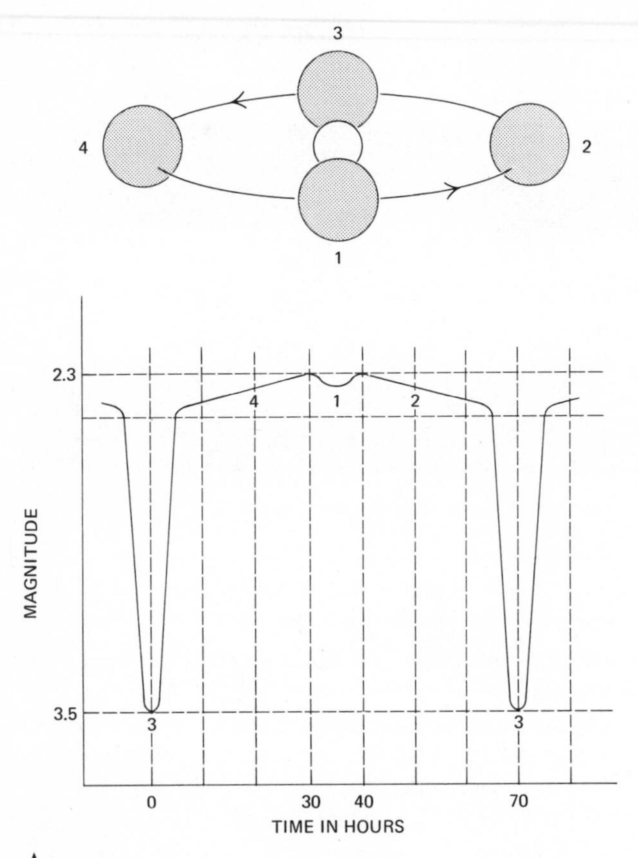

▲
FIG. 12.4
The Algol (β Persei) eclipsing binary and its light curve.

A study of the light curve of an eclipsing binary reveals the binary system's orbital period, the inclination of its orbit, and the size and luminosity of its stars. Figure 12.5(a) represents the light curve of two stars that have approximately the same size and luminosity, and a slight orbital inclination. Their minima points are about equal. Figure 12.5(b) represents the light curve when the luminosity of star 1 is greater than that of star 2. When star 1 eclipses star 2, its minimum is smaller than when star 2 eclipses star 1. The pointed minima indicate that total eclipse lasts for only a moment.

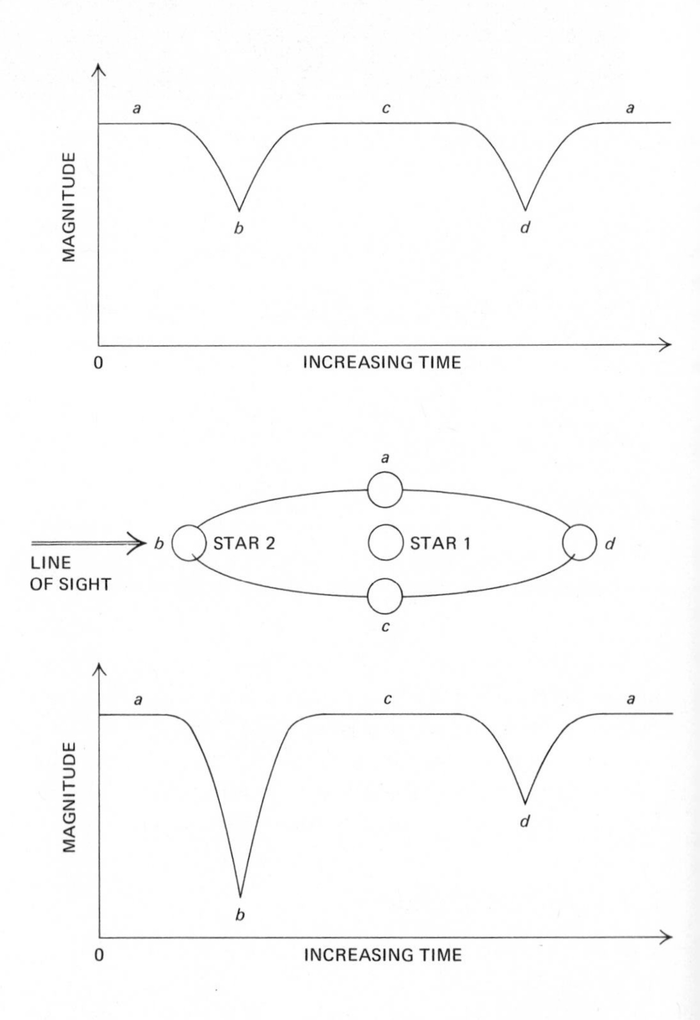

FIG. 12.5
Light curves of eclipsing binaries: (a) two stars of equal size, magnitude, and small orbital inclination; (b) two stars of equal size and small orbital inclination—star 1 is brighter than star 2.

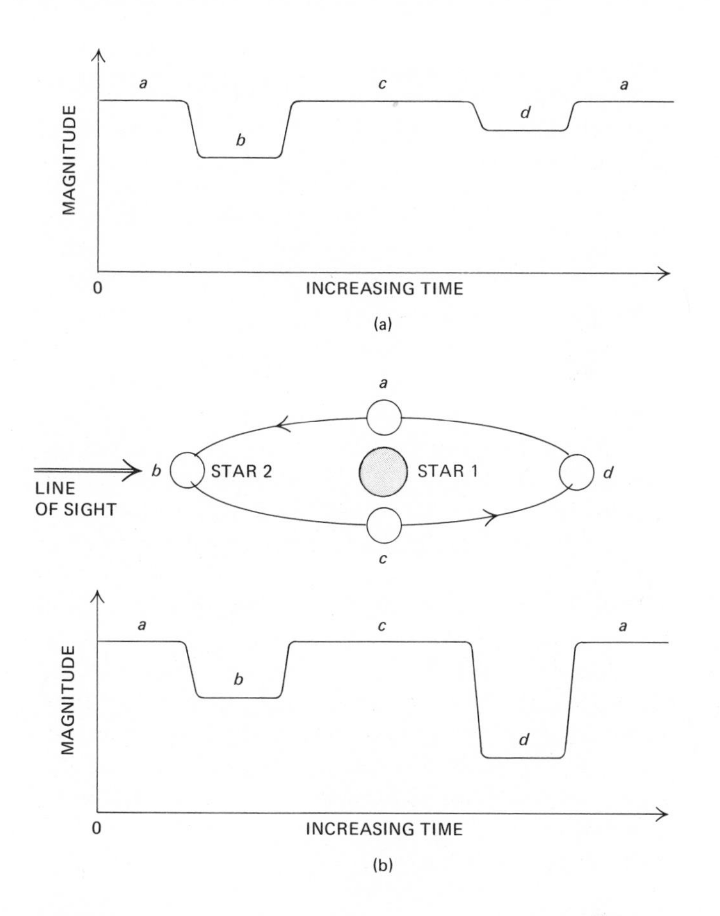

FIG. 12.6
The light curves of eclipsing binaries: (a) star 1 is larger and brighter than star 2; (b) star 1 is larger but dimmer than star 2.

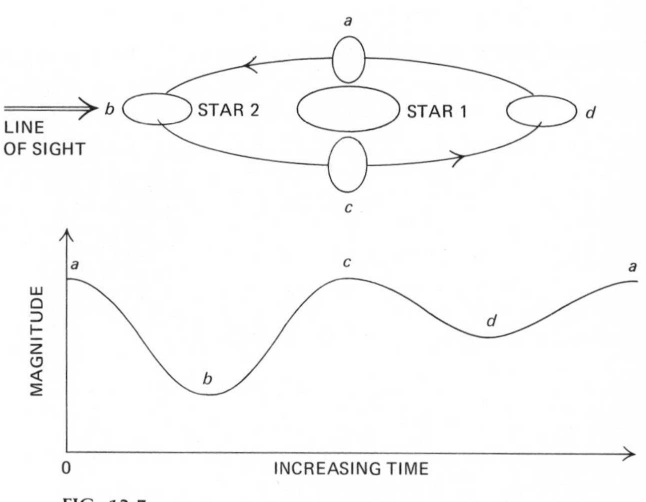

FIG. 12.7
The light curve of an eclipsing binary whose stars are so close together that enormous tides result from their mutual gravitational force, distorting the stars into elliptical shapes and producing a wavy light curve.

Figure 12.6(a) represents the light curve when the size and luminosity of star 1 are greater than those of star 2. When the hotter star 1 eclipses star 2, the minimum is less than when the cooler star 2 eclipses star 1. The minima are flat, because total eclipse is not momentary, but has a definite period of duration. Figure 12.6(b) represents the light curve when star 1 is cooler than star 2. When star 1 eclipses star 2, the minimum is greater than when the hotter star 2 eclipses star 1.

When the two stars in a binary system are very close together, enormous tides result from their mutual gravitational force. This distorts the stars into elliptical shapes and causes the light curve to appear wavy (Fig. 12.7). Between eclipses, the stars are seen lengthwise; just before and after an eclipse, they are seen end-on.

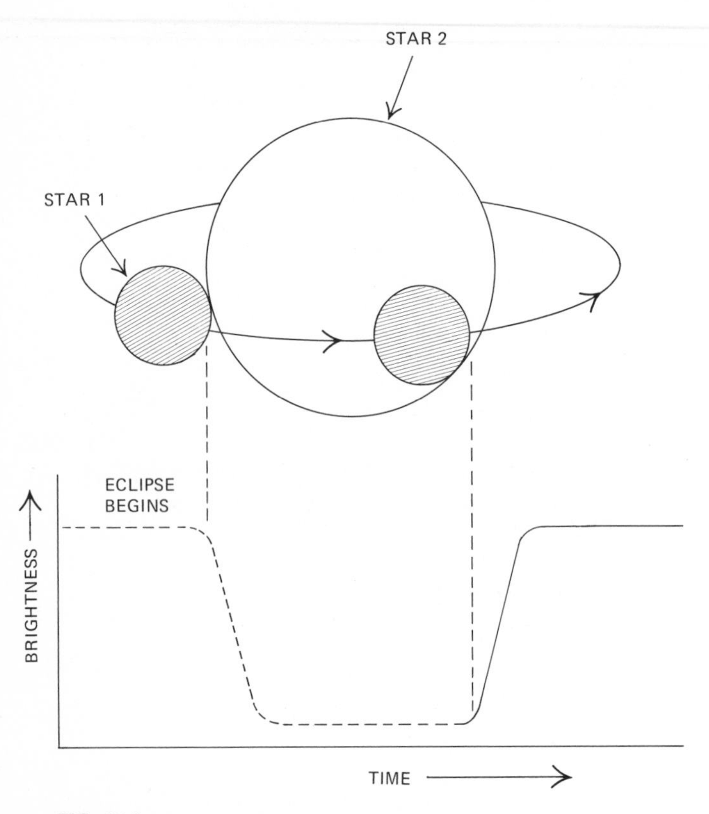

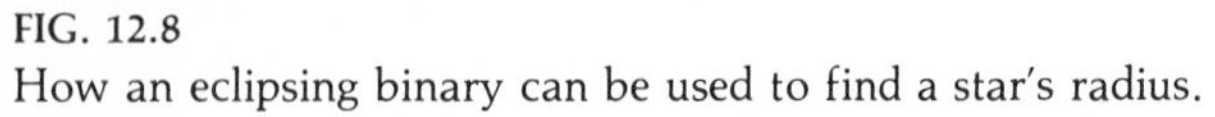
FIG. 12.8
How an eclipsing binary can be used to find a star's radius.

The diameter of the stars in an eclipsing binary can be found if their orbital velocities are known. Figure 12.8 shows that when the edge of the eclipsing star (star 1) first begins to cross the face of star 2, the light curve begins to drop off. The light curve bottoms out when the other edge of star 1 begins to cross the face of star 2. In this interval star 1 must, therefore, have traveled a distance in its orbit equal to its diameter. The orbital velocity when multiplied by the length of the time interval then gives the diameter of star 1. A similar method yields the size of star 2. In the description above we are assuming that star 1 is moving in a circular orbit and star 2 is stationary, although the method can be modified to work for other cases.

12.5 THE MASS OF A BINARY SYSTEM

We learned in Chapter 2 that Newton generalized Kepler's third law of planetary motion as

$$P^2=a^3/(M+m),$$

where P is the body's orbital period in years, a is the semimajor axis of its orbit in astronomical units, and $M+m$ is the sum of the masses of the two bodies as compared to the sun's mass. With this formula it is possible to find the sum of the masses of the two stars in a binary system by observing their period of revolution (P) and the separation of the stars (a). For example, the visual binary of Sirius and its companion, which are designated as Sirius A and Sirius B, has an orbital period P of about 50 years. Suppose we know the binary's angular semimajor axis (7″.7) and its distance from the sun (2.6 parsecs). We find that their product (7.7)(2.6), or 20.0, is the system's semimajor axis (a) in astronomical units, which is the separation of the two stars. Therefore,

$$P^2=a^3/(M+m)$$
$$M+M=a^3/P^2=(20.0)^3/(50)^2$$
$$=3.2 \text{ solar masses.}$$

To determine the individual mass of each star, we must take into consideration their motions with respect to their barycenter. The barycenter lies on a line which joins the two stars, and the distance of each star from the barycenter is inversely proportional to its mass, that is, the more massive body is closer to the barycenter. This is expressed by the relationship

$$m_1/m_2=d_2/d_1.$$

(This situation was discussed in Chapter 7 in relation to the earth-moon system.) As the binary system of Sirius A and Sirius B moves through space, its barycenter follows

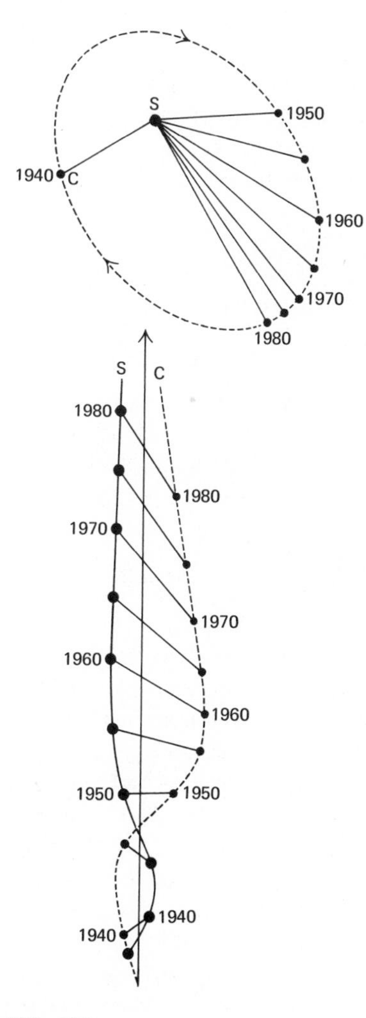

FIG. 12.9
The proper motion of Sirius and companion. The straight line represents the barycenter of the system, the solid line is the path of Sirius, and the dotted line is the path of the companion.

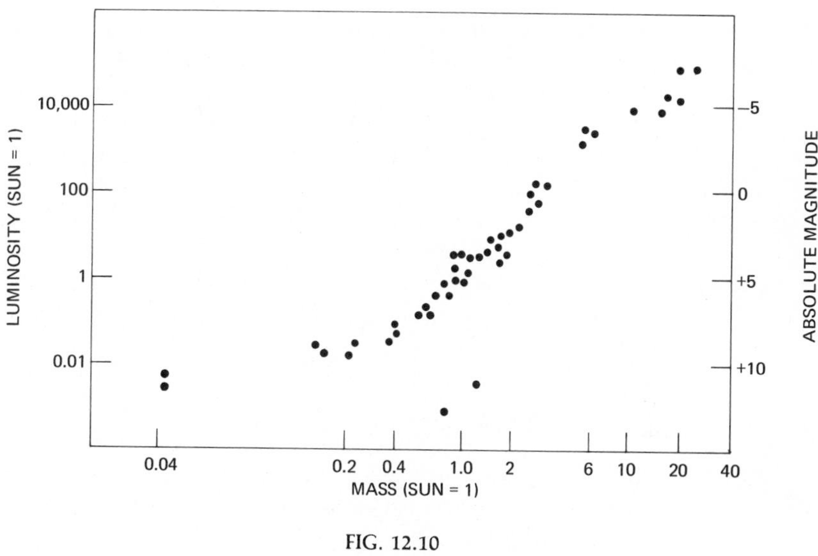

FIG. 12.10
Mass-luminosity relation.

a uniform path, and the two stars oscillate around it in varying degrees. In Fig. 12.9, the paths of Sirius A and Sirius B are shown as wavy lines around a straight line, which represents the path of the barycenter. The ratio of their distances from the barycenter (d_2/d_1) is 2.2, which means that Sirius B is orbiting about 2.2 times farther from the barycenter than is Sirius A. Therefore, the mass of Sirius B is equal to the sun's mass, and the mass of Sirius A is 2.2 times greater.

12.6 THE MASS-LUMINOSITY RELATIONSHIP

In 1924 the English astrophysicist Arthur S. Eddington discovered one of the most important and significant relationships in astronomy—the mass-luminosity relationship (Fig. 12.10). By plotting the mass of stars (in terms of the sun's mass) against their absolute bolometric magnitudes, he discovered that most of the stars plotted were on a narrow band which extended from the lower left to the upper right of the diagram. This indicated that a definite relationship exists between the star's mass and its luminosity and that luminosity increases as mass increases. Roughly, the luminosity increases as the mass cubed ($L \approx M^3$). The mass-luminosity relationship, which holds true for most stars except white dwarfs, indicates that mass must play an important role in determining a star's properties. From this relationship the masses of most stars can be determined approximately from their absolute magnitudes. There is no known star with a mass greater than 100 times that of the sun.

12.7 STAR CLUSTERS

We have learned that stars appear in space independently and in multiple star systems. They also appear in clusters of two major types: galactic (or open) and globular. A *star cluster* is a physically related group of stars bound together by their mutual gravitational attraction. In a cluster each star moves along an orbit determined by adding together the gravitational attraction from each of the separate stars. The typical orbital speed of a star in a cluster is about 1 km per second. When seen from a great distance, the individual stellar motions are hard to detect and the stars appear to converge or diverge from a common point in the sky, similar to meteors in a meteor shower. Their motions suggest that they have a common origin and age. With these characteristics, clusters become very interesting objects for studying the evolution of stars.

12.8 GALACTIC CLUSTERS

Nearly 800 galactic clusters have been discovered. They are located almost exclusively along the spiral arms and in the disk of the Milky Way. It is believed they move on relatively circular orbits about the center of the galaxy, and that tens of thousands of these clusters exist but that they have escaped detection because the dust particles in the plane of the Milky Way are obscuring them. The number of stars in a cluster ranges from about 10 to nearly 1000.

The well-known Pleiades and Hyades in Taurus the Bull are excellent examples of galactic clusters visible to the unaided eye. Photographs of the Pleiades (Fig. 12.11) show that the brighter stars in the cluster are surrounded with wisps of nebulosity. The nebulosity is made visible by reflection of the light from the cluster stars off the dust grains in the interstellar matter in and near the cluster. Seen with the 200-inch Hale telescope, the cluster reveals the presence of several hundred stars. The Hyades appears as a V-shaped arrangement of stars that marks the head of Taurus the Bull (Fig. 12.12). The galactic cluster of the Praesepe in Cancer the Crab is visible to the unaided eye as a very faint, hazy spot of light. It is often referred to as the "Beehive," because it resembles a swarm of bees when seen telescopically. The Hyades and Praesepe are examples of star clusters without nebulosities.

12.9 THE DISTANCE OF GALACTIC CLUSTERS

When the apparent magnitude of the stars in a cluster are plotted according to their color index (Fig. 12.13), we obtain a color-magnitude diagram. It is similar to the H-R diagram. Since the stars in a cluster are at approximately the same distance from the sun, the distance modulus (the difference between the apparent and the absolute magnitudes of each star ($m-M$)) is constant; therefore, star distances can be easily determined from the color-magnitude and H-R diagrams, as will now be explained. Superimpose the color-magnitude diagram on the H-R diagram so that the two color-index scales coincide. Then move the color-magnitude diagram vertically until its main sequence coincides with that of the H-R diagram. With the diagrams in this position, the absolute magnitude of the stars can be obtained; from the distance modulus, their distances can be determined. This method is extremely useful in finding distances to star clusters, because the apparent magnitude and color index can always be found by direct and relatively simple observations.

12.10 ASSOCIATIONS

Since the turn of the century it has been known that O and early B stars occur in the Milky Way in stellar groups in which the stars are farther apart than those in galactic

FIG. 12.11
The Pleiades, NGC 1432, a typical galactic cluster in Taurus, which shows reflection nebulosities. (Courtesy of the Hale Observatories.)

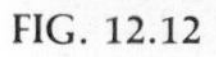

FIG. 12.12
The V-shaped Hyades cluster, a galactic cluster in the constellation of Taurus. (Courtesy of the Lick Observatory.)

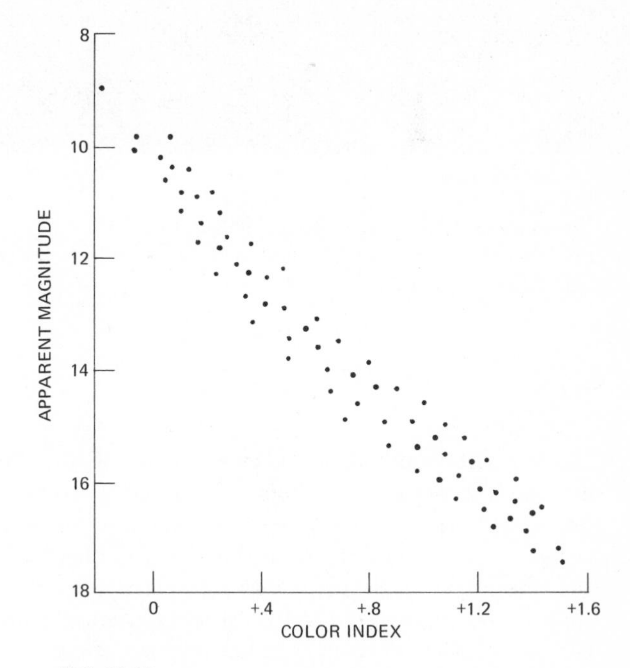

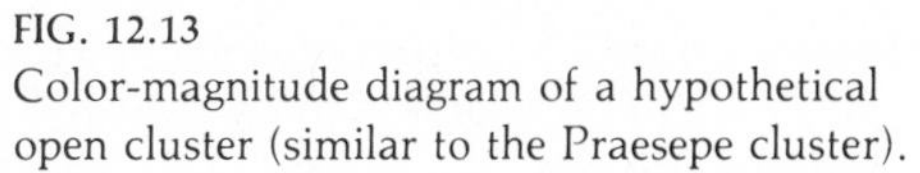

FIG. 12.13
Color-magnitude diagram of a hypothetical open cluster (similar to the Praesepe cluster).

FIG. 12.14
The globular cluster NGC 6205 (Messier 13) in Hercules, photographed with the 200-inch telescope. (Courtesy of the Hale Observatories.)

clusters. After studying the T-Tauri stars* and later the O and early B stars in groupings in the Milky Way, the Soviet astronomer V. A. Ambartsumian in 1949 presented the first explanation of their nature. He observed that the shapes of the groupings are nearly spherical and that the stars are moving at great speeds away from the center of the grouping. Ambartsumian called these groupings *associations*, because his study led him to believe that the stars in a grouping must have had a common origin, that is to say, they did not come together by chance but are associated. The stars are also extremely young, possibly still in the formative stage; otherwise, expansion of the association would have already dispersed them. An excellent example of an association is the cluster of stars in the center of the great nebula in Orion.

* See Chapter 14.

12.11 GLOBULAR CLUSTERS

At present, there are about 120 known globular clusters in the Milky Way. They appear to be distributed throughout a spheroidal volume around the galaxy. They are also concentrated slightly toward the center of the galaxy, which lies in the direction of the constellation Sagittarius. In contrast to galactic clusters, globular clusters are larger, more compact, spheroidal in shape, more distant (the nearest is about 6000 parsecs), and contain hundreds of thousands of stars. The clusters themselves are found to move in and out of the central region of our galaxy at high velocity.

The few globular clusters visible to the unaided eye appear as faint, hazy spots of light. In the southern hemisphere, the two brightest, Omega Centauri and 47 Tucanae are visible; in the northern hemisphere, the brightest and best known is M13 in the constellation of Hercules. A long-exposure photograph of M13 (Fig. 12.14) reveals it to be a spectacular celestial object. Not even the largest telescopes can resolve the stars in the center of M13, which are at a distance of over 6300 parsecs.

12.12 CLUSTERS AS DIAGNOSTICS OF EVOLUTION

The H-R diagrams of star clusters are important in the study of both the composition of clusters and stellar evolution. The diagrams of most of the globular clusters are quite similar, whereas those of the galactic clusters vary considerably. The diagram of a typical globular cluster, M3 in our galaxy (Fig. 12.15), shows that its stars have left the upper part of the main sequence. In general, the very hot, highly luminous, and short-lived stars are

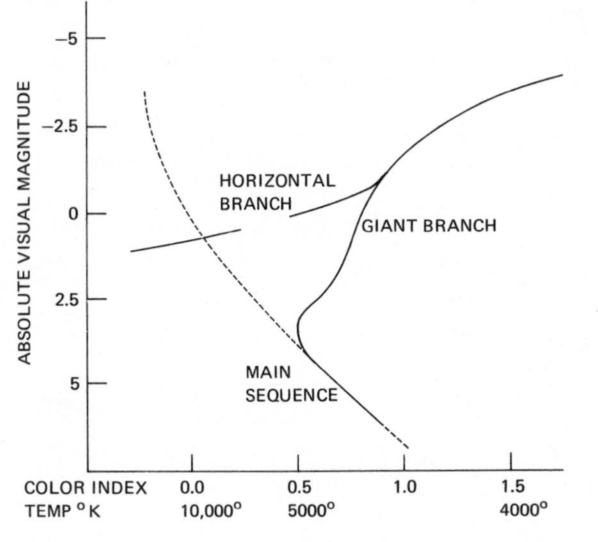

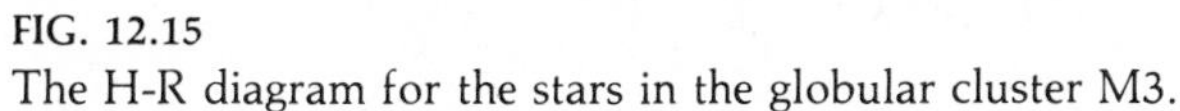

FIG. 12.15
The H-R diagram for the stars in the globular cluster M3.

not present in globular clusters. These stars have aged and have moved to other parts of the diagram, indicating various stages of evolution. Some might have evolved into supernovae and disappeared from the diagram.

The diagram shows the following to be present in a typical globular cluster: small, less luminous, cool stars in the lower part of the main sequence whose life-spans are much longer than those in the upper part of the main sequence; large, red, luminous giants with absolute magnitudes up to −3; and blue, hot stars on the horizontal branch that extends from the giant branch. The gap in the horizontal branch indicates the region of the RR Lyrae varible stars, a type of pulsating star common in globular clusters.

SUMMARY

Stars are very often found in groups. The groups are held together by gravity. Approximately 40% of all stars occur in multiple star systems, in which two or more stars orbit about a common center of mass. Binary stars are classified according to whether they are (a) visual (two or more stars visible telescopically); (b) spectroscopic (spectral lines show doubling); (c) astrometric (companion detectable by its effect on motion of primary); and (d) eclipsing (light output of system varies as stars alternately block each other's light). Larger groupings are called star clusters. There are two main types of star clusters: globular and galactic. Their properties are indicated below.

Globular	*Galactic (or open)*
Old	Generally young
Usually contain red and giant stars	Stars are often blue
Found in a halo around the galaxy	Found in galactic disk and spiral arms
Contain millions of stars	Contain hundreds of stars
Size about 30 light years	Size about 20 light years

Looser groups of young stars are referred to as associations. Binary stars are extremely important because study of the motions (Doppler shift and changing position) can give information about masses of stars. If a binary star system is also eclipsing, the star's radius can be found.

REVIEW QUESTIONS

1. What is the difference between an optical double and a binary?
2. What is a spectroscopic binary? Explain how we can tell from its spectrum that the object is a true physical binary. Why do some spectroscopic binaries show single spectra?
3. Suppose the spectra of a star revealed lines of ionized helium and titanium oxide. What would you conclude about the star? (*Hint*: Refer to Table 11.3)
4. What is an astrometric binary? How can it be detected? Give a famous example of an astrometric binary.
5. Suppose you were to observe the sun from a distance of five light years. What kind of binary star might it appear

to be? Does this suggest a way to detect planets around other stars?

6. What is an eclipsing binary? Draw the light curve of the eclipsing binary Algol and explain the physical characteristics of the two stars and the system which produce the curve.
7. Describe and explain how the different light curves of eclipsing binaries are produced.
8. You are an industrial investigator and note a suspicious car traveling at 30 mph (44 feet per second) alongside a warehouse. It takes the car 10 seconds to get from one end of the warehouse to the other. How long is the warehouse?
9. Suppose you observe an eclipsing binary in which it takes 20 hours for the star to go from the beginning of the eclipse to the point where it flattens into eclipse minimum. From a spectrum you learn that the star's orbital velocity is 40,000 mph. How big is the eclipsed star?
10. Explain how the stellar masses of a binary system are determined.
11. What is the mass-luminosity relation? Who discovered it? Does it hold true for all stars?
12. What are galactic clusters? Give several examples. Where in the Milky Way are they located? How are the stars in the cluster distinguished from those that are not?
13. Explain how the distance to a galactic cluster is determined. Make a tracing of Fig. 11.6 and use it to measure the distance to the star cluster shown in Fig. 12.13.
14. What are stellar associations? Are they stable? Compare them with clusters.
15. What are globular clusters? Compare them with galactic clusters.
16. Explain why clusters are useful in studies of stellar evolution.

CHAPTER 13
STELLAR EVOLUTION: THE AGING PROCESS IN STARS

What goes on four legs in the morning, two legs at noon, and three legs in the evening?

Riddle of the Sphinx

The answer to the ancient riddle is man: crawling on hands and knees as a baby; walking on two legs as an adult; and using a cane when old. In this riddle we see an encapsuled view of the stages of a person's life from birth to death. Aging is a process that all matter undergoes. Stars are no exception to the vagaries of time.

Before we can comprehend the stages in a star's life, we must first recognize certain processes that occur in its interior. Although the interior of a star is inaccessible to direct observation, a great deal of information about its composition, structure, temperature, pressure, and energy has been accumulated from the knowledge of gas laws, hydrostatic and thermal equilibrium, methods of heat transfer, and methods of generating energy. Most stars are in a state of stable equilibrium, which means that at any point within the star, its temperature, pressure, and density are constant so that the star is neither expanding nor contracting.

By its very nature, stellar evolution is not a process that can be studied physically in the laboratory. Thus astronomers have had to rely heavily on mathematical models in deducing what the probable life history of stars must be. The study of stellar evolution has gained new impetus since computers have become available to assist in making calculations. It is believed that the basic physical laws that govern the structure of a star are known. With a large computer one can solve all the relevant equations. A detailed stellar model can be worked out on paper and then, by allowing the model to change as conditions in the star change, one can study how the model star evolves, and whether its radius, surface temperature, luminosity, composition, and mass agree with those of real stars. The H-R diagram plays a crucial role in such comparisons. Not only can one check whether the model star's location in the H-R diagram agrees with the temperature and luminosity of a real star, but one can also compare the number of stars found in a given location on the H-R diagram with how long the model predicts a star should be in that location. Those areas of the diagram where stars spend long periods of time will be most densely populated. Thus, what takes millions of years to occur in a star can be simulated in a few minutes on a fast computer. Nearly all of the information in this chapter has been gained through this indirect method. One should bear in mind, however, that astronomers may well have omitted some important ingredient in their model stars. Ideas of stellar structure and evolution have changed markedly in the last 50 years and will probably be somewhat modified in the future.

13.1 THE GAS LAWS

Since a star is under tremendous pressures and temperatures, it is completely gaseous, and its molecules, which are in constant motion, continually collide with one another. These collisions produce gas pressure, which is proportional to the temperature and density of the gas. The relationship of gas pressure, density, and temperature is summarized in three laws. Boyle's law states that *the pressure of a gas at constant temperature is inversely proportional to its volume, or directly proportional to its density.* Charles' law states that *the pressure of a gas at constant volume is proportional to the temperature.* When these two laws are combined, the result is the *perfect gas law*, which states that *the pressure of a gas is proportional to the products of its temperature and density.* Even though the stellar gas is highly compressed, it reacts like a gas because the atoms or ions are moving so fast at the extreme temperatures that the attractive forces between them (molecular forces) are not able to bind these atoms or ions together.

13.2 HYDROSTATIC EQUILIBRIUM

The mutual gravitation between the masses within a star produces tremendous forces that act inwardly, toward the center of the star. These forces tend to cause the star to collapse toward the center, making it the region of greatest density. For a stable star, this gravitational inward pressure is balanced by an internal force acting outwardly, which is produced by gas pressure and by radiation pressure. The gas pressure is produced by the force exerted by collisions of the gas particles with one another, whereas the radiation pressure is produced by the scattering and absorbtion by atoms and electrons of radiation streaming out from the star's core. A stable star is in hydrostatic equilibrium when its gravitational force inward is balanced by the pressure forces (gas and radiation) outward.

13.3 THERMAL EQUILIBRIUM

Since a star emits energy from its surface and since energy flows from a hot to a cold region, it can be concluded that the star's temperature is highest at its core. If the energy at the core is not replaced, the star's temperature will gradually decrease. However, this energy is replaced by nuclear reactions. A stable star is in thermal equilibrium when the rate at which energy is produced equals the rate at which it is emitted into space.

13.4 TRANSFER OF ENERGY

The transfer of energy within a star can be accomplished by the processes of *conduction, convection,* and *radiation*. The process of conduction, the direct passing of energy from one atom to another, is slow and inefficient except in white dwarfs, where the atoms are very close together. For this reason, conduction is not significant in most stars. Convection, the movement of atoms from a hotter to a cooler region, is an efficient means of heat transfer. For the greater part of a star's life, convection generally occurs only in certain layers where the gas temperatures increase at a faster rate than the pressures. However, during much of the formative stage in a star's evolution, convection plays the predominate role in the transfer of energy.

In most stars energy flows outward by electromagnetic radiation—the transfer of energy from atom to atom by electromagnetic waves. By contrast with convection, radiation is not a very efficient process. When an atom that has become excited by absorbing a photon of energy that is moving outward returns to its previous state, it emits a photon whose chance of moving outward is 50%, since it can be emitted in any direction.

13.5 NUCLEAR REACTIONS

The source of solar energy had been a riddle to scientists for years, because no known source could have kept the sun emitting its steady light for the past several billion years. In 1854 the German physicist H. von Helmholtz presented the first reasonable explanation when he suggested that solar energy is derived from the compression of the sun's gases as they move inward due to gravitational attraction. Although this view was widely accepted, the compression process alone could keep the sun shining at its present rate for only some tens of millions of years. The answer to the riddle was finally provided when the discovery of radioactivity made known the tremendous energy within an atom's nucleus. Stellar energy is derived from a loss of mass that occurs in the process of building elements of heavier atomic weight from elements of lighter atomic weight. The energy release, E, in a reaction is proportional to the mass loss, m, as given by the famous formula

$$E = mc^2,$$

where c is the speed of light.

A star is a nuclear furnace. During most of its lifetime, its energy is derived from two sources: the fusion of hydrogen into helium by the *proton-proton reaction*, and the *carbon cycle*. The proton-proton reaction is the simpler and more direct of the two. The steps are indicated by the following reactions, where the mass loss in units of the proton's mass is shown:

$$_1H^1 + {_1H^1} \rightarrow {_1H^2} + e^+ + \nu + \gamma + 0.001 \text{ (mass loss)}$$
$$_1H^2 + {_1H^1} \rightarrow {_2He^3} + \gamma + 0.006 \text{ (mass loss)}$$
$$_2He^3 + {_2He^3} \rightarrow {_2He^4} + {_1H^1} + {_1H^1}. + 0.014 \text{ (mass loss)}$$

(Note that each of the first two reactions must occur twice in order to produce two $_2He^3$ for the third reaction.)

In this reaction, two protons ($_1H^1$) combine directly to form heavy hydrogen, or deuteron, ($_1H^2$), with the release of a positron (e^+), neutrino (ν), and gamma ray (γ). The deuteron combines with another proton to form a helium isotope ($_2He^3$), with the release of a gamma ray. Two helium isotopes combine to form an ordinary helium nucleus ($_2He^4$) and two protons. The total mass loss is .028 units. The subscript and superscript on the chemical symbol for the element indicate the number of protons and protons plus neutrons, respectively. Thus $_2He^3$ indicates a helium atom with 2 protons and 1 neutron, while $_7N^{13}$ represents a nitrogen atom with 7 protons and 6 neutrons. It should be noted that if the subscript is subtracted from the superscript, the result (the difference) indicates the number of neutrons in that atom.

The concept of the carbon cycle was proposed independently in 1938 by the American astronomer H. Bethe and the German astronomer C. von Weizsacker. The steps are indicated by the following reactions:

Reaction	Mass loss
$_6C^{12} + {_1H^1} \rightarrow {_7N^{13}} + \gamma$	0.0020 (mass loss)
$_7N^{13} \rightarrow {_6C^{13}} + e^+ + \nu$	0.0024 " "
$_6C^{13} + {_1H^1} \rightarrow {_7N^{14}} + \gamma$	0.0080 " "
$_7N^{14} + {_1H^1} \rightarrow {_8O^{15}} + \gamma$	0.0078 " "
$_8O^{15} \rightarrow {_7N^{15}} + e^+ + \nu$	0.0029 " "
$_7N^{15} + {_1H^1} \rightarrow {_6C^{12}} + {_2He^4}$.	0.0053 " "
	0.0284 Total

In the carbon cycle, a carbon nucleus ($_6C^{12}$) combines with a proton to form an unstable, radioactive form of nitrogen isotope ($_7N^{13}$) and the release of a gamma ray. The nitrogen isotope almost immediately releases a positron and a neutrino and is converted into a stable form of carbon ($_6C^{13}$). The stable carbon combines with a proton to form ordinary nitrogen ($_7N^{14}$) and the release of a gamma ray. The ordinary nitrogen combines with a proton to form unstable, radioactive oxygen isotope ($_8O^{15}$) and the release of a gamma ray. The oxygen isotope disintegrates by releasing a positron and a neutrino and forming nitrogen ($_7N^{15}$). The nitrogen combines with another proton to form the original carbon ($_6C^{12}$) and a helium nucleus ($_2He^4$).

In this cycle, four protons have combined to form one helium nucleus. The atomic weight of one proton is 1.008 units. Four protons weigh 4.032 units, whereas one helium nucleus weighs 4.003 units. Thus, in both the proton-proton reaction and the carbon cycle, approximately

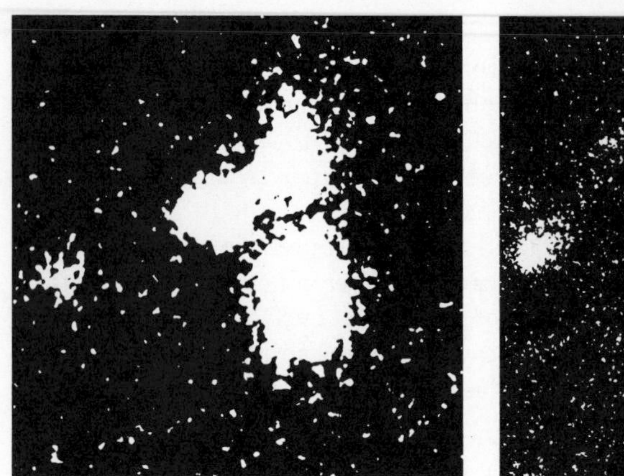

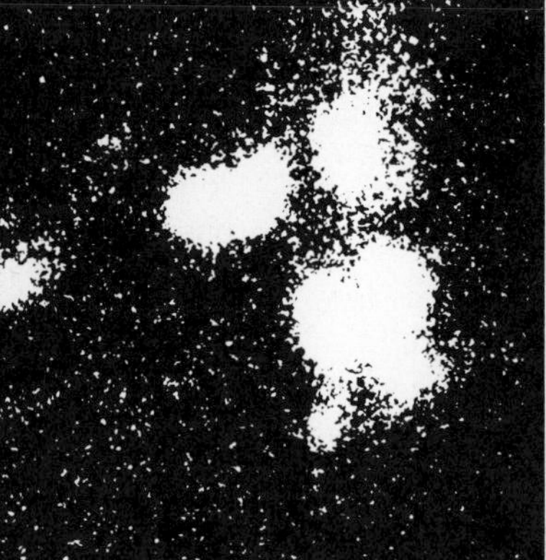

FIG. 13.1
Photographs of Herbig-Haro objects taken at Lick Observatory. The first photograph was taken in 1947; the second was taken in 1960. Notice the changes and the appearance of a new blob. This has been interpreted as a protostar becoming luminous. (Courtesy of the Lick Observatory.)

0.029 mass units are converted into various forms of energy. The carbon cycle, which requires a considerably higher temperature than does the proton-proton reaction, is the main source of energy in the hotter stars, whereas the proton-proton reaction supplies most of the energy in our own sun and in the cooler stars. Note that in both cycles, even though the actual mechanism is different, the overall result is the same. Four hydrogen atoms are converted into one helium atom, and all other elements are restored to their original state. Astronomers often refer to nuclear reactions of this type as "burning." But one should not confuse *nuclear transformations*, in which one element is converted into another, with the ordinary chemical oxidation that occurs in the more conventional meaning of the word "burning."

13.6 THE BIRTH OF STARS

The H-R diagram has proved a valuable tool in the study of the evolution of stars (Chapter 11.11). Although we do not know the complete evolutionary track of a star from its birth to death, we can plot certain observable stages in the development of stars on the H-R diagram, and from these we can formulate an overall picture.

Most of the brightest stars that are visible today can sustain themselves for only several million years at the present rate at which they are using their "fuel." These stars could not have been created five billion years ago when the earth was created. This means that stars are being created continuously.

Young stars are found in clouds of interstellar matter. These clouds are located in the spiral arms of galaxies, in nebulosities, and in dark nebulae. Such locations indicate that interstellar matter might be a prerequisite for the birth of stars. In the early stages of stellar evolution, these clouds of interstellar matter became so dense that the gas and dust particles began to move under the influence of their mutual gravitation to form centers of condensations. Such condensations are called *protostars* and may be related to the objects called *globules*, visible in nebulae such as the Lagoon (Plate 24) and the Rosette (Plate 31). The appearance of a "star" in the Orion association where none was visible in earlier photographs was reported by G. Herbig at the Lick Observatory (Fig. 13.1).

As the protostar contracts, its density, pressure, and temperature increase. The increase in temperature results from the energy released by the gravitational infall of the

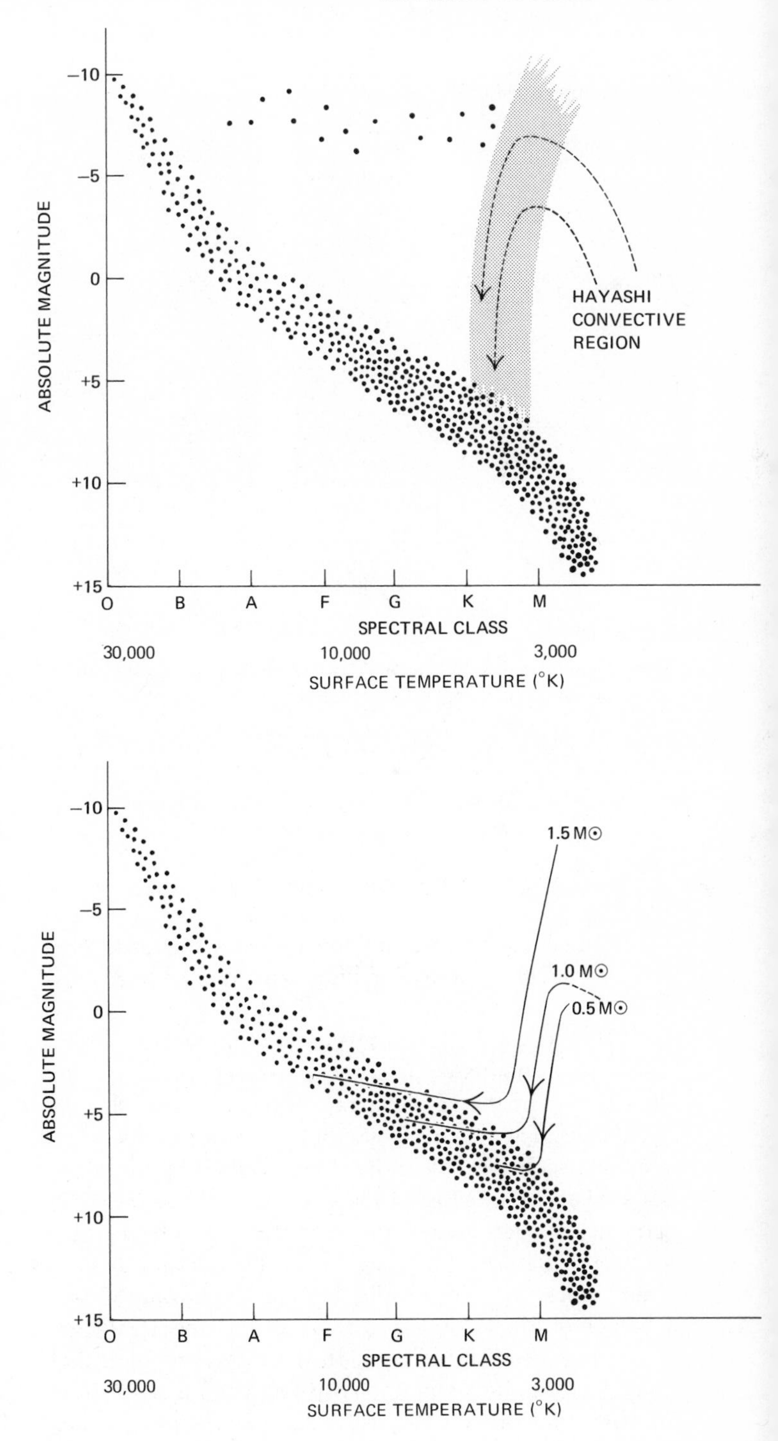

FIG. 13.2 ►
Hayashi convective region, located above the main-sequence stars and to the right of the giants. As a protostar contracts, its density, pressure, and temperature increase (dashed line). As it moves downward, it is completely convective. Protostars to the right of the Hayashi region are unstable.

matter, which is absorbed by the cloud. When thermal equilibrium is established, the temperature is sufficient for the star to shine and be visible. At this stage, according to the Japanese physicist C. Hayashi, the star is completely convective, that is, energy from the interior is carried to the surface by convection. Its position on the H-R diagram is within a nearly vertical, narrow band to the right of the areas occupied by the red giants and supergiants (Fig. 13.2). The Hayashi theoretical evolutionary track for a convective star shows that it moves downward on the H-R diagram, because as the star collapses, its luminosity decreases, with very little change in temperature. During this period, which lasts about 1 million years, the core density increases.

When the transfer of energy in the core changes from convection to radiation, the star turns abruptly and moves horizontally to the left on the H-R diagram. It reaches the main-sequence line when the core temperature is high enough to start and support a nuclear reaction (the synthesis of hydrogen into helium), and this point in the star's evolutionary track is called *zero-age main sequence*. The star's position on the main-sequence line is determined by its mass and chemical composition (Fig. 13.3). A less massive star remains in the convective stage longer; therefore, it moves farther down the Haya-

FIG. 13.3 ►
Evolutionary tracks for stars of 0.5, 1.0, 1.5, and 5 solar masses. The abrupt turn to the left in the Hayashi evolutionary track indicates that radiation is beginning to transport energy in the star rather than convection.

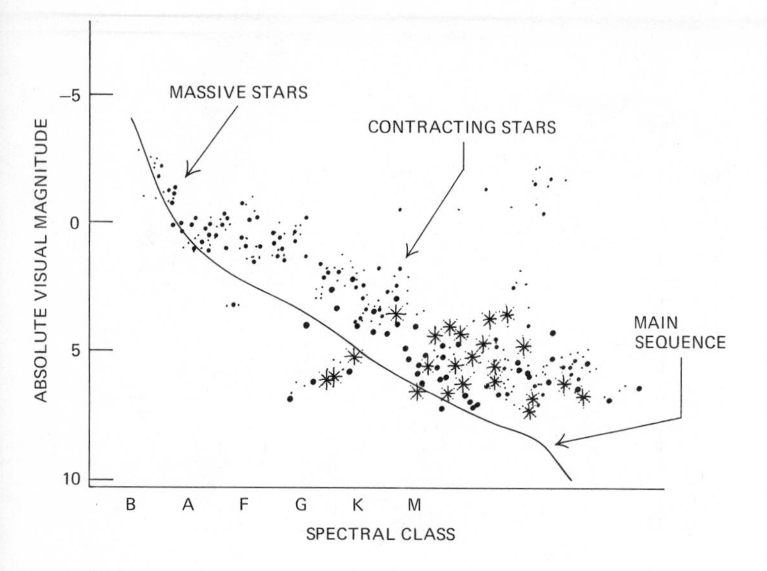

FIG. 13.4
H-R diagram of the young star cluster NGC 2264 (based on work of M. Walker; Lick Observatory)

shi band and joins the main-sequence line at the lower right portion. The time that it takes a star to reach the main sequence is several million years for the most massive and several hundred million years for the least massive.

This general behavior by contracting stars of moving down toward the main sequence is supported by the appearance of the H-R diagram for young star clusters (Fig. 13.4). The massive stars that contract rapidly lie on the main sequence; the more slowly collapsing low-mass stars lie above it. Most of the stars lying above the main sequence in such young clusters are found to be T Tauri stars and are often still surrounded by gas and dust. T Tauri stars vary erratically in their light output (see Chapter 14.5), possibly as a result of their contraction.

There seem to be limits on the masses of stars that can form. As far as is known, there are no stars more massive than about 100 $M_{\odot}$. Likewise, although they would be extremely difficult to see and thus might easily be missed, it appears as if there are few stars less massive than about 0.02 $M_{\odot}$. Any star smaller than this would never be able to raise its internal temperature to the point where nuclear reactions would occur and therefore could not radiate enough energy to be observable. Such faint, cold stars are sometimes referred to as *black dwarfs*. The lack of stars more massive than 100 $M_{\odot}$ is probably due to interruption of their condensation by their own heat. In a collapsing star the core heats up first. For massive stars the heat liberated by the core may stop collapse of the surrounding gases and thereby limit the amount of material that can accrete onto the star.

13.7 MAIN-SEQUENCE STARS

A main-sequence star is in equilibrium because it generates its energy almost exclusively from nuclear reactions in the core at the same rate as its surface radiates energy into space. Most of a star's evolutionary time is spent on the main-sequence line. A more massive and luminous star leaves the main sequence earlier than does a less massive, dimmer star, because it converts hydrogen into helium at a tremendously higher rate. As more hydrogen is converted into helium, the star's core becomes denser and hotter, which causes an increase in the star's nuclear reaction rate and luminosity. This is indicated on the H-R diagram in the star's evolutionary track by a vertical rise within the main-sequence band of less than one magnitude.

The amount of time a star spends on the main sequence depends on how long it takes the star to use up most of the hydrogen in its core. More massive stars have larger amounts of hydrogen to burn, but burn it faster. The main-sequence lifetime is roughly determined by M/L, the ratio of the star's mass to its luminosity. The luminosity, remember, is measured by the star's consumption of hydrogen. From the mass-luminosity law (Chapter 12.6), the luminosity increases rapidly with mass; in fact, L increases proportional to M^3. For example, doubling M increases L by 8 ($=2\times2\times2$). Thus a star's main-sequence lifetime varies as $1/M^2$. A star with ten times the sun's mass has a lifetime 100 times shorter. The main-sequence lifetime of the sun is about 10 billion years.

13.8 ORIGIN OF RED GIANT STARS

When the hydrogen in the core has been almost completely converted into helium, the core of the star begins to contract gravitationally. Core contraction occurs because after hydrogen has been converted into helium, there are fewer particles and thus the core pressure has been reduced. (Recall that for every four protons there is now only one helium atom.) The star's energy is now derived from the nuclear reactions that occur in the hydrogen-rich shell around the core and by the potential energy released by the gases as they collapse gravitationally toward the star's center. Part of the potential gravitational energy and the radiation pressure of the gases in the outer layers of the star force the star to expand. As its size increases, its surface temperature decreases. Since the internal structure of the star has been altered, its evolutionary track shows it moving upward and to the right of the main-sequence line. Eventually, the increase in size and decrease in surface temperature cause the star to become what is known as a red giant or a red supergiant (Fig. 13.5).

In its final stage, as a red giant or supergiant, the star's size has increased tremendously, and its core temperature is about 100 million degrees Kelvin. At this point in its evolutionary track, it is the reddest and most luminous. It is believed that the high temperature triggers an explosion (*Helium flash*), which starts the triple-alpha nuclear reaction process, in which three helium nuclei combine to form one carbon nucleus. The energy released by the triple-alpha process causes the core to expand and thus to reverse the process that led to expansion of the outer layers of the star. The star as a whole contracts, so that its surface temperature increases and causes the star to move horizontally to the left on the H-R diagram in its evolutionary track (Fig. 13.6).

It should be noted that only the lower main-sequence stars go through the exact cycle just described; more massive stars initiate the triple-alpha process quietly, with no sudden rise in luminosity. It should also be kept in mind

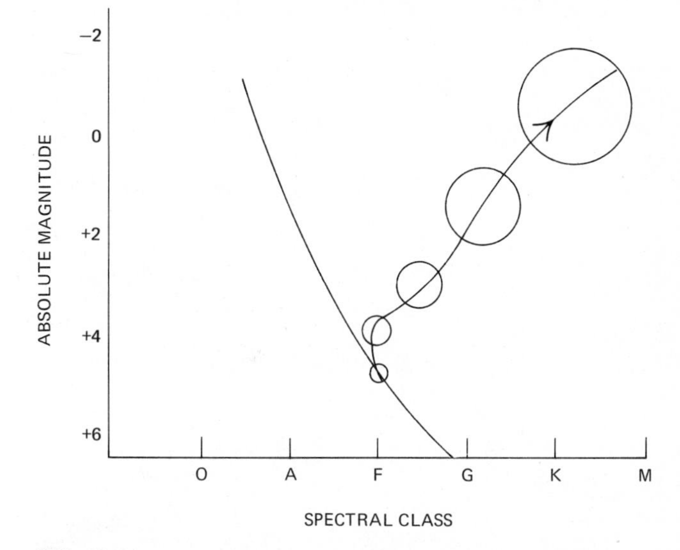

FIG. 13.5
Evolutionary track of a solar mass star from main sequence to red giant. As more hydrogen is converted to helium, the star rises vertically within the main-sequence band in its evolutionary track. When hydrogen in the core is almost completely exhausted, the star moves upward and to the right on the H-R diagram, steadily growing in radius.

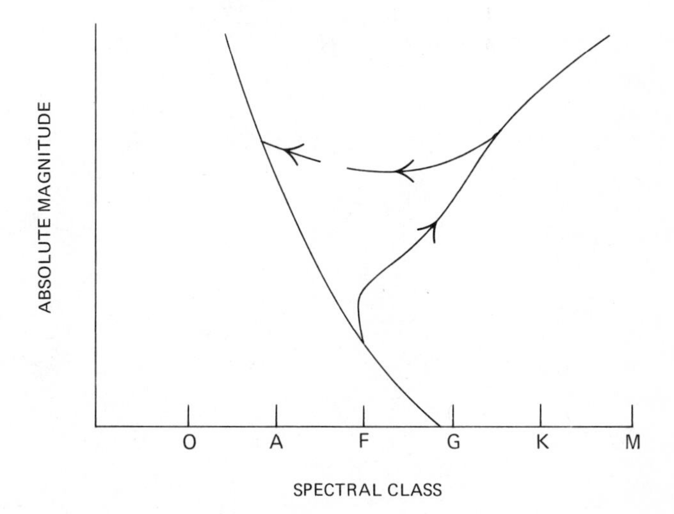

FIG. 13.6
Evolutionary track of a solar mass star from giant along the horizontal branch of the H-R diagram. The "gap" in the horizontal branch represents the location of the RR Lyrae stars.

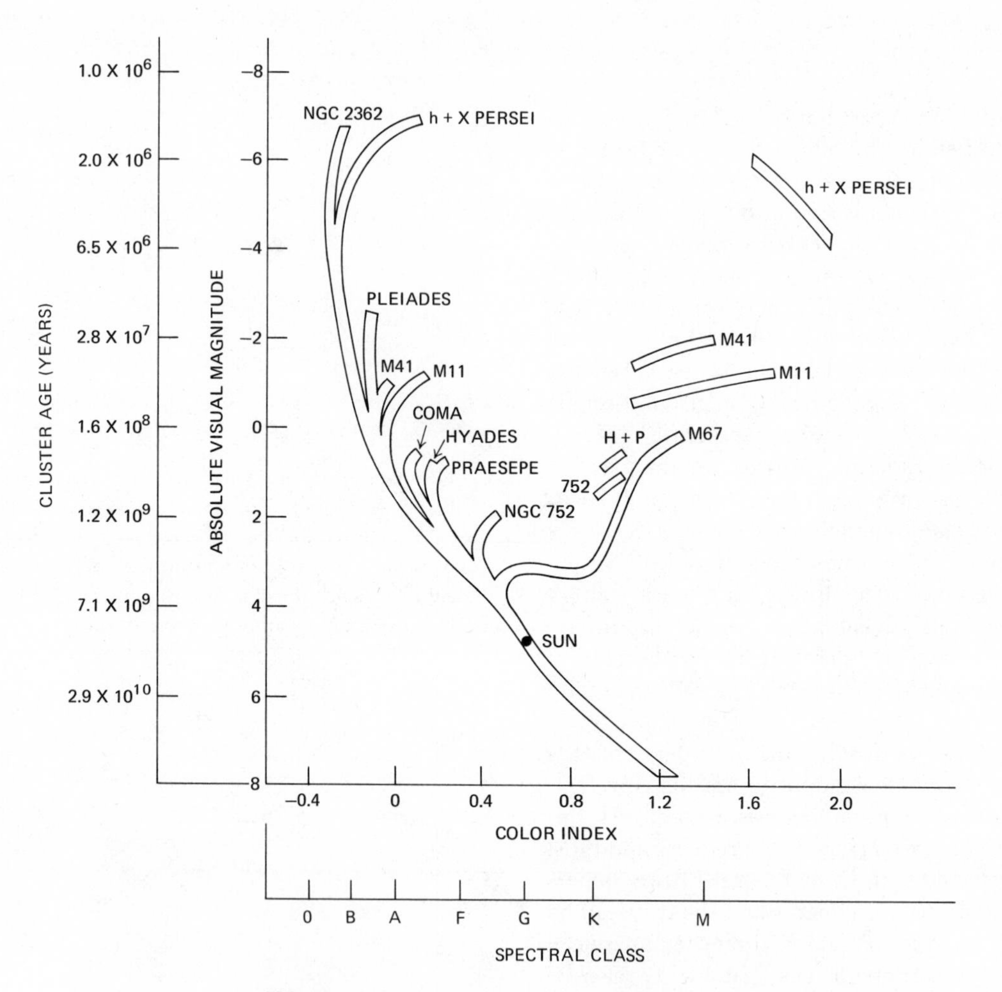

FIG. 13.7
Color-magnitude diagrams of open clusters. The youngest is NGC 2362. The oldest is Messier 67 in Cancer. Where the evolutionary track breaks away from the main-sequence line determines the age of the cluster.

that, at this stage in their evolution, stars may shift their positions in the H-R diagram along very complicated tracks, and in fact, stars may even lose some of their mass by a mechanism similar to the solar wind.

Red giants are very different in structure from main-sequence stars. In main-sequence stars the density, temperature, and pressure change fairly smoothly and gradually from the core to the surface. Low-mass red giants have small and extremely hot, dense cores due to the compression that occurred as hydrogen was exhausted in the previous stage. The core contains one-quarter to one-half of the star's mass, but packs it into only the inner .01% of the star's radius. Surrounding the dense core is a bloated cool atmosphere whose density is less than that of the air in the earth's lower atmosphere.

13.9 COLOR-MAGNITUDE DIAGRAMS

Figure 13.7 shows the superimposed color-magnitude diagrams of several galactic clusters. (Refer to Chapter 11.11 to review color-magnitude diagrams.) Since a cluster is a group of stars of different masses but of one origin, composition, and age, a comparison of their color-magnitude diagrams serves as an indicator of their age and stellar evolution. The youngest cluster shown is NGC 2362 and the oldest is Messier 67 in Cancer. Their ages correspond to the position where the evolutionary track of each cluster breaks away from the main-sequence line.

Color-magnitude diagrams of globular clusters show that most helium-burning stars lie on the horizontal branch of the H-R diagram. The "gap" in the horizontal branch is occupied by the RR Lyrae stars (see Chapter 14.3). It appears that as a star evolves, it must pass through this gap as an unstable variable star. Between the end of the horizontal branch and the main sequence, a star is unstable and pulsates.

13.10 FURTHER EVOLUTION OF RED GIANT STARS

Evolution is not over for a star that has become a red giant. Although a star spends most of its life on the main sequence, as a red giant, its individual destiny depends strongly on the star's mass.

For stars smaller than about 0.5 $M_{\odot}$, once the star has built a core of helium, it can do no more. Exhausted of all fuel, the red giant slowly shrinks down again to the size of a normal star and then, with no nuclear energy source to sustain its interior temperature, it collapses further still, packing its bulk into an even smaller volume. With no internal pressure of any sort, the star might shrivel to something like the size of a golf ball. There are, however, other ways of maintaining pressures than by motion of the core particles. Their explanation takes us from the world of stars to the world of atoms.

A basic principle of atomic physics is that no more than two electrons may occupy the same region of space (a cell) at the same time. This is called the *Pauli exclusion principle*. (Two electrons are allowed because electrons possess a property called *spin*, which can be thought of as measuring their internal rotation. There are two possible spin orientations; if two electrons are of opposite spin they can both occupy the region.) Because of the exclusion principle there is a maximum density to which ordinary matter may be compressed before the electrons begin to find themselves too crowded. In this situation, referred to as *degeneracy*, the electrons respond by moving faster. They are boxed in again, however, by a second effect. Relativity theory says no particle may move faster than the speed of light. Thus, caught between relativity theory and the exclusion principle, the particles react by refusing to be squeezed, that is, they exert a pressure that limits the collapse of the star. The degeneracy pressure becomes important when the density of matter is about 10 million grams per cubic centimeter. (One teaspoon of such matter would weigh several tons.) A 0.5 $M_{\odot}$ star has this density when it is about the size of the earth. We are therefore not surprised to find there exists a family of stars of extremely small size and high density. The compression by which they became so small produces great heat, so they tend to be extremely hot initially, and therefore to appear blue-white. The combination of their color and small size accounts for their name—*white dwarfs*.

The English physicist A. Eddington and the Anglo-Indian astrophysicist S. Chandrasekhar laid the foundation for the theory of the structure of white dwarfs.

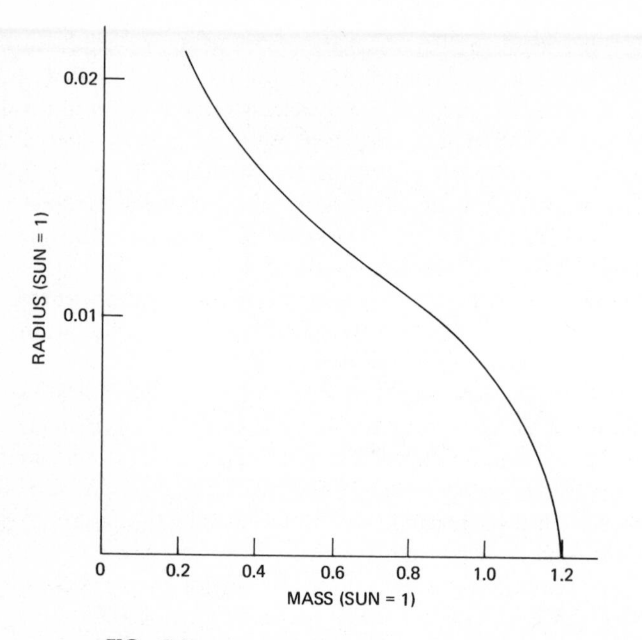

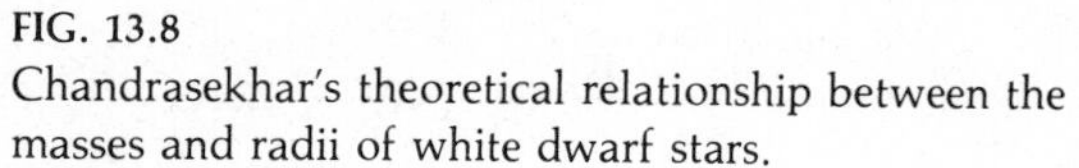
FIG. 13.8
Chandrasekhar's theoretical relationship between the masses and radii of white dwarf stars.

Eddington was the first to show that some of the electrons in white dwarfs have speeds that approach the speed of light, and Chandrasekhar was the first to apply the theory of relativity to the structure of white dwarfs. From his theoretical equations, Chandrasekhar found that the radius of a white dwarf is related to its mass. This theoretical relationship is shown in Fig. 13.8. As the mass of a white dwarf increases, its radius decreases. According to his theory, the maximum mass possible for a white dwarf is 1.2 times the solar mass. More recent calculations show that if the star is spinning, the maximum mass is increased slightly to a few solar masses. Centrifugal forces are then partially able to support the star.

A white dwarf's only source of energy is stored thermal energy. As the velocity of the nuclei of the atoms in a white dwarf decreases, the thermal energy that is released is conducted to the star's surface by the free electrons and then radiated into space. Since a white dwarf cools at an extremely slow rate, its evolutionary track on the H-R diagram is indicated by a diagonal line below and to the left of the main-sequence line.

As described above, white dwarfs represent the end stage of evolution for low-mass red giants. With no fuel source to replenish the energy it emits, the star gradually cools and fades, and its envelope shrinks. Space is probably full of cold remnant white dwarfs, most far too faint to detect.

White dwarfs may originate in other ways as well. For stars whose masses range from about 0.5 $M_\odot$ to about 2 $M_\odot$, heat generated by the contraction of the core as a result of the exhaustion of hydrogen is sufficient to ignite the helium deep in the star's interior. Helium burns into carbon, giving the star an additional fuel source to rely on. For reasons that are still obscure, in association with the conversion of helium to carbon or shortly thereafter, the outer envelope of the star may be blown away in a mild explosion. Pushed outward from the small hot core, the star's atmosphere appears as a huge expanding shell, a *planetary nebula*, illuminated by fierce ultraviolet light from the core (Fig. 13.9). Thereafter the shriveled core gradually cools, becoming a white dwarf; meanwhile, the ejected shell expands into space at about 20 kilometers per second and in 10 to 20 thousand years is dissipated.

For stars more massive than 5 $M_\odot$, the evolution proceeds differently. To support the larger mass, such stars must have a higher core temperature. In addition, the larger gravitational forces of the more massive star hold its outer layers firmly and seem to prevent it from forming a planetary nebula.

The higher core temperature is significant because it allows a wider range of nuclear reactions to occur. Normally, nuclei repel each other because of the charges on the nuclei, but at high temperatures the nuclei move sufficiently fast to overcome these repulsive forces. Thus, while two hydrogen nuclei (each with a charge of 1) can fuse at 10^6 °K, the fusion of helium (charge 2) requires a temperature of 10^8 °K. The fusion of carbon atoms requires still higher temperatures. A few of the possible

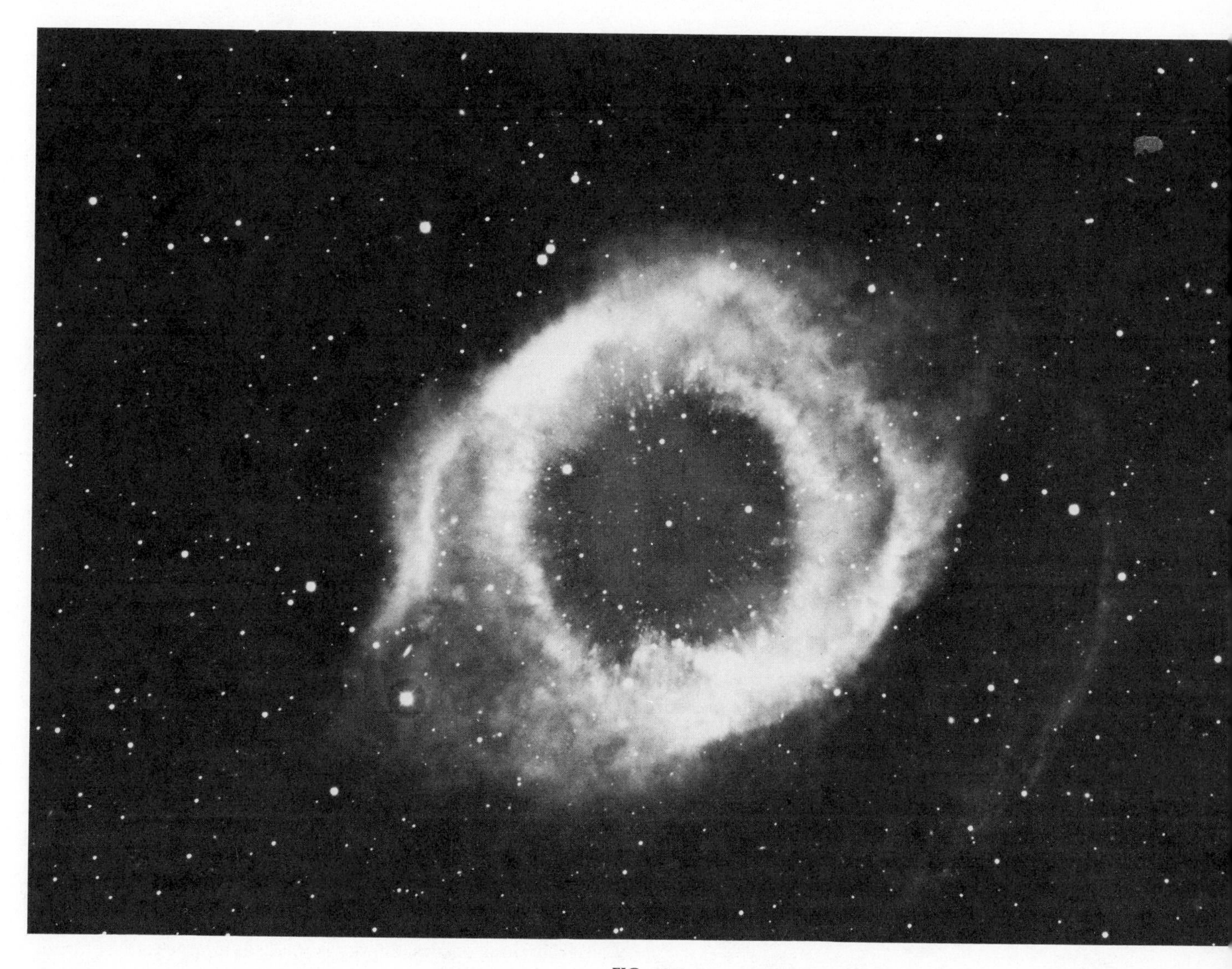

FIG. 13.9
The planetary nebula NGC 7293. (Courtesy of Cerro Tololo Inter-American Observatory.)

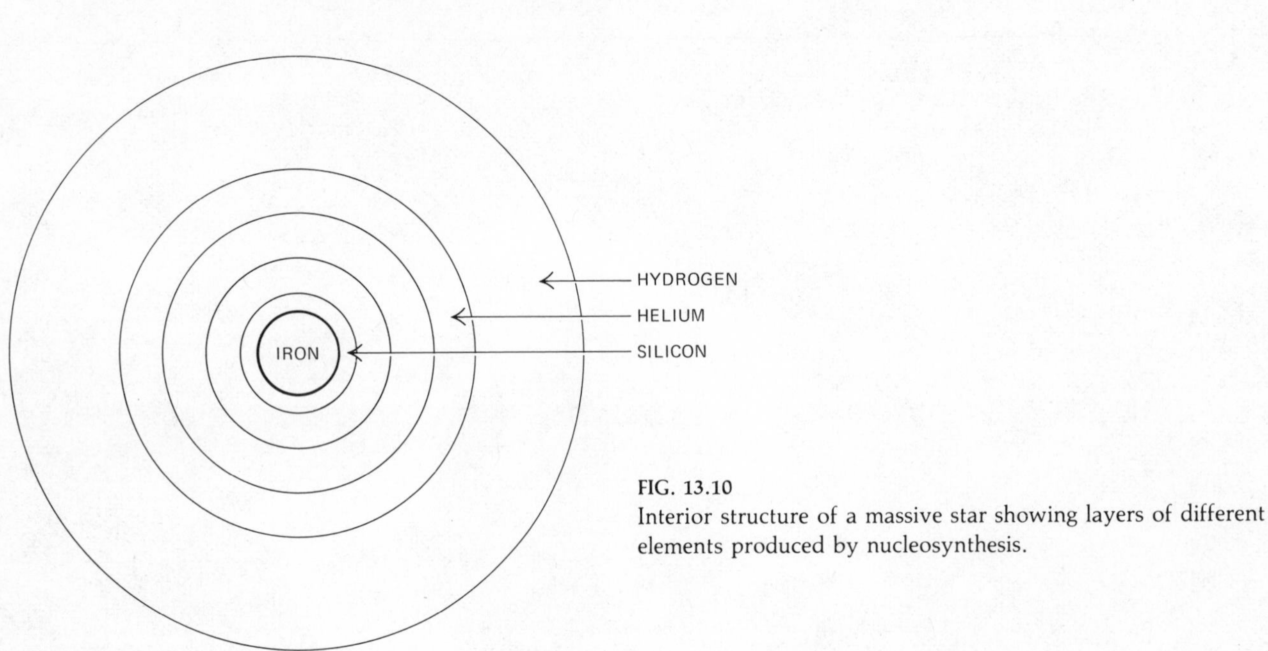

FIG. 13.10
Interior structure of a massive star showing layers of different elements produced by nucleosynthesis.

reactions in massive stars—including the triple-alpha process (helium burning) that we have already discussed—are listed below.

Helium burning
(Triple-alpha process) $3\,{}_2He^4 \longrightarrow {}_6C^{12}$

$${}_2He^4 + {}_6C^{12} \longrightarrow {}_8O^{16} + \gamma$$

Oxygen burning $${}_6C^{12} + {}_8O^{16} \longrightarrow {}_{14}Si^{28}$$

or

$$2\,{}_8O^{16} \longrightarrow {}_{14}Si^{28} + {}_2He^4$$

Silicon burning $$2\,{}_{14}Si^{28} \longrightarrow {}_{28}Fe^{56}$$

All of the above reactions produce energy. All of the reactions also convert a low-mass, low-charge nucleus into a more massive and more highly charged species. As one moves down the list, higher and higher temperatures are required to effect the change. These reactions are sometimes referred to as nuclear transformations or *nucleosynthesis*, because they involve the conversion of low-mass elements into more massive elements.

When a massive star has completed helium burning, the core contracts, the core temperature again rises due to compression, and this time, if the temperature is great enough, the carbon residue begins to burn. When the carbon is consumed and converted to neon, sodium, and magnesium, the core again undergoes contraction and heating, and if enough additional heat is present, oxygen is burned into silicon. Each time a new series of nuclear reactions occurs, energy is made available to the star, but the star rapidly radiates the energy through its surface. The interior of the star thus takes on a layer-cake appearance with an iron core surrounded by a shell of silicon, which in turn is surrounded by a shell of oxygen, and so on (Fig. 13.10).

Once a star has produced an iron core, a fundamental change takes place. In earlier stages, contraction of the core has always liberated energy. However, compression and heating of the iron core causes the iron nuclei to disintegrate into helium atoms and neutrons, a process that is accompanied by a sharp *absorption* of energy. The core is thus momentarily cooled and its pressure falls. The reduced pressure means the layers immediately surrounding the core cannot be supported. They fall in, literally in seconds, pulled by the enormous gravity at the star's center. Falling into the hot core, they ignite explosively, burning the silicon into iron, oxygen into silicon, etc., and releasing a prodigious pulse of energy. The energy travels out through the star's envelope, heating it and tearing it free. The pulse also travels inward toward the center of the star, compressing the remnant core still further. The star disappears in a blinding flash, its atmosphere blasted out into space as a *supernova*.

The occurrence of a supernova has many important consequences. It returns to interstellar space material that has been converted from hydrogen into heavier elements. All of the carbon atoms in our bodies, the oxygen atoms we breathe, and the silicon atoms of the ground beneath our feet were once in a star. A supernova also heats the interstellar medium and may, in fact, trigger the formation of new stars in adjacent regions by compressing the interstellar gas (Fig. 13.11). The supernova process is also responsible for creating two of the strangest astronomical objects yet discovered: *neutron stars* and *black holes*.

FIG. 13.11
Star formation triggered by a supernova explosion. The expanding supernova shell has compressed a cloud causing it to collapse. (National Geographic Society—Palomar Observatory Sky Survey. Courtesy Hale Observatories.)

13.11 NEUTRON STARS

The compression of the core of a supernova is so extreme that negatively charged electrons, normally far removed from the nucleus, are forced into the nucleus, where they combine with positively charged protons to form neutrons. The remnant of the supernova becomes an aggregate of neutrons or a *neutron star*.

Neutron stars possess a number of remarkable properties. They are extremely small, with radii of only 10 kilometers, approximately. They are extremely dense. The stellar core, which may contain several solar masses of material, reaches a density of 10^{14} gm per cc when compressed into a sphere only 10 km across. A teaspoon full of neutron star matter weighs as much as a mountain of solid rock half a mile high and a mile in circumference.

Neutron stars possess strong magnetic fields. The compression involved in the collapse of the star's core intensifies any magnetic field it has. Neutron stars also rotate rapidly. As we have discussed earlier, decreasing the size of a rotating object makes it spin faster if its angular momentum is conserved. A star like the sun,

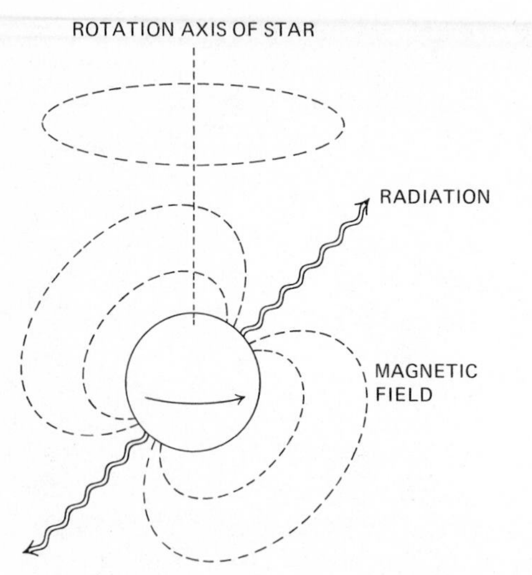

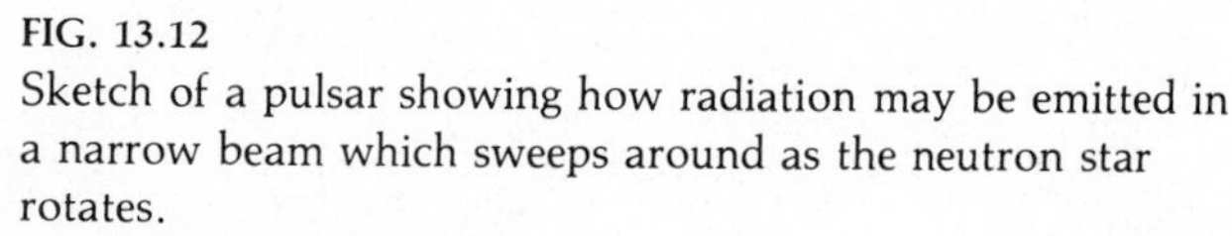

FIG. 13.12
Sketch of a pulsar showing how radiation may be emitted in a narrow beam which sweeps around as the neutron star rotates.

rotating once a month, would spin 10 times a second if compressed to neutron star size without losing any angular momentum.

By a mechanism not yet thoroughly understood, the rapid rotation of the neutron star and its strong magnetic field cause it to send out nonthermal radiation in two narrow beams, one along each of its magnetic poles (Fig. 13.12). This nonthermal radiation (also called *synchroton radiation*) may be emitted at virtually all wavelengths, but usually it is strongest at radio wavelengths. The rotation of the star causes these two beams to sweep across space like the beacon of a lighthouse (or the more familiar blinker on a police car). If the beams happen to sweep across the earth, we can detect them. Each time the neutron star turns we see a pulse of radio waves. Since an object as massive as a star turns at a very constant rate, the repetition of pulses is extremely precise. The discovery of the first radio-pulsating neutron star or *pulsar* was made in 1967 by a young English astronomer, Jocellyn Bell. The observed properties of pulsars are described further in Chapter 14.

13.12 BLACK HOLES

A massive star may be destined for a fate other than becoming a neutron star. It may collapse completely at the end of its lifetime and become instead what is called a *black hole*. For all the stars discussed thus far, a balance is achieved between gravitational forces attempting to crush the star and internal pressure attempting to support the star. For ordinary stars, gas pressure is enough to balance the gravitational forces. For white dwarfs, with their higher density and greater gravitational forces, electron degenerate pressure is strong enough to support the star, and for neutron stars, neutron degenerate pressure suffices. But can one always rely on a new type of pressure to intervene at the last moment and save a star against its own collapse? Recall that pressure is a measure of internal energy. However, the theory of relativity relates energy to an equivalent mass. At extremely high pressures, the gravitational field acting to compress matter in stars of more than a certain mass is enhanced by the internal energy acting to support the star. At a critical density set by relativity theory, the star cannot be supported against gravity. It will therefore collapse totally and become a black hole.

Black holes are one of astronomy's most curious phenomena. Black holes emit no radiation: no visible light, no radio waves, no particles. It should not be surprising that a black hole emits no light if we recall the discussion of escape velocity in Chapter 8.2. It can be shown that the square of the escape velocity depends on the ratio of an object's mass to its radius. If the mass increases or if the radius decreases, the escape velocity becomes larger. For a star like the sun the escape velocity is about 0.001 the speed of light (c). For a white dwarf, with its much smaller radius, the escape velocity is 0.01 c. For a neutron star it is 0.1 c. Suppose now that an object has either so large a mass or so small a radius that the escape velocity equals the speed of light. In that case no light or radiation of any kind can escape from it.

From the expression for the escape velocity we can write an expression for the radius of a black hole. If the escape velocity is c, the speed of light, then $c^2 = 2GM/R$, where G is the gravitational constant and M and R are the mass and radius of the hole. Solving for R, we have $R = 2GM/c^2$. This radius is called the *Schwarzschild radius.* For one solar mass, R is 3 km. Thus neutron stars are not quite small enough to be black holes, but they are close enough in size to make the compression of a supernova core into a black hole a believable possibility. Black holes may also have been produced in the early stages of the evolution of the universe, but this is far from certain.

It is believed that black holes can possess only mass, charge, and angular momentum. The mass of a black hole is what determines its gravitational pull. Despite being collapsed, black holes exert a gravitational force on their surroundings identical to that of an ordinary-size body containing the same amount of material. Thus, were the sun to become a black hole, the earth would continue to orbit it as before. Black holes do *not* pull everything in their vicinity into them, any more than does the sun. However, if anything does fall in, it can never get out, because escape would require motion faster than light, which is impossible. Black holes also have the ominous property that as material falls in, the hole's mass and therefore its radius are increased. A larger radius allows it to collect material faster . . . and so on.

Do black holes exist? The evidence is so far very uncertain. There would be no direct way of seeing one if it existed. The hole itself would be just what its name suggests—black. It would be far too small to be seen silhouetted against background stars or nebulosity. Material that is falling in toward a black hole, however, might well be visible. As we will see in Chapter 14.10, black holes may be strong x-ray sources.

There is still much to be learned about black holes. For example, a theory recently developed by S. Hawking, a young English mathematician, suggests that black holes can create high-energy radiation at the expense of their mass. The energy escapes and the black hole's mass is slightly reduced. However, this evaporation is extremely slow for objects as large as those of stellar mass.

13.13 A REVIEW OF STELLAR EVOLUTION

The success of stellar models in accounting quite accurately for the properties of such diverse objects as red giants and white dwarfs suggests that our basic ideas about stellar structure and evolution are correct. Our understanding of the hows and whys of stellar deaths is not on such firm ground. The nature of neutron stars, supernovae, and black holes is still clouded with uncertainty. New observations using x-ray telescopes, radio telescopes, and all of the other new techniques of the space age add to our understanding of these exotic objects only piece by piece. It is this search that makes astronomy so exciting to the astronomer.

A summary of the evolutionary track of a low-mass star is shown in Fig. 13.13. It is believed that the dust and gas particles within a dense region of interstellar matter begin to contract under their mutual gravitational attraction to form condensations. A condensation visible as a dark "globule" is called a protostar. Thermal equilibrium is established at point *a* on the diagram, when the temperature is sufficient for the star to shine and be visible. Since the energy at this stage is transferred convectively to the star's surface, the star moves downward on the diagram, maintaining a fairly constant temperature, but gradually decreasing its luminosity. When the star begins to transfer energy within the core by radiation and its core temperature is sufficient to start and maintain a nuclear reaction, the star's evolutionary track abruptly changes and moves to the left (point *b*). It hooks onto the main-sequence line at point *c* when the star becomes stable, that is, when it generates energy at the same rate at which it emits it into space. This point is called zero-age main-sequence. Practically its entire life is spent as a stable, main-sequence star. As it converts hydrogen into helium, the star moves about one magnitude vertically within the main-sequence band. When almost all of the hydrogen in the core has been exhausted, it begins to contract gravitationally. Part of the gravitational potential

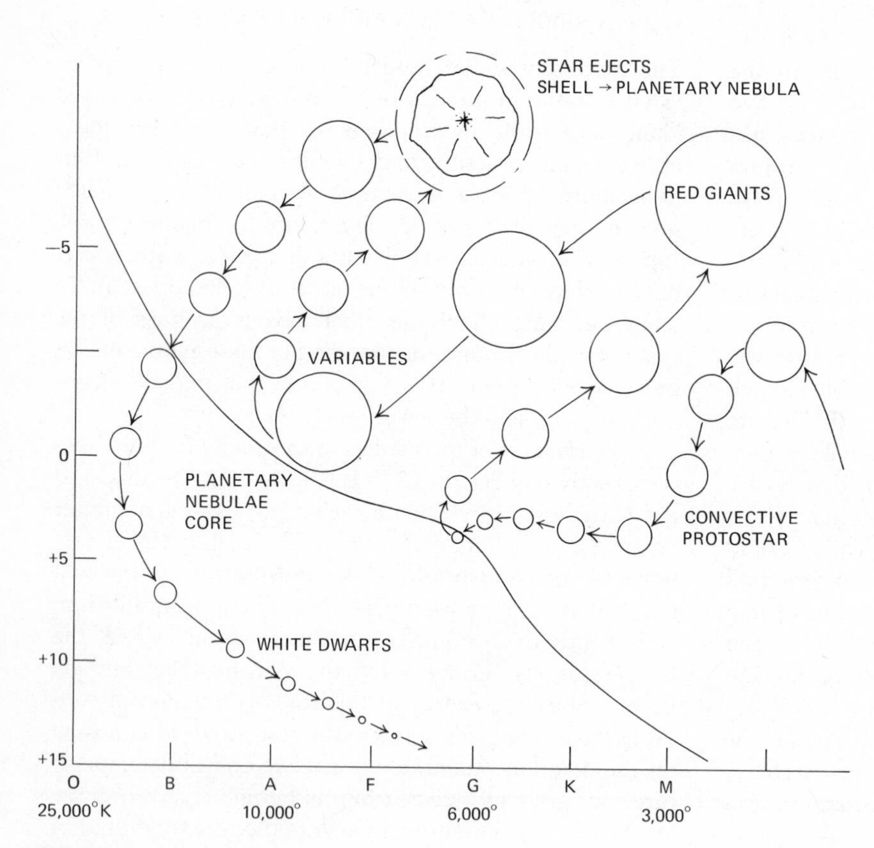

FIG. 13.13
The evolutionary track of a star slightly more massive than the sun, from protostar to white dwarf.

TABLE 13.1
Stellar evolution

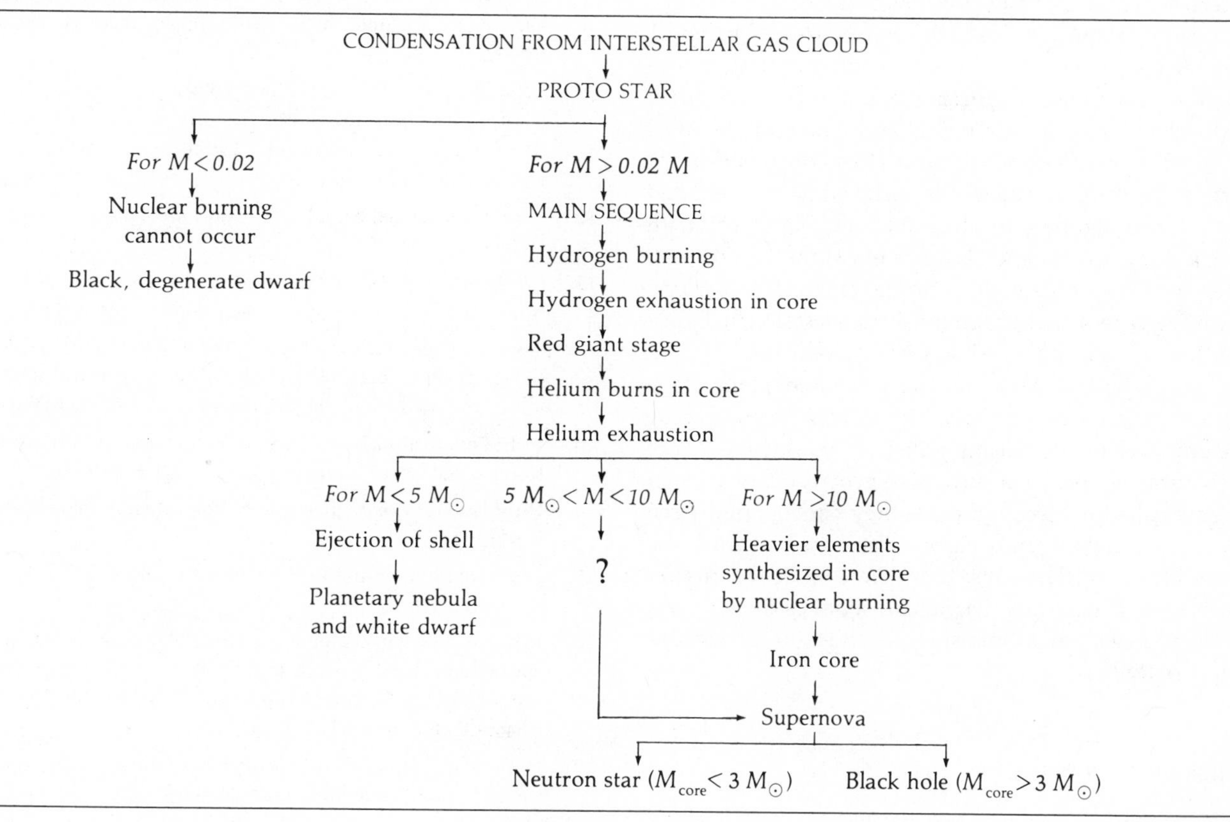

energy is used to expand the outer layers of the star. Since the star's internal structure has been changed, it begins to move above and to the right of the main-sequence line and eventually becomes a red giant, with a core temperature of about 100 million degrees Kelvin. This enormous temperature triggers an explosion which starts the *triple-alpha nuclear reaction*—the conversion of three helium nuclei into one carbon nucleus. This reverses the expansion process, and the star begins to contract. During this stage, the star becomes unstable (a variable star) and fluctuates considerably in size and luminosity. The final stage in the star's evolutionary track is a white dwarf. Stars of higher mass may end their lives as neutron stars or black holes.

SUMMARY

The structure of a star is determined by its need to be in both hydrostatic (mechanical) and thermal equilibrium. Hydrostatic equilibrium requires that the gravitational force inward balance the pressure force outward. Thermal equilibrium requires that the star's energy loss rate be balanced by a supply of energy from the interior. In almost all stars energy is supplied by nuclear reactions in which atoms of a light element (such as hydrogen) are converted to atoms of a heavier element (such as helium). This fusion process is sometimes referred to as "burning." As a star consumes its hydrogen, it can no longer maintain either mechanical or thermal equilibrium, and therefore its structure (e.g., its radius and temperature) must change. For main-sequence stars the consumption of hydrogen in the core causes the surface layers to expand and cool, turning the star into a red giant. The speed with which a star evolves depends on its mass; the support of a higher mass requires a higher internal temperature, which in turn leads to a higher luminosity and faster fuel consumption. Thus more massive stars evolve faster.

The evolution of a star past the red giant stage depends critically on its mass. Lower-mass stars become planetary nebula by ejecting their outer layers. More massive stars engage in a successive generation of heavier elements called nucleosynthesis—they convert hydrogen into helium, helium into carbon, and carbon into still heavier elements. The subsequent ejection of this matter into space in a supernova explosion supplies the universe with heavy-element atoms such as those of which our bodies are built.

REVIEW QUESTIONS

1. What is meant by saying that a stable star is in (a) hydrostatic equilibrium (b) thermal equilibrium?
2. List and describe the three mechanisms by which energy is transferred within a star. In what part of the star is each the dominant mechanism?
3. Describe the proton-proton reaction in the production of stellar energy.
4. What is being converted in the carbon cycle? What is being produced? What part does the carbon nucleus play in this reaction?
5. In what part of the star does the proton-proton reaction and the carbon cycle occur? Why?
6. Why is it not surprising that young star clusters are often surrounded by interstellar dust and gas?
7. Explain what is indicated by a change in a star's position on its evolutionary track on the H-R diagram.
8. Explain what has occurred within a star, according to Hayashi, when it starts to shine and becomes visible. Where is it located on the evolutionary track on the H-R diagram?
9. What has occurred when the evolutionary track of a star abruptly starts to move horizontally on the H-R diagram toward the main sequence? What occurs when it has reached the main sequence?
10. Explain what determines the star's position on the main sequence.
11. What is a main-sequence star? In the evolution of a star, how much of its life is spent as a main-sequence star?
12. Two stars reach the main-sequence line at the same time. If one is spectral class K and the other is spectral class B, which one will leave the main-sequence line first? Why?
13. In the evolutionary process of a star from main sequence to red giant, what happens to its surface temperature?
14. Why do stars of higher mass evolve faster than those of lower mass?
15. Why are high temperatures necessary for nuclear reactions to occur?
16. What is the triple-alpha reaction? At what point in the star's evolutionary track does it occur? What happens to the star as the result of the triple-alpha process? How is this indicated on the H-R diagram?
17. Why is the main sequence bent over toward the right in the H-R diagram of older star clusters?
18. What do T Tauri stars in a cluster indicate about the cluster's age?
19. What is the difference between the carbon cycle and carbon burning?
20. What are white dwarf stars? Explain what is meant by degenerate gas.
21. What is meant by the limiting mass of a white dwarf?
22. Explain how the color-magnitude on the H-R diagram is used in estimating cluster distances.
23. What is a neutron star?
24. What is believed to cause supernova explosions?
25. What determines whether a supernova leaves a black hole or a neutron star as its end product?
26. Describe the evolution of a 15 $M_\odot$ star from its beginning to its end.

CHAPTER 14
VARIABLE STARS

In the previous chapter we saw that most stars lead uneventful lives apart from their birth and death pangs. Like the sun, the average star spends most of its lifetime on the main sequence, slowly converting hydrogen into helium in a measured way. The night sky bears this out. Most stars seem to shine with light of an almost chilling constancy. However, a very small fraction of them undergo changes in their light output on time scales that human beings can perceive. Just as aberrant human behavior is often more interesting than normal behavior, so these aberrant stars have a particular fascination.

Stars whose light output changes are called *variable stars*. Some vary in a rhythmic and predictable way, others undergo explosive and destructive outbursts. We have already discussed binary stars—two stars that revolve around their barycenter under the influence of their mutual gravitational force. Eclipsing binary stars are not true variables: their variation in brightness is caused extrinsically by the periodic eclipsing of one star by the other as viewed from the fixed vantage point, earth. On the other hand, the variation in brightness of intrinsic variable stars is caused by actual physical changes that occur within the stars. These are the true variables (Figs. 14.1, 14.2).

Two general types of intrinsic variable stars can be identified: *pulsating variables* and *eruptive* (explosive) *variables*. Most of the pulsating variables are giants, and a few are supergiants of spectral classes B to N. Some show irregular fluctuations in brightness, but most of them have a fairly regular rhythmic period which may be as little as one hour or as much as three years. The pulsating variables include the *Cepheid variables,* the *RR Lyrae stars,* and the *red irregular variables.*

The eruptive variables are stars that display sudden, unexpected outbursts of energy. In novae stars, the increase in brightness may be as much as 12 magnitudes, with a duration of about two weeks. In supernovae stars, the increase is so enormous that the star may sometimes be clearly visible in the daytime sky for a period of several months.

Variable stars are designated by single Roman letters or by combinations of letters within each constellation. The first variable discovered in a constellation is assigned the letter R. Thus the first star discovered in the constella-

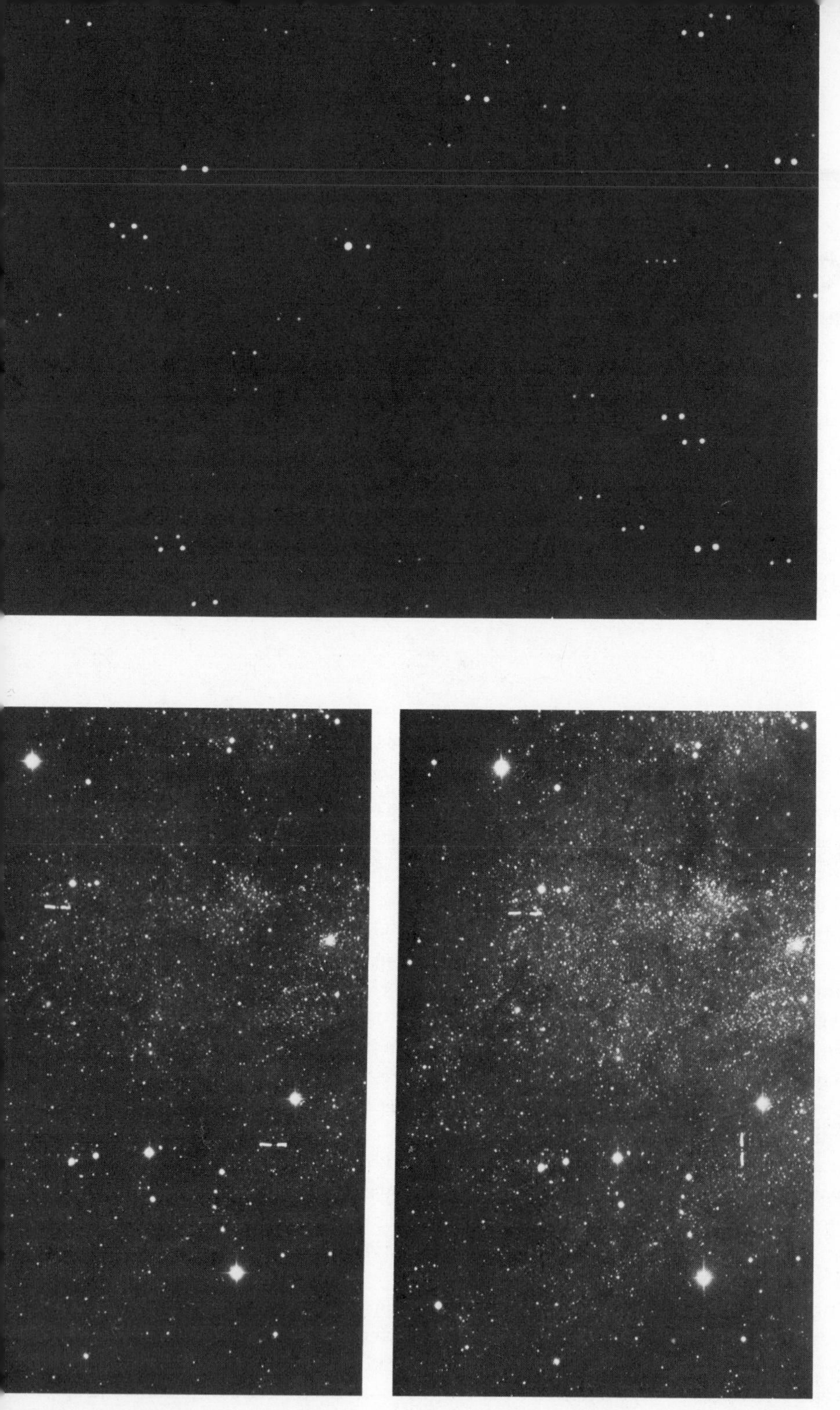

FIG. 14.1
WW Cygni, eclipsing Algol-type variable star, photographed at maximum and minimum brightness. This is a double-exposure photograph with a slight displacement. The variable is seen in the center of the photograph. (Courtesy of the Hale Observatories.)

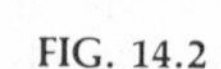

FIG. 14.2
A field of variable stars in the Andromeda Galaxy (NGC 224) with two variables marked. Photographed with the 200-inch telescope. (Courtesy of the Hale Observatories.)

tion of Corona Borealis is designated as R Corona Borealis. Subsequent discoveries are assigned letters S, T, . . . , Z; then two-letter combinations, RR, RS, . . . , RZ; SS, ST, . . . , SZ; and so on to ZZ; then AA, AB, . . . , AZ; BB, BC, . . . , BZ; and so on to OZ, with the letter J omitted. This system accommodates 334 variable stars in each constellation. With additional discoveries, the letter V, followed by Arabic numbers, is used, and the variable is designated V335, . . . , and so on. In Cygnus, for example, the 335th variable is designated as V335 Cygni.

14.1 CEPHEID VARIABLES

Why do stars pulsate? The reason is actually fairly straightforward. The atmospheres of many giant stars have the property that, when hot and compressed, they trap the energy liberated in the core. When the star is cool and distended, energy can flow freely. If energy cannot flow through the atmosphere, the outer layers block the radiation seeking to escape and bottle it up. As more and more energy is trapped, the atmosphere swells up. However, once the atmosphere expands, the trapped energy can escape freely into space and the atmospheric gases settle back to their original state. Compressed again, the gases start trapping heat once more and the cycle repeats. It is a bit like blowing up a balloon with a small tear in it. As you puff air into the balloon it expands. Eventually you put in so much air that the tear opens up and all the air rushes out. You then have to start blowing it up again, and the whole cycle repeats.

The prototypes of the pulsating variables are the Cepheids, which derive their name from the star Delta Cephei, the first to be discovered. Nearly 600 Cepheids have been discovered in our galaxy, and all lie very close to the galactic plane within the layer of dust and gas. They are the yellow supergiants of spectral classes F or G at maximum brightness, with periodic light variations that range from 1 to 50 days (average period is about 6 days).

The light curve of Delta Cephei, a typical Cepheid, is smooth and asymmetrical (Fig. 14.3). It shows that the star brightens very rapidly from an apparent magnitude of 4.4 to 3.7, then fades at a slower rate back to 4.4. This light curve is typical for the Cepheids with periods up to about five days. Those with longer periods show a definite hump on the fading portion of the curve, which is probably caused by shock waves as the star pulsates. The average variation in brightness for the Cepheids is one magnitude. Polaris, the north star, is a Cepheid variable with a short period of four days and a variation in brightness of only 0.1 magnitude, which can be observed with binoculars.

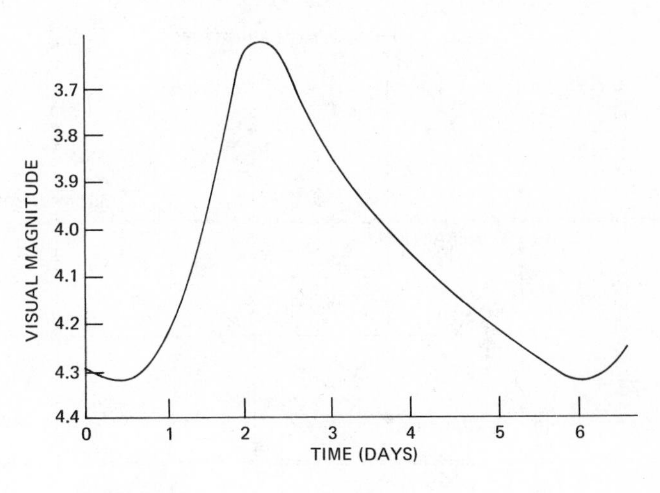

FIG. 14.3
The light curve of Delta Cephei, the prototype Cepheid.

A spectroscopic analysis of the light from a Cepheid variable shows that changes occur in the radial velocity of the star and in the spectral class with a variation in

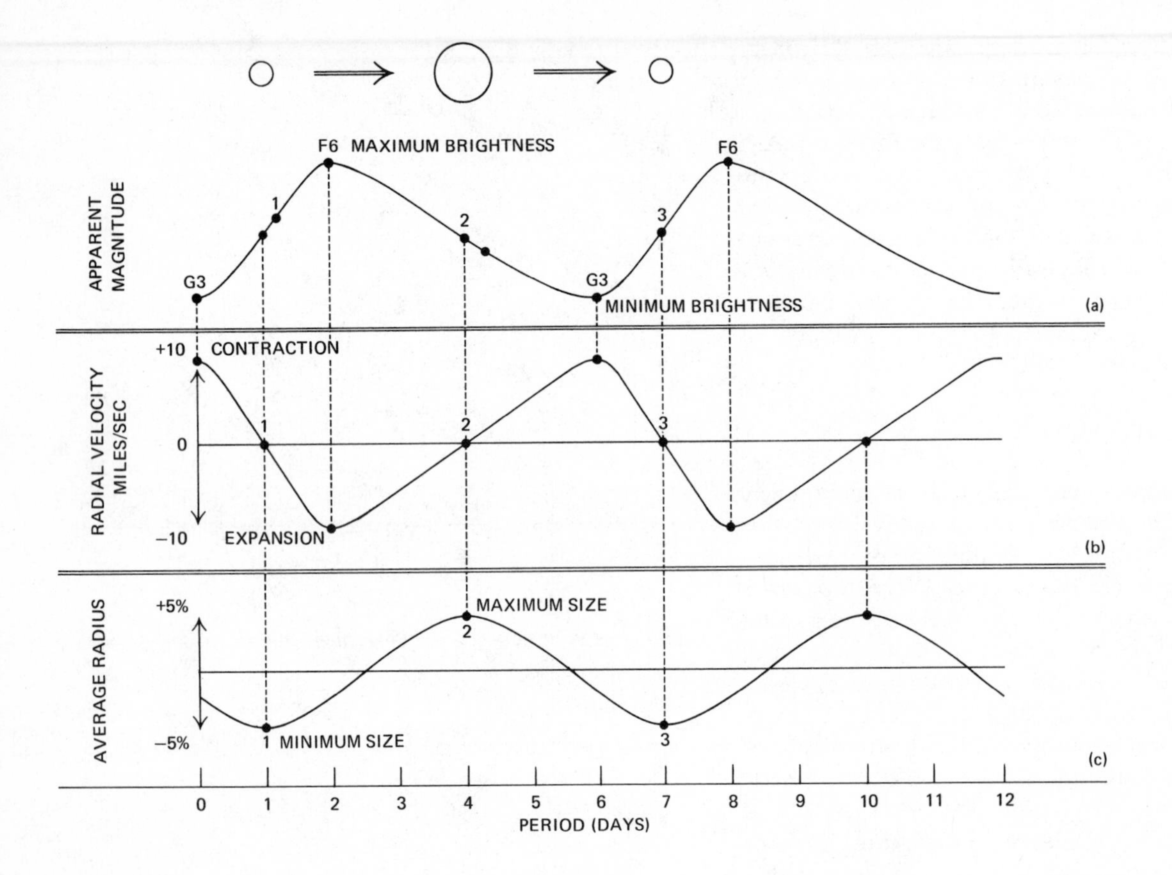

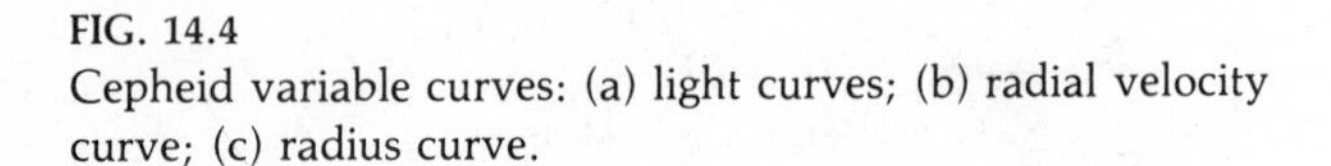

FIG. 14.4
Cepheid variable curves: (a) light curves; (b) radial velocity curve; (c) radius curve.

brightness. Figure 14.4 shows the relationship between the Cepheid's light curve, spectral class, radial velocity, and size. The variation in brightness is interpreted as a periodic rise and fall (pulsations) of the star's radiating surface evidenced by the Doppler shift in the spectrum. The radial velocity curve is a mirror-image of the light curve; for a typical Cepheid, its total range is about 24 miles (39 km) per second.

When the star is in position 1 on the light curve, its radial velocity is zero (no Doppler displacement), and its size is minimum. As the star expands, its spectral lines shift to the violet, because the star's surface approaches the observer, and its size reaches a maximum (2) when the radial velocity becomes zero. During the expansion period, the total change in the star's radius from minimum to maximum is about 10% of the average radius of a typical Cepheid. As the star contracts, its spectral lines shift to the red, because the star's surface recedes from the observer, and its size reaches a minimum (3) when the radial velocity becomes zero again.

If a Cepheid reacted like a typical star, part of the energy released during the contraction period would radiate into space, and the rest would increase the kinetic energy of the star's atoms, resulting in increased luminosity, with maximum occurring when the star's size became minimum. However, the light curve of a Cepheid does not follow this pattern. Maximum luminosity occurs when the star is expanding at its maximum radial velocity and after it has reached its minimum size. Minimum luminosity occurs when the star is contracting at its maximum radial velocity and after it has reached its maximum size. A variation in the brightness of a Cepheid produces a change in the spectral class from G at minimum brightness to F at maximum brightness, thus indicating that maximum energy may be produced when the star's size is minimum, but its effect does not reach the star's surface until later because of a time lag.

14.2 THE PERIOD-LUMINOSITY RELATION

When Henrietta Leavitt of Harvard College Observatory plotted the periods of the Cepheids in the Small Magellanic Cloud against their apparent magnitudes in 1912, she discovered a significant relationship, as shown in Fig. 14.5. The Cepheids of longer period have greater luminosity. Since a star's apparent magnitude depends on its absolute magnitude and distance, she assumed that the stars in the Cloud are at approximately the same distance from the earth, because the size of the Cloud in relation to its distance from the earth is relatively small; therefore, the difference between the absolute and apparent magnitudes of each star is constant. Thus the apparent magnitudes of the Cepheids could be taken as a measure of their absolute magnitude, thereby establishing the period-luminosity relationship.

The period-luminosity relationship is extremely important because it provides the tool for measuring the distances to nearby galaxies. This is accomplished by observing the apparent magnitude and the period of the Cepheid variable. With the period, the absolute magnitude is obtained from the period-luminosity diagram. When the apparent and absolute magnitudes are known, the Cepheid's distance can be determined from the inverse square law (Chapter 11.7).

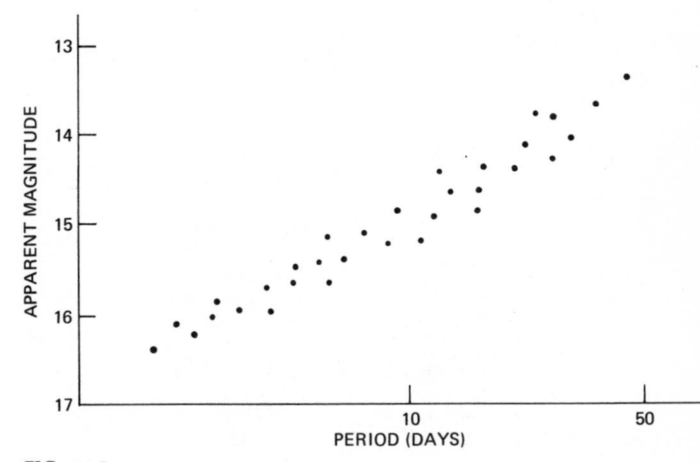

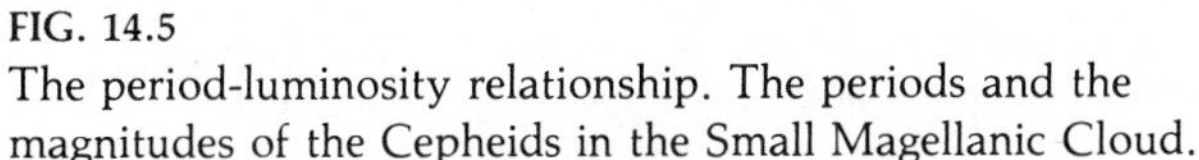
FIG. 14.5
The period-luminosity relationship. The periods and the magnitudes of the Cepheids in the Small Magellanic Cloud.

14.3 RR LYRAE STARS

Another group of pulsating stars contains the RR Lyrae stars, which were first observed in globular clusters. They are the blue-white giants with absolute magnitude of zero that lie on the horizontal branch of the H-R diagram. Their periods of light variations range from about one hour to one day. At maximum brightness most of them are spectral class A; at minimum brightness, spectral class F. Most RR Lyrae stars belong to Population II (the old stars found in globular clusters around the galactic center and in the galactic plane), although a few of them belong to Population I, young stars found in the galactic disk. (Stellar populations are discussed in Chapter 15.3.)

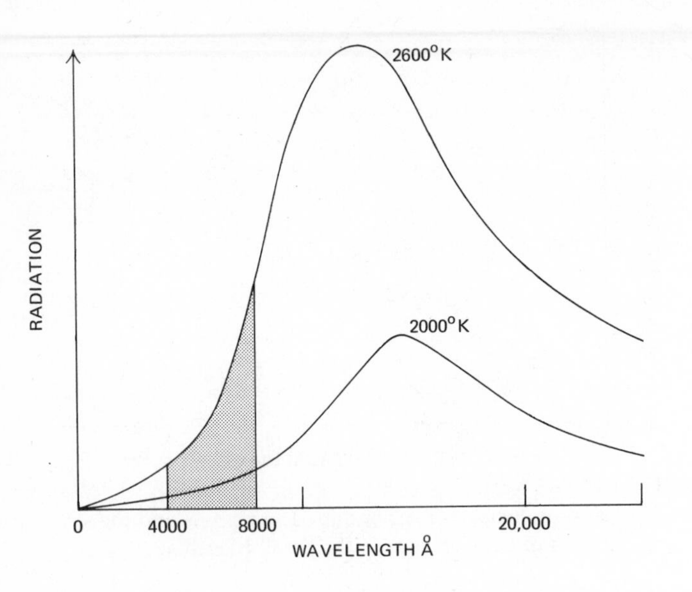

FIG. 14.6
Energy-distribution curves for irregular variable Mira Ceti. The area under the temperature curve represents total radiation. The shaded area represents the visible radiation, which increases at a faster rate than the total radiation between maximum and minimum brightness.

14.4 RED VARIABLE STARS

The red variables are the most common variable stars in our galaxy. These red giants and supergiants are of two types—long-period variables (Mira-type), and irregular variables. Both types show variability in brightness and surface temperature.

The *long-period variables* have low densities, are of spectral classes M, R, N, or S, have magnitude increases that range from less than 2 to 7, and periods of 90 to 700 days. The prototype of the long-period variables is the remarkable red supergiant Mira Ceti, which is about 300 times larger than the sun. During an average period of 331 days, its visual magnitude changes from between 3 and 5 at maximum brightness to between 8 and 10 at minimum brightness, and its spectral class varies from M6 at maximum to M9 at minimum. Its surface temperature varies from 2600°K at maximum to 2000°K at minimum.

Stefan's law states that the total energy radiated by a unit area of a star's surface is proportional to the fourth power of the temperature (Chapter 3.8). Therefore, since the ratio of Mira's maximum and minimum surface temperatures (2600/2000) is 1.3, one square centimeter of its surface should radiate $(1.3)^4$, or nearly three times more energy at maximum brightness than at minimum. Actually, with a five-magnitude change in brightness, Mira radiates 100 times more energy at maximum brightness than at minimum. This large increase in radiation with only a 600° rise in temperature can be partially accounted for and explained by the energy-distribution curves for Mira at 2600°K and 2000°K (Fig. 14.6). The radiation in visible light (represented by the shaded area under each curve) increases at a greater rate than the total radiation; therefore, a considerably larger amount of light is visible at the higher temperature than at the lower. Also, during pulsations the diameter of Mira increases by about 20%, which results in a 50% increase in its surface area.

Among the most irregular of the red variables are the R Corona Borealis stars, which remain at maximum brightness for many months and sometimes years, then suddenly decrease their brightness by two to seven magnitudes. They remain at minimum brightness for a very brief period, then return slowly and erratically back to maximum brightness. The spectra of the R Corona Borealis stars reveal that they are rich in carbon and are highly luminous.

14.5 ERUPTIVE VARIABLE STARS

Eruptive variables are stars whose light output changes sporadically, usually undergoing a rapid increase and then returning to a faint level. Some eruptive variables undergo repeated outbursts; others may undergo only a single cataclysmic explosion that destroys the star. Eruptive variables tend to be either newly formed stars or dying stars.

Stars still in the process of contracting toward the main sequence need to undergo continuous readjustment of their structure. These adjustments seem to trigger light outbursts. There are two major groups of stars in this category, the T Tauri stars and the flare stars.

T Tauri stars are found in or near dense dust and gas clouds and display erratic fluctuations in brightness. They often are surrounded by nebulosity and are strong infrared sources. It is believed that they are ejecting material and are sweeping away the remaining dust and gas out of which they formed. The infrared radiation is believed to come from a shell or disk of dust around the star, which is perhaps a forerunner of a planetary system.

Flare stars are generally low-mass main-sequence stars that display sudden outbursts of energy that last only minutes and often double the brightness of the star. It is believed that the flares represent localized outbursts of energy (small surface areas) from below the star's surface similar to the flares that occur on the sun's surface. The prototype of flare stars is UV Ceti. Its outbursts occur at about 1½-day intervals, with increases in brightness of about two magnitudes. Other examples of flare stars are Krueger 60 and Proxima Centauri, the star nearest the earth.

In 1572, Tycho Brahe observed a star that had never been seen before. By the following day, it had become as bright as Venus, so that it was clearly visible in the daytime. This convinced him that it was a *nova stella* (new star). Today, we know that a nova is not a new star, but a faint, preexisting subdwarf, smaller and denser than the sun, which unexpectedly and abruptly increases its brightness up to 12 to 13 magnitudes (60,000 to 150,000) in less than two days. As it slowly returns to its prenova brightness, it experiences several irregular fluctuations. When these have disappeared, the nova appears to have returned to being a faint star.

During the very brief period when the nova is increasing in brightness, its spectrum shows absorption lines greatly displaced toward the violet, indicating that the star's outer shell is moving rapidly outward and toward the observer. After maximum brightness has been attained, the outer shell is no longer opaque, so that the light from all parts of the shell is visible. The spectrum is most complex, the bright lines from the back side are shifted toward the red, the lines from the front are shifted toward the violet, and the lines from the sides are not displaced. Figure 14.7 shows Nova DQ Hercules 1934, a *slow nova*, which remained at maximum brightness for three months.

Most novae have been seen to erupt only once. The few that have been observed to erupt several times are called *recurrent novae* and are similar to novae except for their change in brightness, which is only about nine magnitudes. After each explosion, the nova appears to return to its former condition, except for a slight loss in its mass. A prominent recurrent nova is T Coronae Borealis, a faint, blue star in a binary system which erupted once in 1866 and reached a brightness of magnitude 2, and erupted again in 1946 and reached a magnitude of 3. Prior to the 1946 outburst, its brightness was observed to vary slightly and erratically. At the present time, the star is showing similar activity.

Dwarf novae are stars that exhibit some of the basic characteristics of a nova star, but to a lesser degree. Their light curves show that the outbursts occur with a very rapid increase in brightness every few months, followed by a slower decline. During minimum brightness, many small but constant fluctuations have been observed. The range of brightness is about two to four magnitudes. Although their spectra are similar to those of the novae stars, they do not show the displacement of the bright lines at minimum brightness in the direction of the

(a)

FIG. 14.7
Nova Hercules 1934, showing a large increase in brightness in two months: (a) 10 March 1935; (b) 6 May 1935. (Courtesy of the Lick Observatory.)

(b)

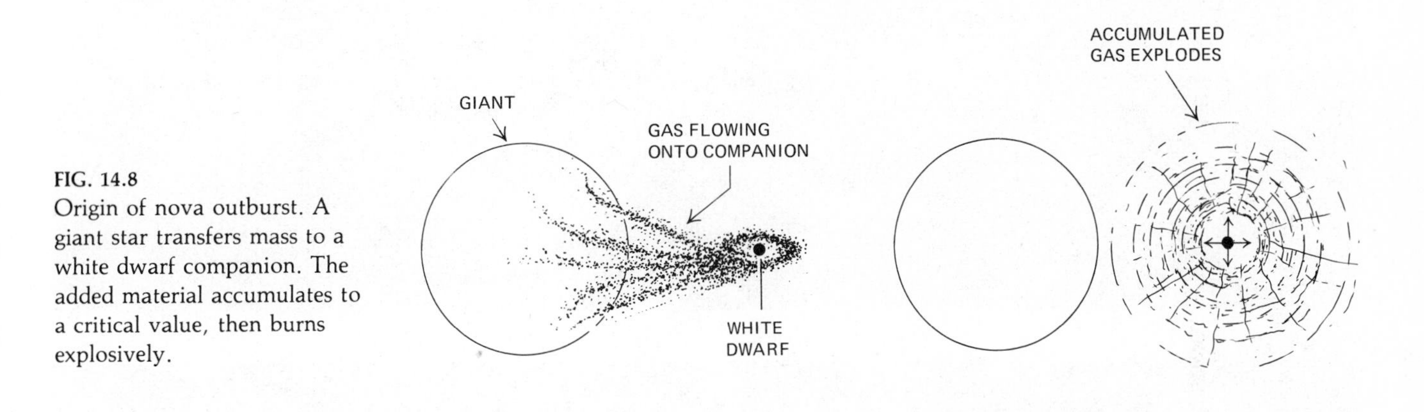

FIG. 14.8
Origin of nova outburst. A giant star transfers mass to a white dwarf companion. The added material accumulates to a critical value, then burns explosively.

observer, which suggests that rather than being ejected into space, their outer shells expand and contract. Two well-known dwarf novae are SS Cygni and U Geminorum.

It is now believed that the nova outburst of a star is intimately related to its binary nature. Virtually all novae have been found to occur in close binary systems in which one of the stars is a white dwarf. As the white dwarf's companion star evolves and expands, matter pours out of its atmosphere and falls onto the surface of the white dwarf (Fig. 14.8). This accreted matter eventually becomes hot enough to explode violently, expelling a shell of gas from the system. Matter again begins to accumulate and in a number of years another outburst occurs. The different varieties of novae are presumably due to differences in the spacing of the binaries, which affects the rate at which matter can be transfered from one star to the other.

14.6 SUPERNOVAE

The most spectacular eruptive variable star is the supernova. As we saw in Chapter 13.10, a supernova represents the death throes of a star and may result in ejection of a solar mass or more of material. Cataclysmic explosions hurl gaseous material into space at speeds above 3000 miles (4800 km) per second. The brightness of a supernova can increase millions of times and reach an absolute magnitude of −17 (Fig. 14.9).

Although more than 200 supernovae have been recorded in other galaxies, they are rare in our own galaxy. The average rate of their appearance is still being debated. According to F. Zwicky, one supernova appears in each galaxy about once every 300–400 years, a conservative estimate. However, since many are so remote, detection is impossible. I. S. Shklovsky's estimate is one every 30 years.

The explosion of a supernova creates a rapidly expanding shell of gas moving out through the space around the star. As the ejected shell moves out through the surrounding interstellar gas, it sweeps up gas and dust atoms, as well as any magnetic field that may be nearby. The swept-up matter forms a dense shell, called a *supernova remnant*, inside which the hot ejected matter resides. The hot residual gas from the explosion radiates x-rays, and recent observations from spacecraft have detected a number of supernova remnants by their x-ray radiation. The swept-up matter produces strong radio emission (discussed below) so that with a radio telescope one can detect a loop surrounding the site of the

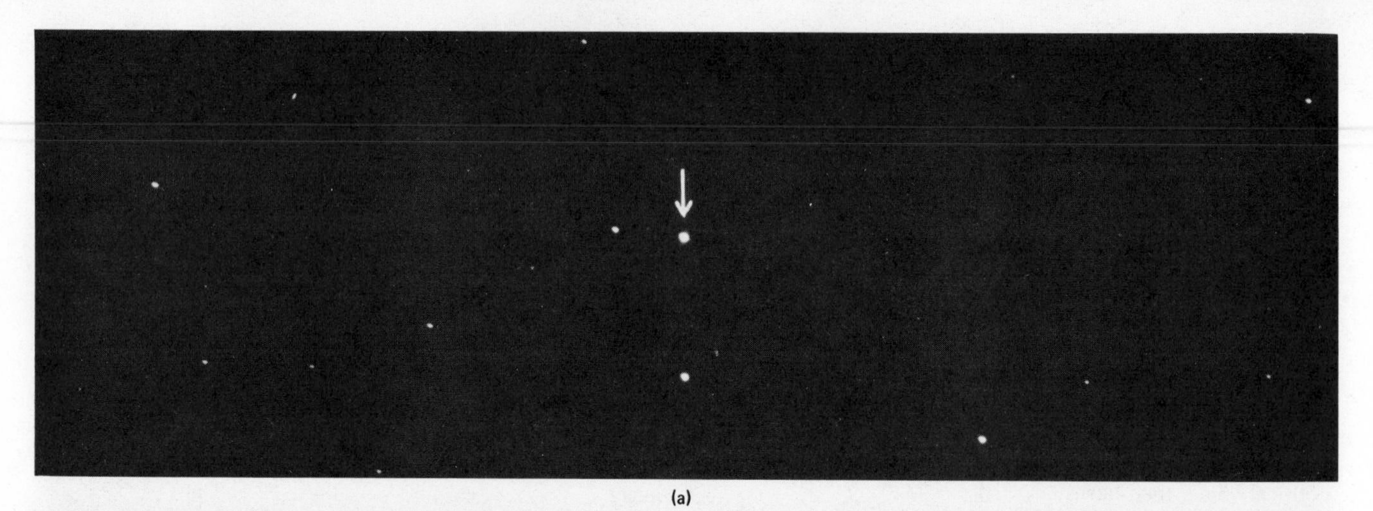

(a)

(b)

(c)

FIG. 14.9
A supernova in IC 4182, a galaxy in Virgo, photographed with the 100-inch telescope: (a) 23 August 1937, maximum brightness, 20-minute exposure; (b) 24 November 1938, faint, 45-minute exposure; (c) 19 January 1942, too faint to observe, 85-minute exposure. (Courtesy of the Hale Observatories.)

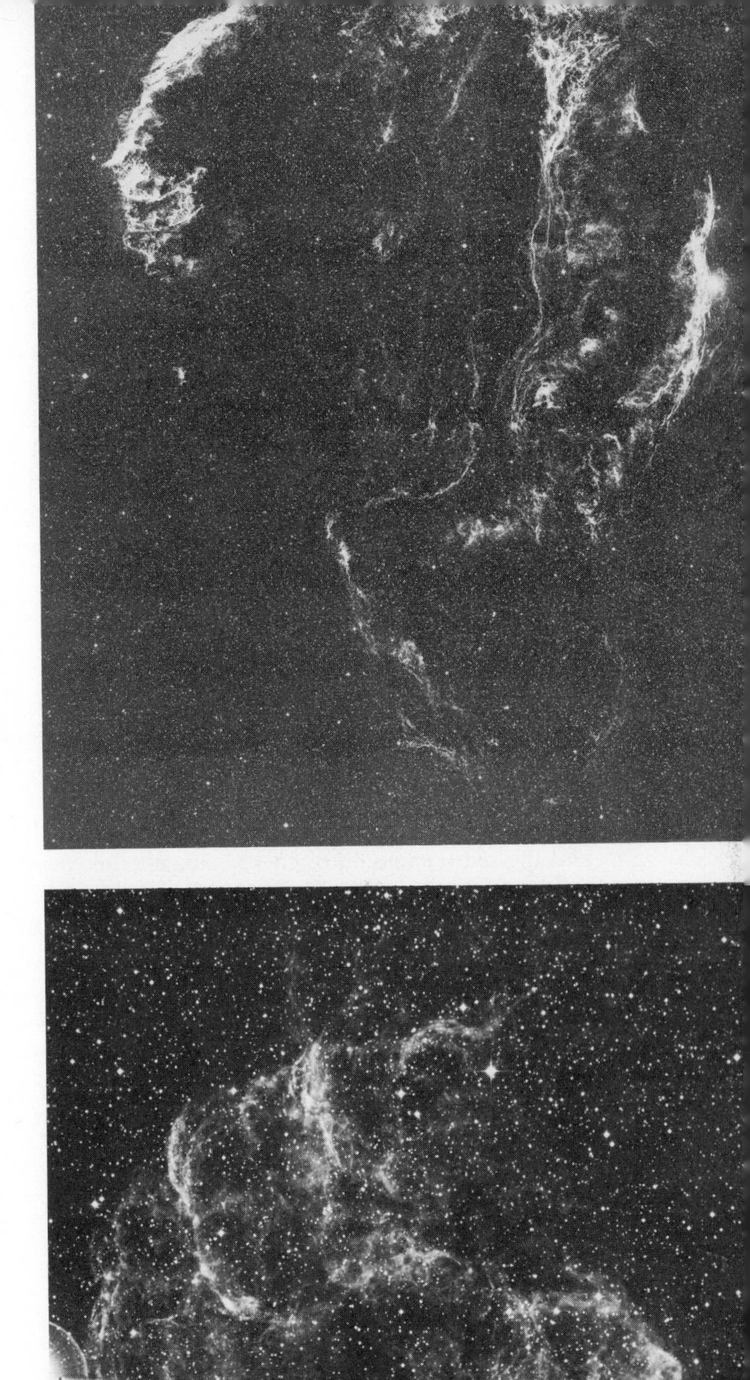

explosion. The dense swept-up gas may also emit visible light, producing an optical supernova remnant (Fig. 14.10 and Plate 18). The supernova remnant continues to expand, but eventually slows down and dissipates. The total lifetime of the remnant may be a few hundred thousand years. There are many optical supernova remnants in our galaxy and still more revealed by radio telescopes, which are able to see through the dust layer of the galactic disk.

Three definite supernova outbursts have been recorded in our galaxy. The first appeared in 1054 in the constellation of Taurus and was recorded by the Chinese and Japanese as a "guest star." It became several times brighter than Venus and was clearly visible in the daytime. The remnant of this supernova is visible today as an expanding nebulosity (Crab Nebula) in the constellation of Taurus. The second supernova was observed in 1572 by Brahe in the constellation of Cassiopeia. It was visible for nearly two years and at maximum was brighter than Venus. The third supernova was observed in 1604 by Kepler in the constellation of Ophiuchus. Comparisons of ancient astronomical records from around the world

FIG. 14.10 ►
Supernova remnants. The material ejected from the star has swept up the surrounding interstellar matter, compressing it into a shell or segments of a shell. The remnant on the top is in Cygnus. The remnant on the bottom is in Gemini. (Courtesy of the Hale Observatories.)

with radio maps of supernova remnants suggest there may have been another half dozen or so supernova outbursts in our galaxy in the last 2000 years.

Supernovae appear to fall into two definite groups: Type I and Type II. Type I supernovae become brighter by about two magnitudes and show a more rapid decline in brightness than do Type II supernovae. Their spectra show that hydrogen is prominent in Type II and rather weak in Type I. An interesting and prominent feature of their spectra is the extremely broad emission lines—an indication of the violent explosion that has occurred in the star. An analysis of these broad emission lines is most difficult, and only a few of them have been identified.

14.7 THE CRAB NEBULA

An excellent example of the remnant of a supernova is the Crab Nebula in Taurus, which consists of a homogeneous, shapeless, central mass surrounded by a complex network of fine filaments (Plate 18). Its central mass, which shows a continuous spectrum of strongly polarized light, and the radio emission of energy suggest *synchrotron radiation,* that is, radiation emitted by extremely high-velocity electrons revolving in a magnetic field. At the present time the nebula has a radius of about 180″, and its present rate of expansion is about 0″.21 per year. The outer parts of the nebula are moving away from the center at about 800 miles (1300 km) per second. The nebula is a strong source of both radio and x-ray emissions. The spectrum of the filaments shows emission lines of hydrogen, neutral and ionized helium, and other elements. Near the center of the nebula lies the Crab Pulsar (Chapter 14.9).

14.8 PLANETARY NEBULAE

Many stars eject material with no visible outbursts, in a less violent and more continuous manner than the novae and supernovae. These are extremely hot stars with extended atmospheric shells. The presence of the shell is revealed by the rather broad emission lines, or bands, that are superimposed on a continuous spectrum. The continuous spectrum is produced by the central star, and the broad emission lines are believed to be produced by the gases in the semitransparent shell—the gases nearest the observer are approaching and show a Doppler shift toward the violet, whereas those on the opposite side are receding and show a Doppler shift toward the red. Examples of stars with extended atmospheric shells are the P Cygni, the Wolf-Rayet, and the planetary nebulae. The spectra of the P Cygni and Wolf-Rayet stars show lines of ionized nitrogen and silicon in various levels of ionization—their surface temperatures range up to 60,000°K. The large Doppler shifts in the spectral lines of the Wolf-Rayet stars indicate that the gases in the shells are moving out at speeds up to 2000 miles (3200 km) per second.

As we saw in Chapter 13.10, planetary nebulae are stars whose outer envelope has been ejected, leaving an extremely hot central star surrounded by a giant expanding shell of gas. The gas is heated by the ultraviolet energy radiated by the central star. The shell may contain as much as 0.2 $M_{\odot}$ of material. Planetary nebula are believed to be an important link in the chain by which material is cycled from the interstellar medium to stars and then back to the interstellar medium.

Planetary nebulae derive their name from their resemblance to the telescopic disks of planets. Over 500 planetary nebulae are known; many of the faint ones were discovered by R. Minkowski. The studies of central stars of planetary nebulae indicate that they are small, dense, and extremely hot. Some have temperatures of nearly 200,000°K, which places them among the hottest stars known. The gas shells of planetary nebulae have diameters that range up to 200,000 astronomical units. Since the material in a shell is expanding at the average speed of 20 miles (30 km) per second, the gases would eventually become too tenuous for the shell to be visible. Excellent examples of planetary nebulae are the "Ring" in Lyra (Plate 19), the Planetary Nebula in Aquarius (Plate 20), the "Owl" in Ursa Major (Fig. 14.11), and the "Dumbbell" in Vulpecula (Plate 21).

14.9 PULSARS

In our discussion of stellar evolution in Chapter 13, we argued that following a supernova explosion the stellar remnant may become a neutron star. We also said that a

FIG. 14.11
The Owl Nebula in Ursa Major, photographed with the 60-inch telescope. (Courtesy of the Hale Observatories.)

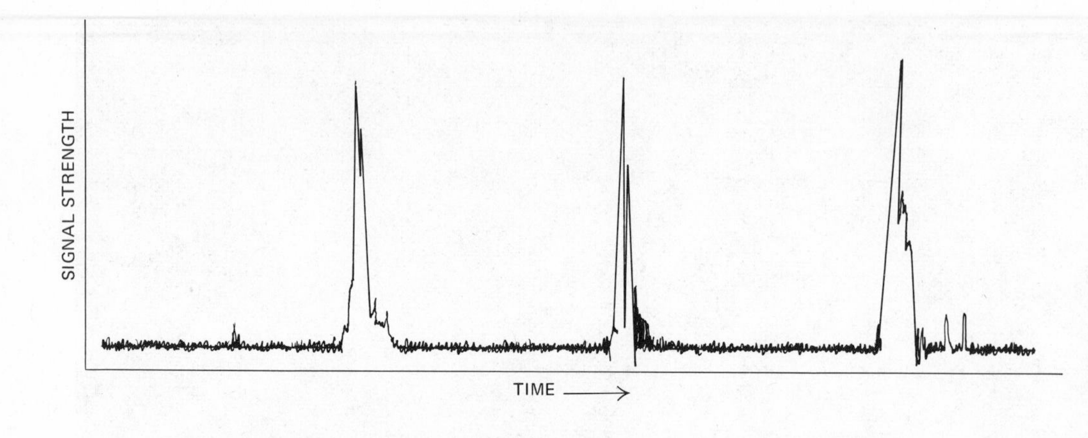

FIG. 14.12
Radio pulses from the Crab Nebula pulsar.

neutron star may make itself visible as a pulsar. The story of the discovery of pulsars and their subsequent identification with stellar remnants is an interesting example of how scientific discoveries are sometimes made. In 1939 two physicists, J. R. Oppenheimer and G. Volkoff, concluded from purely theoretical arguments that there should be stars even denser than white dwarfs, and that these would be composed of neutrons. However, detecting a neutron star seemed impossible at the time. It would be far too faint to be detected optically. Twenty-eight years passed. Then, in the course of studying extragalactic radio sources, the young English astronomer Jocellyn Bell, working for Anthony Hewish, noted an object that seemed to fluctuate with an extraordinarily short and regular period. The source was not exceptionally weak, but because it varied so rapidly (roughly one pulse every 1.33 second), its variation might have gone unnoticed without her clear perception. Subsequent observations by Bell and Hewish showed that the pulsating radio source (soon shortened to "pulsar") emitted radio waves at time intervals spaced with fantastically high precision (far better than most terrestrial clocks). Some people speculated that this was the long-awaited first indication of intelligence on other planets. However, other pulsars were discovered soon after, and it was determined that they all were slowing down and that none showed Doppler shifts such as might be expected if the source were located on a planet orbiting a star. What, then, could be the explanation? The suggestion that pulsars were the neutron stars first conceived of more than 20 years earlier was made by T. Gold of Cornell. The subsequent discovery by D. Staelin and E. Reifenstein III of a pulsar in the center of the Crab Nebula (a known supernova remnant) seemed to confirm the idea that pulsars were neutron stars, because the period of one-thirtieth of a second was too fast to be the pulsation of a white dwarf.

In the intervening years nearly 300 pulsars have been discovered. They are all characterized by extraordinarily precise periods of repetition of their radio emission (Fig. 14.12). The shape of the pulse varies from pulsar to pulsar, and even among the pulses of a single star. Each pulsar, however, has a well-defined average pulse shape, so that if any hundred pulses in a sequence are averaged, the pulse shape looks like that formed by averaging the next hundred pulses.

As described in Chapter 13.10, pulsars are now believed to be rapidly spinning magnetized neutron stars which emit nonthermal radiation along the direction of their magnetic poles. Pulse periods (and by inference the rotation periods of the stars) vary from 0.033 seconds to about 4 seconds. A pulsar's period is believed to be determined by its age. The rotational energy of the neutron star powers the emission in a way as yet not completely

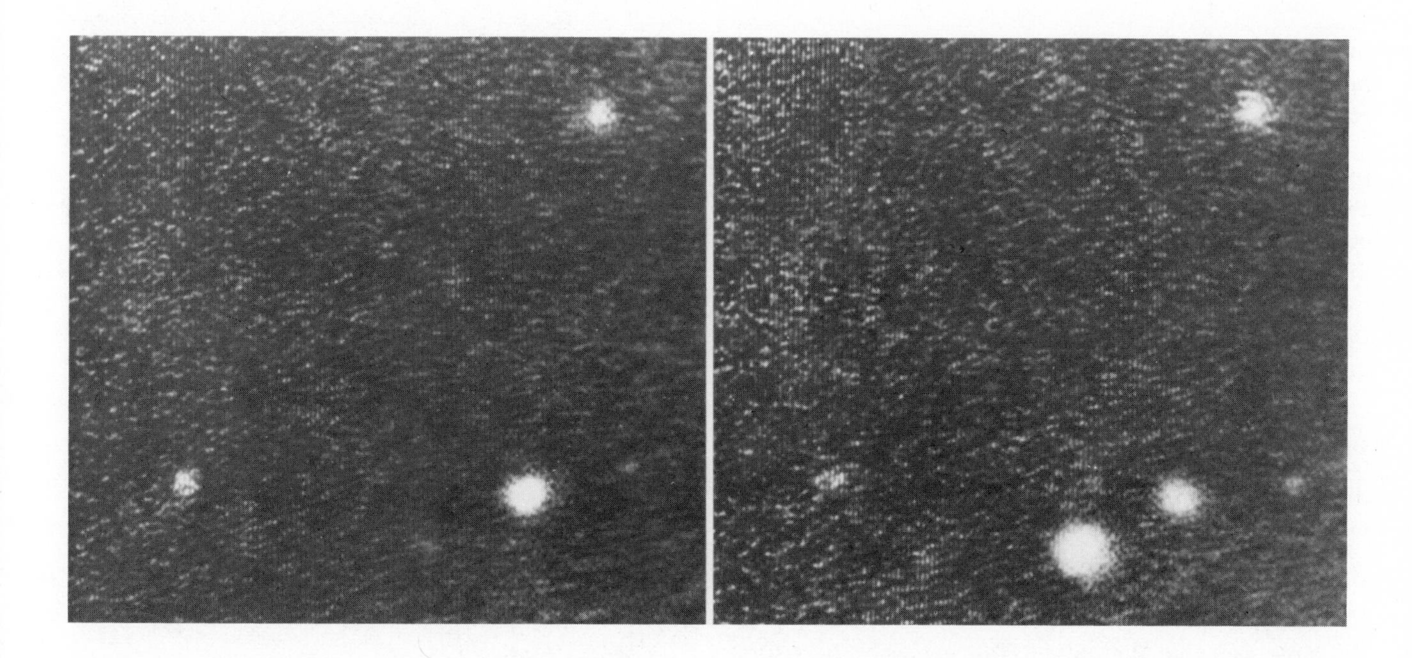

FIG. 14.13
Pulsar NP 0532 in the Crab Nebula, photographed with the 120-inch telescope with a rotating disk. In the photograph on the left, the pulsar is almost invisible; in the photograph on the right, the pulsar is near maximum brightness. (Courtesy of the Lick Observatories.)

understood. When the pulsar is created it is rotating perhaps a hundred times a second. As it ages it slows down, eventually rotating so slowly that it ceases to radiate.

The pulsar in the Crab Nebula (Fig. 14.13) is the fastest known pulsar, while one in a supernova remnant in Vela is the third fastest. The location of these pulsars in visible supernova remnants argues for their youth. A recent origin is also suggested by the fact that both the Crab and Vela pulsars emit optically visible light as well as even more energetic radiation in the x-ray and gamma-ray part of the spectrum. Astronomers at the University of Arizona first detected optical flashes from the Crab pulsar, and recently a team of Anglo-Australian astronomers (one of whom was also involved in the detection of optical variations from the Crab pulsar) measured faint flickers from Vela. The reason only young pulsars emit visible light is presumably related to the fact that, being a higher-energy radiation, visible light is harder to produce in a more slowly rotating (and thus older) neutron star.

Observations of the time of arrival of pulses over a few years has shown that pulsars are moving through space. Astronomers at the University of Massachusetts have found that many pulsars are traveling at hundreds of miles per second through space, much faster than ordinary stars and, in fact, sufficiently fast to escape from the galaxy. One theory, developed by E. R. Harrison and E. Tademaru at the University of Massachusetts, suggests that the pulsar, when it forms, emits more radiation from one magnetic pole than from the other. The pulsar is then propelled through space by this "light-rocket" effect, which gives it the high velocity observed.

Of the nearly 300 pulsars known, only one, detected by J. Taylor and R. Hulse at the University of Massachusetts, is a member of a binary star system. Since it is a component of a binary star, the limits on the mass of the pulsar can be determined. It is known that the sum of the two masses in the system must be less than 2.8 $M_{\odot}$. Interestingly, the object around which the pulsar is orbiting may also be a neutron star (but one that is not pulsing). Thus the pulsar itself probably has a mass of no more than 1.6 $M_{\odot}$. The binary pulsar may also allow tests of the general relativistic theory of gravity, because the gravitational field in which the stars are moving is extremely intense due to their small radii and separation.

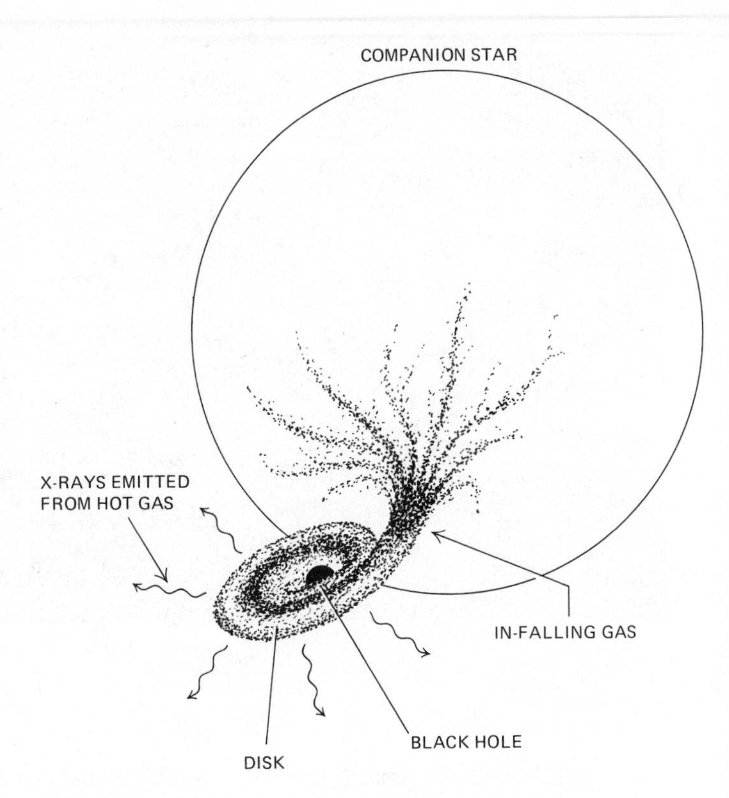

FIG. 14.14
Accretion disk around a black hole. Gas is heated as it swirls in toward the black hole.

14.10 X-RAY VARIABLES

We have seen that the pulsation or explosion of a star leads to marked optical variability and that rapid rotation of a neutron star can lead to radio variability. An even more remarkable kind of variation—x-ray variability—is associated with the orbital motion of a collapsed star in a binary system. New instrumentation often leads to new discoveries, and the discovery by orbiting observatories of x-ray emission from binary stars is no exception. X-ray variables are thought to offer the first concrete evidence for the existence of black holes.

We saw in Chapter 14.5 that in close binary star systems the leakage of mass from one star onto a companion white dwarf may lead to a nova outburst. If the companion star is a neutron star or black hole, the mass transfer is believed to lead to a very different outcome. Both neutron stars and black holes are so small and have such strong gravitational fields relative to other stars that as matter spills over toward them the matter behaves a bit like water flowing down a drain. Water moving toward an open drain normally develops a circulating motion immediately around the opening, where the water swirls faster and faster before vanishing down the opening. Likewise, matter streaming around a black hole or neutron star is believed to develop a circulating motion and to form a flattened cloud of material called an *accretion disk* (Fig. 14.14). Perhaps a million miles across and one-tenth that thickness, gases near the center of the accretion disk swirl furiously in the intense gravitational field of the collapsed star. The high velocity of the moving gas heats it to temperatures of millions of degrees. At such high temperatures, the gas emits x-rays copiously. The accretion disk is very tiny compared with the star it is orbiting. It thus may pass out of view behind

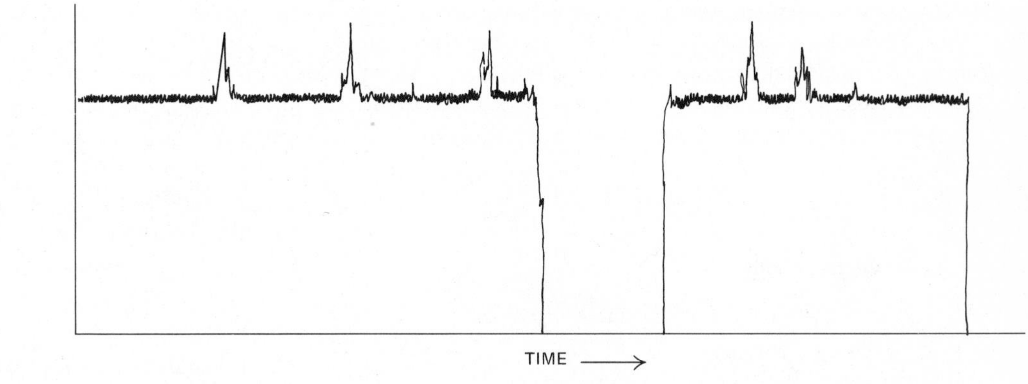

FIG. 14.15
X-ray energy from a binary x-ray source. Note periods when there is no radiation. This occurs as the companion eclipses the black hole and its accretion disk.

the disk of its neighbor. This causes the x-ray emission to be cut off for the duration of the eclipse, exactly as in normal eclipsing binaries (Chapter 12). In addition, variations in the gas density and fluctuations in the motions in the accretion disk may lead to a fluctuating x-ray output. Both the flickering output and the total occultation of the x-ray emission (Fig. 14.15) are characteristic of these strange variables. As we saw in Chapter 13.10, only very massive stars can form black holes. The binary nature of the x-ray sources has in fact made it possible to measure their masses, and these are in several cases so large that it is difficult to believe that anything but flow into a black hole is responsible for the x-ray emission.

SUMMARY

The light output of some stars is observed to vary with time. A variable star can be classified on the basis of its light curve (a plot of its changing brightness against time) as either a pulsating or an eruptive variable.

Pulsating variables are stars that periodically expand and contract as their deep atmospheres store and then release the energy generated in the interior. Periods range from fractions of a day to many months but obey the period-luminosity law, according to which intrinsically brighter stars have longer periods. Some of the more important pulsating variables, in order of increasing period, are RR Lyrae stars, Cepheids, and Mira variables. The pulsating variables are for the most part yellow or red giants and are found above the main sequence near the middle of the H-R diagram.

There are many types of eruptive variables. In general, eruptive variables are either at the beginning or at the end of their lifetimes. Novae, stars that flare up dramatically in a few days and slowly over months fade back to their pre-outburst brightness, are probably white dwarf stars in binary systems. The accretion of mass from the companion is a source of new fuel and energizes the outburst.

T Tauri stars, on the other hand, are very young stars that are still contracting. Their flaring may be due to changes in their structure brought about by their slow collapse.

Planetary nebulae, although they are not strictly variable stars, are often compared with novae because both represent stars late in their evolution and both have ejected shells of gas.

Supernovae represent a shattering explosion in which a star blows off nearly all of its envelope of gases. They are the most energetic stellar phenomena known, their luminosity becoming a trillion times that of the sun at their maximum brightness. Supernovae may leave as remnants either neutron stars or black holes. Neutron stars are extremely dense, rapidly rotating stars with strong magnetic fields. They produce a narrow beam of radiation that sweeps around them as they rotate. If the beam of radiation crosses the earth, we see a pulse of energy, and the neutron star is called a pulsar. If a supernova forms a black hole, there may be variable x-ray emission as matter falls in toward the black hole.

REVIEW QUESTIONS

1. What is a variable star? Define the two main types of variable stars.
2. Describe the light curve of a Cepheid variable.
3. What is the relationship between the period and luminosity of a variable star? Who first established this relationship?
4. How is the period-luminosity relation used to measure stellar distances?
5. What is the difference between an extrinsic and intrinsic variable? Give examples.
6. Explain what happens to the spectrum of a variable star as its brightness changes.
7. Explain how the light and velocity curves of a Cepheid variable show that it is not an eclipsing binary.
8. Why are x-ray variables believed to be binary stars?
9. Describe the characteristics of irregular variable stars. Where are they usually found? How does the environment in which they are located explain their irregular variability?
10. What are flare stars?
11. What kind of a star may become a nova? What are dwarf novae? Give examples. What are recurrent novae? Give examples.
12. What are the differences between novae and supernovae?
13. What is the Crab Nebula?
14. What are planetary nebulae? What are the characteristics of the central star? Why were these nebulae named planetary? Give examples of planetary nebulae.
15. Distinguish between pulsating stars and pulsars.
16. Describe the pulsar that was found in the Crab Nebula.
17. Why are the fastest varying pulsars believed to be the youngest?
18. Why are some x-ray sources believed to be black holes?

CHAPTER 15
GALAXIES

15.1 THE MILKY WAY

When the constellation of Cygnus the Swan is directly overhead on a moonless summer evening in the northern midlatitudes, a diffusely glowing, narrow band of light appears that completely encircles the sky. This band of light is an immense assemblage of stars (our own solar system is a minor member) known as the Milky Way. The Milky Way passes from the northeastern horizon through Cassiopeia, Cepheus, and Cygnus, where it divides into two branches. The eastern branch, which is wider and brighter, moves through Aquila, Sagittarius, and Scorpius to the southwestern horizon. The western branch passes through Lyra, almost completely disappears in Ophiuchus, and reappears in Scorpius. The dark region between the two branches, called the *Great Rift,* is a dust and gas cloud that obscures the many stars that are located in this part of the sky. Since the earth is located within the Milky Way, no single complete picture of the Milky Way can be obtained. Figure 15.1 shows a mosaic of several photographs of the Milky Way between the constellations of Cassiopeia and Sagittarius.

When the constellation of Auriga the Charioteer is directly overhead on a winter evening in the northern midlatitudes, the Milky Way passes from the northwestern horizon through Cepheus, Cassiopeia, Auriga, Taurus, Gemini, Orion, and Canis Major, to the southeastern horizon. In winter the Milky Way appears narrower, dimmer, and less spectacular than it does in summer. In the southern hemisphere in winter, the Milky Way, with its brilliant star clouds in Norma and Carina, presents a most impressive sight. The Milky Way passes through the constellations of Sagittarius, Scorpius, Centaurus, and Crux (Southern Cross).

The Milky Way is our home galaxy, an immense wheel of stars, gas, and dust particles. Although we cannot obtain a complete picture of its structure, by combining observations made with optical and radio telescopes and by examining neighboring galaxies we have been able to construct a rough model of our own system. As far as is known, galaxies are the largest coherent structures in the universe and possess a variety of forms. Our own is a flat disk surrounded by a spheroidal halo and containing about 100 billion stars. The word "galaxy"

FIG. 15.1
Mosaic of the Milky Way from Sagittarius to Cassiopeia.
(Courtesy of the Hale Observatories.)

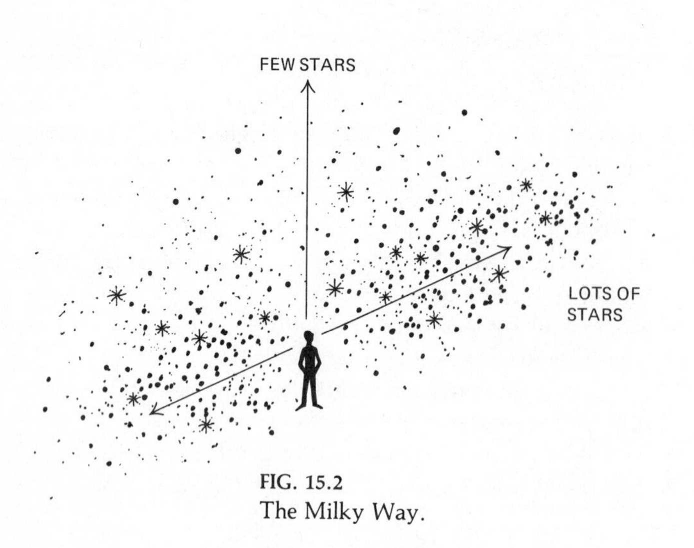

FIG. 15.2
The Milky Way.

actually derives from the name Milky Way (the Greek word for milk is *galactos*). William Herschel in 1785 was among the first to recognize that, because the Milky Way appears as a band across the sky (that is, more stars are seen in the band along the Milky Way than outside it), we must live in a flattened star system. Figure 15.2 illustrates this effect. Looking along the disk one sees far more stars than looking up through and out of it. It remained, however, to Harlow Shapely, in 1917, to recognize that we are located, not near the center of the system, but near the edge. Shapely, by studying the distribution of globular clusters, was able to show that they formed a giant spheroidal system about 100,000 light years across, with the sun located about 30,000 light years from the center.

The view of our galaxy at optical wavelengths is severely hampered because of interstellar matter. Dust particles are spread in a thin layer across the entire galaxy. The sun is unfortunately embedded in this layer and, because the dust absorbs starlight strongly, we can see only a few thousand light years into the plane of the disk. Thus the center of our galaxy cannot be seen in visible wavelengths, and the nucleus and more distant parts are revealed only by radio radiation, which is able to penetrate freely through the absorbing layer.

15.2 THE STRUCTURE OF THE MILKY WAY

By combining both optical and radio data, it has been possible to map out much of the structure of our galaxy. The Milky Way, which as we have said is about 100,000 light years across, consists of a giant disk containing stars, gas, and dust. In the disk the stars and interstellar matter move in roughly circular orbits about a central bulge of stars called the *nucleus*. The nucleus is about 10,000 light years across. In the vicinity of the sun the stars move at about 150 miles (250 km) per second along their orbits and thus complete about one revolution every 200 million years.

The main disk of the galaxy is embedded in a giant halo of stars in which the globular clusters are located (Fig. 15.3). The halo is spheroidal and the stars and globular clusters in it move in orbits that plunge downward toward the galactic nucleus, pass by the nucleus, swing out on the other side of the galaxy, and then loop back (Fig. 15.4).

The disk is 1000 light years thick, and its proportions are similar to those of an ordinary phonograph record. The disk is broken into spiral arms, as revealed both by optical studies of young O and B stars and by the radio emission produced by the cold gas clouds. The exact shape of the spiral arms is still unclear. The sun appears to be in a spiral arm and slightly above the main galactic plane.

It is not fully understood why our galaxy has spiral arms. Current theories suggest that the arms are ripple-like structures in which gas and dust accumulate. The higher density of gas then triggers the formation of stars, making the ripples glow with the light of the newly formed stars. If the stars in our galaxy all moved around the nucleus in the same time, so that the galaxy rotated like a disk, disturbances in the distribution of stars and gas would produce ripples concentric with the center like those seen in a cup of coffee when the side is tapped gently. However, stars at different distances from the galactic nucleus revolve around it at different speeds. This phenomenon is referred to as *differential rotation*; in our own solar system we have seen examples of it in both Jupiter and the sun. The effect of differential rotation on the galaxy is to cause a slight twisting of the ripples in the star distribution, so that they are arranged in spirals rather than rings.

We saw in Chapter 12.5 that the motion of binary stars could be used to find the masses of the stars in a binary system. The motion of stars in our galaxy can be used in a similar way to measure the mass of our galaxy. If we assume that most of the mass of the galaxy is concentrated in the center of the galaxy and that the sun revolves around it, the galaxy's mass can be determined by substituting the sun's distance from the galaxy's center (30,000 light years) and its orbital period around it (200 million years) in Kepler's third law of planetary motion as modified by Newton:

$$(M_{galaxy}+m_{sun})\,(P_{sun})^2=(a_{sun})^3.$$

Although this method is oversimplified, it produces a fairly good approximation of the galaxy's mass of about 200 billion times that of the sun. When more precise

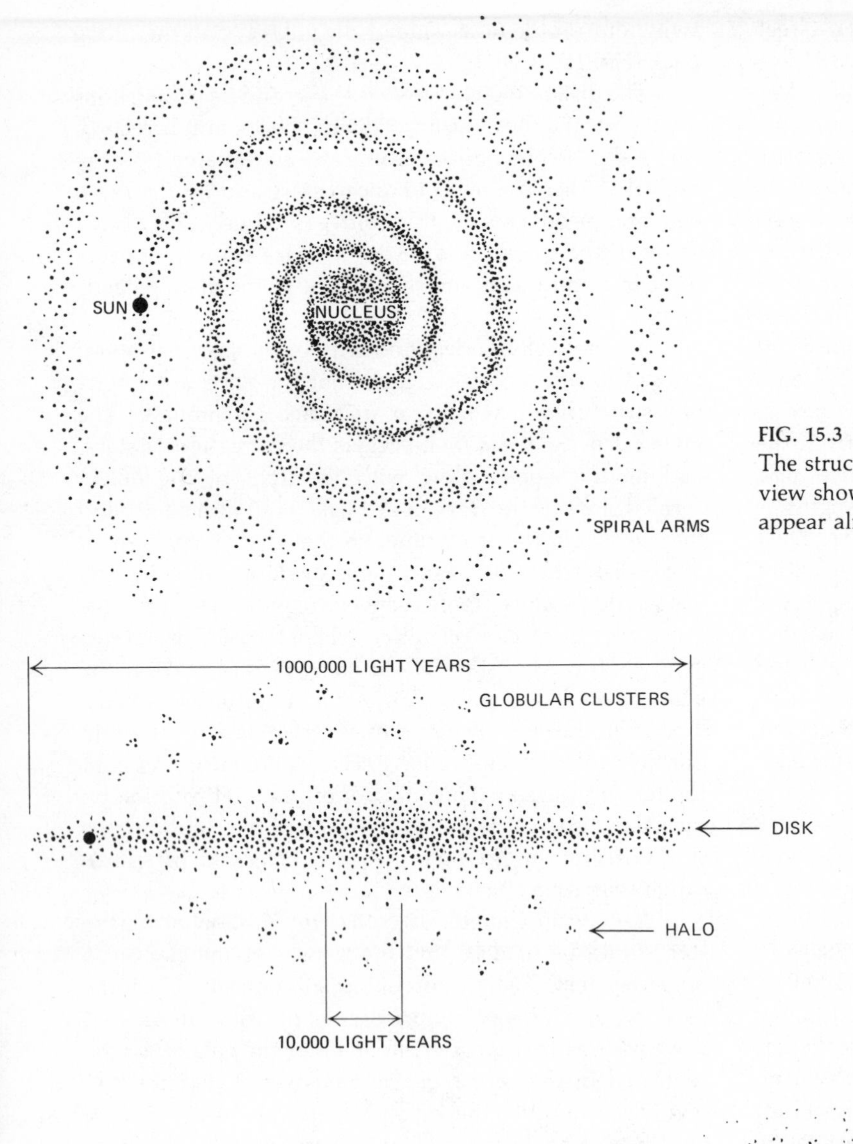

FIG. 15.3
The structure of the Milky Way, schematic model. Edge-on view shows spirals to be rather thin. Spiral arms in top view appear almost circular.

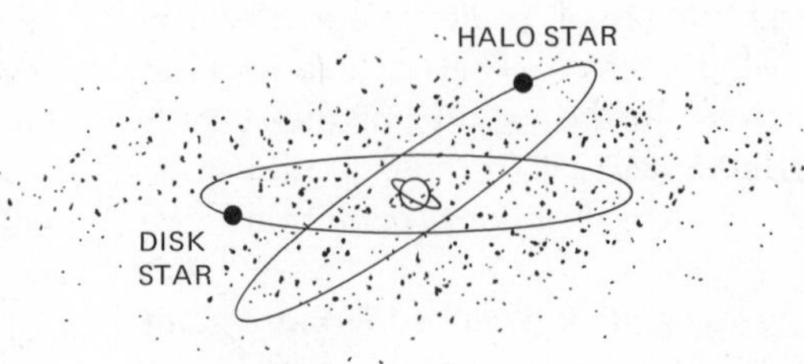

FIG. 15.4
Star motions in the Milky Way.

values are used for the sun's motion, the galaxy's mass is about 150 billion times that of the sun.

15.3 STELLAR POPULATIONS

In the last days of World War II, during blackouts in the Los Angeles area, Walter Baade obtained a number of photographs of our neighboring galaxy M31, the Great Nebula in Andromeda, as well as its satellite galaxies. These pictures, taken under the unusually dark sky conditions of the blackout, revealed some extremely faint stars. Baade noted that there were two distinct populations of stars that were characterized by their color as well as their location in the galaxy. He named them Population I and II or, for short, Pop I and Pop II. Pop I stars are the bright blue objects found in the disk of the galaxy and concentrated in the spiral arms. Pop II stars are the red stars distributed more uniformly throughout the galaxy and found in the halo and globular clusters. Subsequent observations of stars, both in our own galaxy and in other star systems, show that the breakdown into stellar populations arises from differences in their ages. Pop I stars are young, while Pop II stars are old. Other distinguishing features are chemical composition and orbital motion around the galaxy. Pop I stars contain ten to one hundred times the amount of heavy metals that Pop II stars do. Pop I stars move on roughly circular orbits around the galaxy, while Pop II stars move on highly eccentric, plunging orbits. We will see in Chapter 15.20 why these differences occur.

15.4 INTERSTELLAR MATTER

In addition to the stars, galaxies (including our Milky Way) contain matter distributed between the stars, referred to as interstellar matter. Interstellar matter consists of gas and dust. In our own galaxy, about 15% of the mass of the whole system is gas and dust. There are thus roughly 15 billion solar masses of interstellar matter in the Milky Way. Of this, roughly 1% is dust. The gas is composed mostly of hydrogen and helium as are stars. Heavy elements make up only 1% or so. The gas exists in many different forms, but typically it occurs in clouds about 10 light years across and containing a few hundred solar masses of matter. The density of the gas in the clouds is incredibly small by terrestrial standards—the clouds contain roughly one hydrogen atom per cubic centimeter. By contrast, the air you are now breathing has a density of about 10 quintillion (10^{19}) molecules per cubic centimeter. (And this in turn is about the relative abundance of one grain of sand to all the sand on a beach a few hundred yards long.) The gas is also usually very cold. Temperatures of $-200\,^{\circ}C$ ($-350\,^{\circ}F$) are typical.

The number of dust grains in a given volume of space is also very small, with perhaps only 100 particles per cubic mile! The grains are extremely tiny (about 10^{-5} cm, close in size to the wavelength of visible light).

Interstellar clouds are often called *nebulae*, from the latin word for cloud. Nebulae that are made luminous by the light from nearby stars are called *bright nebulae*. When the gas and dust particles are so dense that light from background stars and luminous gas is blocked, they are called *dark nebulae*. Most of the gases in the nebulae, which are cold and nonluminous, become visible by the process of emission or line absorption of starlight. The solid particles reveal themselves by the general obscuration, reddening, reflection, and polarization of starlight.

15.5 DIFFUSE NEBULAE

On a clear, very dark evening, the Great Nebula in Orion is visible to the unaided eye as a very faint haze around the middle star in Orion's sword. Telescopically, this diffuse nebula is one of the most beautiful of celestial bodies (Plate 22). It is a mass of interstellar gases made luminous by the excitation of groups of extremely hot, luminous O- and B-type stars imbedded in the nebula. This is an excellent example of a bright-emission nebula. The hydrogen gas in the nebula is ionized by the ultraviolet radiation from the hot stars, and light is emitted when the protons recombine with the electrons. The regions of interstellar space around stars that contain hot ionized hydrogen are called *H-II regions*, whereas colder clouds that contain neutral hydrogen are called *H-I regions*.

The Orion emission nebula, at a distance of approximately 1500 light years, has a diameter of about 15 light years. The central region appears bluish green because the photograph has been overexposed to reveal the faint outer details. The blue light is caused by ionized oxygen, and the red light is caused by hydrogen. Other excellent examples of diffuse emission nebulae are the Lagoon Nebula in Sagittarius (Plate 23)* and the North America Nebula in Cygnus.

The collision of hydrogen atoms, if sufficiently energetic, can also produce luminescence in a nebula. The Veil Nebula in Cygnus (a supernova remnant) is an excellent example (Plate 24). No star has been found near the nebula to account for its luminosity. The nebula is expanding at about 60 miles (100 km) per second and the emission of light occurs when the expanding filaments of gas collide with the low-density gas of the interstellar medium.

When a star near a nebula is cooler than type A0, it does not produce enough ultraviolet light to make a visible H-II region. However, the dust particles in the cloud scatter the starlight, and the nebula becomes visible just as dust particles are made visible in a sunbeam. It is then called a *reflection nebula*. Although *emission nebulae* also reflect starlight, the amount of light reflected in comparison to the amount emitted is negligible. The nebulosities around the bright stars of the Pleiades are excellent examples of reflection nebulae (Plate 25).

A cloud of interstellar matter that is too far away from a bright star to emit or reflect light is called a *dark nebula*. Some are so dense that they are clearly visible as dark objects obscuring background stars. The average dark nebula is irregular in shape and is less than 30 light years in diameter. An excellent example of a dark nebula is the Horsehead in Orion (Plate 26). The dark cloud at the left obscures the emission nebula which is in the background. Only a few of the bright stars are visible through the dark nebula. The outline of the horse's head is simply an extension of the dark nebula. Other examples of dark nebulae are the Great Rift, which runs lengthwise in the Milky Way, dividing it into two parts, and the Coal Sack in both the Northern Cross and the Southern Cross. The smallest of the dark nebulae are the *globules*, about 10,000 astronomical units in diameter, that are found in several nebulae such as the Lagoon. They may be stars in the formative stage.

* Plates 23–30 appear following p. 296.

15.6 GENERAL OBSCURATION

In addition to being concentrated in the dark nebulae, the dust particles are more or less uniformly distributed throughout the spiral arms of galaxies, where their presence causes the stars to appear dimmer and more distant than they actually are. Another evidence of general obscuration is the apparent distribution of galaxies in space. This effect is similar to the dimming of starlight when the star is observed near the horizon. The dimming is proportional to the distance through which the light passes. E. P. Hubble at Mount Wilson Observatory discovered that it is almost impossible to observe galaxies in the plane of the Milky Way. The number of observable galaxies increases as the distance increases from the plane of the Milky Way. Hubble called the region in the plane of the Milky Way the *zone of avoidance*. Today, we know that galaxies are difficult to detect in the zone of avoidance because of the increase in the concentration of the dust and gas particles and because of the increase in the absorption of light.

15.7 INTERSTELLAR REDDENING

The dust particles responsible for interstellar obscuration do not absorb all wavelengths of light equally. As in our own atmosphere, the shorter wavelengths are scattered more efficiently. This means that more of the shorter wavelengths have been absorbed by the interstellar material from a beam of visible light, so that the starlight appears to be redder. The reddening effect can be seen strikingly when the spectrum of what looks like a dim red star reveals it to be in actuality a bright, hot, blue star of spectral class B.

15.8 INTERSTELLAR POLARIZATION

Starlight is polarized by interstellar matter. In this process the dust particles are more effective than the gas particles. Polarization has been verified in a number of ways. It is not present in the light from stars that are near and unobscured, whereas it is present to varying degrees in the light from stars that are more distant, which has been affected by interstellar absorption and reddening. Also, the stars that are observed through the same cloud display about the same degree of polarization and in the same direction. These observations indicate that polarization is produced by the interstellar matter rather than by the stars themselves. It has also been observed that the light from stars near the plane of the Milky Way is polarized parallel to the galactic plane. The fact that the interstellar matter polarizes starlight indicates that the interstellar particles must be elongated, that is, needle-shaped. Also, the alignment parallel to the galactic plane suggests the presence of a magnetic field parallel to the galactic plane that aligns the needle-shaped particles uniformly.

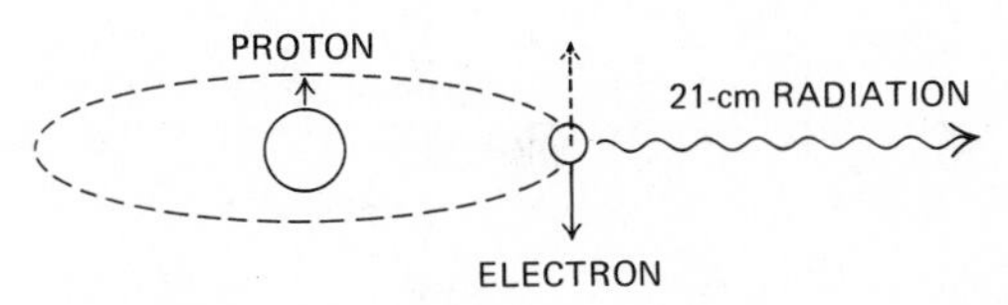

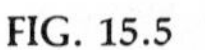

FIG. 15.5
21-cm radiation is produced when the electron in a hydrogen atom "flips over," lowering the atom's energy slightly.

15.9 RADIO EMISSION FROM INTERSTELLAR CLOUDS

We saw in Chapter 3 that the characteristic wavelength at which a substance radiates is determined by its temperature. Objects with temperatures of thousands of degrees radiate strongly at visible wavelengths. As temperatures drop, the radiation emitted is at longer and longer wavelengths. Under the frigid conditions of interstellar clouds, it should be no surprise that most of the radiation emitted is at far infrared and radio wavelengths.

Interstellar gas emits radio waves by a variety of mechanisms. We have already seen (Chapter 14.6) that supernova remnants emit nonthermal radio waves with a continuous spectrum. The ionized hydrogen associated with H-II regions emits continuous radiation at radio wavelengths as well, but with a very different spectrum and by a different mechanism. The motion of the free electrons in the ionized gas produces the radio emission from H-II regions. H-I regions emit radio waves as well, but neutral hydrogen clouds produce radiation almost exclusively at one wavelength, 21 cm. This radiation is produced, not by a transition of the electron from one orbit to the next, but by the flipping over of the electron in its lowest orbit (Fig. 15.5). Observations of the 21-cm radiation of the cold hydrogen has been one of the most powerful tools used to date in studying the structure of the Milky Way. Traveling across the galactic disk, basically unaffected by the presence of the intervening gas or dust, radio waves from the cold clouds have allowed precise measurements of the distribution and motions of the gas layer in our galaxy. From these observations, it has been possible to learn of the arrangements of the spiral arms in the Milky Way and to make far more accurate measurements of the size and mass of our galaxy.

There is still another important type of radio emission from interstellar clouds. In dense, dark gas clouds the atomic hydrogen combines to form molecular hydrogen, which does not emit 21-cm radiation. For years astronomers despaired of observing the formation of stars because, at the high densities needed for stellar condensation to occur, there is no 21-cm radiation and the dust prevents visible light from getting out. However, in 1963 radio emission was detected from the interstellar molecule OH. In 1968 ammonia (NH_3) and water molecules were observed at radio wavelengths. Since then a flood of other molecules have been discovered, including formaldyhyde, ethynyl and ethyl alcohol, carbon monoxide, and even more complicated species of molecules (Fig. 15.6). These molecules radiate by converting their rotational energy into electromagnetic waves, most of which are at centimeter and millimeter wavelengths. Typically formed only in the densest clouds, interstellar molecules have allowed astronomers to peer into the dark clouds where stars and planetary systems are born (Fig. 15.7).

Interstellar molecules have proved to be an additional surprise. Most of them are what would be termed

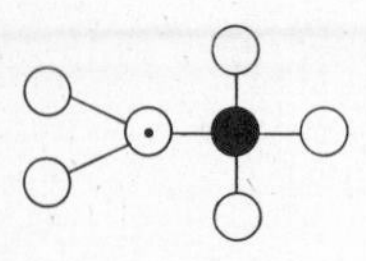

METHYLAMINE
CH_3NH_2

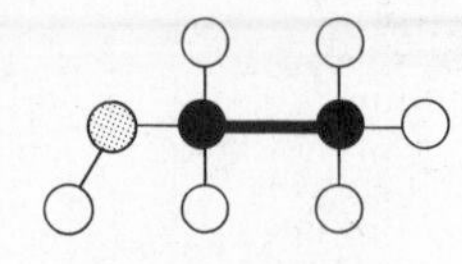

ETHYL ALCOHOL
(C_2H_5OH)

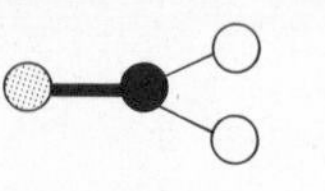

FORMALDYHYDE
H_2CO

FIG. 15.6
A few of the many interstellar molecules discovered by radio astronomy.

organic molecules. In fact, two of the types detected, if combined, would form an amino acid, one of the building blocks of proteins and thus of life. The wide distribution and high abundance of the relatively complex molecules suggests that some of the initial processes in the formation of life may be occurring in many places throughout the galaxy.

15.10 IDENTIFYING GALAXIES

The few, faint, luminous patches of light located in the sky away from the Milky Way that were observed before the invention of the telescope were believed to be interstellar clouds (nebulae) within the Milky Way. With the introduction of the telescope, more of these supposed nebulae were discovered, and by 1781 they had been catalogued by the French astronomer Charles Messier. His famous list of nebulae made no distinction between actual nebulae, star clusters, and galaxies. For example, the Andromeda galaxy was listed as the Andromeda nebula. There were a few men who disagreed with Messier's catalog. The German philosopher Immanuel Kant and the English astronomer William Herschel, for example, speculated that these nebulae were actually star systems like our own Milky Way, that is, "island universes."

The breakthrough occurred in 1924, when Edwin Hubble at Mount Wilson Observatory identified Cepheid variables in the Andromeda nebula. By measuring their apparent brightnesses and their periods of light fluctuations, and by using the period-luminosity relation, the present, corrected value for the distance to the Androm-

FIG. 15.7
Giant molecular cloud and region of star formation in the constellation Monoceros. The young star cluster NGC 2264 and the Madonna nebula are in the lower left. A dark dust and molecular cloud partially obscures the luminous material to the right and above the cluster toward the picture center. (Photograph courtesy of Hale Observatories, copyright National Geographic Society—Palomar Observatory Sky Survey.)

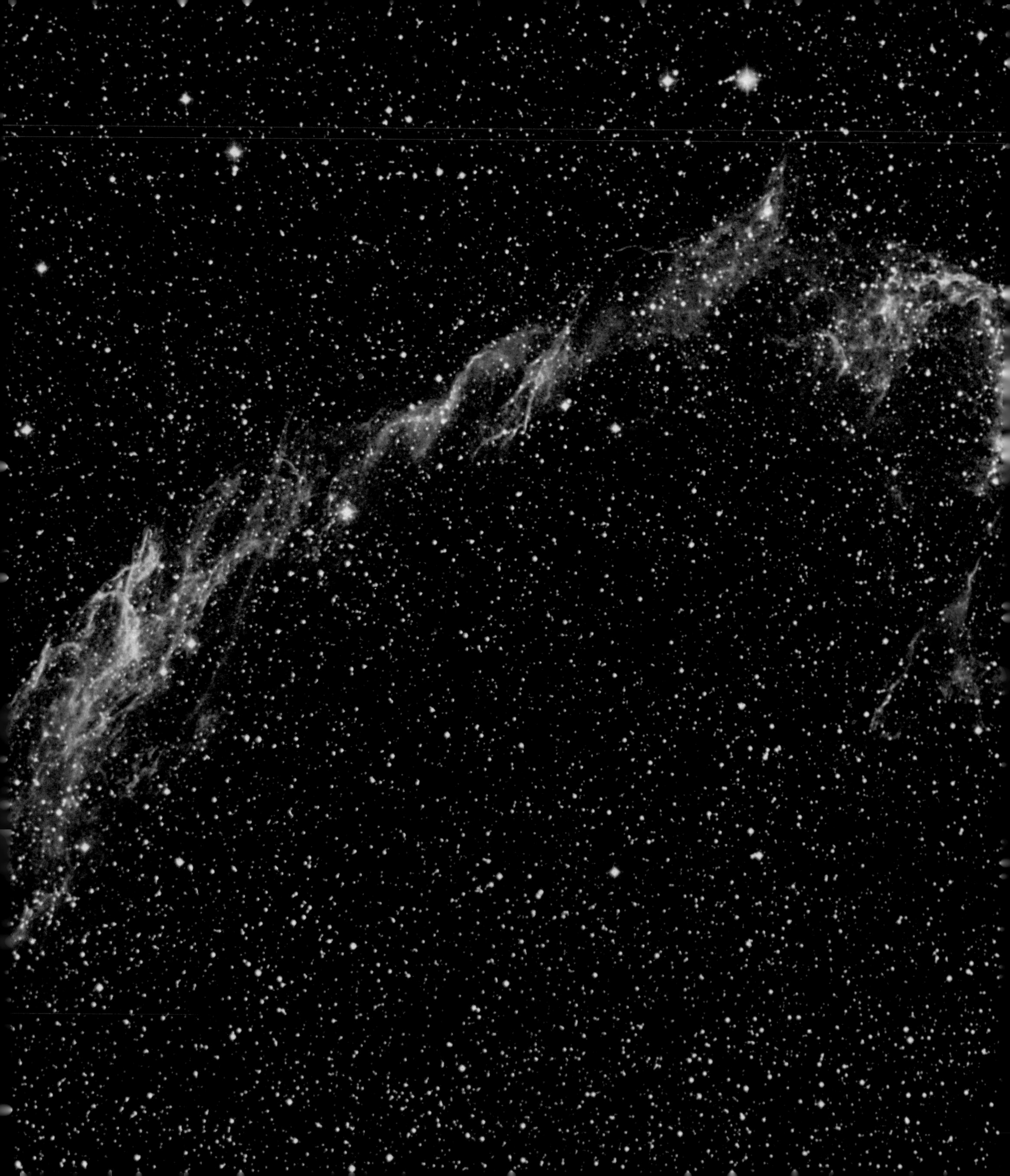

PLATE 23 (first page this insert) The Lagoon Nebula in Sagittarius, an emission nebula. Photographed with the 200-inch telescope. (Photograph from the Hale Observatories)

PLATE 24 (left) The Veil Nebula in Cygnus, photographed with the 48-inch Schmidt telescope. (Photograph from the Hale Observatories)

PLATE 25 (above) The Pleiades in Taurus is a typical galactic cluster. The stars are hot, blue, and young. The six brightest stars are visible with the unaided eye. The dust surrounding the stars reflects the starlight, producing the beautiful nebulosity. Photographed with the 48-inch Schmidt telescope. (Photograph from the Hale Observatories)

PLATE 26 (next page) The Horsehead Nebula in Orion. The nebula in the upper part of the photograph is bright because its luminosity is derived from the energy of the nearby stars. The nebula in the lower part appears opaque to light because the interstellar material is denser and the region contains no stars. The stars visible in the dark nebula are between the observer and the nebula. The outline of a horse's head is produced by the projection of a cool, dark cloud of dust against the background of the bright nebula. Photographed with the 48-inch Schmidt telescope. (Photograph from the Hale Observatories)

PLATE 27 (preceding page) The Andromeda Galaxy. This great, spiral galaxy, about two million light-years away, is the nearest to the Milky Way. Two elliptical satellite galaxies, NGC 205 and NGC 221, are visible near the Andromeda Galaxy; one is below it, and the other is above and to the right of it. Photographed with the 48-inch Schmidt telescope. (Photograph from the Hale Observatories)

PLATE 28 (above) Irregular Galaxy in Ursa Major, NGC 3034, Messier 82. This type of galaxy appears to lack symmetry, contain large amounts of interstellar dust, and stars (if present) too faint to be resolved. (Photograph from the Hale Observatories)

PLATE 29 (right) The Trifid Nebula in Sagittarius appears to be divided into three parts by interstellar dust. The nebula contains very small dark spots (globules) which are believed to be stars in the formative stage. (Photograph from the Hale Observatories)

PLATE 30 (overleaf) The Rosette Nebula in Monoceros, NGC 2237. (Photograph from the Hale Observatories)

FIG. 15.8 ►
Hubble's famous "tuning fork" diagram for classifying galaxies according to their appearance.

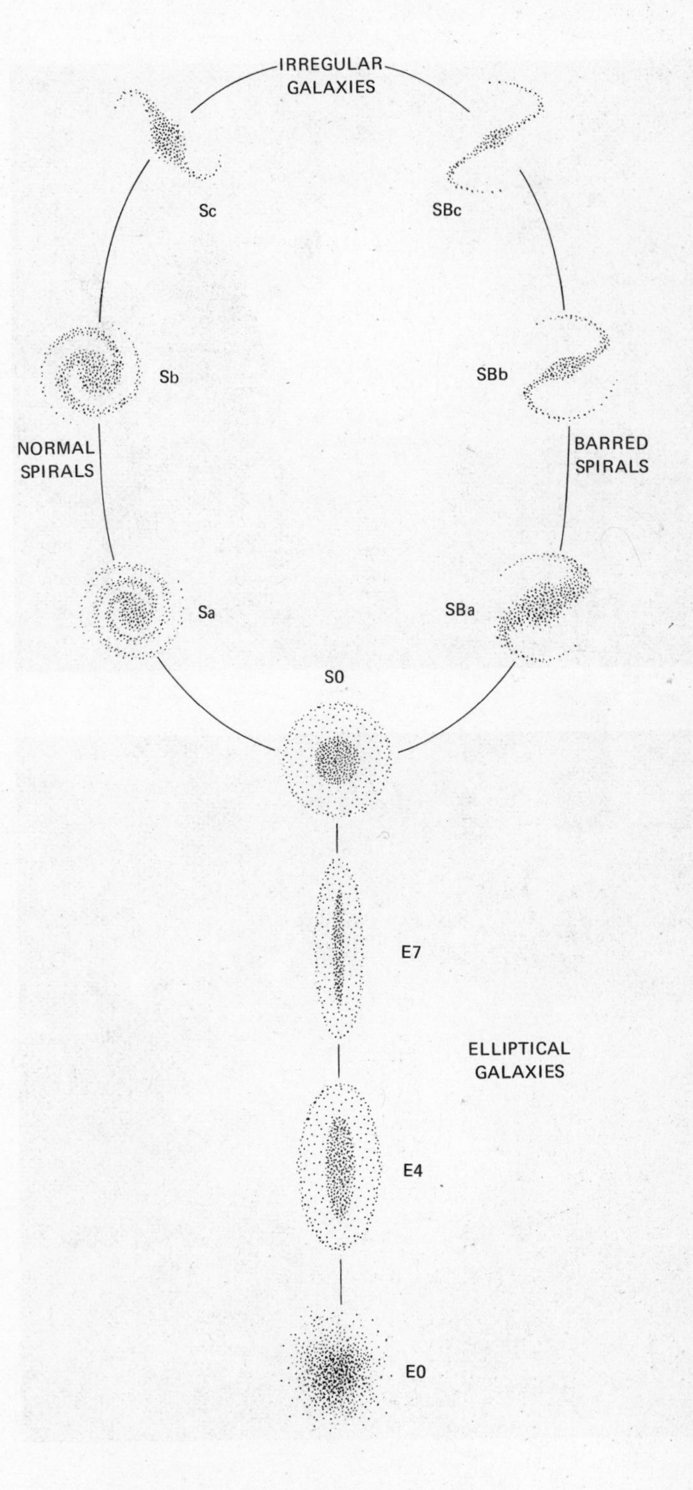

eda galaxy was established as slightly over two million light years. Their discovery established the "nebulae" as individual galaxies far beyond the limits of the Milky Way.

In 1926 Edwin Hubble classified the galaxies according to their apparent rather than actual shapes: elliptical, spiral, and irregular, which he presented in his famous "tuning fork" diagram (Fig. 15.8). Hubble noted that there appears to be a progression of galactic forms from the spheroidal galaxies through flatter and flatter types in which arms are eventually visible.

15.11 GALAXY NOMENCLATURE

Because of their extreme faintness, the distant galaxies, unlike stars and planets, do not have traditional names. Instead, they have only numerical designations taken from catalogs. The first catalog to include galaxies was that compiled by Messier (Appendix 9). Galaxies in the Messier catalog are denoted by "M" numbers, for example, M87. A far more complete list of galaxies is given in the *New General Catalog* (NGC), which was begun by Herschel, and in its supplement, the *Index Catalog* (IC). Many of the galaxies that have been studied recently are too faint to appear even as NGC objects. The *Third* and *Fourth Cambridge Catalogs* (3C and 4C), which are compiled from lists of radio sources, include a number of galaxies of this type.

15.12 ELLIPTICAL GALAXIES

The apparent forms of elliptical galaxies range from spheres to ellipsoids whose lengths are about three times longer than their widths. To indicate that a galaxy is elliptical, the designation for the type of galaxy begins

FIG. 15.9
The NGC 4486 (Messier 87) galaxy in the Virgo cluster, a source of radio radiation, photographed with the 200-inch telescope. (Courtesy of the Hale Observatories.)

with the letter E, followed by a number which indicates the galaxy's degree of ellipticity. There are eight classes that range from the spherical (E0, Fig. 15.9) to the elliptical (E7, Fig. 15.10). They all have bright centers, with luminosities that gradually diminish toward the edges. They are generally free of dust, and their stars are members of Pop II.

15.13 SPIRAL GALAXIES

About three-fourths of the bright galaxies are spirals. Generally, they consist of a bright nucleus and two spiral arms that extend from opposite ends of the nucleus and wind around it. They are classified as normal spirals (S) and barred spirals (SB), according to the shape of their nuclei. The spirals are further divided into three classes (a,b, and c) according to the size of their nuclei and the tightness with which their spiral arms are wound around the nuclei. The (a) designation indicates a large nucleus, with thin spiral arms that are wound tightly around the nucleus: the (c) designation indicates a small nucleus, with wide spiral arms that are wound loosely around the nucleus. Examples of each type are shown in Fig. 15.11. S and SB galaxies, unlike E galaxies, contain moderate

FIG. 15.10
Elliptical galaxy NGC 205, E5 type, a satellite of the Great Galaxy in Andromeda. Taken in red light with the 200-inch telescope. (Courtesy of the Hale Observatories.)

FIG. 15.11 ►
Classification of galaxies: (a) normal spiral galaxies; (b) barred spiral galaxies. (Courtesy of the Hale Observatories.)

NGC 1201 TYPE S0

NGC 2811 TYPE Sa

NGC 488 TYPE Sab

NGC 2841 TYPE Sb

NGC 3031 M81 TYPE Sb

NGC 628 M74 TYPE Sc

(a)

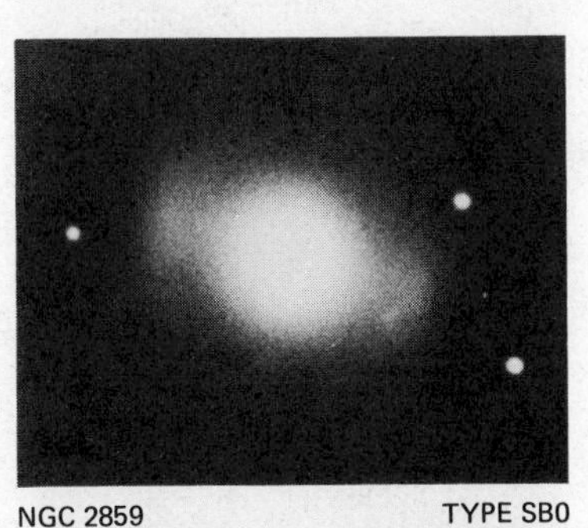
NGC 2859 TYPE SB0

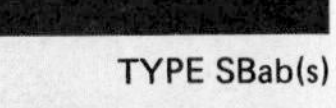
NGC 175 TYPE SBab(s)

NGC 1300 TYPE SBb(s)

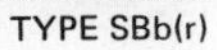
NGC 2523 TYPE SBb(r)

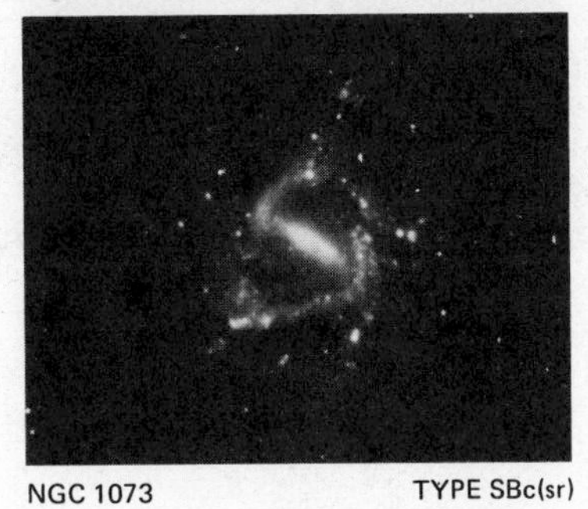
NGC 1073 TYPE SBc(sr)

NGC 2525 TYPE SBc(s)

(b)

FIG. 15.12
The Large Magellanic Cloud, a satellite of our Milky Way. It is a typical small irregular type galaxy. (Courtesy of the Lick Observatory.)

amounts of dust and gas, perhaps 15% of their total mass, and have mixed stellar populations. Pop I stars are found in the spiral arms. They are the blue giants and supergiants of spectral classes O and B. Pop II stars are found in the nucleus, in the halo, and especially in the globular clusters.

The Andromeda galaxy, at a distance of over two million light years, is probably the best known and most famous of all the spiral galaxies. It was observed by the ancient astronomers and as early as A.D. 964 was recorded as another "fixed star" by the Persian stargazer Al Sufi. Andromeda is clearly visible to the unaided eye as a small, elongated nebulosity of fourth magnitude located to the west of the Andromeda constellation. When it is seen through binoculars, this galaxy appears as a small disk, smaller than the moon; when seen through a telescope (Plate 27), it appears as a great elliptical object with several spiral arms. Studies reveal that it is larger and more massive than the Milky Way, with spiral arms that appear to be tightly wound. It is interesting to note that to an observer in the Andromeda galaxy, the Milky Way would appear much as the Andromeda galaxy appears from the earth.

15.14 IRREGULAR GALAXIES

The irregular galaxies, which are few in number, display no symmetry of form. They are rich in dust and gas and in general contain mostly Pop I stars. The stars are distributed around a central cloud whose form is most chaotic. The Large and Small Magellanic Clouds, typical irregular galaxies of type I, contain luminous stars of spectral class O and B (Pop I stars), some globular clusters, and RR Lyrae stars (Pop II stars). The *Large Magellanic Cloud*, located in the constellation of Dorado, appears as a detached portion of the Milky Way (Fig. 15.12). Its actual diameter is about 30,000 light years, but at an estimated distance from the earth of 160,000 light years, its apparent diameter is about 12°. Although it is classified as irregular, it shows a relatively dense central bar which consists of stars surrounded by large, irregular clusters of bright stars. The cloud also contains large quantities of interstellar matter. The *Small Magellanic Cloud*, located in the constellation of Tucana, is about 180,000 light years away and has an actual diameter of about 25,000 light years. Both the Large and Small Magellanic clouds are believed to be satellite galaxies of the Milky Way.

Another type of irregular galaxy recognized by Hubble is M82 (Plate 28). M82 is a typical type II irregular galaxy in that it appears basically featureless. It seems to have undergone a violent explosion in its nucleus and may have ejected huge plumes of gas and dust out of its disk.

The S0-type galaxy was not included in the original "tuning-fork" diagram, but was later introduced at the point where the elliptical galaxies end and the two branches of the normal and barred spirals begin. Hubble described the new type as being *lenticulars*, that is, disk-shaped spirals with a nucleus and without spiral arms. Galaxies of this type were regarded as old, elliptical galaxies that are located in rich clusters: they are now considered to be in a class all their own. They may arise from spiral galaxies that have had all their interstellar gas and dust "combed out" as they move through the cluster.

15.15 PROPERTIES OF GALAXIES

Although the three basic galaxy types differ greatly in external appearance, there are few systematic differences in their fundamental properties, such as size and mass. Just as our sun is a rather typical star, the Milky Way is a rather typical galaxy. The greatest range in properties seems to occur among the ellipticals. Ellipticals have masses ranging from a few tens of millions of solar masses for the dwarf galaxies up to several trillion solar masses for the giants that may dominate an entire galaxy cluster. The ellipticals likewise vary in size, but they are typically a hundred thousand light years across. Spiral and irregular systems have properties similar to those of an average elliptical, and do not exhibit the extreme range found among ellipticals. For example, there do not appear to be either dwarf or giant spirals.

One way in which galaxies do differ among the classes is in their spectra. Ellipticals, composed of Pop II stars, have spectra of class G or K. Irregulars, dominated by Pop I stars, often have spectra of classes as early as A. Spirals have intermediate spectral classes.

The spectrum of a galaxy is a composite of the spectra of its billions of component stars. Galaxies with many A stars will show in their spectra those features seen in A stars. Galaxies with numerous H-II regions or clouds of ionized hot gas will show emission lines due to this matter. The spectra of galaxies give additional information about the star system, and in particular their state of motion. Rotation of the galaxy can be measured by studying the tilt of the spectral lines (recall the discussion of Saturn's rings). Most galaxies, including ellipticals, show rotation.

In addition to rotation, all but the nearest galaxies exhibit a strong red shift of their spectral lines, indicating that they are receding from us. While we will explore this point in more detail in Chapter 16, here we should note that the red shift of a galaxy has proved to be a powerful tool in measuring its distance. In fact, except for nearby galaxies in which one can observe Cepheid variables, supernova, and H-II regions, the red shift is the only reliable way yet devised to determine distance to galaxies.

15.16 PECULIAR GALAXIES

Just as there are peculiar stars, there are peculiar galaxies. Unfortunately, so little is known about how galaxies evolve that many galaxies classed as peculiar may in fact represent merely short-lived stages found in the development of ordinary galaxies. Nevertheless, there are several important types of peculiar galaxies, which we will briefly discuss below.

In Chapter 7 we saw how the moon, by its gravitational field, is able to raise tides on the earth. In Chapter 9 we mentioned that at one time it was thought a passing star might have torn matter off the sun to form the planets. Tidal forces produced by the close approach of one galaxy to another can produce strangely contorted galaxies looking totally unlike any of the classes we have described so far. As was first suggested by F. Zwicky and recently studied in detail by A. Toomre and J. Toomre, galaxy-galaxy encounters result in enormous tides that can pull a plume of stars out of one galaxy and spew them across space to form a narrow bridge or tail (Fig. 15.13).

15.17 RADIO GALAXIES

Among the most peculiar galaxies known are those objects which, because of their prodigious production of radio waves, are called radio galaxies. We have noted that the gases in the interstellar space in the Milky Way emit radio waves. To a distant observer, the Milky Way would emit weak but detectable amounts of radio energy. However, the radio energy from the Milky Way is only about one one-millionth the amount of energy radiated in the visible part of the spectrum by its stars; thus an observer detects our galaxy primarily by the light from its stars, not by radio waves from its gases (Fig. 15.14a). For radio galaxies the energy emitted as radio waves may be nearly equal to that emitted as starlight (Fig. 15.14b) and millions of times that produced in a galaxy like the Milky Way.

The first radio galaxy was found by R. Minkowski and W. Baade in 1951 when they noted that a strong radio source lying in the direction of the constellation Cygnus was coincident with a faint smudge of light. The radio source, denoted Cygnus A because of its strength and location, was soon identified by Minkowski and Baade as a peculiar galaxy at an enormous distance from us. It was at first believed that the source represented two galaxies in collision, but this explanation was discarded after the discovery of many other sources and as it became clear that the radio energy was coming, not from the patch of light, but from two areas on either side of the galaxy.

Radio galaxies are still not fully understood. However, they share many general features that have led to a model (discussed in Chapter 15.19) capable of explaining many of their properties. Nearly all strong radio galaxies are elliptical galaxies. The radio emission comes from two sources located at equal or nearly equal distances on either side of the optical image of the galaxy (Fig. 15.15). Radio galaxies often show peculiarities in their optical properties as well. They may show lanes of dust, wisps of ejected gas, or emission lines from hot gas in their nuclei.

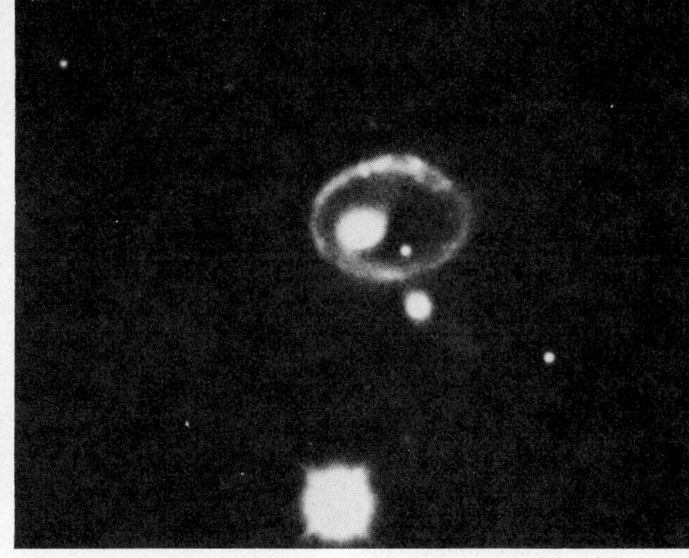

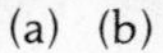

▲
FIG. 15.13
Two pairs of tidally disturbed galaxies. The long tails seen in the pair on the left are formed of stars torn from the main body of the galaxy. The two galaxies on the right have collided and one has "punched a hole" through the other. (Courtesy of Kitt Peak National Observatory.)

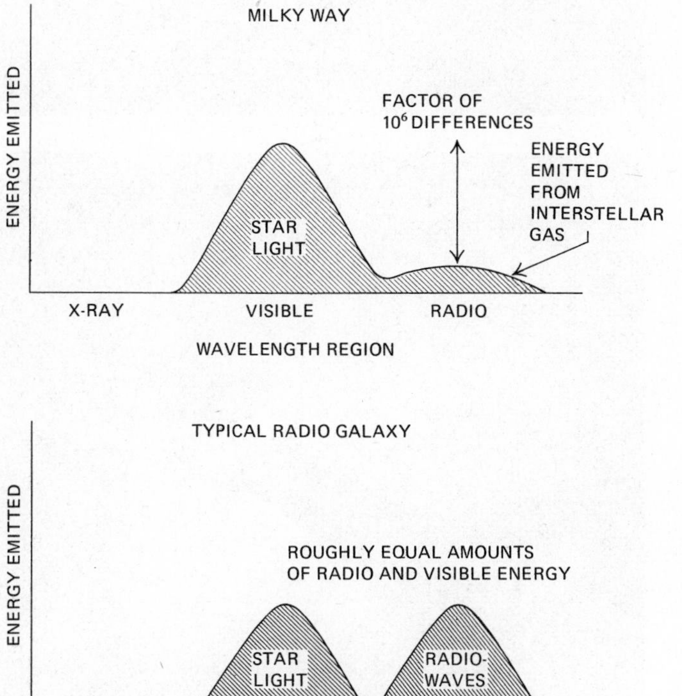

◄ FIG. 15.14
Comparison of radiated energy in a normal galaxy (the Milky Way) and a radio galaxy.

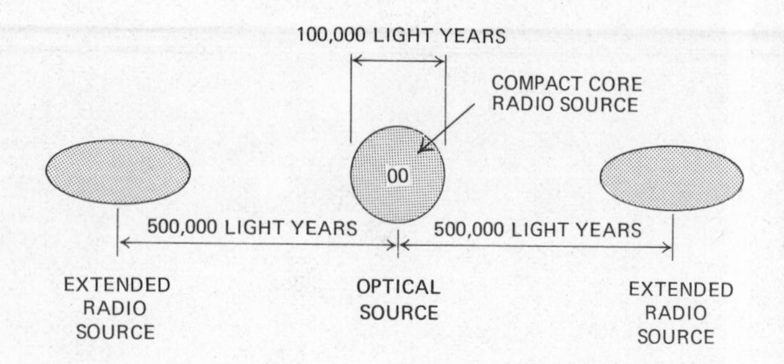

FIG. 15.15
Schematic picture showing relationship of radio-emitting regions to the optical position of a radio galaxy.

FIG. 15.16 ►
Radio galaxies: the elliptical galaxy NGC 4486 (Messier 87) showing the "jet" in polarized light, and the unusual galaxy NGC 5128, associated with the bright radio source Centaurus A. Photographed with the 200-inch telescope. (Courtesy of the Hale Observatories.)

Figure 15.16 illustrates some of these peculiarities. The radio emission from radio galaxies is nonthermal radiation and is produced by high-speed electrons spiraling around magnetic field lines in the galaxy. Often the core of the galaxy is a strong radio source; in the galaxy NGC 5128 three sets of double sources can be seen, all aligned in the same direction: one tiny pair at the center, a slightly larger pair a little further out, and finally the giant lobes of gas and magnetic field outside the main body of the galaxy. The size of the radio-emitting regions in the core may be smaller than a light year, while the huge external lobes are in some cases a million light years across.

In Chapter 14 we saw that nonthermal radiation was associated with supernova explosions. The presence of the nonthermal emission, the radio-emitting regions outside the galaxy, and the optical peculiarities of the galaxy all point to the occurrence of a violent event in the galaxy's core.

15.18 SEYFERT GALAXIES

In 1943 Carl Seyfert at Mount Wilson Observatory discovered a class of galaxies (Seyfert galaxies) similar to spirals, yet definitely different. Their distinguishing feature is the small, sharp, intensely bright nucleus, which appears almost star-like, as shown in a print of the Seyfert galaxy NGC 4151 (Fig. 15.17). Several of the Seyfert galaxies are strong radio sources; some emit tremendous amounts of energy in the infrared region of the spectrum, and others emit greater amounts of ultraviolet radiation than is normal. Their spectra reveal very broad emission lines, indicating that the radiating atoms are moving extremely fast. This motion could result from either a high temperature or, more probably, turbulence and mass motions. Velocities of 13,000 km per second have been found in one Seyfert galaxy. In the spectrum of NGC 4151, lines produced by highly ionized iron have been observed. This is a notable feature, because identical lines are observed in the spectrum of the solar corona, a fact suggesting that the core of this galaxy contains enormously hot gas.

FIG. 15.17
A Seyfert-type galaxy, NGC 4151, photographed with the 200-inch telescope. (Courtesy of Hale Observatories.)

15.19 QUASARS

Seyfert and radio galaxies are in many ways complementary objects. Seyferts are spiral galaxies with strong optical emission and radio galaxies are elliptical systems with strong radio emission. Many of the properties seen in both radio and Seyfert galaxies are combined in a third group of objects that are not at present classed as galaxies. These are the enigmatic *quasi-stellar objects*, referred to as quasars or QSO's. The first ones discovered were strong radio sources that were found to be associated with faint star-like objects, hence they were called quasi-stellar radio sources, or simply quasars.

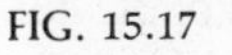

The first quasar was discovered in 1960 by Allan Sandage at Palomar Observatory. It was found in a

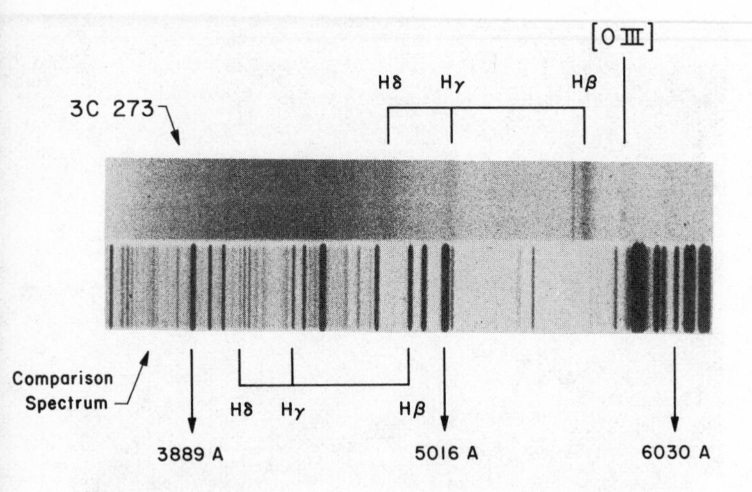

FIG. 15.18
Spectrum of the quasar 3C 273. The lower spectrum consists of hydrogen and helium lines and serves to establish the scale of wavelengths. The upper part is the spectrum of the quasar, a star-like object of magnitude 13. The Balmer lines Hβ, Hγ and Hδ in the quasar spectrum are at longer wavelengths than in the comparison spectrum. The redshift of 16 percent corresponds to a distance of two billion light years in the expanding universe. (Courtesy Hale Observatories and M. Schmidt.)

search attempting to locate optically certain radio sources in the *Third Cambridge Catalog* of radio sources. This source was designated 3C 48. In 1963 the spectra of 3C 48 and a similar object, 3C 273, were obtained by Maarten Schmidt, also at Palomar Observatory. Initially the spectra (Fig. 15.18) were a complete riddle. Although they showed very strong emission lines, it was not at first recognized that the lines were from hydrogen, because they were red-shifted by an amount never before observed. The Balmer lines were shifted by 16% toward the red end of the spectrum. Subsequently, several hundred quasars have been found. Not all of them share the properties of the first two, but all have large red shifts, which suggests that they are at enormous distances from us. Despite the large red shift, several astronomers believe they may be relatively nearby objects that are moving at very high speed, perhaps ejected from the nucleus of a nearby galaxy.

The nature of quasars is still very obscure. Their properties are numerous and, unfortunately, are not shared by all objects that are so designated. As was indicated above, all quasars have large red shifts and many are strong radio sources. Both the optical and the radio emissions vary in what so far seems an erratic fashion. The rapidness of the variation that is observed in many cases implies that the radiating gas is localized in a very small volume. To increase its light output on a time scale of months, a source could not very well be any larger than a few light months in diameter. The light from the far side of the object must travel further to get to the observer than must the light from the near side; thus it will take as many light months for the source to turn on as it is light months across. The very small size of the emitting regions has also been confirmed by the techniques of radio interferometry (Chapter 4.14).

The radio emission and some of the optical emission is nonthermal, which indicates the presence of high-speed electrons and magnetic fields. In many quasars faint wisps of matter are seen (Fig. 15.19). In many the light is highly polarized. Most show extremely strong emission lines of hydrogen, similar to those found in Seyfert galaxies, as well as lines from neon, oxygen, and nitrogen. The most dramatic property of quasars, however, is their energy production. The quasar 3C 273 emits roughly one thousand times the energy of an ordinary galaxy.

A recently discovered group of objects, apparently very similar to quasars, show no spectral features at all. Called *Lacertids*, after their prototype, the object BL Lacertae, they often show a faint halo of light around a very bright core (reminiscent of Seyfert galaxies). The halo seems to be composed of stars, and many astronomers believe Lacertids are elliptical galaxies with extremely bright nuclei. The similarity between quasars,

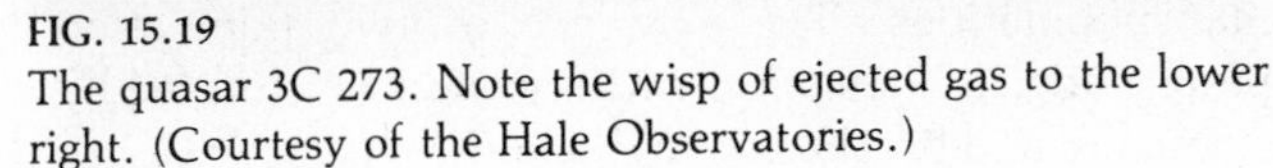
FIG. 15.19
The quasar 3C 273. Note the wisp of ejected gas to the lower right. (Courtesy of the Hale Observatories.)

radio galaxies, Seyfert galaxies, and Lacertids suggests strongly that quasars may after all be a kind of peculiar galaxy. Perhaps they are galaxies in the process of collapse, perhaps galaxies whose nuclei are undergoing some catastrophic disruption. But many other theories have also been suggested to account for their properties—nearly as many theories as there are astronomers studying quasars. The presence of nonthermal emission and the variability of the sources has suggested to some astronomers that the activity in all the peculiar galaxies and quasars is produced by supernova outbursts. Perhaps in the nuclei of galaxies, where stars are far closer together, collisions between stars occur. These collisions might produce effects similar to those exhibited by a supernova outburst or, if stars are permanently fused together by such collisions, they might eventually become so massive as to form supernova. Another suggestion is that the gas ejected from stars (planetary nebulae, nova outbursts, etc.) settles toward the nucleus of a galaxy. There it forms supermassive stars, perhaps of millions of solar masses. These monster stars then explode to energize the quasar or Seyfert galaxy.

One of the most fantastic ideas to explain quasars and their kin is that the centers of galaxies contain giant black holes. With so many stars in the nucleus, it is reasonable to suppose that eventually one would turn into a black hole as a result of its normal evolution. In the closely packed environment of the nucleus there would always be some gas that gets a little too close to the black hole and is "eaten." The added mass would make the black hole larger and would increase its gravitational pull. As it grew, it would eventually get large enough to swallow entire stars. Matter moving toward the black hole would be heated and would accumulate in a giant shell just outside the black hole before it is finally pulled in to oblivion. By this theory, what is detected as a quasar, or on a smaller scale as the core of a Seyfert or radio galaxy, is this shell of gas poised on the edge of a black hole. Eventually the black hole would devour all the available stars in the galactic center. With nothing to feed it, it would no longer be active, and to a distant observer the activity of the nucleus would appear to die away. But the hole would still lurk there, and occasionally an unfortunate star or gas cloud would be snared. The galaxy would then appear basically normal except for some radio activity at its nucleus. Our own Milky Way is such a galaxy—there is a strong radio source at its center, and at some time in the past hundred million years there seems to have been some event that caused the ejection of gases outward from the nucleus. Is there a black hole there now? Perhaps the whole theory (so to speak) is wrong. It is presented here as an example of how mysterious quasars are and to what extremes theories have been carried in order to explain their peculiarities.

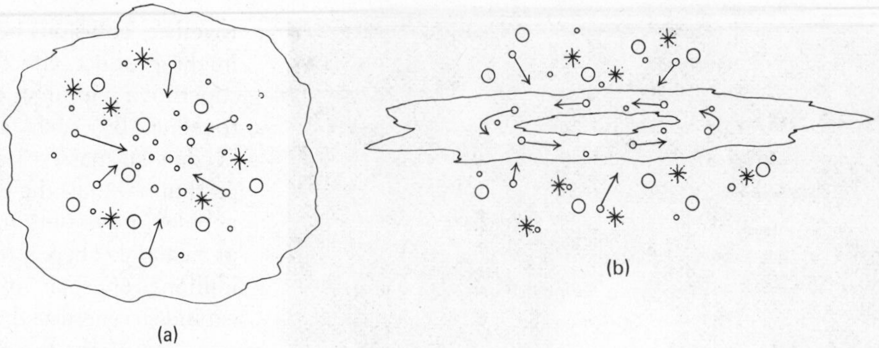

FIG. 15.20
(a) Initial collapse and fragmentation of the Milky Way, with formation of Pop II stars. (b) Formation of Pop I stars in a disk surrounded by a halo of older Pop II stars.

Nevertheless, unlikely as the preceding scenario may appear, a group of astronomers in California and Arizona have recently discovered evidence that a giant dark mass may lie in the center of the elliptical galaxy M87 (Fig. 15.9). Measurements of the variation of brightness across the galaxy show that at the center the brightness sharply increases, as if the stars have been pulled into an exceptionally dense cluster. Also, the velocity with which the stars are moving rises rapidly, as though they feel the pull of some enormous mass. It is estimated that in the central core of M87 there are 5 billion solar masses of matter in an invisible form. M87 is the nearest radio galaxy to us, and may now be the first to reveal the existence of a massive black hole.

15.20 ORIGIN AND EVOLUTION OF GALAXIES

The origin and evolution of galaxies ranks as one of the major unsolved problems in astronomy today. We do not even know for certain whether the different types of galaxies (E, S, and Irr) represent evolutionary phases of a single object or whether the form of a galaxy is a permanent characteristic dating from its birth. One current (although not universally accepted) view is that a galaxy's history is determined mainly by its angular momentum (initial amount of rotation)—that is, the formation of a galaxy and its subsequent evolution depend critically on the rotational speed of the primordial matter from which the galaxy forms.

This theory of galaxy evolution and formation can best be studied with reference to our own Milky Way. We saw that the Milky Way contains two disparate stellar populations. The existence of two population types is understandable if the Milky Way was formed by the collapse and break-up of a massive pregalactic cloud, much as stars are thought to have condensed from interstellar clouds. According to this theory, the matter in the primordial intergalactic clouds was initially hot (10,000°C or about 20,000°F) and very rarefied (1 atom per ten cubic meters), and consisted of a mixture of 75% hydrogen and 25% helium atoms, with no heavy elements at all. Our galaxy probably formed from a cloud containing a few tens of billions of solar masses of matter and having a diameter of perhaps a million light years. It is thought that in galaxy formation, gravity pulls the matter together and compresses it. Contraction amplifies any initial rotation that the cloud has and causes it to flatten. During the initial collapse of the pregalactic cloud, stars form (Fig. 15.20a) within the in-falling matter and evolve, and the more massive ones explode as supernova, returning gases enriched with heavy elements into what up to that time had been a pure hydrogen and helium cloud. Star formation is not 100% efficient and some of the primordial gas is not condensed into stars. In systems with a high amount of rotation, this uncondensed gas forms a disk through which the orbital paths of the first stars plunge (Fig. 15.20b). Gas in the disk becomes contaminated by the heavy elements formed by the first generation of stars. Eventually a second generation of stars forms from the disk gas. These stars contain heavy elements and move in roughly circular orbits because they were formed from gas moving in that fashion. Galaxies with moderate amounts of rotation will thus contain two stellar populations—one group moving on plunging orbits and deficient in heavy elements (Pop II), the other group moving in circular orbits and containing moderate amounts of heavy elements (Pop I).

The preceding description accounts for the formation of spiral galaxies such as our own. Galaxies with essentially no rotation do not form a gaseous disk. Matter falling inward toward their cores does so until it becomes dense enough for stars to form. Having little rotation, these galaxies are unflattened. Since they have consumed all their gas in the first stages of collapse, there is no interstellar matter available for a second generation of stars. Such galaxies are identified with ellipticals. Galaxies that are formed from matter rotating very rapidly behave in the opposite way. They are so turbulent at their birth that in their initial collapse essentially no stars can form. They retain large amounts of gas to form multitudes of second-generation stars, but they contain few old stars. Their rapid rotation prevents them from developing the regular structure associated with spiral or elliptical galaxies. Thus they become irregular galaxies.

It is important to bear in mind that the above scenario of galactic origin and evolution is very tentative. There is still much to be learned about the early days of any galactic system.

15.21 GALAXY CLUSTERS

We have seen that astronomical objects seldom occur in isolation. Planets are grouped around stars; stars are grouped into clusters and into galaxies. Galaxies show the same tendency, being in their turn grouped into galaxy clusters. The clustering of galaxies is evident on wide-area photographs of the sky. Once the obscuring effect of the dust in our own galaxy is taken into account, galaxies are found not to be distributed randomly. The number of galaxies in a cluster varies from a handful to several thousands. The Milky Way belongs to a fairly small cluster called, modestly, the *local group*. The local group is an example of an irregular galaxy cluster. It contains 17 moderate to large member galaxies (including the great nebula, M31 in Andromeda) and an as-yet-undetermined number of small satellite galaxies. Both the Milky Way and M31 have two conspicuous satellite systems, but M31 may contain many far smaller ones.

Galaxy clusters are classed as *irregular* or *regular*. The regular clusters often contain thousands of member galaxies, most of which are ellipticals, and usually possess a roughly spherical structure. The irregular clusters contain fewer members, include many spiral galaxies, and possess little symmetry. The local group and the Hercules cluster (Fig. 15.21) are examples of irregular clusters. The Coma cluster is an example of a regular cluster (Fig. 15.22). A galaxy cluster is typically a few million light years in diameter. Clusters are usually separated from each other by tens of millions of light years. It is not known whether galaxies form first and then group into clusters, or whether a cluster forms and then breaks up into galaxies. It is also not known whether the galaxy clusters themselves cluster. H. Shapley and, more recently, G. deVaucouleurs have argued persuasively that the nearby clusters form a super-cluster, and they cite the seemingly regular arrangement of the local group, the Virgo cluster, the Coma cluster, and several other groups of galaxies as evidence. Seeing structure in the universe on this scale (millions of light years) is not easy, however, and not all astronomers agree with this interpretation.

Another unresolved problem is the amount and state of matter outside of galaxies but inside of galaxy clusters. In our own galaxy, we speak of the interstellar matter, or matter between stars. There is growing evidence that there is also matter between galaxies. This material, referred to as intergalactic matter, is extremely hot (millions of degrees) and very rarefied. It has been observed in at least two different ways. First, the high temperature causes the gas to emit x-rays. X-ray radia-

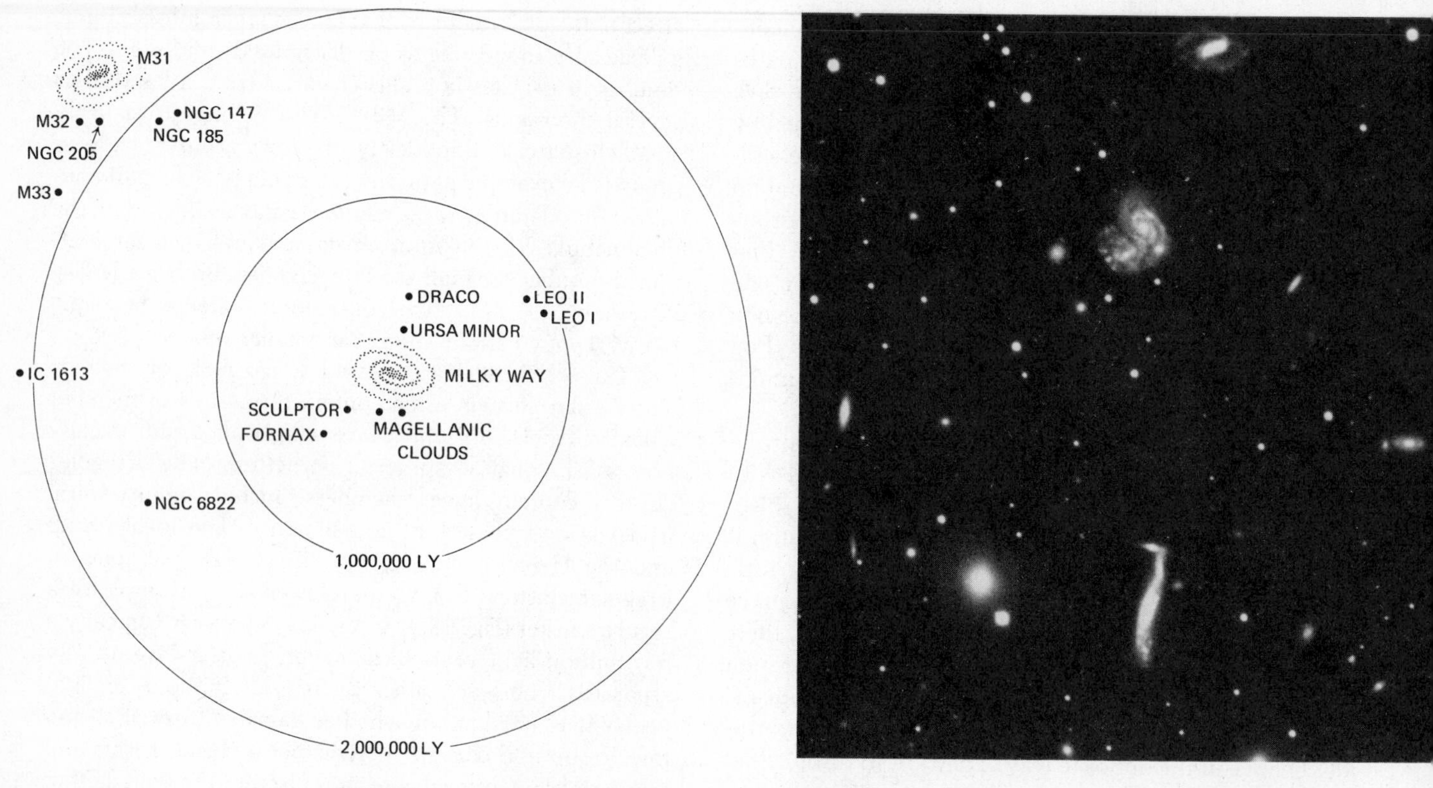

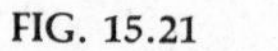

FIG. 15.21
Two irregular galaxy clusters: (left) a sketch of the local group and (right) the large cluster in Hercules. (Photo courtesy of Kitt Peale National Observatory.)

FIG. 15.22
The giant galaxy cluster lying in the direction of the constellation Coma Berenices. (Courtesy of Kitt Peale National Observatory.)

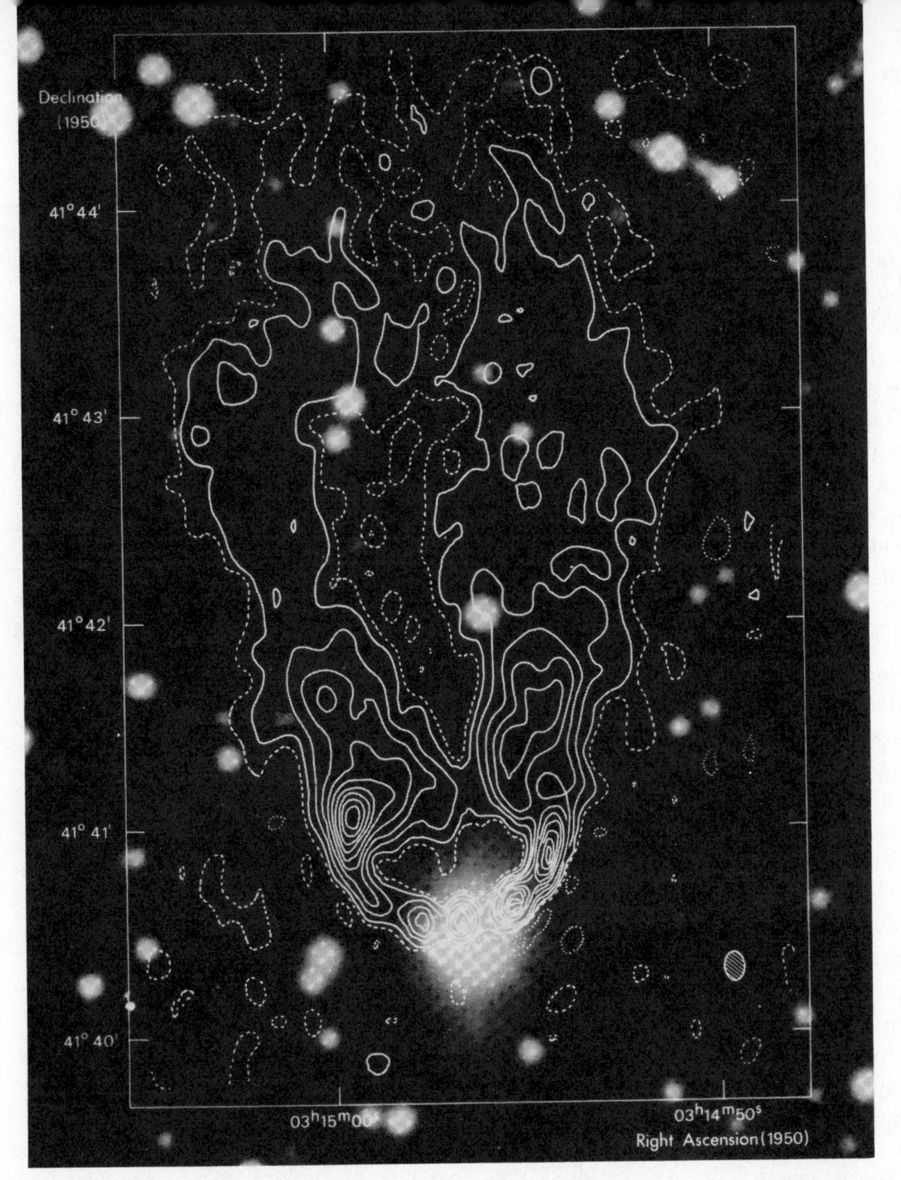

FIG. 15.23
Peculiar radio galaxy with wake of gas produced, perhaps, by the galaxy's motion through intergalactic gas. (Copyright Netherlands Foundation for Radio Astronomy, published with permission of Leiden Observatory.)

tion has recently been detected from a number of regular clusters, called the "rich" clusters, in which the galaxies are moving very rapidly and are very numerous. Intergalactic matter also seems to have an influence on radio sources inside of galaxy clusters. The most striking example of this is the galaxy NGC 1265 in the Perseus galaxy cluster. This radio source has been observed with the radio interferometer at Westerbork, in the Netherlands, to be trailing a wake of radio-emitting matter behind it (Fig. 15.23).

SUMMARY

Stars and star clusters are grouped into even larger aggregates called galaxies. A typical galaxy may contain billions of stars as well as gas and dust (interstellar matter). The sun belongs to a disk-shaped galaxy roughly 100,000 light years across called the Milky Way. From the earth the Milky Way appears as a faint band of stars running around the celestial sphere. The basic sections of the Milky Way galaxy are its nucleus, its disk (which includes the spiral arms), and its halo. The stars in the disk tend to be Pop I while those in the halo and nucleus are Pop II. Interstellar matter is found mainly in the disk and is observable in the following ways: obscuration (blocking background light), reddening of distant starlight, emission (when heated by young stars), reflection (when near luminous stars) and through a variety of radio emissions, including 21-cm radiation and interstellar molecular radiation.

Galaxies can be divided into four basic categories: spiral, barred spiral, elliptical, and irregular. Apart from their appearance, the galaxy types differ in their stellar populations. In general, spiral galaxies possess mixed populations, elliptical galaxies contain mostly Pop II stars, and irregular galaxies contain mostly Pop I stars.

Galaxies often show peculiar activity in their nuclei. Radio galaxies appear to have ejected enormous clouds of high-energy particles and magnetic fields. Seyfert galaxies have clouds of extremely hot gas in their nuclei; the nucleus occupies a volume perhaps only a few light years across, but emits nearly as much energy as all the rest of the galaxy. Quasi-stellar objects (quasars) seem to be an even more extreme example of activity in galactic nuclei. The energy radiated from these peculiar systems may originate from matter falling into an enormous black hole.

Galaxies, like stars, are believed to have formed from the collapse of gas clouds. The differences between the various galaxy types may depend on the amount of rotation that the cloud has: slowly rotating clouds become elliptical galaxies, while more rapidly spinning clouds become spiral systems.

REVIEW QUESTIONS

1. Why does the Milky Way appear as a faint band of light across the sky?
2. Draw a sketch of the galaxy (top and side view). Label the different parts and indicate the scale in light years. Indicate the position of the sun.
3. What is meant by Pop I and Pop II stars? How are they different?
4. How long does it take the Milky Way galaxy to rotate? How do we know it rotates?
5. How much of the Milky Way galaxy can be seen from the earth with optical telescopes? Why?
6. Why do astronomers believe the Milky Way is a spiral galaxy?
7. What is 21-cm radiation? How is it produced? Why is it useful to astronomers?
8. What other sources of radio emission are there in the Milky Way? How might the Milky Way look with a radio telescope as seen from the Andromeda galaxy?
9. List five reasons astronomers give for believing that interstellar matter exists. Is there any evidence that is visible with the naked eye?
10. How are emission and reflection nebulae illuminated?
11. Why might you expect not to see any reflection nebulae illuminated by class O stars?
12. Sketch the Hubble tuning fork diagram of galaxy types. Label the galaxy types on it.
13. Which type of galaxy is richest in interstellar matter? Which type seems to have the least interstellar matter? Is that consistent with their stellar populations?
14. What are the two major types of galaxy clusters? Are clusters themselves grouped? Why are astronomers unsure about the existence of super-clusters?
15. About how many galaxies are in the local group? Is the Milky Way a member of the local group? Are there any giant elliptical galaxies in this group? Are there *any* elliptical galaxies in it?
16. Does the Milky Way galaxy have any satellites? What are they called? Can they be seen from the United States?
17. How does the model of galaxy formation described in the text help to explain the origin of and the differences between Pop I and Pop II stars?
18. What is a Seyfert galaxy? Is the Milky Way a Seyfert galaxy? Why?
19. What are the characteristics of a radio galaxy?

20. What are the characteristics of a Lacertid?

21. What is a QSO?

22. Describe some of the mechanisms suggested for producing the energy of quasars.

23. Are radio galaxies often spiral galaxies?

24. At one time it was argued that dark nebulae were "tunnels" passing through the galaxy in which no stars existed. What would this suggest about the orientation of the tunnels with respect to the sun?

CHAPTER 16
COSMOLOGY

16.1 THE MYSTERIOUS

The night sky has always held a strange, disquieting fascination for humanity. Standing alone, away from the artificial lighting of modern civilization, one looks into a darkness that is broken by a single, illuminated disk and thousands of tiny, flickering lights. If one allows the imagination to run free, one can "fall" into the immensity of space. Rushing onward, stars, star systems, galaxies, and interstellar matter speed by, but the end is never reached. Even the wildest imagination is incapable of visualizing the ultimate conclusion of such a trip. How can one ever reach a point beyond which it is impossible to step? How can nothingness begin? Refusing to be cowed by the seeming futility of the search, humanity has never ceased in the attempt to discern the structure of the universe. Astronomy has painstakingly developed an impressive body of knowledge about the structure of that part of the universe nearest the earth. However, astronomers have not been content merely to establish the laws that govern the celestial bodies; they also want to discover the ultimate nature of the universe itself. In coming to grips with the universe as a whole, the astronomer must face several basic questions. What is the extent and shape of the universe? What is its age? What will be its final end? What, if anything, lies beyond the universe? Although these questions touch at the very core of existence, the answers may never be known.

Many of the early ideas about the origin and nature of the universe were woven together in the religions of humanity. One of these is found in the Bible in the book of Genesis, which tells us that the physical universe is a product of God's mind:

> *In the beginning God created the heaven and the earth; and the earth was without form and void; and darkness was upon the face of the deep and the spirit of God moved upon the face of the waters. And God said, "Let there be light," and there was light. And God saw the light, that it was good. (Gen. 1:1–3).*

Since the Christian church was the dominant force in Europe during the medieval period, its beliefs about the origin and nature of the universe were generally accepted by philosophers, who simply tried to show that these beliefs were just and reasonable. The Christian universe was finite and bounded. Beyond the physical universe was the infinity of God. This concept began to crumble with Copernicus' suggestion that the placement of the sun at the center of the cosmos would better explain the movements of the planets. In 1576 Thomas Diggs took the step that Copernicus had avoided and wrote that the infinity of the universe was a distinct possibility. Humanity's conception of the universe was beginning to expand. Giordano Bruno, who was burned at the stake in 1600, believed that there was not just one world, but an infinity of worlds. In the middle of the eighteenth century, the German philosopher Immanuel Kant conceived of an infinity of space populated by "island universes." With the advent of new instruments and methods, *cosmology*—the study of the origin, structure, and evolution of the universe—became an important branch of astronomy.

Elaborate scientific theories have been postulated in an attempt to answer the basic questions about the universe. While scientists try to remain within the safe pale of objectivity and base their theories on facts, cosmology is of necessity still a speculative area. Within it, science makes contact again with philosophy and religion. Cosmology may well be the limit beyond which humans will never be able to extend their knowledge. Yet the probe will assuredly continue, for as Albert Einstein wrote:

> *The fairest thing we can experience is the mysterious. It is the fundamental emotion which stands at the cradle of true art and true science. He who knows it not and can no longer wonder, no longer feel amazement, is as good as dead, a snuffed-out candle . . . Enough for me the mystery of the eternity of life, and the inkling of the marvelous structure of reality, together with the single-hearted endeavor to comprehend a portion, be it ever so tiny, of the reason that manifests itself in nature.**

* Albert Einstein, *The World As I See It*, translated by Allan Harris, New York: Philosophical Library, 1949, p. 5.

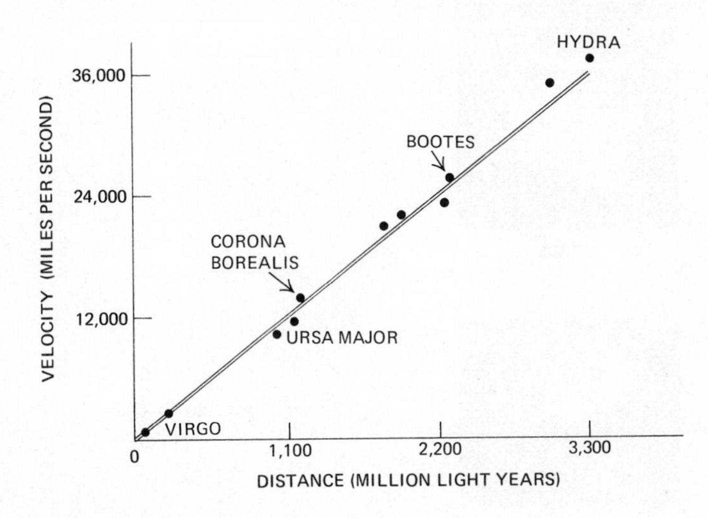

FIG. 16.1
The linear velocity-distance relationship for clusters of galaxies illustrates the relationship between the red shifts of galaxies and their distances.

16.2 OUR VIEW OF THE UNIVERSE

Our observations show that the distant galaxies and clusters of galaxies are roughly uniformly distributed in space. The red shifts noted in their spectral lines, if due to the Doppler effect, indicate that the galaxies are moving away from us at a rate which increases with distance. In 1912, V. M. Slipher first observed and measured the red shift in the spectrum of a galaxy. By 1929, Hubble had discovered that the radial velocities of galaxies are in direct proportion to their distances from the observer. From this, he established the velocity-distance relationship, which is known as the *law of the red shifts,* or Hubble's law. This linear relationship is shown in Fig. 16.1 and is expressed algebraically as

$$V = Hr,$$

where V is the velocity of the galaxy, H is Hubble's constant, and r is the galaxy's distance. Using Hubble's

GALAXIES NEBULA IN	DISTANCE IN LIGHT YEARS	RED SHIFTS
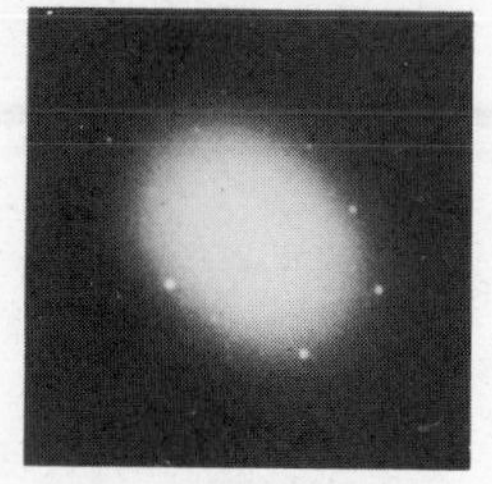 VIRGO	78,000,000	H+K 750 MI/SEC
 URSA MAJOR	980,000,000	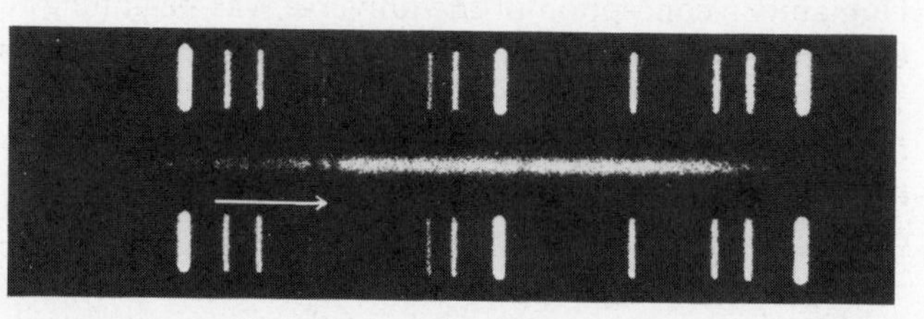 9,300 MI/SEC
 CORONA BOREALIS	1,400,000,000	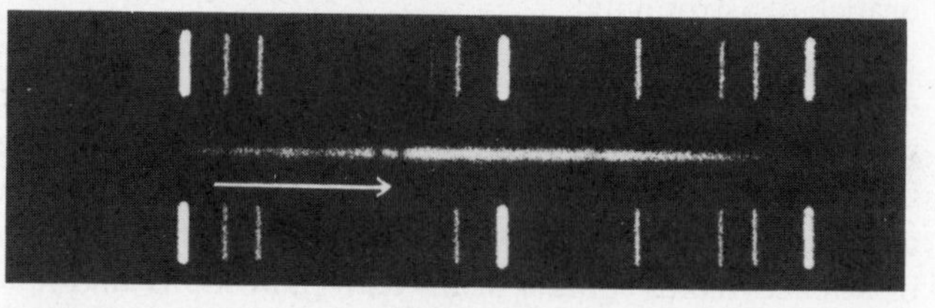 13,400 MI/SEC
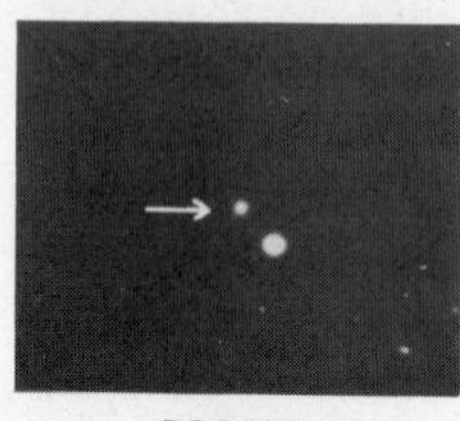 BOOTES	2,500,000,000	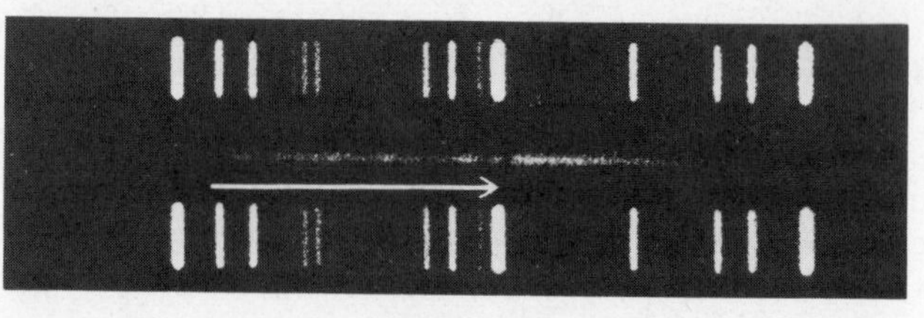 24,400 MI/SEC
 HYDRA	4,000,000,000	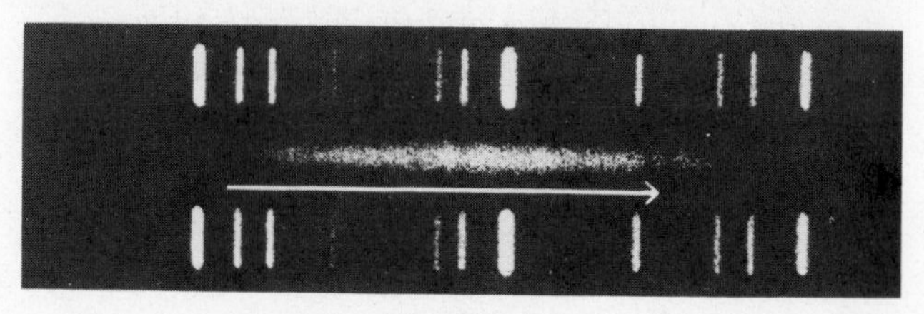 38,000 MI/SEC

◄ FIG. 16.2
Relation between red shift and distance for galaxies. Each galaxy shown on the left is a member of a cluster of galaxies in the constellation indicated. The red shift of the two absorption lines of ionized calcium in their spectra shown on the right is indicated by arrows. The red shifts are expressed in miles per second. The more distant galaxies show greater red shifts. (Courtesy of the Hale Observatories.)

constant, which is about 11 miles (17 km) per second per million light years, astronomers have been able to determine the distance of galaxies. The photographs and spectra of five galaxies with distances that range up to about two billion light years are shown in Fig. 16.2.

Recessional motion is the only verifiable cause that can account for the observed red shifts in the spectra of distant galaxies. It is therefore generally accepted as observational evidence that the universe is expanding. However, this does not imply that the earth is at the center of the universe, nor does it imply that the individual galaxies are expanding. It is the space in which the galaxies are embedded that is expanding. The red shift arises because light waves traveling across the expanding void are stretched and thus reddened.

Some scientists, rejecting the interpretation that the red shift indicates an expanding universe, have suggested that it is instead caused by the loss of energy in the photons as light travels across great distances of space. This "tiring" causes the increase in the light's wavelength. Others contend that something not yet discovered is causing the red shift. However, none of the suggestions that have been proposed are supported by either theoretical or observational evidence.

16.3 COSMOLOGICAL MODELS

In our earlier discussions of planets, stars, and galaxies, we have in each case tried to use the available observations in order to construct a model that will explain the shape, structure, and evolution of the object in question. In studying the universe as a whole, it is especially difficult to construct such a model for the following reasons: (a) There are only a few relevant observations that can now be made. (b) There is by definition one and only one universe, and it encompasses all that there is. Thus one cannot appeal to other "universes" to help clarify the model. (c) The observer is inextricably inbedded inside the universe, preventing the sort of external view so useful in studying stars and galaxies. Nevertheless, it is possible to devise representations of the universe that at least crudely describe what is observed (such as the shape, size, and evolution of the universe). These representations are called *cosmological models.*

Cosmological models are based on mathematical equations because mathematics is the language most suitable and most widely accepted for adequately describing the complicated cosmological situation. The equations are based on an assumption called the *cosmological principle*, which states that *the universe will appear the same in all directions to all observers*, regardless of their location in space. When the principle is extended to uniformity in time—that is, to an observer anywhere in space the universe will appear the same regardless of time—it is called the *perfect cosmological principle.* There is strong evidence, which will be discussed in Section 16.6, that the perfect cosmological principle is not valid. However, the less restrictive cosmological principle will be assumed true in the remainder of the discussion of cosmology.

In Einstein's theory of relativity, the cosmological equations predict the possibility of many types of universes, depending on the values assigned to the several variables. The two most important variables measure the amount of matter the universe contains (the density) and its rate of expansion. There are two classes of models that have appeared since Einstein presented his theory of relativity: *evolutionary* and *steady-state* models. Both have accepted the theory that the red shifts indicate an expanding universe, and each will be discussed in turn. However, before we turn to a comparison of the evolutionary and steady-state theories, a short description of the features they have in common is appropriate.

Cosmological models, or model universes, do not attempt to explain the fine details of the universe such as stars or planets. In fact, even galaxies and galaxy clusters are reduced in cosmological models to the role of *marker particles*, which allow astronomers to see the structure of the universe but which play no role by themselves. Model universes that have been devised so far are described in terms of a smoothed average density, which can be thought of as what the universe would be like if the galaxies and stars were all spread out into their component atoms and distributed uniformly throughout all space. This smoothed distribution of material is characterized by a density, ρ, measuring the number of grams per cubic centimeter. On the basis of the matter directly observable in the universe, the smoothed density has the value of 2×10^{-31} gm/cm^{-3}, equivalent to about 1 hydrogen atom per ten cubic meters. The amount of matter in the universe plays a critical role in governing its history. Since each atom attracts every other atom via its gravitational force, every segment of the universe is in principle able to "feel" the others by means of gravity. Since gravity always acts to draw matter closer together, it tends to hold the universe together. It is a bit like a balloon full of water set down on a table. Without the balloon enclosing it, the water would flow freely across the table surface in random pathways. The elastic skin of the balloon, however, confines it and gives it form. So, too, does matter through its gravity give form to the universe.

The shape of the universe is determined in principle by the distribution of the galaxies or galaxy clusters, treated as the marker particles referred to earlier. The galaxies may be thought of in this context as if they were snowflakes, leaves, or motes of dust borne along by the wind and thereby making the motion of the unseen air visible.

Astronomers describe cosmological models in terms of whether they are infinite or finite, "flat" or "curved." Sometimes the terms "open" and "closed" are used to denote infinite and finite universes, respectively. For most humans, understanding the size and shape of the universe is as difficult as it probably was for a medieval philosopher to understand that the world is round. For example, how does one visualize a universe with no edges or outside, and yet at the same time, a universe that extends forever? In order not to turn the discussion into one laden with technicalities about four-dimensional spaces, it may be helpful to appeal to a few analogies.

Suppose an ant is crawling on a sheet of paper. If it were an intelligent ant, it might be able to figure out, by crawling slowly around the edge, that the center is the point equidistant from the edges (Fig. 16.3). Now suppose the sheet of paper is somehow cut and pieced together to form a sphere, and taped so that there are no edges (Fig. 16.4). The sphere is so large that the ant is unaware the paper is curved. The crawling ant no longer finds a boundary. The surface of the paper ball on which it moves has no center on the surface. The center is located "inside" the ball at a point the ant cannot reach, and which it perhaps cannot even imagine. To human beings able to think and to visualize things in three dimensions, the center of the ball is an easy concept; for the ant, a prisoner of the two-dimensional surface on which it must crawl, this third dimension leading to the inside of the sphere is alien and unimaginable.

Suppose now it is possible to "bend" the three-dimensional universe into a curved four-dimensional object. If the curvature or bending of the universe is small, it will not be noticed any more than the ant realized its piece of paper was bent. Just as the ant cannot conceive of the "center" of the sphere on which it moves, so do humans have difficulty visualizing a center to the universe. The ant needs to imagine a third dimension to discover the center of its universe; humans need to envision a fourth dimension to visualize the center of their universe. Just as there is no point on the piece of paper that is the center for the ant, there is no point in the universe that is its center for us. In fact, the question may be posed incorrectly. Since the fourth dimension is equivalent to what is normally perceived as time, perhaps one should ask not where the center is, but *when* it is. It is a point in time from which the galaxies are expanding, not a point in space. There is nothing "outside" the universe because "inside" and "outside" refer in a four-

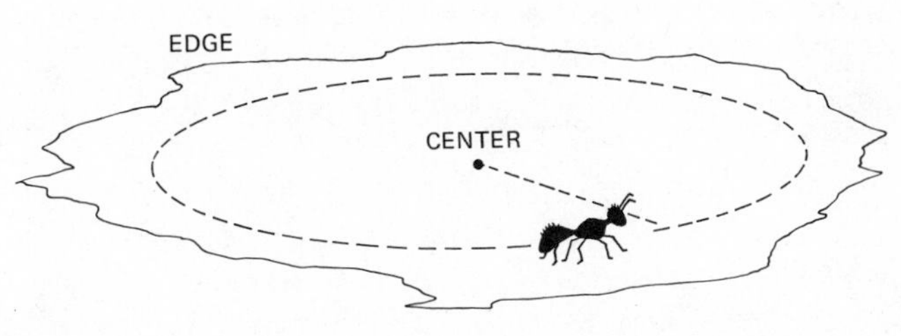

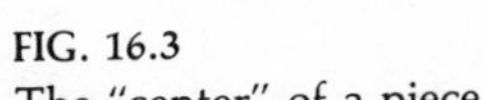
FIG. 16.3
The "center" of a piece of paper as measured by a crawling ant.

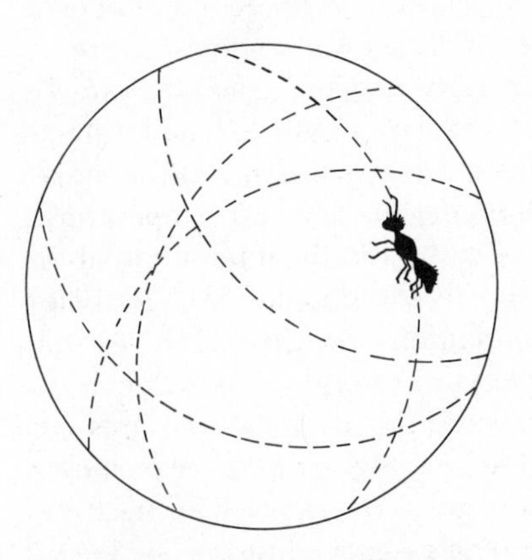

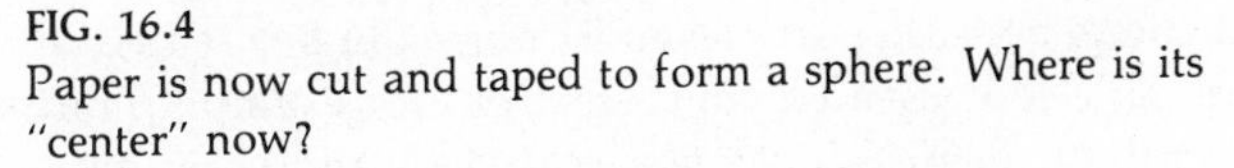
FIG. 16.4
Paper is now cut and taped to form a sphere. Where is its "center" now?

dimensional space to the past and future, and since the future has not happened yet, there cannot be anything there.

16.4 THE EVOLUTIONARY (BIG BANG) MODEL

The evolutionary cosmological model was first proposed by the Belgian Abbé Georges Lemaitre in 1927. He suggested that the universe originated from an explosion of a *primeval nucleus*, which contained all the matter in the universe within a volume of space equal to the diameter of the earth's orbit around the sun. The mechanics of how the universe was started from the primeval nucleus was developed by the American physicist George Gamow in his "big bang" theory (Fig. 16.5). (Steady-state theory is shown in Fig. 16.6 for comparison.)

According to Gamow, since the galaxies are receding from one another now, they must in the past have been closer together. If on the basis of the expansion now, one asks how the universe looked billions of years ago, a little reflection shows it must have been far denser and hotter. Gamow suggested that under the high densities and temperatures of the early universe, all matter was in the form of neutrons. Since the sphere of neutrons was at a tremendous temperature and under enormous pressure, an explosion occurred, ejecting material into space at great velocities. During this expansion period, the neutrons decayed into protons and electrons. As the expansion continued, the temperature dropped sufficiently to allow the protons to capture neutrons and to form deuterons. Gamow theorized that the elements were built from the neutron capture process within 30 minutes.

Although the theoretical curve based on the neutron capture is in general agreement with the curve based on the observed relative abundances of the elements, some investigators doubt that the elements were formed in this manner, because the process would have come to a halt with the formation of unstable helium isotope with an atomic weight of 5. Since the unstable helium isotope immediately decays to the stable helium element with an atomic weight of 4, how were the heavier elements produced? Gamow's explanation was that since 98% of

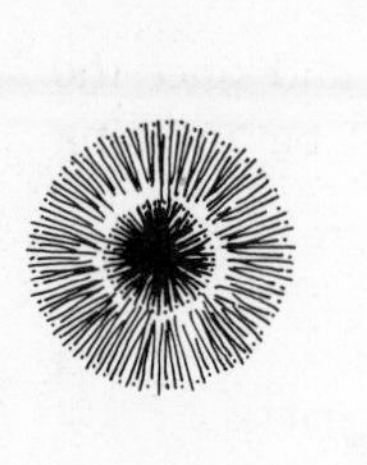
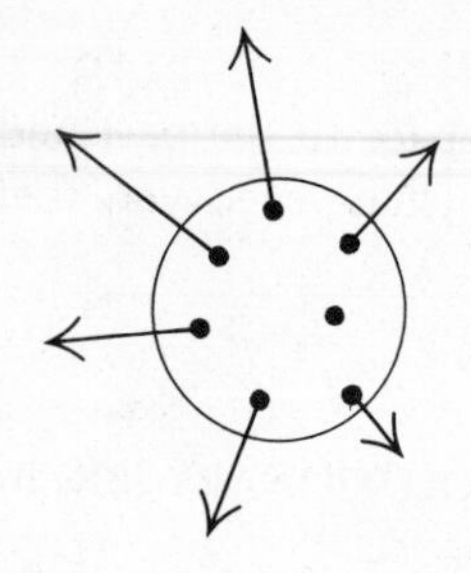
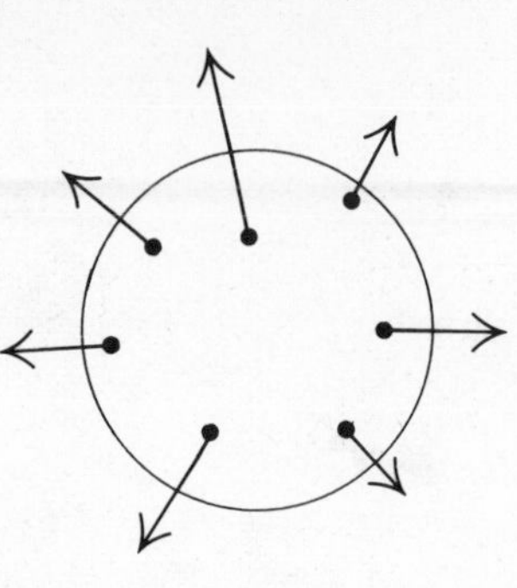

FIG. 16.5
According to the "big bang" theory, the universe originated from a tremendous explosion. The galaxies are moving away from the hypothetical center. The universe is expanding, and the space between galaxies is increasing.

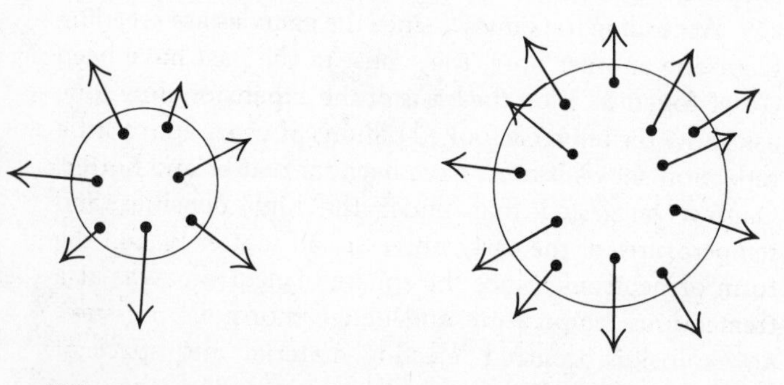

FIG. 16.6
According to the steady-state theory, the universe always has been and always will be as it appears today, that is, its density is the same, regardless of time. Since the universe is expanding, new matter is continuously being created.

the universe consists of the low atomic-weight elements hydrogen and helium, these two elements were formed immediately after the explosion by the neutron-capture process. As seen in Chapter 13, the heavier elements were and still are being synthesized within the interior of stars that are evolving, a suggestion originally made by Hoyle.

Over the past few decades a new view has emerged that differs somewhat from Gamow's original picture. It is now believed that the early universe consisted largely of extremely high-energy radiation with a small admixture of subatomic particles. The temperature of the radiation was perhaps a trillion degrees. At this temperature, atoms themselves cannot exist, and the intense radiation field could spontaneously create a population of particles entirely alien to the conditions existing now. At the temperatures of the everyday world, it is difficult to create matter from radiation because it requires an amount of energy equal to $2mc^2$, where m is the particle's mass and c is the speed of light. To form even an electron, this means an enormous energy expenditure. (The factor of 2 enters because particles must normally be created in pairs, an electron and an anti-electron, for example.) But at a trillion degrees, "strange particles" and their anti-particles would be copiously created. This primordial gruel would be rich in neutrinos, muons, pions, lamda particles, and other particles that now reveal themselves mainly in atomic accelerators or cosmic rays.

It is thought that about 15 billion years ago, for reasons that are not known, the matter and energy began to expand. Expansion led to cooling. Cooling lowered the energy of the radiation and made it hard for the strange particles to be created. By the time expansion had caused the temperature to drop to a few billion degrees, the major constituents of the universe were protons, neutrons, and electrons plus the radiation. As expansion continued the temperature dropped still lower. Most of the neutrons decayed to form protons and electrons, while a few combined with protons to form small amounts of deuterium and helium, but not heavier elements. Expansion and cooling continued still further and, when the temperature reached about 10,000°K, hydrogen atoms began to form by the recombination of electrons and protons. At this time matter and radiation began to behave independently. The universe was for the first time able to form smaller subcondensations. These primordial eddies in the expanding gas are believed to have been the ancestors of galaxies. At this time the universe was probably about 10,000 years old.

It is very hard today, looking out at the night sky, to imagine what the early universe was like. By the big-bang theory its creation involved an explosion the likes of which perhaps will never occur again. The early expansion and cooling occurred very rapidly, but in the period from the first formation of galaxies to the present the universe appears to have been relatively quiet; the galaxies continue to race away from one another, but with a deliberate orderedness that perhaps can be imagined as an apology for their cataclysmic beginnings.

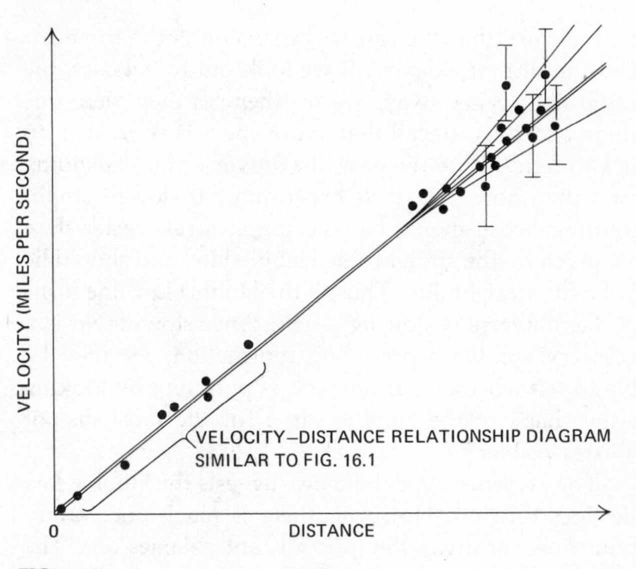

FIG. 16.7
The velocity-distance relationship extended to the more distant cluster galaxies and radio galaxies.

16.5 THE PULSATING UNIVERSE

What is the fate of the expanding universe? Does it expand forever, the stars burning out one by one until the universe is completely dark, containing only dead stars, black holes, and other inhabitants of the astronomical graveyard? One of the major questions in astronomy today centers on whether the expansion stops and reverses. A universe that expands to maximum "size" and then contracts under the gravitational pull of its component galaxies is called a *pulsating universe*. In such a universe contraction leads to a merging of all the matter in the universe. A new primeval nucleus is created. A new explosion follows. The universe of stars and galaxies is reborn.

In a pulsating universe, there is no beginning and no end. The universe pulsates forever, cycle after cycle. It is obvious that for pulsation to occur, the galaxies' mutual recession must slow and ultimately stop. Whether this is possible depends on how much matter the universe contains. More matter means more gravity, which means more deceleration. The density of matter required to halt expansion is called the *critical density*, ρ_c, and has been estimated to be about 10^{-29} gm/cm^{-3}. This is about one hundred times the observed density (see Chapter 16.3). Thus there appears to be too little material to stop the expansion. This material not seen, yet required to "close" the universe and stop its expansion, is referred to as *missing matter*. Recently, however, evidence from x-ray telescopes in orbit above the atmosphere suggests there may be huge clouds of extremely hot gas in many distant galaxy clusters. It is still too early to say whether this matter is abundant enough to close our universe and prevent it from perpetually expanding.

In a more direct way, we might look at the motion of galaxies to see if there is any evidence that their mutual recession is slowing down. Such a test can be performed using the Hubble relation shown in Fig. 16.7. A straight line indicates that the rate of expansion is the same for near and distant galaxies. If we look out to galaxies one billion light years away, we see them as they were one billion years ago (recall that *out* in space is equivalent to *back* in time). If, in the past, the universe was expanding faster than now (i.e., if its expansion has slowed down) remote galaxies should be receding at a rate faster than that given by the straight-line Hubble law, and should lie *above* the straight line. Thus, if the Hubble law line turns *up*, the universe is *slowing down*. Since slow-down is a necessary condition preceding contraction, we may be able to tell whether our universe is pulsating by looking at the shape of the Hubble curve for the most distant galaxies visible.

The evidence available now suggests the Hubble law line does turn up. However, there is much uncertainty about how far away the most distant galaxies are. The bars shown in Fig. 16.7 reveal the uncertainty in distance and thus show we are not yet able to determine whether the universe is or is not pulsating.

16.6 THE STEADY-STATE MODEL

The steady-state model of the universe, proposed by H. Bondi, T. Gold, and F. Hoyle, was developed because they questioned whether the universe had, as the evolutionary model required, a definite beginning and a finite age. In the steady-state model, the universe has always been and will always remain as it appears today, that is, its density remains the same, regardless of time. In this theory, the universe has no beginning and will have no end. Because the expansion of the universe was established by the red shift, proponents of the steady-state model had to explain how the density of the universe could nevertheless remain constant. Hoyle accomplished this by introducing his famous continuous-creation-of-matter hypothesis, which states that matter is created continuously at a sufficient rate to replace the matter that is moving outward. This controversial hypothesis is the basic weakness of the steady-state theory, because it postulates the creation of matter from nothing.

One evidence against the validity of the steady-state theory was produced in 1965 by Arno Penzias and Robert Wilson of the Bell Telephone Laboratories when they discovered, by using conventional radio telescopes, low-energy cosmic radio radiation coming from all directions. It appeared as though the entire universe was filled with this radiation, which was characteristic of the radiation emitted by a *black body*, a perfect radiator, at a temperature of about 3°K.

This universal cosmic radiation had been predicted by the big-bang theorist George Gamow. As early as 1948, Gamow said that if the universe started from an explosion of a primeval nucleus, one should be able to detect radiation from it at the present time. The primeval nucleus was originally extremely hot and dense. As it expanded, it cooled and eventually permeated the entire universe. Gamow inferred from his theory that the present-day temperature of the radiation should be about 3°K, which is what it actually turned out to be. The discovery of this radiation was a devastating blow to the steady-state theory. Since the steady-state universe never existed in a dense state, it could never have created the black-body radiation.

The steady-state theory presents other problems as well. Recall our discussion (Chapter 15) of quasi-stellar objects with large red shifts. If these objects are at the enormous distances indicated by their red shifts, they are seen the way they were a billion years ago. (An object one light year away is seen the way it was one year ago, etc.). Since QSO's are found *only* at large distances, they must represent a class of objects that existed only long ago. All QSO's near the Milky Way galaxy have long since died away. Thus the universe now seems to be different from the universe many billion years ago. It cannot be a steady-state universe because large-scale changes have occurred.

16.7 AGE OF THE UNIVERSE

In order to estimate the age of the universe, we must assume that the universe had a definite beginning. Keeping this assumption in mind, we can try to determine the lapse of time between its beginning and its present state. We know that the oldest stars, located in globular clusters, have an age greater than 10^{10} years. Moreover,

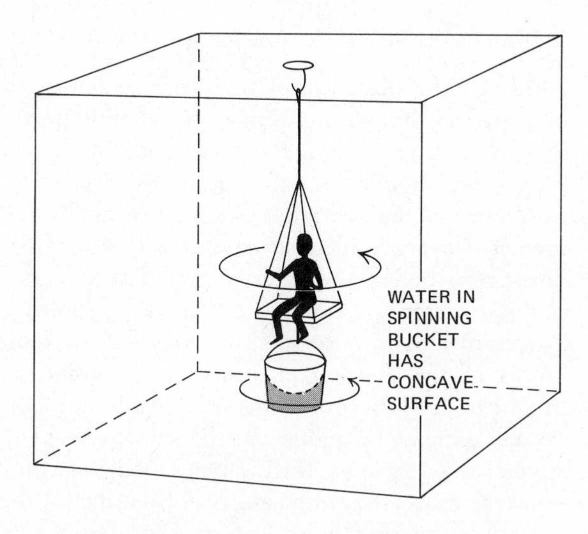

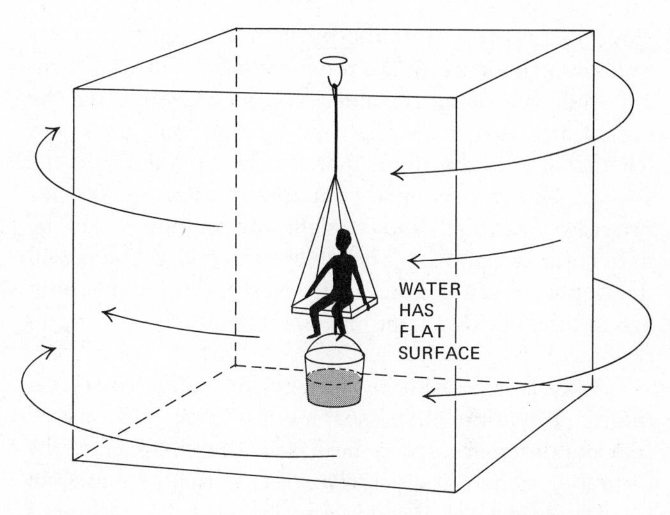

FIG. 16.8
A swing in a room. Can you tell whether you are being spun in the room or the room is being spun around you?

red-shift studies yield a comparable time for the expansion of the universe. Thus, with our present knowledge, the universe appears to be about 15 billion years old.

16.8 SOME COSMOLOGICAL MISCELLANY

It is easy to come to believe that cosmology is a remote and impersonal subject. However, just as all life forms on earth are woven into an interrelated pattern studied under the heading of ecology, a phenomenon that may at first appear to occur only on a cosmic scale is often reflected in the world close at hand. There are almost certainly profound connections between what happens on the scale of the receding galaxies and what occurs on the human scale. Some of these connections are well understood, others are fragmentarily known, and still others are speculative. In concluding the discussion of cosmology, it may be appropriate to consider two examples of possible connections between cosmology and the world around us. One connection is partially understood, the other is purely speculative.

An example of the relationship between the universe in the large and the world close at hand that is partially understood is *Mach's principle*. Mach was a nineteenth-century German scientist and philosopher whose work appears to have strongly influenced Einstein. To understand Mach's principle, it is necessary first to digress briefly on the topic of relative motion. The problem of detecting relative motion is in fact familiar to anyone who has driven a car. Imagine being stopped on a slight uphill slope at a traffic light. As you look out the window at the car in front, all of a sudden you may perceive yourself to be rolling backward. Jabbing at the brake seems to have no effect. When you look again, however, you realize that the car in front is creeping *forward*, and that your car has remained stationary. In general, an observer cannot readily distinguish between self-motion and motion by the surroundings as long as the motion occurs in a straight line at constant velocity. Now, however, imagine sitting on a swing hung from the ceiling in a closed room (Fig. 16.8). If the swing is set spinning you are quickly made aware of the rotation by the dizzy, queasy sensation you experience. If the room were somehow set turning while the swing remained fixed, you would *see* the same thing as if the swing were turning, but you could tell that it was the room that was moving and not the swing. Newton pointed out the curious difference between rotational and straight-line motion when he noted that a spinning bucket of water in a room will develop a concave surface. (Obviously, turning the room around the bucket would have no effect on the water in the bucket.)

Why is it possible to distinguish relative rotational motion, but not relative linear motion? Mach conjectured that the difference occurs because of the presence of the distant stars and galaxies. The distribution of matter in the far reaches of the universe establishes a reference frame that allows the immediate detection of any rotating motion. If there were no distant stars, neither the sense of dizziness associated with spinning nor physical effects like centrifugal force and the Coriolis force would exist. Thus, since the wind blows in large measure because of the Coriolis force, when a flower sways in the breeze, it is moving in response to a force established by the distant stars.

A more speculative cosmological question deals with why the universe is the way it is. Why is the universe expanding rather than contracting? Why are the stars spread sparsely across the sky instead of appearing close together? Indeed, why are there any stars at all? Why is the force of gravity weak while the force between electrically charged particles is strong? B. Carter, a young English mathematician and physicist, has speculated that many of the known properties of the universe exist because if the universe were any different, there would be no intelligent life to perceive it. The existence of human life requires planets, which in turn require heavy elements, which in turn require that some stars have died, etc. Thus human life can exist in the universe only because the universe has evolved to the set of conditions that now prevail. It is therefore meaningless to ask why the universe is expanding. The very existence of the questioner requires that the universe have all those properties which it now has.

16.9 CONCLUSION

Although the observational evidence, such as the 3°K background radiation, points to the abandonment of the steady-state model and the acceptance of the big-bang model, we should not conclude that the latter model is the definitive one. Cosmological models are like all other models—they are reconstructions of the universe along lines prescribed by the observational data available. Cosmology presents a number of difficult problems to an astronomer who attempts to construct a reasonable model of the universe. Because of the great distances involved, facts become somewhat tenuous. As long as doubts remain, no model can be said to be proved. In spite of the problems, the big-bang model presently provides the most efficient means of accounting for the given data. Continuing studies into quasars, remote galaxies, and intergalactic matter may radically alter the situation in the near future. The "mysterious" still beckons.

SUMMARY

Observation of distant galaxies shows they are receding from one another and that the recession velocity increases with distance (the Hubble law). It appears that the universe was created about 15 billion years ago in a cataclysmic explosion referred to as the "big-bang." In addition to the observed expansion, the big-bang theory accounts for the presence of the 3°K background radiation. Intense radiation is believed to have characterized the hot early phases, but as the universe expanded the radiation cooled. Current evidence suggests the expansion will continue for many billions of years more, but that ultimately the universe will recollapse.

REVIEW QUESTIONS

1. What is the Hubble law?
2. It is observed that M31—the great galaxy in Andromeda, and a member of the local group of galaxies—is *approaching* us. Does this mean the Hubble law is wrong? Why?
3. The quasi-stellar object 3C 47 has a red shift of .425. What is its distance from the earth according to the Hubble law?

4. Does the perfect cosmological principle seem a sensible hypothesis to you? If the universe has an edge, will all observers regardless of location see the same thing? Why?
5. What information does the Hubble law give about the age of the universe? Suppose the Hubble constant were suddenly discovered to be five times larger than is now believed. How would that affect our estimate of the age of the universe?
6. Suppose the velocity-distance curve for galaxies turned down from a straight line at large distances. What would that mean about how fast the universe expands?
7. Describe a current theory for the formation of galaxies.
8. What is meant by missing mass?
9. Draw four galaxies along a straight line, all separated by the same distance and each receding from its neighbor at 100 km/sec. Show that a person on *any* of the four will see a Hubble-type law of expansion.
10. Does it interest or bother you that the universe may expand forever?

APPENDIXES

APPENDIX 1
CONSTELLATIONS

Few people notice the magnificent pageant that unfolds in the sky every night. Yet many hours of enjoyment and pleasure are readily available to anyone who is willing to do one simple thing—look. The task of learning about and understanding the stars is not as formidable as it may appear, because although the number of stars in the universe has been compared to the number of grains of sand on all the coasts of the earth, the stars visible to the unaided eye number less than 3000.

The initial step in becoming acquainted with the stars is to scan the sky and observe that the stars differ considerably in their color and brightness (Chapter 11). The colors range from blue (hottest) to red (coolest), and the brightnesses range from −1 magnitude (brightest) to +6 magnitude (just barely visible). For simplicity in observing the stars with the unaided eye, the following symbols and their magnitudes have been adopted.

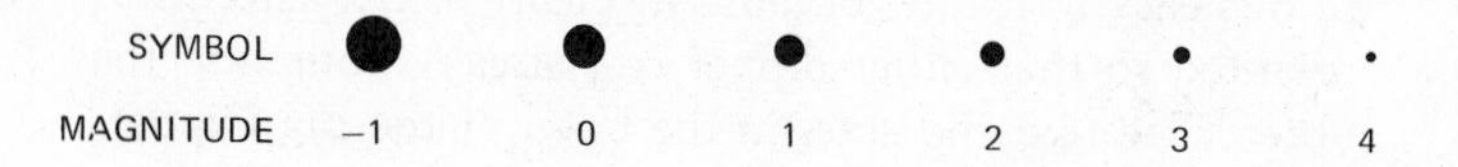

There are only about 100 stars that are either bright enough or of sufficient interest to be identified by a proper name. Three of these are Polaris, the north pole star, which is almost directly above the earth's north pole; Sirius, the brightest star in the sky; and Vega, the fourth brightest star, which marks the general direction in space toward which the sun and the solar system are moving at about 12 miles (19 km) per second.

At first glance, the stars appear to be randomly distributed; on closer scrutiny, some of them appear to occur in geometrical patterns, such as straight lines, triangles, squares, and rectangles; others appear to be in definite groupings. Ancient peoples perceived these shapes and groupings—the constellations—as the outlines of the people, animals, and physical objects that were significant in their religions. Although the ancient Greeks recognized 48 constellations and gave many of them Greek names, they were known several thousand years earlier to the people living in the Tigris and Euphrates river valleys. Most of the star groupings are not even fair approximations of the constellation pictures

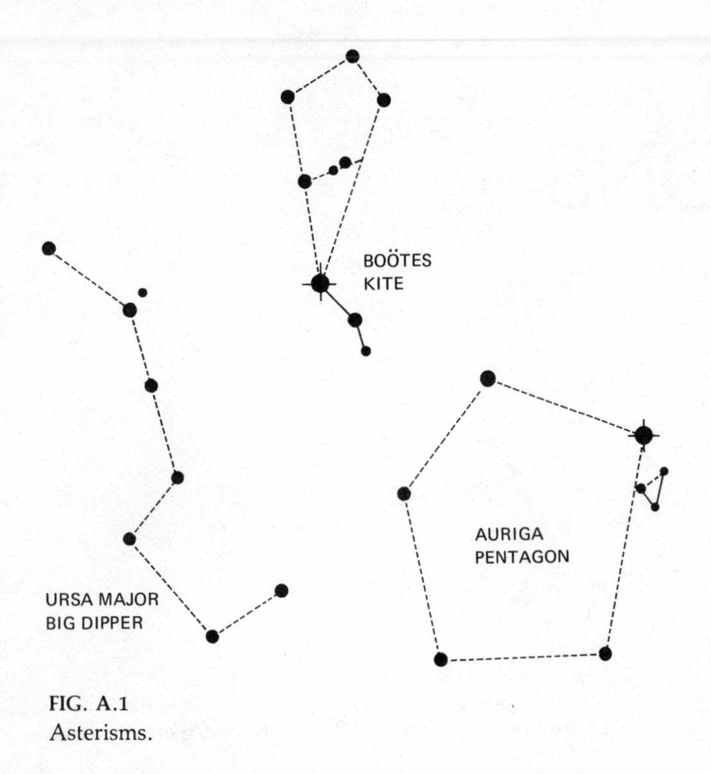

FIG. A.1
Asterisms.

they are supposed to represent; therefore, recognition of the supposed person or object from the groupings is not easy. Constellation identification was simplified when the few bright stars within the constellations were connected by straight lines to form *asterisms*—simple pictures which everyone can easily recognize. In Fig. A.1, the asterism for Ursa Major the Big Bear is the Big Dipper; for Boötes the Bear Driver, it is the Kite; and for Auriga the Charioteer, it is the Pentagon. The pictures associated with the constellations can be visualized by first locating their asterisms, then (with a little imagination) tracing the constellation pictures around the asterisms.

Constellations acquired a new meaning in 1927 when astronomers divided the entire sky into definite, irregular areas around the 88 constellations and assigned the name of the constellation to the respective area in which it is located. This produced a map of the sky on which the celestial bodies and events can be easily and quickly located by reference to the constellations.

1. THE NORTH CIRCUMPOLAR CONSTELLATIONS (CHART A.1)

The starting point for locating the constellations in the northern hemisphere is Polaris, the north star. To the unaided eye it is the nearest star to the celestial north pole (CNP), an imaginary point that marks the projection of the earth's north pole to the celestial sphere. At the present time, Polaris is less than one degree from the celestial north pole. When the observer faces the north point on the horizon, which is directly below Polaris, the stars appear to move in a counterclockwise direction around Polaris, which appears to remain stationary.

Five constellations revolve around Polaris—Ursa Major the Big Bear, Ursa Minor the Little Bear, Cepheus the King, Cassiopeia the Queen, and Draco the Dragon. These are called *circumpolar constellations* because they are visible every night at any time to observers located north of about 40° north latitude.

To locate the five circumpolar constellations, refer to Chart A.1, in which Polaris is in the center and the five constellations are around it. The circumference of the circle is divided into 12 equal parts to represent the 12 months of the year. It is further divided into 24 equal parts to represent the hours in each day. To orient the chart with the sky for any evening at 8:00 P.M., face Polaris and hold the chart so that the proper month appears on the bottom.

Polaris, which marks the end of the handle of the Little Dipper and the tip of the tail of the Little Bear, locates the constellation of Ursa Minor. This constellation is easy to locate, because its clear, bright asterism is oriented so that either dipper can always pour into the other. The two end stars in the bowl of the Big Dipper,

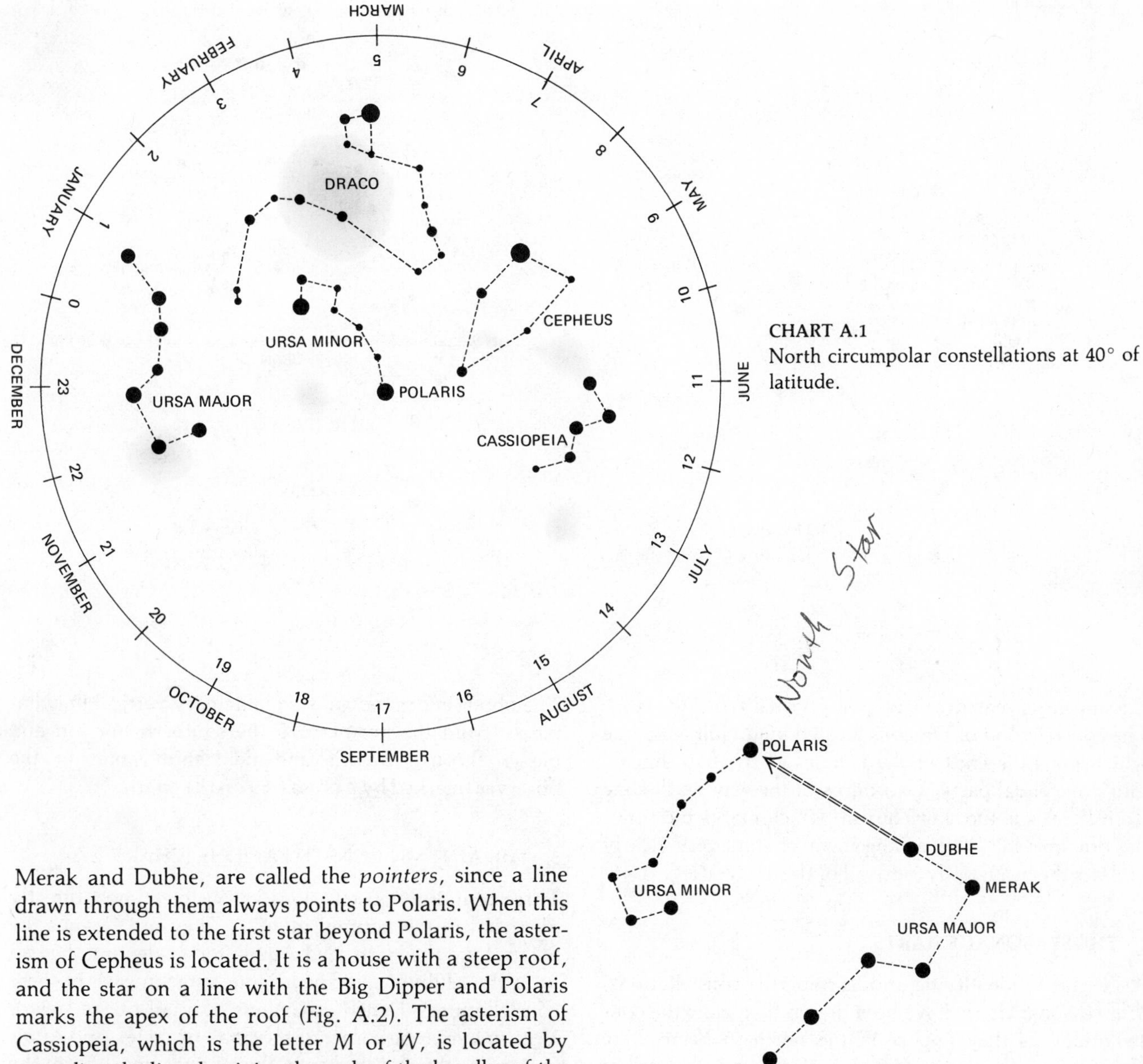

CHART A.1
North circumpolar constellations at 40° of latitude.

FIG. A.2
Locating Polaris in Ursa Minor.

Merak and Dubhe, are called the *pointers*, since a line drawn through them always points to Polaris. When this line is extended to the first star beyond Polaris, the asterism of Cepheus is located. It is a house with a steep roof, and the star on a line with the Big Dipper and Polaris marks the apex of the roof (Fig. A.2). The asterism of Cassiopeia, which is the letter *M* or *W*, is located by extending the line that joins the ends of the handles of the two dippers to the first star beyond Polaris. This star marks one end of the asterism. When another star is included in the asterism, the outline of the chair on which

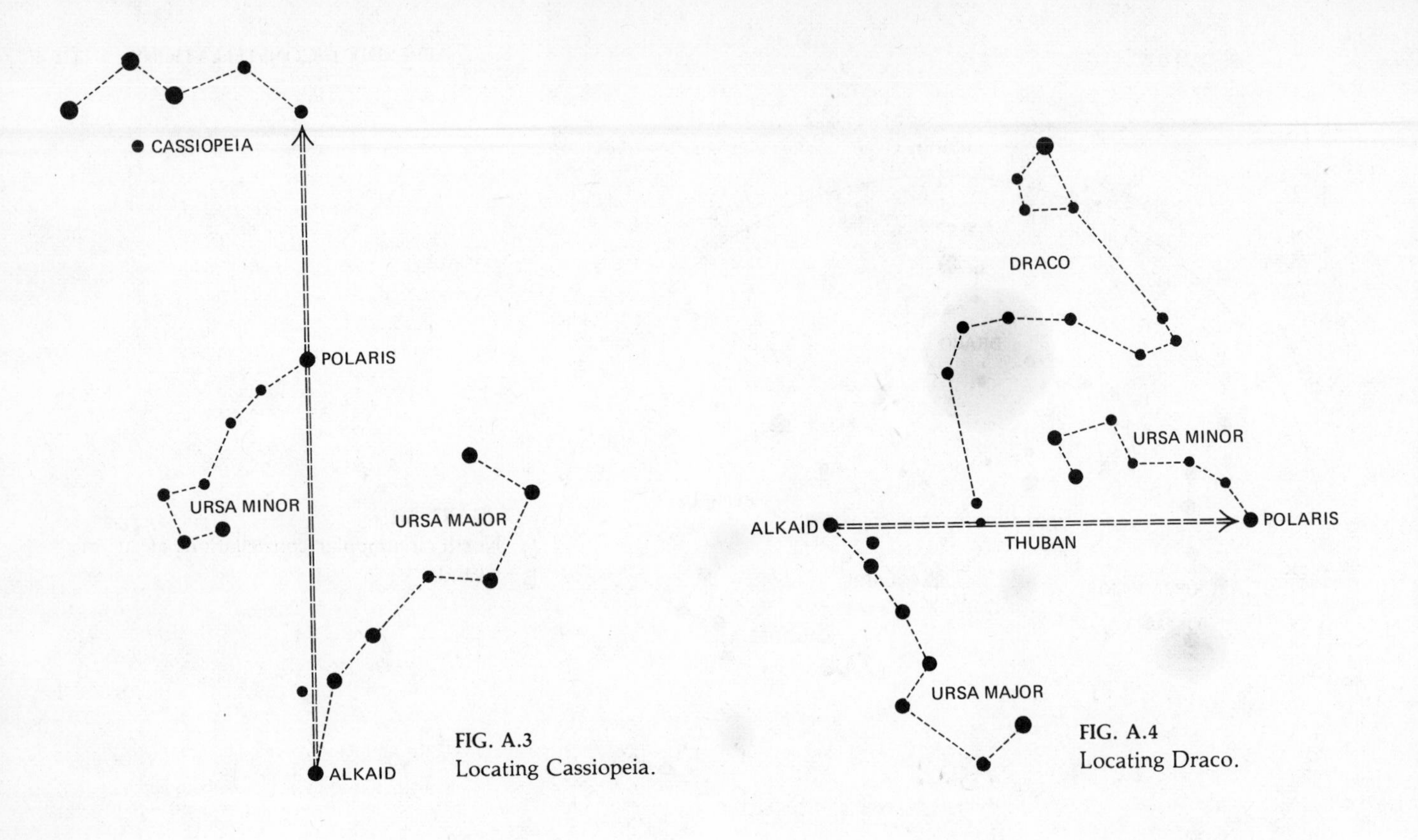

FIG. A.3
Locating Cassiopeia.

FIG. A.4
Locating Draco.

Cassiopeia is seated can be easily visualized (Fig. A.3). The constellation of Draco is located by dividing the line which joins the ends of the handles of the two dippers into three equal parts. Two-thirds of the way on the line from Polaris is the star Thuban, which marks the tip of the dragon's tail. The arrangement of stars very clearly outlines the head, body, and tail of the dragon (Fig. A.4).

2. THE SEASONAL CHARTS

As an aid to identifying and learning the constellations, four seasonal charts have been drawn that show the constellations as they appear in the northern hemisphere about 9:00 P.M. on the first day of autumn (September 21), winter (December 21), spring (March 21), and summer (June 21). When using a chart, face one of the horizons, hold the chart vertically, and turn it so that the direction in which you are facing appears at the bottom. The observer's meridian is an imaginary vertical line that passes from the northern to the southern horizon and passes through Polaris and the zenith point of the observer (marked by a cross (+) on the chart).

3. THE AUTUMN CONSTELLATIONS (CHART A.2)

Three beautiful constellations are visible almost directly overhead in the autumn months. They are Cygnus the Swan, Lyra the Harp, and Aquila the Eagle, and they all lie in the Milky Way. The brightest star in each of these constellations—Deneb in Cygnus (the tail of the swan), Vega in Lyra (one corner of the small triangle), and Altair in Aquila (the middle star in the straight line)—form the summer triangle, which is almost a right triangle. The asterisms are the Northern Cross for Cygnus, a rectangle and triangle for Lyra, and a straight line of three stars for Aquila.

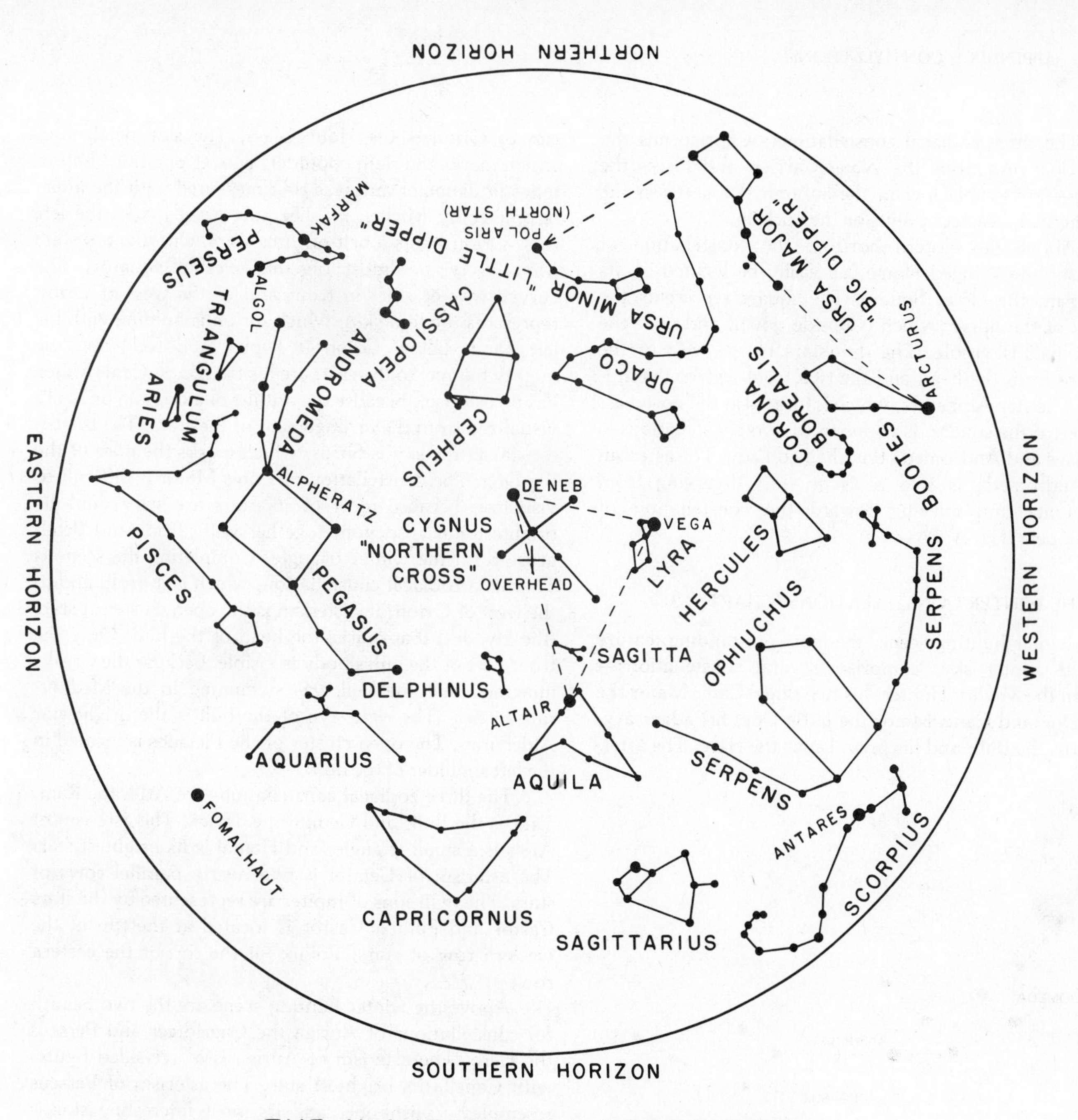

THE NIGHT SKY IN SEPTEMBER

CHART A.2
The autumn constellations. (Courtesy of the Griffith Observatory.)

The three zodiacal constellations—Capricornus the Sea Goat, Aquarius the Water Carrier, and Pisces the Fishes—are visible low in the autumn sky between the southern and eastern points on the horizon.

Above the eastern horizon, the constellation of Pegasus the Winged Horse can easily be located by its asterism, the Great Square. The square represents the body of the horse, which is upside down, and only the front half is visible. The dim stars to the west of the square form the head and the two front legs of the animal. The star Alpheratz, which is located at the northeast corner of the square, is common to two constellations—Pegasus and Andromeda the Chained Lady. The asterism of Andromeda is two rows of stars diverging from Alpheratz and curving toward the constellation of Cassiopeia (Fig. A.5).

4. THE WINTER CONSTELLATIONS (CHART A.3)

The winter hunting scene, the most outstanding feature of the winter sky, comprises several constellations—Orion the Mighty Hunter; his two dogs, Canis Major the Big Dog and Canis Minor the Little Dog; his adversary, Taurus the Bull; and his prey, Lepus the Hare. The asterism of Orion is the Hour Glass. The star Betelgeuse, which marks the right shoulder, is over one-third billion miles in diameter and was first measured with the interferometer by Michelson. The star Rigel marks the left foot. Orion wears a belt of three stars, with the top star, Mintaka lying almost on the celestial equator. The curved row of stars in front and to the west of Orion represents the lion skin, which Orion is holding with his left hand. Below Orion is Lepus, and following the mighty hunter, to the east, are his two dogs. Canis Major has no asterism, because the outline of a dog can be easily visualized from the arrangement of the stars. The brightest star in the sky is Sirius, which marks the nose of the Big Dog. The constellation of Canis Minor is difficult to visualize, because most of its stars are very dim. Its brightest star, Procyon, together with Sirius and Betelgeuse form the winter triangle. Completing the scene is Taurus, a zodiacal constellation, which is in front and to the west of Orion. Its asterism is the open cluster of stars (the Hyades) that marks the head of the bull. Only the front part of the bull's body is visible, because the Greeks imagined that the bull was swimming in the Mediterranean Sea. The right eye of the bull is the bright star Aldebaran. The open cluster of the Pleiades is located in the left shoulder of the bull.

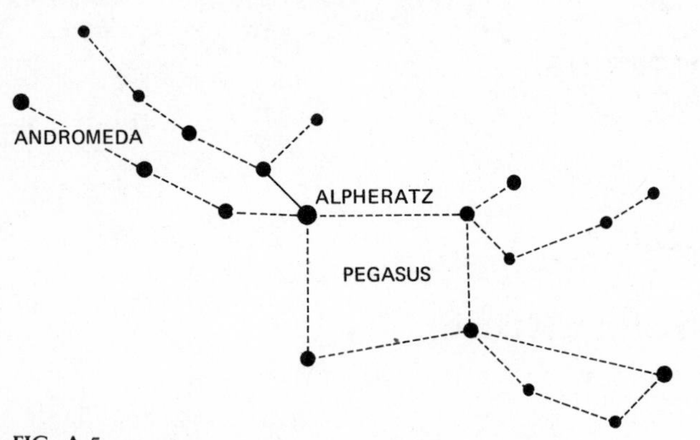

FIG. A.5
The constellations of Pegasus and Andromeda.

The three zodiacal constellations are Aries the Ram, Taurus the Bull, and Gemini the Twins. The asterism of Aries is a small triangle, and Hamal is its brightest star. The asterism of Gemini is two nearly parallel rows of stars. The twin sons of Jupiter are represented by the stars Castor and Pollux. Castor is located at the top of the western row of stars; Pollux, at the top of the eastern row.

Above the winter hunting scene are the two beautiful constellations of Auriga the Charioteer and Perseus the Hero. The asterism of Auriga is a five-sided figure, with Capella its brightest star. The asterism of Perseus resembles a wishbone, and its most interesting star is Algol, the "Blinking Demon." In the southern part of the sky are two long, indistinct constellations that can be located and traced on a clear moonless evening. These are Eridanus the Po River and Cetus the Whale. Eridanus meanders in the region between Cetus, Orion, and Lepus.

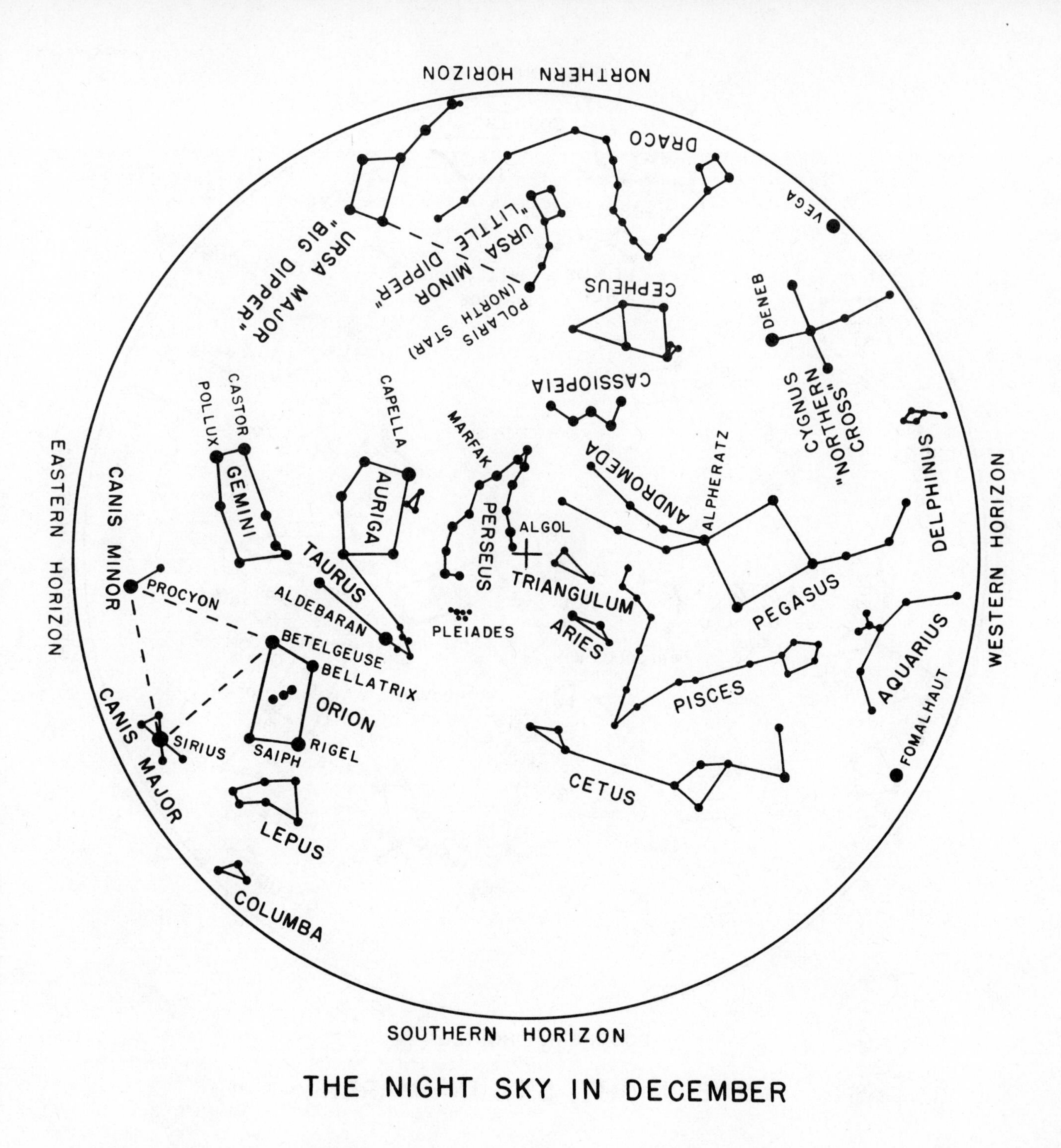

THE NIGHT SKY IN DECEMBER

CHART A.3
The winter constellations. (Courtesy of the Griffith Observatory.)

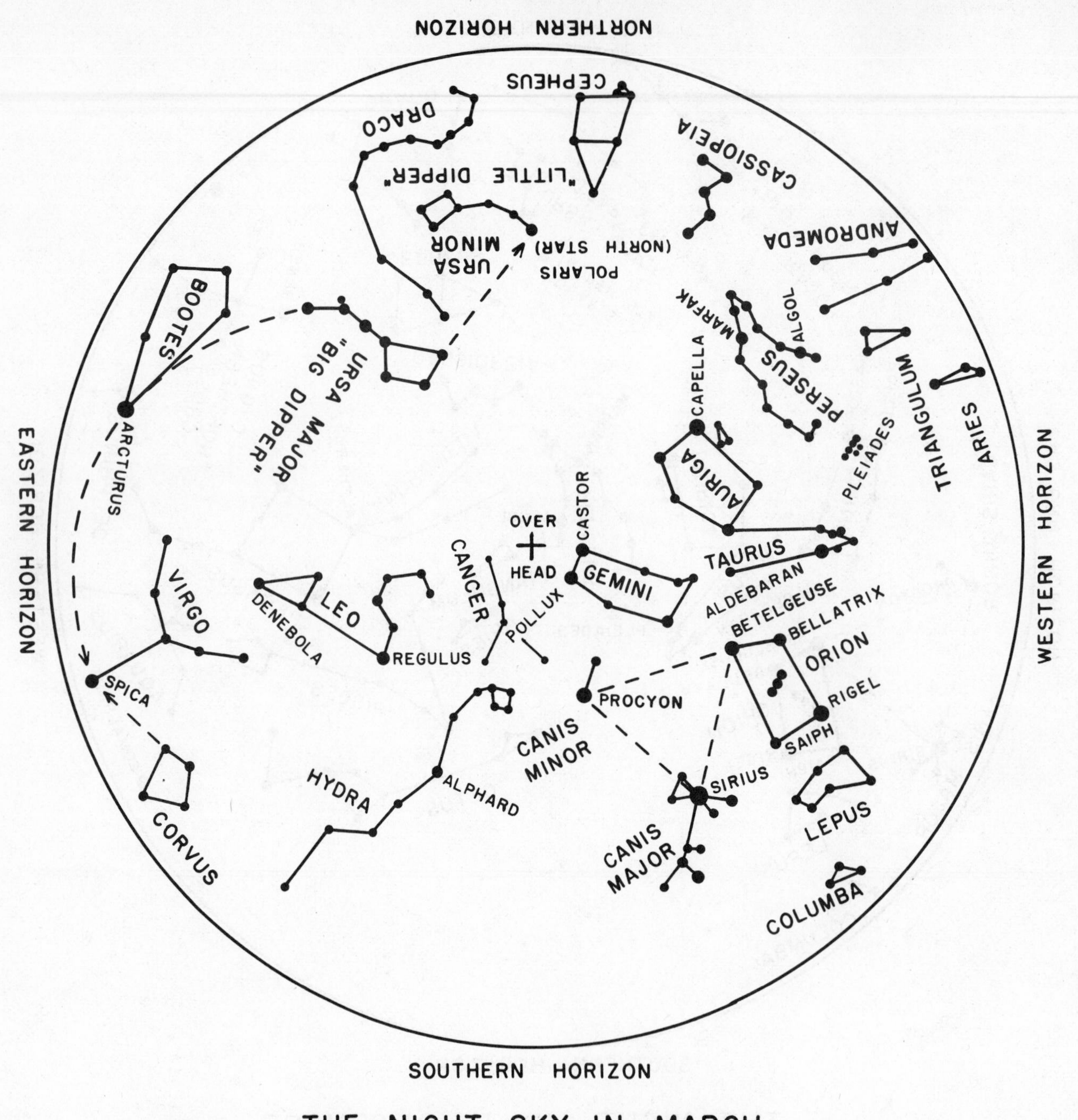

THE NIGHT SKY IN MARCH

CHART A.4
The spring constellations. (Courtesy of the Griffith Observatory.)

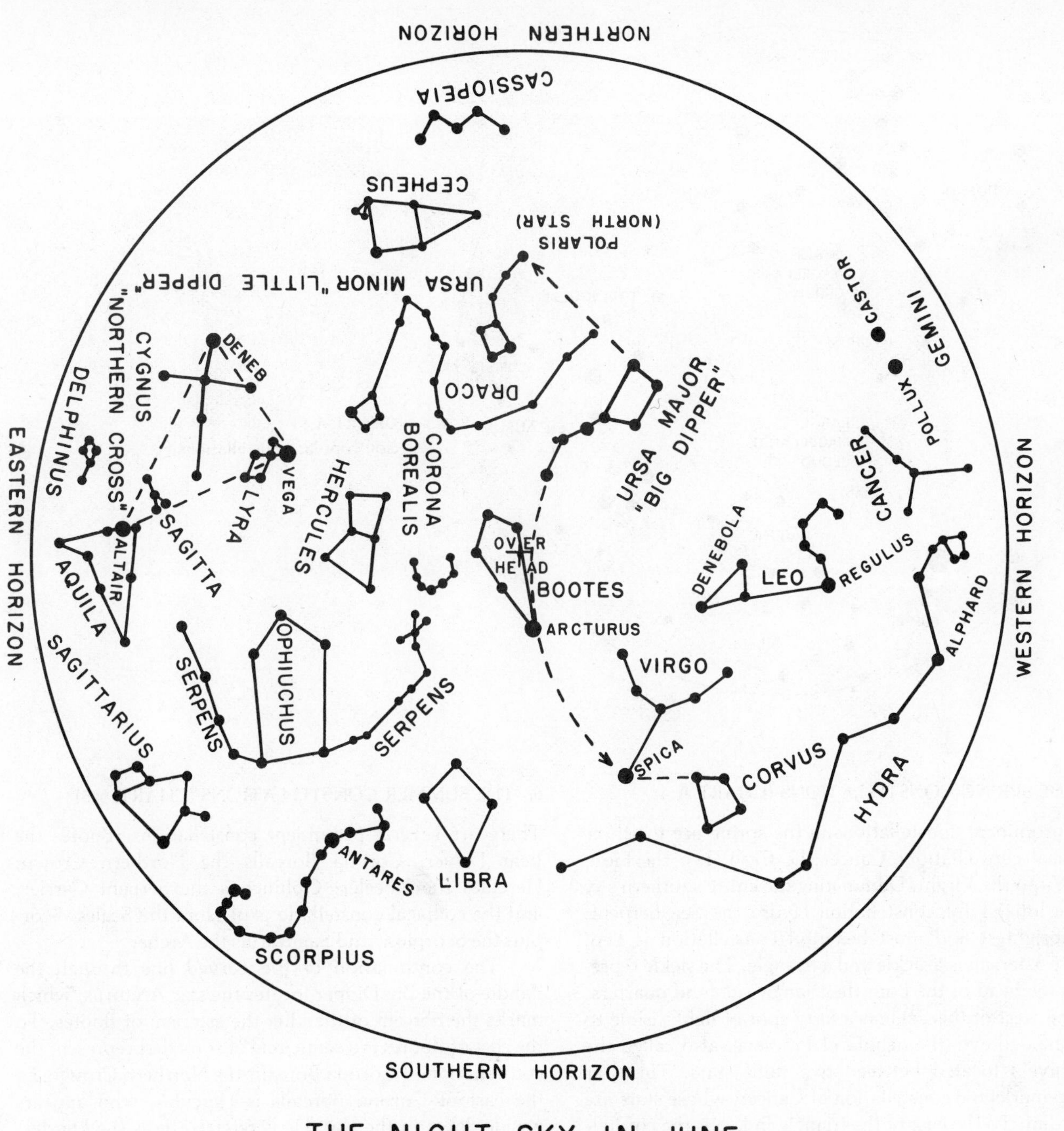

CHART A.5
The summer constellations. (Courtesy of the Griffith Observatory.)

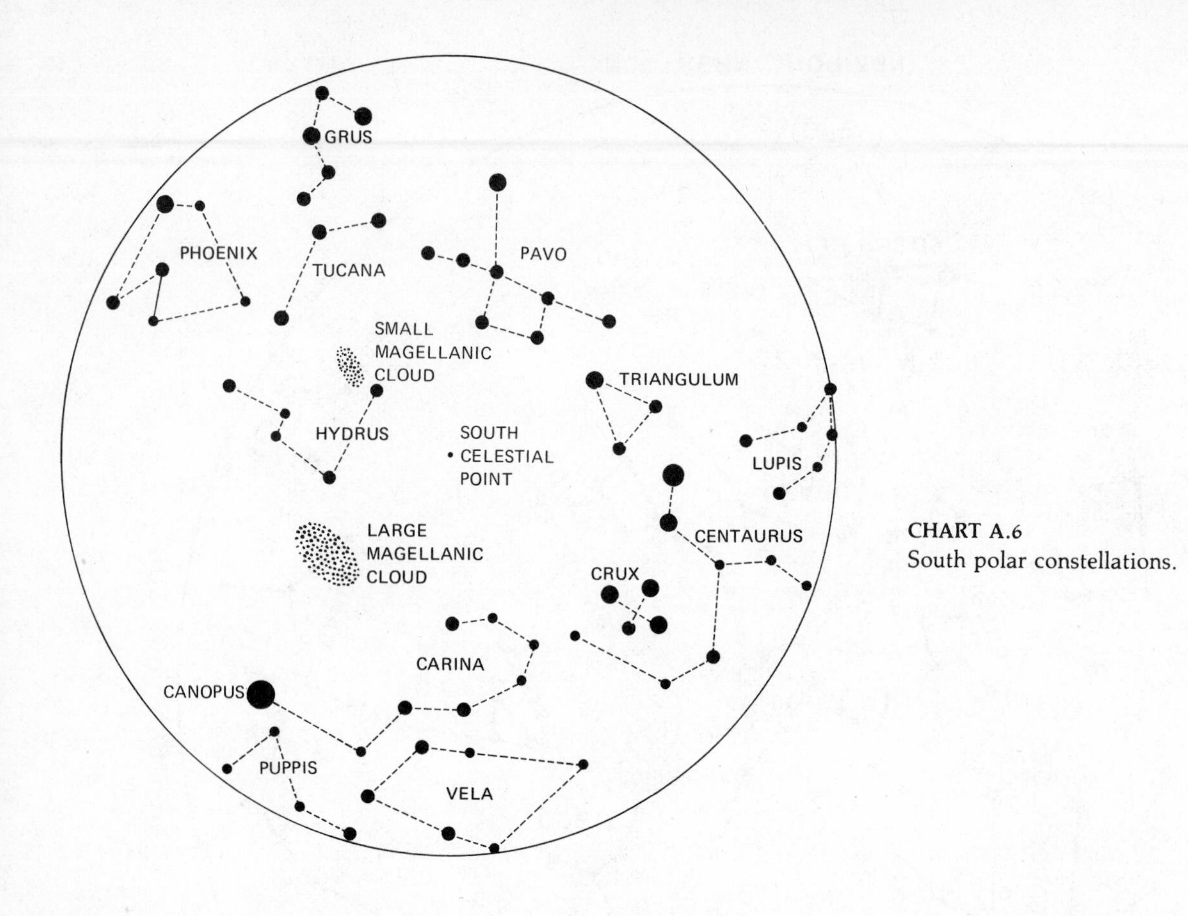

CHART A.6
South polar constellations.

5. THE SPRING CONSTELLATIONS (CHART A.4)

The prominent constellations in the spring are the three zodiacal constellations Cancer the Crab, Leo the Lion, and Virgo the Virgin. Dominating the entire southern sky is the long, faint constellation Hydra the Sea Serpent. The brightest and most beautiful constellation is Leo, whose asterism is a sickle and a triangle. The sickle represents the head of the lion; the triangle, its hind quarters. To the west of the sickle is a hazy spot of light visible to the unaided eye (the nebula of Praesepe, also called the "Beehive") located between two faint stars. This star group marks the constellation of Cancer, whose stars are very dim. To the east of the triangle in Leo is the constellation of Virgo. The location of this constellation is also marked by the continuation of the curved line through the handle of the Big Dipper, through the bright star Arcturus in Boötes, and to the bright star Spica in Virgo. Spica represents the sheaf of wheat in Virgo's left hand.

6. THE SUMMER CONSTELLATIONS (CHART A.5)

There are several prominent constellations: Boötes the Bear Driver; Corona Borealis the Northern Crown; Hercules the Kneeler; Ophiuchus the Serpent Carrier; and the zodiacal constellations of Libra the Scales, Scorpius the Scorpion, and Sagittarius the Archer.

The continuation of the curved line through the handle of the Big Dipper locates the star Arcturus, which marks the bottom of the kite, the asterism of Boötes. To the east of Boötes is a semicircle of stars that represent the constellation of Corona Borealis the Northern Crown. To the east of Corona Borealis is Hercules, who appears upside down in the sky. He is referred to as the kneeler, because he appears in a kneeling position. The asterism of Hercules is the letter *H*. The asterism of Ophiuchus is a triangle on top of a large vertical rectangle that represents the body of Ophiuchus. Around his body is the constellation Serpens the Serpent. The most prominent of the three zodiacal constellations is Scorpius. Its arrangement of stars clearly outlines the head, body, and tail of the scorpion. Its brightest star, Antares, marks the heart of the scorpion. In the late summer, Sagittarius, looking like a teapot with handle and spout, is conspicuous in the sky to the south.

7. THE SOUTH POLAR CONSTELLATIONS (CHART A.6)

The south polar constellations are shown in Chart A.6. At the present time, there is no south celestial pole star. In fact, the area in comparison to that shown in Chart A.1 contains fewer stars and less prominent constellations. However, the area contains several brilliant stars: Canopus, which is one of the most beautiful, as well as two bright stars in Centaurus that are close together and point toward the Southern Cross.

The largest, most interesting southern constellation is the ship Argo, named after the famous ship that carried Jason and his crew to Colchis in search of the golden fleece. Since there are so many stars in this constellation, it has been divided into several smaller constellations that represent various parts of the ship—the constellation Carina is the keel, Puppis is the stern, and Vela is the sails. The brilliant star Canopus marks Argo's rudder.

A famous, fabled constellation is Crux Australis, whose asterism is the Southern Cross. The asterism, which looks more like a diamond than a cross, appears to many to be less impressive than the large, beautiful Northern Cross.

8. THE 88 CONSTELLATIONS

The 88 constellations are listed in the following table by their Latin and English names and their approximate position in the sky.

Constellation Table

LATIN NAME	ENGLISH NAME	SKY POSITION R.A. (HOURS)	SKY POSITION DECLINATION (DEGREES)
Andromeda	Princess of Ethiopia	1	+40
Antila	Air pump	10	−35
Apus	Bird of paradise	16	−75
Aquarius	Water carrier	23	−15
Aquila	Eagle	20	+ 5
Ara	Altar	17	−55
Aries	Ram	3	+20
Auriga	Charioteer	6	+40
Boötes	Bear driver	15	+30
Caelum	Graving tool	5	−40
Camelopardus	Giraffe	6	−70
Cancer	Crab	9	+20
Canes Venatici	Hunting dogs	13	+40
Canis Major	Big dog	7	−20
Canis Minor	Little dog	8	+ 5
Capricornus	Sea goat	21	−20
Carina	Argo's keel	9	−60
Cassiopeia	Queen of Ethiopia	1	+60
Centaurus	Centaur	13	−50
Cepheus	King of Ethiopia	22	+70
Cetus	Whale	2	−10
Chamaeleon	Chameleon	11	−80
Circinus	Compasses	15	−60
Columba	Noah's dove	6	−35
Coma Berenices	Berenice's hair	13	+20
Corona Australis	Southern crown	19	−40
Corona Borealis	Northern crown	16	+30
Corvus	Crow	12	−20
Crater	Cup	11	−15
Crux	Southern Cross	12	−60
Cygnus	Swan	21	+40
Delphinus	Dolphin	21	+10
Dorado	Swordfish	5	−65
Draco	Dragon	17	+65

(continued)

Constellation Table (cont.)

LATIN NAME	ENGLISH NAME	SKY POSITION R.A. (HOURS)	SKY POSITION DECLINATION (DEGREES)
Equuleus	Little horse	21	+10
Eridanus	Po River	3	−20
Fornax	Furnace	3	−30
Gemini	Twins	7	+20
Grus	Crane	22	−45
Hercules	Hercules (Zeus' son)	17	+30
Horologium	Clock	3	−60
Hydra	Sea serpent	10	−20
Hydrus	Water snake	2	−75
Indus	Indian	21	−55
Lacerta	Lizard	22	+45
Leo	Lion	11	+15
Leo Minor	Little lion	10	+35
Lepus	Hare	6	−20
Libra	Balance	15	−15
Lupus	Wolf	15	−45
Lynx	Lynx	8	+45
Lyra	Harp	19	+40
Mensa	Table mountain	5	−80
Microscopium	Microscope	21	−35
Monoceros	Unicorn	7	− 5
Musca	Fly	12	−70
Norma	Level	16	−50
Octans	Octant	22	−85
Ophiuchus	Serpent carrier	17	0
Orion	Mighty hunter	5	+ 5
Pavo	Peacock	20	−65
Pegasus	Winged horse	22	+20
Perseus	Hero	3	+45
Phoenix	Phoenix	1	−50
Pictor	Easel	6	−55
Pisces	Fishes	1	+15
Piscis Austrinus	Southern fish	22	−30
Puppis	Argo's stern	8	−40
Pyxis	Argo's compass	9	−30
Reticulum	Net	4	−60
Sagitta	Arrow	20	+10
Sagittarius	Archer	19	−25
Scorpius	Scorpion	17	−40
Sculptor	Sculptor's tools	0	−30
Scutum	Shield	19	−10
Serpens	Serpent	17	0
Sextans	Sextant	10	0
Taurus	Bull	4	+15
Telescopium	Telescope	19	−50
Triangulum	Triangle	2	+30
Triangulum Australe	Southern triangle	16	−65
Tucana	Toucan	0	−65
Ursa Major	Big bear	11	+50
Ursa Minor	Little bear	15	+70
Vela	Argo's sail	9	−50
Virgo	Virgin	13	0
Volans	Flying fish	8	−70
Vulpecula	Fox	20	+25

REVIEW QUESTIONS

1. Give two definitions for the term "constellation." What is an asterism?
2. What star is nearest the north celestial pole? How can it be located? What is its magnitude? What type of star is it?
3. What are circumpolar constellations? Name those at your latitude.
4. What is the significance of the pointers in the asterism of the Big Dipper?
5. What is the asterism of the following constellations: (a) Auriga, (b) Crux, (c) Cygnus, (d) Leo, (e) Pegasus?
6. What stars make up (a) the winter triangle, (b) the summer triangle?
7. Explain how the vernal equinox can be located.

8. In what constellation is the Hyades located? What does this group of stars represent? Is the bright star Aldebaran which appears in this group part of the Hyades?
9. What are the Pleiades? Where are they located? How many stars are visible to the naked eye? How many have been observed in the Pleiades?
10. In what constellation is each of the following located: (a) the eclipsing binary Algol, the "blinking demon," (b) the Andromeda galaxy, (c) the Ring Nebula, (d) Capella?
11. What is the star of greatest apparent brightness in the sky?
12. Name the five brightest stars, with their magnitudes, that appear in the winter sky at about 9:00 P.M. Name the two brightest stars that appear in the early hours in the summer sky.
13. The following examples represent three methods for designating stars: (a) Algol, (b) Beta Persei, (c) Kruger 60. Explain.
14. What constellations appear in the Milky Way in the summer sky? Where is the Coal Sack located? What is it? Describe and explain the Great Rift.
15. What constellations appear in the Milky Way in the winter sky? Compare the appearance of the Milky Way as to size and luminosity in summer and winter. Explain the difference with a diagram.
16. The constellations of Canis Major and Gemini appear to cross the observer's meridian at the same time. Do they rise on the eastern horizon at the same time? Explain.

APPENDIX 2
SCIENTIFIC NOTATION

In astronomical work, very large and very small numbers are used. For example, the earth's distance from the sun is approximately 93,000,000 miles, or 15,000,000,000,000 cm, and the wavelength of an x-ray is approximately 0.00000001 cm. Since the many zeros make it inconvenient to write these numbers, a simpler system involving the powers of ten is used. This system is called *scientific notation*.

Powers of ten

$10^3 = 10 \times 10 \times 10$	=	1000
$10^2 = 10 \times 10$	=	100
$10^1 = 10$	=	10
$10^0 = 1$	=	1
$10^{-1} = 1/10$	=	0.1
$10^{-2} = 1/10 \times 1/10 = 1/100$	=	0.01
$10^{-3} = 1/10 \times 1/10 \times 1/10 = 1/1000$	=	0.001

In scientific notation, a number is usually written as the product of a number from 1 to 9.99 . . . and a power of ten. The power of ten (exponent) indicates the number of places the decimal point must be moved to restore the number. If the power is positive, the decimal point must be moved to the right; if negative, it must be moved to the left.

Number	*Scientific notation*
15,000,000,000,000	1.5×10^{13}
93,000,000	9.3×10^{7}
69,600,000,000	6.96×10^{10}
0.00000001	1.0×10^{-8}
0.00000872	8.72×10^{-6}

Scientific notation is an important and useful technique because it expresses bulky numbers in simple terms so that they can be easily read and compared. It also simplifies the multiplication and division processes in calculations.

When multiplying two numbers expressed in powers-of-ten notation, multiply together the significant figures; then simply add the exponents. Thus

2.5×10^4 times $3 \times 10^5 = 7.5 \times 10^9$.

When dividing, first divide significant figures and then subtract one exponent from the other. Thus

$8.8 \times 10^5 / 2.2 \times 10^3 = 4 \times 10^2$.

Numbers may also be raised to powers easily:

$(2 \times 10^4)^3 = 2^3 \times 10^{12} = 8 \times 10^{12}$.

Here one multiplies the exponent by the power to which it is being raised. As another example, take the square root (½ power) of 4×10^{10}:

$(4 \times 10^{10})^{1/2} = 4^{1/2} \times 10^{(10)1/2} = 2 \times 10^5$.

APPENDIX 3
FUNDAMENTAL METRIC AND ENGLISH UNITS

Fundamental units	*Metric system*	*English system*
Length	1 meter (m)	1 yard (yd)
Mass	1 kilogram (kg)	1 pound (lb)
Time	1 second (sec)	1 second (sec)

Units of length

1 kilometer (km) = 1000 meters (m) = 0.6214 mile (mi)
1 meter (m) = 0.001 km = 100 centimeters (cm)
= 1.094 yd = 39.37 inches (in.)
1 centimeter (cm) = 0.01 m = 10 millimeters (mm)
= 0.3937 in.
1 millimeter (mm) = 0.001 m = 0.03937 in.
= 1000 microns (μ)
1 micron (μ) = 0.000001 m = 3.3937×10^{-5} in.
= 10,000 Angstroms (Å)
1 statute mile (mi) = 5280 feet (ft) = 1.6093 km
1 inch (in.) = 2.5400 cm

Units of mass

1 metric ton = 1000 kilograms (kg) = 1.102 English tons
1 kilogram (kg) = 1000 grams (gm) = 2.2046 lb
1 gram (gm) = 0.0022046 lb = 0.0353 ounce (oz)
1 pound (lb) = 16 oz = 453.6 gm
1 ounce (oz) = 28.3495 gm

APPENDIX 4
USEFUL PHYSICAL CONSTANTS AND ASTRONOMICAL QUANTITIES

Pi (π) = 3.14159 . . . .
Radian (R) = 360°/2π = 57°.3 . . . . = 206265 sec
Angstrom unit (Å) = 10^{-8} cm
Astronomical unit (a.u.) = 1.496×10^{8} km
= 9.3×10^{7} mi
Velocity of light (c) = 2.99793×10^{5} km/sec
= 1.86×10^{5} mi/sec
Light year (l.y.) = 9.461×10^{12} km = 5.9×10^{12} mi
Parsec (pc) = 206265 a.u. = 3.26 l.y.
Constant of gravitation (G) =
6.670×10^{-8} dyne • cm²/gm²
Solar constant (S) = 1.95 calories/cm² • min
= 1.36 kilowatts/m²
Sun radius ($R_{\odot}$) = 6.960×10^{5} km = 4.32×10^{5} mi
Earth radius (equatorial) ($R_{\oplus}$) = 6378.16 km
= 3963.20 mi
Sun mass ($M_{\odot}$) = 1.99×10^{33} gm
Earth mass ($M_{\oplus}$) = 5.977×10^{27} gm
Earth velocity of escape = 11.2 km/sec = 6.95 mi/sec
Proton mass (m_p) = 1.67×10^{-24} gm
Electron mass (m_e) = 9.11×10^{-28} gm
Hydrogen atom mass (m_H) = 1.673×10^{-24} gm
Planck's constant h = 6.62×10^{-27} erg sec
Stefan-Boltzmann constant (σ) =
5.67×10^{-5} erg cm^{-2} deg^{-4}s^{-1}
Boltzmann constant (k) = 1.38×10^{-16} erg deg^{-1}
Tropical year = 365.24 days = 3.156×10^{7} sec

APPENDIX 5
TEMPERATURE SCALES AND CONVERSIONS

The different temperature scales have been chosen rationally, although at a first glance it may not appear that way. The Celsius scale (named for its originator and until recently referred to as degrees centigrade) is based on the important temperatures where water changes its state from liquid to solid (0°C) and from liquid to gas (100°C). The Kelvin scale (named for the nineteenth century scientist Lord Kelvin, who studied heat) is based on the principle that temperature measures the energy content of a substance. Since a substance cannot have negative energy, there can be no negative degrees. Zero degrees Kelvin, also referred to as absolute zero, is a temperature at which all motion, even the vibration of atoms, ceases. There is thus nothing colder. Kelvin temperatures are important to scientists as they can be used to measure directly such things as the velocity of atoms in a gas ($v^2 = 3kT/2m$, where k is a constant called *Boltzmann's constant* and has the value 1.38×10^{-16} ergs per degree and m is the atom's mass in grams). For convenience in converting to degrees Celsius, Kelvin degrees are chosen to have the same size as degrees Celsius (i.e., there are also 100 Kelvin degrees between the freezing and boiling points of water).

The Fahrenheit scale (named for Gabriel Fahrenheit) at first appears to be completely irrational. Fahrenheit, however, based his scale on the coldest temperature he could produce in his laboratory (that of a mixture of crushed ice and salt), which he defined as having a temperature of 0°F, and the human body temperature, which he chose as 96°F. Unfortunately, the boiling point of water did not come out as a whole number of degrees. Thus he rounded off the boiling point of water to 212°F and redefined his temperature scale accordingly.

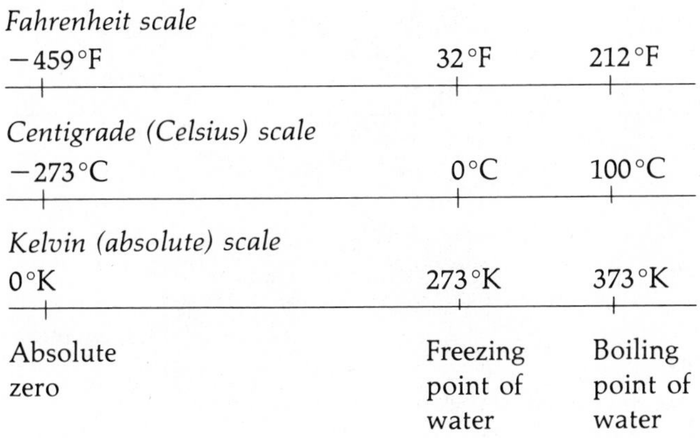

Fahrenheit scale

−459 °F	32 °F	212 °F

Centigrade (Celsius) scale

−273 °C	0 °C	100 °C

Kelvin (absolute) scale

0 °K	273 °K	373 °K
Absolute zero	Freezing point of water	Boiling point of water

Conversions

$°C = 5/9(°F - 32°)$

$°F = 9/5(°C) + 32°$

$°K = 5/9(°F + 459°)$

$°C = °K - 273°$

$°F = 9/5(°K) - 459°$

$°K = °C + 273°$

APPENDIX 6
PHYSICAL AND ORBITAL DATA FOR THE PLANETS

SYMBOL	MERCURY ☿	VENUS ♀	EARTH ⊕
Mean distance from sun (a.u.)	0.387	0.723	1.000
Mean distance from sun (miles)	3.6×10^7	6.7×10^7	9.3×10^7
Mean orbital speed (km/sec)	47.9	35.0	29.8
Synodic period (days)	115.9	583.9	—
Sidereal period (d, days; y, years)	88 d	224.7 d	365.26 d
Orbital eccentricity	0.206	0.007	0.017
Inclination of orbit to plane of the ecliptic	7°.0	3°.4	0°(by definition)
Inclination of equator to the orbital plane	2° (roughly)	3°	23°.5
Equatorial diameter (km)	4,880	12,104	12,756
Equatorial diameter (miles)	3,025	7,526	7,927
Diameter (earth=1)	0.38	0.95	1.00
Mass (earth=1)	0.055	0.815	1.000
Density (gm/cm³)	5.46	5.23	5.52
Escape velocity (km/sec)	4.2	10.3	11.2
Period of rotation	59 days	243 days retrograde	23 hours 56 minutes
Albedo	0.06	0.72	0.39

MARS ♂	JUPITER ♃	SATURN ♄	URANUS ♅	NEPTUNE ♆	PLUTO ♇
1.524	5.203	9.539	19.182	30.058	39.439
1.42×10^8	4.83×10^8	8.86×10^8	1.78×10^9	2.79×10^9	3.67×10^9
24.1	13.1	9.6	6.8	5.4	4.7
779.9	398.9	378.1	369.7	367.5	366.7
687.0 d	11.86 y	29.46 y	84.01 y	164.79 y	247.69 y
0.093	0.048	0.056	0.047	0.009	0.250
$1^\circ.9$	$1^\circ.3$	$2^\circ.5$	$0^\circ.8$	$1^\circ.8$	$17^\circ.2$
24°	3°	27°	98°	32°	?
6.787	142,800	120,000	51,800	49,500	> 5,800 ?
4,218	88,700	74,600	34,700	30,800	> 3,600 ?
0.53	11.19	9.47	4.37	3.88	> 0.31
0.108	318.0	95.2	14.6	17.2	.002
3.93	1.33	0.69	1.20	1.7	?
0.51	61	37.0	21.1	23.6	?
24 hours 37 minutes	9 hours 50 minutes	10 hours 14 minutes	24 ± 3 hours	22 ± 4 hours	6.4 days
0.16	0.7	0.75	0.9	0.82	0.1 ?

APPENDIX 7
THE NEAREST STARS

STAR	RIGHT ASCENSION (1970) H M	DECLINATION (1970) ° ′	DISTANCE (L.Y.)	PROPER MOTION (SEC)	RADIAL VELOCITY (KM/SEC)	APPARENT VISUAL MAGNITUDE	ABSOLUTE VISUAL MAGNITUDE	SPECTRAL CLASS
Sun			0.000016 (from Earth)			−26.8	+ 4.8	G2 V
α Centauri	14 37	−60 43	4.3	3.68	− 23	0.1	+ 4.4	G2 V
Barnard's star	17 56	+04 36	6.0	10.30	−108	9.5	+13.2	M5 V
Wolf 359	10 55	+07 13	7.6	4.84	+ 13	13.5	+16.8	M6 V
Lalande 21185	11 02	+36 10	8.1	4.78	− 86	7.5	+10.5	M2 V
Sirius	06 44	−16 41	8.6	1.32	− 8	− 1.5	+ 1.4	A1 V
Luyten 726	01 37	−18 07	8.9	3.35	+ 29	12.5	+15.4	M6 V
Ross 154	18 48	−23 51	9.4	0.74	− 4	10.6	+13.3	M5 V
Ross 248	23 40	+44 01	10.3	1.82	− 81	12.2	+14.7	M6 V
ϵ Eridani	03 32	−09 34	10.7	0.97	+ 15	3.7	+ 6.1	K2 V

STAR	RIGHT ASCENSION (1970) H M	DECLINATION (1970) ° ′	DISTANCE (L.Y.)	PROPER MOTION (SEC)	RADIAL VELOCITY (KM/SEC)	APPARENT VISUAL MAGNITUDE	ABSOLUTE VISUAL MAGNITUDE	SPECTRAL CLASS
Luyten 789-6	22 37	−15 31	10.8	3.27	− 60	12.2	+14.9	M6 V
Ross 128	11 46	+01 01	10.8	1.40	− 13	11.1	+13.5	M5 V
61 Cygni	21 06	+38 36	11.2	5.22	− 64	5.2	+ 7.5	K5 V
ϵ Indi	22 02	−56 55	11.2	4.67	− 40	4.7	+ 7.0	K5 V
Procyon	07 38	+05 18	11.4	1.25	− 3	0.3	+ 2.7	F5 V
Σ 2398	18 42	+59 35	11.5	2.29	+ 8	8.9	+11.1	M4 V
Groom. 34	00 17	+43 51	11.6	2.91	+ 13	8.1	+10.3	M1 V
Lacaille 9352	23 04	−36 02	11.7	6.87	+ 10	7.4	+ 9.6	M2 V
τ Ceti	01 43	−16 06	11.9	1.92	− 16	3.5	+ 5.7	G8 V
BD+5°1668	07 26	+05 28	12.2	3.73	+ 26	9.8	+11.9	M4 V

APPENDIX 8
THE BRIGHTEST STARS

STAR	RIGHT ASCENSION (1970) H M	DECLINATION (1970) ° ′	DISTANCE (L.Y.)	APPARENT VISUAL MAGNITUDE	ABSOLUTE VISUAL MAGNITUDE	SPECTRAL CLASS	PROPER MOTION (SEC)	RADIAL VELOCITY (KM/SEC)
Sun				−26.73	+4.84	G2 V		
Sirius	06 43.8	−16 41	8.7	− 1.42	+1.45	A1 V	1.324	−07.6
Canopus	06 23.3	−52 41	98.0	− 0.72	−3.1	FO lb-II	0.025	+20.5
α Centauri	14 37.6	−60 43	4.3	+ 0.01	+4.39	G2 V	3.676	−24.6
Arcturus	14 14.3	+19 20	36.0	− 0.06	−0.3	K2 III	2.284	−05.2
Vega	18 35.9	+38 45	26.5	+ 0.04	+0.5	AO V	0.345	−13.9
Capella	05 14.5	+45 58	45.0	+ 0.05	−0.6	G8 III	0.435	+30.2
Rigel*	05 13.1	−08 14	900.0	+ 0.14	−7.1	B8 Ia	0.001	+20.7
Procyon	07 37.7	+05 18	11.3	+ 0.37	+2.7	F5 IV-V	1.250	−03.2

Note: la, most luminous supergiant; lb, least luminous supergiant; II, bright giant; III, normal giant; IV, subgiant; V, main sequence; *, variable star.

STAR	RIGHT ASCENSION (1970) H M	DECLINATION (1970) ° ′	DISTANCE (L.Y.)	APPARENT VISUAL MAGNITUDE	ABSOLUTE VISUAL MAGNITUDE	SPECTRAL CLASS	PROPER MOTION (SEC)	RADIAL VELOCITY (KM/SEC)
Betelgeuse*	05 53.5	+07 24	520.0	+ 0.41	−5.6	M2 Iab	0.028	+21.0
Achernar	01 36.6	−57 23	118.0	+ 0.51	−2.3	B3 V	0.098	+19.0
β Centauri	14 01.7	−60 13	490.0	+ 0.63	−5.2	B1 III	0.035	−12.0
Altair	19 49.3	+08 47	16.5	+ 0.77	+2.2	A7 IV-V	0.658	−26.3
Aldebaran	04 34.2	+16 27	68.0	+ 0.86	−0.7	K5 III	0.202	+54.1
Spica*	13 23.6	−11 00	220.0	+ 0.91	−3.3	B1 V	0.054	+01.0
Antares*	16 27.6	−26 22	520.0	+ 0.92	−5.1	M1 lb	0.029	−03.2
Pollux	07 43.5	+28 06	35.0	+ 1.16	+1.0	KO III	0.625	+03.3
Fomalhaut	22 56.0	−29 47	22.6	+ 1.19	+2.0	A3 V	0.367	+06.5
Deneb	20 40.4	+45 10	1600.0	+ 1.26	−7.1	A2 Ia	0.003	−04.6
β Crucis	12 46.0	−59 32	490.0	+ 1.28	−4.6	BO III	0.049	+20.0
Regulus	10 06.8	+12 07	84.0	+ 1.36	−0.7	B7 V	0.248	+03.5
α Crucis	12 24.9	−62 56	370.0	+ 1.39	−3.9	B1 IV	0.042	−11.2

APPENDIX 9
THE MESSIER CATALOG OF NEBULAE AND STAR CLUSTERS

NUMBER M	NGC*	RIGHT ASCENSION (1970)	DECLINATION	CONSTELLATION	APPARENT VISUAL MAGNITUDE	DESCRIPTION
1	1952	5 32.7	+22 01	Taurus	11.3	"Crab" nebula; remains of supernova 1054
2	7089	21 31.9	−00 57	Aquarius	6.27	Globular cluster
3	5272	13 40.8	+28 32	Canes Venatici	6.22	Globular cluster
4	6121	16 21.8	−26 26	Scorpio	6.07	Globular cluster
5	5904	15 17.0	+02 13	Serpens	5.99	Globular cluster
6	6405	17 38.1	−32 11	Scorpio	6.0	Open cluster
7	6475	17 51.9	−34 48	Scorpio	5.0	Open cluster
8	6523	18 01.8	−24 23	Sagittarius		"Lagoon" nebula; diffuse nebula
9	6333	17 17.5	−18 29	Ophiuchus	7.58	Globular cluster
10	6254	16 55.5	−04 04	Ophiuchus	6.40	Globular cluster
11	6705	18 49.5	−06 19	Scutum	7.0	Open cluster
12	6218	16 45.6	−01 54	Ophiuchus	6.74	Globular cluster
13	6205	16 40.6	+36 31	Hercules	5.78	Globular cluster
14	6402	17 36.0	−03 14	Ophiuchus	7.82	Globular cluster
15	7078	21 28.6	+12 02	Pegasus	6.29	Globular cluster
16	6611	18 17.2	−13 48	Serpens	7.0	Open cluster with nebulosity

* New General Catalog

NUMBER M	NGC*	RIGHT ASCENSION	(1970) DECLINATION	CONSTELLATION	APPARENT VISUAL MAGNITUDE	DESCRIPTION
17	6618	18 19.1	−16 12	Sagittarius	7.0	"Swan" or "Omega" nebula; diffuse nebula
18	6613	18 18.2	−17 09	Sagittarius	7.0	Open cluster
19	6273	17 00.7	−26 13	Ophiuchus	6.94	Globular cluster
20	6514	18 00.6	−23 02	Sagittarius		"Trifid" nebula; diffuse nebula
21	6531	18 02.8	−22 30	Sagittarius	7.0	Open cluster
22	6656	18 34.6	−23 56	Sagittarius	5.22	Globular cluster
23	6494	17 55.1	−19 00	Sagittarius	6.0	Open cluster
24	6603	18 16.7	−18 27	Sagittarius	6.0	Open cluster
25	4725†	18 29.9	−19 16	Sagittarius	6.0	Open cluster
26	6694	18 43.6	−09 26	Scutum	9.0	Open cluster
27	6853	19 58.4	+22 38	Vulpecula	8.2	"Dumbbell" nebula; planetary nebula
28	6626	18 22.6	−24 52	Sagittarius	7.07	Globular cluster
29	6913	20 22.9	+38 25	Cygnus	8.0	Open cluster
30	7099	21 38.6	−23 18	Capricornus	7.63	Globular cluster
31	224	0 41.1	+41 06	Andromeda	3.7	Andromeda galaxy (Sb)
32	221	0 41.1	+40 42	Andromeda	8.5	Elliptical galaxy companion to M31
33	598	1 32.2	+30 30	Triangulum	5.9	Spiral galaxy (Sc)
34	1039	2 40.1	+42 40	Perseus	6.0	Open cluster
35	2168	6 07.0	+24 21	Gemini	6.0	Open cluster
36	1960	5 34.3	+34 05	Auriga	6.0	Open cluster
37	2099	5 50.4	+32 33	Auriga	6.0	Open cluster
38	1912	5 26.6	+35 48	Auriga	6.0	Open cluster
39	7092	21 31.1	+48 18	Cygnus	6.0	Open cluster
40	—	12 20.0	+59 00	Ursa Major		Double star
41	2287	6 45.8	−20 42	Canis Major	6.0	Loose open cluster
42	1976	5 33.9	−05 24	Orion		Orion nebula; diffuse nebula
43	1982	5 34.1	−05 18	Orion		Northeast portion of Orion nebula
44	2632	8 38.2	+20 06	Cancer	4.0	Praesepe; open cluster
45	—	3 45.7	+24 01	Taurus	2.0	The Pleiades; open cluster
46	2437	7 40.4	−14 45	Puppis	7.0	Open cluster
47	2422	7 35.1	−14 26	Puppis	5.0	Open cluster
48	2548	8 12.0	−05 41	Hydra	6.0	Open cluster
49	4472	12 28.3	+08 10	Virgo	8.9	Elliptical galaxy
50	2323	7 01.5	−08 18	Monoceros	7.0	Loose open cluster
51	5194	13 28.6	+47 21	Canes Venatici	8.4	"Whirlpool" galaxy; Spiral galaxy (Sc)

† Index Catalog number (IC)

NUMBER M	NGC*	RIGHT ASCENSION (1970)	DECLINATION	CONSTELLATION	APPARENT VISUAL MAGNITUDE	DESCRIPTION
52	7654	23 22.9	+61 26	Cassiopeia	7.0	Loose open cluster
53	5024	13 11.5	+18 20	Coma Berenices	7.70	Globular cluster
54	6715	18 53.2	−30 31	Sagittarius	7.7	Globular cluster
55	6809	19 38.1	−31 01	Sagittarius	6.09	Globular cluster
56	6779	19 15.4	+30 07	Lyra	8.33	Globular cluster
57	6720	18 52.5	+33 00	Lyra	9.0	"Ring" nebula; planetary nebula
58	4579	12 36.2	+11 59	Virgo	9.9	Spiral galaxy (SBb)
59	4621	12 40.5	+11 50	Virgo	10.3	Elliptical galaxy
60	4649	12 42.1	+11 44	Virgo	9.3	Elliptical galaxy
61	4303	12 20.3	+04 39	Virgo	9.7	Spiral galaxy (Sc)
62	6266	16 59.3	−30 04	Scorpio	7.2	Globular cluster
63	5055	13 14.4	+42 11	Canes Venatici	8.8	Spiral galaxy (Sb)
64	4826	12 55.2	+21 51	Coma Berenices	8.7	Spiral galaxy (Sb)
65	3623	11 17.3	+13 16	Leo	9.6	Spiral galaxy (Sa)
66	3627	11 18.6	+13 10	Leo	9.2	Spiral galaxy (Sb); companion to M65
67	2682	8 49.5	+11 56	Cancer	7.0	Open cluster
68	4590	12 37.8	−26 35	Hydra	8.04	Globular cluster
69	6637	18 29.4	−32 23	Sagittarius	7.7	Globular cluster
70	6681	18 41.3	−32 19	Sagittarius	8.2	Globular cluster
71	6838	19 52.4	+18 42	Sagittarius	6.9	Globular cluster
72	6981	20 51.8	−12 41	Aquarius	9.15	Globular cluster
73	6994	20 57.3	−12 46	Aquarius		Open cluster
74	628	1 35.1	+15 38	Pisces	9.5	Spiral galaxy (Sc)
75	6864	20 04.3	−22 01	Sagittarius	8.31	Globular cluster
76	650	1 40.3	+51 25	Perseus	11.4	Planetary nebula
77	1068	2 41.1	−00 07	Cetus	9.1	Spiral galaxy (Sb)
78	2068	5 45.3	+00 02	Orion		Small diffuse nebula
79	1904	5 22.9	−24 33	Lepus	7.3	Globular cluster
80	6093	16 15.2	−22 55	Scorpio	7.17	Globular cluster
81	3031	9 53.4	+69 12	Ursa Major	6.9	Spiral galaxy (Sb)
82	3034	9 53.6	+69 50	Ursa Major	8.7	Irregular galaxy
83	5236	13 35.3	−29 43	Hydra	7.5	Spiral galaxy (Sc)
84	4374	12 23.6	+13 03	Virgo	9.8	Elliptical galaxy
85	4382	12 23.8	+18 21	Coma Berenices	9.5	(SO) type galaxy
86	4406	12 24.6	+13 06	Virgo	9.8	Elliptical galaxy
87	4486	12 29.2	+12 33	Virgo	9.3	Elliptical galaxy
88	4501	12 30.4	+14 35	Coma Berenices	9.7	Spiral galaxy (Sb)
89	4552	12 34.1	+12 43	Virgo	10.3	Elliptical galaxy
90	4569	12 35.3	+13 19	Virgo	9.7	Spiral galaxy (Sb)
91?	—	—	—	—	—	—

NUMBER M	NGC*	RIGHT ASCENSION (1970)	DECLINATION (1970)	CONSTELLATION	APPARENT VISUAL MAGNITUDE	DESCRIPTION
92	6341	17 16.2	+43 11	Hercules	6.33	Globular cluster
93	2447	7 43.2	−23 48	Puppis	6.0	Open cluster
94	4736	12 49.6	+41 17	Canes Venatici	8.1	Spiral galaxy (Sb)
95	3351	10 42.3	+11 52	Leo	9.9	Spiral galaxy (SBb)
96	3368	10 45.1	+11 59	Leo	9.4	Spiral galaxy (Sa)
97	3587	11 13.1	+55 11	Ursa Major	11.1	"Owl" nebula; planetary nebula
98	4192	12 12.2	+15 04	Coma Berenices	10.4	Spiral galaxy (Sb)
99	4254	12 17.3	+14 35	Coma Berenices	9.9	Spiral galaxy (Sc)
100	4321	12 21.4	+15 59	Coma Berenices	9.6	Spiral galaxy (Sc)
101	5457	14 02.1	+54 30	Ursa Major	8.1	Spiral galaxy (Sc)
102?	—	—	—	—	—	—
103	581	1 31.2	+60 32	Cassiopeia	7.0	Open cluster
104	4594	12 37.4	−11 21	Virgo	8.3	Spiral galaxy
105	3379	10 45.2	+13 01	Leo	9.7	Elliptical galaxy
106	4258	12 16.5	+47 35	Canes Venatici	8.4	Spiral galaxy
107	6171	16 29.7	−12 57	Ophiuchus	9.2	Globular cluster

The Messier catalog was originally compiled as a guide to what *not* to look at. Messier was an avid seeker of new comets. Since newly discovered comets often have a nebulous appearance, causing them to look like many galactic nebulae, star clusters, or galaxies, it was not easy to tell in a given case whether he had a new comet or only some nebula. The comet ultimately reveals itself by its motion through the heavens, but to avoid wasting time on the faint galaxies and star clusters, Messier made a list to warn him away from the unwanted noncometary objects.

GLOSSARY

GLOSSARY

Aberration of starlight The apparent shift in the position of a star due to the orbital motion of the earth.

Ablation The vaporization of the surface material of a meteroid due to the heat produced by friction as the body passes through the earth's atmosphere.

Absolute magnitude The apparent visual magnitude a body would have at a distance of 10 parsecs (32.6 light years).

Absolute zero The lowest temperature achievable; a condition of zero energy, at which all molecular motion stops (−273°C, −459°F, or 0°K).

Absorption spectrum Dark (absorption) lines superimposed on a continuous spectrum.

Acceleration A change in the velocity or direction of motion of a body.

Achromatic lens A lens system of two or more components used to correct chromatic aberration—the spreading of the spectrum into its constituent colors.

Active sun The state of the sun when it has many centers of activity—sunspots, prominences, flares, and other phenomena.

Accretion The process in which a star, planet, or other object gains mass by sweeping up surrounding gas or matter.

Accretion disk A disk of matter surrounding a star or black hole, formed of matter pulled in toward the object by its gravity.

Airglow The fluorescence of the earth's upper atmosphere.

Albedo The reflecting power of a body. The fraction of the incident light that a body reflects.

Almagest The celebrated book of Claudius Ptolemy which summarized the astronomical work up to his time and presented his theories of planetary motions.

Alpha particle The nucleus of a helium atom. A positively charged particle consisting of two protons and two neutrons. Also a product of radioactive decay.

Altitude The angular distance above the horizon of a celestial body measured along its vertical circle.

Angle of incidence The angle between the incoming ray of light and the normal (perpendicular) to the reflecting or refracting surface.

Angstrom unit A unit of length equal to 10^{-8} cm which is used to measure the wavelength of light. Its symbol is Å.

Angular diameter The angle subtended by the diameter of a body.

Angular momentum The momentum of a body moving about an axis or a point. It is the product of its mass, linear velocity, and distance from the center of motion.

Annular eclipse A solar eclipse in which the sun's apparent disk is larger than the moon's apparent disk so that a ring (the sun's disk) appears to surround the moon.

Antimatter Matter composed of antiparticles such as positrons (antielectrons) and antiprotons. Antiparticles differ from ordinary particles most usually by having the opposite electric charge, but are otherwise identical. When a particle and an antiparticle collide, they totally annihilate each other, leaving no matter and releasing an amount of energy equivalent to their joint mass.

Aperture The diameter of the objective lens or mirror of a telescope.

Aphelion The point in the elliptical orbit of a planet or comet at which the body is at its greatest distance from the sun.

Apogee The point in the elliptical orbit of the moon or artificial satellite at which the body is at its greatest distance from the earth.

Apparent magnitude A measure of the visual brightness of a celestial body as observed from the earth.

Apparent noon The time when the sun's center is on the observer's meridian.

Apparent solar day The period between two successive transits of the sun's center across the observer's meridian.

Apparent sun The true sun.

Appulse A penumbral lunar eclipse.

Ascending node The point in the orbit of a celestial body at which it crosses the reference plane—either the ecliptic or the celestial equator—from south to north.

Association A loose cluster of stars with physical characteristics that indicate a common origin.

Asteroid A small body whose orbital path around the sun is usually between the orbits of Mars and Jupiter. A synonym for a minor planet or planetoid.

Astrology The pseudoscience that claims there is an influence of celestial bodies on human affairs and that allegedly foretells terrestrial events by the positions of the stars and planets.

Astronomy The science that describes the celestial bodies according to their locations, sizes, motions, constitutions, and evolution.

Astronomical unit A unit of length defined by international agreement as the mean distance of the earth from the sun. Its value is about 93,000,000 miles, and its symbol is a.u.

Astrophysics A branch of astronomy in which the methods, tools, laws, and theories of physics are applied to the study of celestial bodies.

Atom The smallest particle of an element that exhibits all the properties of the element.

Atomic number The number of protons in the nucleus of an atom or the number of electrons around the nucleus when the atom is in its normal state.

Atomic weight The mass of the atom.

Aurora The beautiful streamers of light visible in the arctic regions, produced when solar particles interact with atmospheric gases. Aurora borealis are the northern lights, and Aurora australis are the southern lights.

Autumnal equinox The intersection of the celestial equator and the ecliptic at the point where the sun crosses from north to south. Autumn begins when the sun reaches the autumnal equinox.

Azimuth The angle measured from the north point, eastward along the horizon to the vertical circle that passes through the celestial body.

Baily's beads During a total solar eclipse, "beads" of sunlight that are visible around the moon just before and immediately after totality. These are caused by the sunlight passing through the moon's irregular limb.

Balmer lines A series of emission (bright) or absorption (dark) lines in the visible portion of the spectrum produced by electronic transitions in the hydrogen atom from a higher level to the second level (bright lines) or from the second level to a higher level (dark lines).

Barred spiral A spiral galaxy with a bright bar through its nucleus. The arms extend from the two ends of the bar.

Barycenter The center of mass (gravity) around which two bodies orbit.

Beta particle A negatively charged particle emitted by a radioactive atom; an electron.

"Big bang" theory A cosmological theory which states that the universe evolved from a primeval explosion.

Binary star A double star system held together by mutual gravitation with its components revolving around their barycenter.

BL Lacertae A peculiar variable galaxy showing no strong spectral features but emitting radio waves. It was originally mistaken for a star.

Black body A theoretically perfect radiator—a body that absorbs and reemits all the radiation that falls on it.

Black dwarf A star whose mass is too small for nuclear reactions to occur. It is therefore nonluminous.

Black hole An object that has suffered gravitational collapse. The gravitational attraction at its "surface" or horizon is so large that not even light is able to escape. Matter falling into a black hole is similarly unable to escape, and loses its identity. Black holes may be remnants of supernovae.

Blink microscope An instrument in which two photographs of the same region of the sky taken at different times may be viewed alternately. The two photographs are arranged so that corresponding star images are superimposed. As the two photographs are viewed alternately, the body will appear to jump back and forth ("blink") if it has changed its position relative to the other bodies during the period in which the two photographs were taken.

Bode-Titius relationship A sequence of numbers (empirical relationship) which gives the approximate mean distances of the planets in astronomical units.

Bolide A very bright meteor that often breaks up with a loud sound; a fireball.

Bolometer An instrument which records radiant energy over a wide range of wavelengths of the spectrum.

Bolometric magnitude Stellar magnitude based on the radiation emitted at all wavelengths from the star. The measurement is based on the radiation of the entire electromagnetic spectrum corrected for the effect of the earth's atmosphere.

Bright-line spectrum A series of bright colored lines on a dark background, produced by an incandescent, low-pressure gas.

Brightness A measure of the luminosity of a body.

Calorie The unit of heat required to raise the temperature of one gram of water 1°C.

Carbon cycle A series of nuclear reactions in which the carbon nucleus acts as the catalyst for converting hydrogen into helium.

Cardinal points The four principal points of the compass: north, south, east, and west.

Cassegrain focus A telescope arrangement in which a convex hyperboloidal secondary mirror reflects the beam back through an opening in the primary mirror, where the image is formed and viewed. Since the image is formed at the back of the telescope, it is easier to mount and use equipment.

Cassini division The 4800 km (3000 mi) gap between the outer and the middle ring of Saturn.

Celestial equator The projection of the earth's equator on the celestial sphere. It is a great circle whose points are 90° from the north and south celestial poles.

Celestial horizon The great circle on the celestial sphere whose points are 90° from the zenith and nadir.

Celestial mechanics A branch of astronomy that deals with the gravitation and motions of bodies in space.

Celestial meridian The great circle on the celestial sphere that passes through the zenith and nadir and crosses the observer's horizon at 90°.

Celestial poles The points on the celestial sphere established by the extension of the earth's axis of rotation.

Celestial sphere An apparent sphere of infinite radius at whose center the observer is located and on whose surface the celestial bodies appear to be located.

Center of mass (center of gravity) One point in a single body or within a system of several bodies that remains fixed or moves uniformly through space as though the entire mass of the body or system were concentrated at that point.

Centrifugal force A term used to describe the resistance of a particle to being accelerated along a curved path.

Centripetal force A center-directed force which diverts a body from a straight-line path into a curved path.

Cepheid variable A yellow giant or supergiant pulsating star whose brightness varies because it expands and contracts periodically.

Ceres The first and largest asteroid (minor planet) discovered.

Chromatic aberration The failure of a simple lens to bring visible light to a single focus because it refracts the wavelengths of visible light differently, producing a dispersion.

Chromosphere The nearly transparent, hot layer of the sun's atmosphere that lies directly above the solar visible surface, or photosphere.

Circumpolar stars Stars whose angular distances from the north or south celestial poles are less than the latitude of the observer, so that they always remain above the horizon.

Cluster of stars A group of stars with a common origin, bound together by mutual gravitation, and moving in space along nearly parallel paths at about the same speed.

Cluster variables A class of high-velocity pulsating stars (RR Lyrae stars) with periods of less than one day. They are very numerous in the Milky Way and in globular clusters.

Coelostat An instrument that permits the sun's image to be viewed in a fixed position throughout the day. This is accomplished by use of a telescope with an equatorially mounted, clock-driven plane mirror that follows the sun and a fixed mirror that directs the sunlight to the objective of a fixed tower telescope.

Collimator A lens system that renders a diverging beam of light into a parallel beam.

Color index The difference between the photographic and photovisual magnitudes of a star. $C.I. = m_{pg} - m_{pv}$.

Conduction The direct transfer of energy from atom to atom.

Comparison spectrum The spectrum of a vaporized element (often iron) photographed next to a stellar spectrum for comparison and identification of their spectral lines.

Coma (comet) The diffuse gaseous shell around the nucleus. The nucleus and coma comprise the head of the comet.

Coma (optical) An optical defect in which light rays that strike a lens or a mirror at an angle to its axis are not focused at the same place.

Configuration A special orientation of a planet or the moon with respect to the sun.

Conic sections The curves formed by the intersection of a plane and a right circular cone. These are circles, ellipses, parabolas, and hyperbolas.

Conjunction A configuration of two or more celestial bodies that are in line longitudinally and have the same right ascension.

Constellation A configuration of stars with which the ancients associated the name and shape of a person, animal, or object in their mythology. Today, it designates the definite area in the sky in which the configuration is located.

Continuous spectrum The spectrum of light that appears as a continuous blend of colors emitted by an incandescent solid, liquid, or gas under pressure.

Convection The transfer of energy by currents in the medium that contains the energy.

Coordinates Angular distances that are used to locate points on the terrestrial sphere (longitude and latitude) and positions of celestial bodies on the celestial sphere (right ascension and declination).

Core The central part of any celestial body.

Coriolis effect The deflection with respect to the earth's surface of a moving body, air, or water, due to the earth's rotation. The deflection is to the right in the northern hemisphere and to the left in the southern hemisphere.

Corona The tenuous outer layer of the sun's atmosphere that is visible during a total solar eclipse or with a coronograph.

Coronograph An instrument for photographing the sun's corona whenever the sun is visible.

Corpuscular radiation Charged particles, such as atomic nuclei and electrons. The solar wind and cosmic rays are sometimes so termed.

Cosmic rays High-energy particles (mostly protons) from space that strike the earth's atmosphere and produce secondary particles. They are believed to originate in supernova explosions, although some are produced in solar flares.

Cosmogony The study of the creation and evolution of the universe.

Cosmological principle The assumption that the universe at any time is the same from any position and in any direction.

Cosmology The study of the design, character, and extent of the universe.

Coudé focus A secondary optical system in a reflecting telescope that directs the reflected light down the telescope's polar axis to a focus that remains fixed regardless of any movement of the telescope.

Dark nebula A nonluminous cloud of interstellar dust and gas that obscures the light of the stars behind it.

Declination The angular distance north or south of the celestial equator measured along the hour circle that passes through the body.

Deferent The circumference of the large circle in the Ptolemaic system on which the center of the epicycle of a planet moves.

Deflection of starlight The bending of a beam of starlight as it passes close to a massive body, which was predicted by Einstein's theory of relativity.

Degenerate gas A state of matter often found at high density in which the atoms are packed so closely that the normal repulsive forces are replaced by subatomic repulsive forces.

Density The amount of mass (matter) of a body in a unit of volume, usually expressed in grams per cubic centimeter.

Descending node The point in the orbit of a celestial body at which it crosses the reference plane—either the ecliptic or the celestial equator—from north to south.

Deuterium **("heavy hydrogen")** A type (isotope) of hydrogen whose nucleus contains one proton and one neutron. Discovered by H. Urey in 1932.

Diffraction The deflection, or spreading out, of light as it passes the edge of an opaque body or through a small opening.

Diffraction grating A series of narrow parallel lines equally spaced on a metal or glass plate that utilizes the principle of diffraction to produce a spectrum.

Diffraction pattern A series of alternating bright and dark lines (fringes) produced by the interference of the light rays.

Diffuse nebula A cloud of interstellar dust and gas dense enough to be seen as a dark nebula or close enough to nearby stars to either reflect light (reflection nebula) or emit light (emission nebula).

Direct motion The normal motion of a planet against the stars as a background—from west to east.

Dispersion The separation of visible (white) light into its component colors by its wavelengths being refracted or diffracted by different amounts.

Distance modulus The difference between the apparent and absolute magnitudes of a star. A quantity used to express the star's distance d in parsecs, mathematically equal to $5 \log d - 5$.

Diurnal circle The daily apparent path of a celestial body that is parallel to the celestial equator.

Diurnal motion The daily apparent motion of a celestial body caused by the daily rotation of the earth.

Doppler effect The apparent change in the frequency of radiation caused by the relative motion between the radiation source and the observer.

Dwarf star A main-sequence star comparable to the sun in size, mass, and luminosity.

Dyne The unit of force in the metric system that will accelerate a one-gram mass at one cm/sec^2.

Earthshine The sunlight reflected from the earth that dimly illuminates the dark portion of the moon.

Eccentricity The measure of the degree of flattening of an ellipse. It is the ratio of the distance between the foci to the length of the major axis. A circle has zero eccentricity.

Eclipse The cutting off of the light, partially or totally, from one body when another body passes in front of it.

Eclipse limit The maximum angular distance from either side of the node at which an eclipse can occur.

Eclipsing binary A binary star whose orbital plane is seen nearly edgewise so that alternately one star appears to eclipse the other.

Ecliptic The annual apparent path of the sun among the stars. It is the projection of the earth's orbit on the celestial sphere.

Electromagnetic radiation Radiation consisting of wave phenomena associated with electrical and magnetic fields whose wavelengths include gamma rays, x-rays, ultraviolet, visible light, infrared, and radio.

Electromagnetic spectrum The entire family of electromagnetic radiation.

Electron A fundamental subatomic particle that is negatively charged, has an extremely small mass, and revolves around the nucleus of the atom.

Element A basic substance that cannot be reduced to simpler substances by ordinary chemical processes.

Ellipse A conic section. A closed curve formed by the intersection of a nonhorizontal plane and a circular cone.

Elliptical galaxy An ellipsoidal galaxy that lacks spiral arms and any appreciable interstellar material.

Elongation The apparent angle between the sun and a planet as seen from the earth. The difference between the longitude of the sun and a planet.

Emission line A discrete bright line in a spectrum.

Emission nebula A bright gaseous nebula whose light is derived from the ultraviolet light of a star within or near the nebula.

Emission spectrum A series of bright lines (emission lines) produced by a low-pressure incandescent gas.

Energy levels of atoms The possible energies that an atom can have due to its electrons moving in orbits around the nucleus.

Ephemeris A table that gives the positions of celestial bodies at regular intervals of time.

Epicycle The small circle in the Ptolemaic system whose center moves along the circumference of the larger circle (deferent) while the planet moves along the circumference of the small circle.

Epoch An arbitrary date selected as a point of reference to which astronomical observations are referred.

Equation of time The difference between the apparent and mean solar time.

Equator A great circle on the terrestrial sphere whose points are 90° from the north and south poles.

Equatorial mount A telescope mount with one axis parallel to the earth's axis which rotates at the same rate as the earth by means of a clock drive.

Equinox One of two intersections of the ecliptic and the celestial equator.

Erg A unit of energy in the metric system. The amount of work accomplished by a force of one dyne moving a body a distance of one centimeter.

Escape velocity The initial speed that a body must attain to overcome the gravitational pull of another body and escape into space.

Exosphere The top layer of the earth's atmosphere where the molecules readily escape from the earth's gravitational pull.

Extragalactic Outside or beyond the Milky Way.

Faculae Extended bright regions ("little torches") seen near the sun's limb.

Filar micrometer An instrument attached to the eyepiece of a telescope to measure the angle (separation) between two stars and their relative positions.

Fireball An unusually bright meteor.

Flare A very bright flash of light over a small area of the sun's surface, especially near an active sunspot, which is due to an intense outburst of energy.

Flare star A star that suddenly and unpredictably increases its brightness for a short period of time.

Flash spectrum The spectrum of the sun's limb that is a bright-line spectrum of the chromosphere visible for an instant just before totality in a solar eclipse.

Flocculi (plages) Bright regions in the chromosphere in the magnetic fields around sunspots visible in spectroheliograms in monochromatic light.

Fluorescence The absorption of radiation of one wavelength (especially ultraviolet) and its re-emission in another wavelength (visible light).

Focal length The distance from the center of a lens or mirror to the focal point.

Focal ratio The "f" number, or speed. The ratio of the focal length of a lens or mirror to its diameter.

Forbidden lines Spectral lines not usually obtainable under laboratory conditions because their emissions result from atomic transitions that are most improbable.

Force That which can change the speed or the direction of a body.

Fraunhofer line An absorption (dark) line in the spectrum of the sun or a star.

Frequency The number of waves that pass a given point in a unit of time.

Full moon The phase of the moon when it is in opposition (opposite side of the earth from the sun) and its entire visible hemisphere is illuminated.

Fusion The process by which heavier atomic nuclei are created from lighter ones.

Galactic cluster An open cluster of stars found in the spiral arms of the Milky Way.

Galactic equator The plane of the great circle on the celestial sphere that locates the center line of the Milky Way.

Galactic poles The north and south galactic poles that are 90° from the galactic equator.

Galaxy A large assemblage of stars and interstellar material held together by gravitation. A galaxy usually contains from millions to hundreds of billions of stars. The galaxy in which our solar system is located is the Milky Way.

Gamma ray The most energetic form of electromagnetic radiation yet discovered, and the one possessing the shortest known wavelength.

Gegenschein A faint diffuse patch of light (counterglow) seen opposite the sun in the sky. It is probably caused by the reflection of sunlight from very small particles in space around the earth.

Giant star A star of large radius and luminosity.

Gibbous A phase of the moon or a planet during which more than half but less than the whole disk appears illuminated.

Globular cluster A large spherical cluster of stars located in the halo which surrounds the Milky Way and other galaxies.

Globule A small, dark, relatively dense nebula that may be a protostar.

Granules Small, bright spots in the sun's photosphere that give it a mottled appearance similar to rice grains. They are produced by hot gases rising from below the photosphere.

Gravitation The property of matter by which one mass exerts a force of attraction on another.

Great circle The curve formed on the surface of a sphere where a plane that passes through the sphere intersects its center. It divides the sphere into two equal parts.

Greenwich meridian The meridian that passes through a point at the Royal Greenwich Observatory in England, the reference or "prime meridian" for determining longitude.

Gregorian calendar The calendar in common use today, which was introduced by Pope Gregory XIII in 1582.

Ground state The lowest possible energy level of an atom.

H I region A region in space of neutral hydrogen.

H II region A region in space of ionized hydrogen.

Halo A region that surrounds the nucleus of a galaxy and contains globular clusters and stars.

Harmonic law Kepler's third law of planetary motion, which states that the squares of the sidereal periods of planets are proportional to the cubes of the semimajor axes of their orbits: $P^2 = a^3$.

Harvest moon The full moon nearest to the time of the autumnal equinox, which rises after sunset on successive nights with the minimum delay.

Hayashi lines The theoretical evolutionary track on the Hertzsprung-Russell diagram along which a star contracts in its early stages of evolution.

Head of comet The principal part of a comet, which contains the nucleus and coma.

"Heavy" elements In astronomy, (usually) the elements whose atomic numbers are greater than that of helium.

Heliocentric system In models of the planetary system, any system that is centered around the sun.

Helium flash An explosion-like ignition of the helium in the core of a red giant star, which starts the triple-alpha nuclear process.

Hertzsprung-Russell (H-R) diagram A plot of the absolute magnitudes of stars against their spectral class, color index, color, or temperature.

Horizon system A system in which the coordinates azimuth and altitude are used to establish the position of a celestial body.

Hour angle The angle measured from the local (observer's) meridian westward along the celestial equator to the hour circle that passes through the body.

Hour circle A great circle on the celestial sphere that passes through the celestial poles and crosses the celestial equator at 90°.

Hubble law The velocity-distance relation for extragalactic objects, arising from the expansion of the universe. ($V = dH$, where H is the Hubble constant)

Hubble constant A number that relates the rate of recession of a galaxy to its distance. Its reciprocal measures the age of the universe.

Hyperbola A conic section. An open curve formed by the intersection of a plane parallel to the axis of a circular cone with the cone's surface.

Image The optical representation of an object produced by the reflection or refraction of the light rays by a lens or a mirror.

Inclination of an orbit The angle between the orbital plane of a body and usually either the plane of the celestial equator or the ecliptic.

Index Catalog (I.C.) A supplement to the New General Catalog (NGC) of star clusters, nebulae, and galaxies.

Index of refraction The ratio of the speed of light in a vacuum to its speed in a given transparent substance.

Inertia The property of matter by which a body resists a change in its state of motion.

Inferior conjunction The planetary configuration of an inferior planet (Mercury and Venus) when it is between the earth and the sun.

Inferior planet A planet whose mean distance from the sun is less than the earth's. Its orbit lies between the earth and the sun.

Infrared radiation Electromagnetic radiation whose wavelength is longer than that of visible red light and shorter than that of radio waves.

Insolation The amount of the sun's radiation that falls on a unit area of the earth's surface in a unit of time.

Interferometer An optical instrument that utilizes the principle of the interference of light waves to measure small angles, for example, the diameter of stars.

International date line An arbitrary line, nearly coinciding with the 180° meridian, across which the date changes by one day.

Interstellar lines Absorption lines (dark spectral lines) on stellar spectra produced by diffuse gas in space.

Interstellar matter Microscopic dust particles and diffuse gases in space between the stars.

Ion An electrically charged atom produced by either the loss or gain of one or more electrons.

Ionization The process by which atoms lose or gain electrons.

Ionosphere A layer of the earth's atmosphere containing many ionized atoms, which reflect some wavelengths of radio waves.

Irregular galaxy A galaxy whose shape is not symmetrical—it is neither an elliptical nor a spiral galaxy.

Irregular variable A star whose energy output is not periodic.

Isotope Any of two or more forms of a given element, whose atoms have the same atomic number (number of protons) but different atomic weight (mass) by virtue of containing a different number of neutrons.

Jovian planet One of the four large, relatively low-density planets: Jupiter, Saturn, Uranus, and Neptune.

Julian calendar A solar calendar introduced by Julius Caesar.

Julian day calendar A calendar based on the system of the continuous numbering of the days beginning with January 1, 4713 B.C.

Kepler's laws The three basic laws of planetary motion discovered by Johannes Kepler.

Kinetic energy The energy of motion of a body. It is expressed as one-half the product of its mass and the square of its velocity: K.E. $= \frac{1}{2} mv^2$.

Kirchhoff's laws The three laws which explain the formation of continuous, emission (bright-line), and absorption spectra.

Kirkwood's gaps The gaps in the spacing of the asteroids.

Lacertid A type of peculiar extragalactic radio source showing a starlike image but usually no emission or absorption lines in its spectra.

Latitude The angular distance on the terrestrial sphere measured from the equator, north or south, along the meridian which passes through the place.

Law of areas Kepler's second law of planetary motion, which states that the radius vector (line joining the planet and sun) sweeps equal areas in its orbital plane in equal intervals of time.

Law of the red shift The radial velocity of a distant galaxy, which is measured by the red shift, is proportional to its distance; therefore, the red shift is a measure of the galaxy's distance. (See also *Hubble law.*)

Leap year A calendar year of 366 days, which occurs in the years divisible by four except in century years not divisible by 400. 1900 was not a leap year. 2000 will be a leap year.

Libration A real or apparent oscillation of a body that permits the observer on the earth to see more than one hemisphere of the body during a given period of time.

Libration (latitudinal) The libration due to the moon's equator being inclined about $6\frac{1}{2}°$ to its orbital plane, which permits the observer to see about $6\frac{1}{2}°$ beyond the north and south poles of the moon during one lunar month.

Libration (longitudinal) The libration due to the moon's constant rotational speed and its variable orbital speed, which permits the observer on the earth to see over $7\frac{1}{2}°$ beyond the east and west limbs of the moon during one lunar month.

Light An electromagnetic radiation visible to the eye.

Light curve A plot of the variation of the magnitude of a variable star or an eclipsing binary against time.

Light-gathering power of a telescope The amount of light a telescope collects, which is proportional to the area of its objective.

Light year The distance light travels in one year in space (vacuum), which is approximately 6×10^{12} miles, or 9.7×10^{12} km.

Limb The apparent edge of the sun, moon, or planet.

Limb darkening Term describing the fact that the sun's limb appears darker than the center of its disk, because at the disk's center the observer sees into deeper and hotter layers of the sun's photosphere.

Line broadening The phenomenon that increases the width of spectral lines.

Line of apsides The major axis of the elliptical orbit of a body.

Line of nodes The line that connects the ascending and descending nodes of an orbit and that intersects a reference plane such as the ecliptic.

Local apparent time The local hour angle of the apparent sun plus 12 hours.

Local group The cluster of galaxies, including the Milky Way, that appears to form a group.

Local mean time The local hour angle of the mean sun plus 12 hours.

Local standard of rest The coordinate system in which the motions of the stars in the neighborhood of the sun average zero, that is, they appear to be at rest within the system.

Longitude The angular distance on the terrestrial sphere measured from the Greenwich meridian, east or west along the equator to the meridian that passes through the place.

Luminosity The rate at which a star emits electromagnetic radiation into space. It is usually expressed in terms of the sun's luminosity.

Luminosity function The relative number of stars of various absolute magnitudes in a unit volume of space.

Magnifying power The apparent increase in the size of a body when seen through a telescope over its size when seen with the unaided eye.

Magnitude A number that designates the brightness of a body. It is a measure of the light received from a body.

Main sequence A narrow band on the H-R diagram on which the majority of the stars lie. The band runs from the upper left to the lower right of the diagram. Stars on the main sequence are burning hydrogen. (See Fig. 11.6)

Mantle A layer that lies between the crust and the core in the earth's interior.

Mare A large, smooth, circular basin such as those found on the moon. The name comes from the Latin word for sea.

Mass-luminosity relation An empirical relationship that states that the luminosity of a star, primarily a main sequence or a giant, depends on its mass. The more massive stars are the more luminous.

Mean solar day The time between successive crossings of the observer's meridian by the mean sun.

Mean sun An imaginary body that moves eastward along the celestial equator at a uniform rate and completes its circuit in the sky in the same period as the apparent sun.

Meridian The great circle on the terrestrial sphere which passes through the observer's position and the earth's north and south poles. The great circle on the celestial sphere which passes through the observer's zenith and the celestial north and south poles.

Meson A short-lived subatomic particle with a mass between that of a proton and an electron. Mesons play a role in holding nuclei together.

Messier catalog A catalog of nebulae, star clusters, and galaxies compiled by Charles Messier in 1787. The bodies are designated by M and a number, e.g., M31 for the Andromeda galaxy.

Meteor The bright streak of light that is visible when a meteoroid passes through the earth's atmosphere and is heated by friction between it and the air molecules.

Meteor shower Many meteors that appear to radiate from a point in the sky, due to the earth's passing through a swarm or a stream of meteoroid particles.

Meteorite A meteoroid that has survived its flight through the earth's atmosphere and has struck the earth's surface.

Meteoroid The stony or metallic particle that produces a meteor when it passes through the earth's atmosphere.

Micrometeorite An extremely small meteoroid whose size causes it to move very slowly through the earth's atmosphere so that it does not burn.

Milky Way A faint, diffuse band of light that completely encircles the sky and consists of a vast number of stars and interstellar material.

Mohorovicic discontinuity Moho for short. It is the boundary between the earth's crust and the mantle. Named for the Yugoslav scientist Andrja Mohorovicic.

Molecule The smallest unit of a substance that retains the chemical properties of the substance. It is a combination of two or more atoms.

Monochromatic Consisting of one color or wavelength.

n-body problem The problem of determining the motion of a body that is interacting with two or more other bodies under their mutual gravitational attraction.

Nadir The point on the celestial sphere that is 180° from the zenith and directly below the observer.

Neap tides The lowest tides that occur each month, when the moon is near the first- or third-quarter phase.

Nebula A cloud of interstellar dust or gas.

Neutrino A particle with zero mass when at rest and no electric charge, which carries away energy when it is emitted from a nuclear reaction.

Neutron A subatomic particle without a charge and with a mass approximately equal to that of a proton.

Neutron star An extremely small, dense star of approximately one solar mass but only 10 km in diameter, composed of neutrons. Believed to arise from supernova explosions. (See also *Pulsar.*)

New General Catalog (NGC) A catalog of star clusters, nebulae, and galaxies, compiled by J. Dreyer, which succeeded the Messier Catalog.

New moon The phase of the moon that occurs when the longitude of the sun and the moon are the same and the moon's dark hemisphere is toward the earth.

Newtonian focus A telescope arrangement in which a secondary plane mirror is placed near the top of the tube to divert the light rays from the primary mirror and bring them to a focus at the side of the tube, at right angles to the direction of the telescope.

Node The intersection of the orbital path of a body with the reference plane, e.g., the celestial equator or the ecliptic. (See *Ascending node, Descending node.*)

Nova A star that experiences some kind of violent explosion, increasing its brightness several magnitudes, then fading gradually.

Nuclear fusion See *Fusion.*

Nucleus (atom) The central part of the atom, which contains protons and neutrons that comprise almost the entire mass of the atom.

Nucleus (comet) The swarm of solid particles in the head of a comet.

Nucleus (galaxy) The center of the galaxy, where the star density is the greatest.

Nutation Name for the phenomenon that, as the earth's polar axis precesses, the earth's pole wobbles about nine seconds of arc around its mean position.

Objective The primary, light-gathering, image-forming lens or mirror of a telescope.

Oblate spheroid A solid formed by rotating an ellipse about its minor axis. A spherical body that has been flattened by rotation.

Oblateness A measure of the amount of flattening of a sphere. It is the ratio of the difference between the equatorial and polar diameters of the spheroid to the equatorial diameter.

Obliquity of the ecliptic The 23½° angle between the planes of the ecliptic and the celestial equator.

Obscuration (interstellar) The absorption of starlight by interstellar dust.

Occular An eyepiece.

Occultation The phenomenon in which one body is temporarily hidden from view by the passage of another body in front of it, as in the occultation of a star by the moon or the occultation of Jupiter's satellites by the planet.

Opacity The property of a body that allows it to stop the passage of light. The light-absorbing ability of a nebula or region in a star.

Open cluster A loose, unsymmetrical cluster containing from tens to several thousands of stars, located in the disk or spiral arms of the Milky Way or other galaxy.

Opposition The configuration of a planet when it is on the opposite side of the sun as viewed from the earth, that is, when its elongation is 180°. This occurs only for planets whose orbits are larger than the earth's.

Optical binary Two stars that appear to be close together, although they are neither in the same region of space nor gravitationally associated.

Ozone Oxygen molecules composed of three atoms that are formed by the action of the sun's ultraviolet radiation on the oxygen molecules of two atoms found in the stratosphere.

Parabola A conic section. An open curve formed by the intersection of a circular cone and a plane parallel to its side (surface of the cone).

Paraboloid The surface generated by the rotation of a parabola about its axis. The shape of the surface of the primary mirror in most reflecting telescopes.

Parallax (stellar) The apparent angular displacement of a nearby star with respect to the more distant stars due to the earth's orbital motion around the sun. It is the angle that subtends the radius of the earth's orbit (1 a.u.) at the star's distance.

Parsec A unit of distance. The distance of a body when its stellar *par*allax is one *sec*ond of arc. 1 parsec=3.26 light years.

Penumbra The region within a body's shadow that is partially illuminated. It is the transition region between total obscuration in the umbra and total illumination outside the shadow.

Penumbral eclipse A lunar eclipse that occurs when the moon passes through only the penumbra of the earth's shadow.

Perfect gas A gas in which the pressure is related to the density multiplied by the temperature. The pressure arises from collisions between the atoms or molecules of the gas.

Periastron The point in the orbit of a star in a binary star system that is closest to the other star.

Perigee The point in the orbit of an earth satellite that is closest to the earth.

Perihelion The point in the orbit of a body revolving around the sun that is closest to the sun.

Period-luminosity relation The empirical relationship between the absolute magnitude and the period of light variation of cepheid-variable stars.

Perturbation The deviation in the orbital path of a body produced by a third body or an external force.

Phases The progressive changes in the shape of the illuminated hemisphere of the moon or planet as seen from the earth.

Photoelectric cell A vacuum tube in which electrons are ejected from the surface of a light-sensitive substance (cathode) when exposed to light and then are accelerated to the anode. This flow of electrons is an electric current, which can be used to determine the amount of light that strikes the cathode.

Photographic magnitude The magnitude of a body as determined by a photographic plate sensitive to blue and violet light.

Photomultiplier A photoelectric cell in which the number of ejected electrons is increased by having the electrons strike in succession a series of charged grids. This amplifies the electric current generated so that it can be measured more easily and accurately.

Photon A discrete parcel of electromagnetic energy.

Photosphere The apparent luminous visible solar surface; the region from which nearly all of the sun's light is emitted.

Photovisual magnitude The brightness of a body as determined by a photographic plate sensitive to green and yellow light; the spectral region to which the human eye is most sensitive.

Plage (flocculi) A bright region in the chromosphere, above and around a sunspot in its magnetic field, visible in a spectroheliogram.

Planetarium An optical instrument that projects the celestial bodies visible to the unaided eye on a domed ceiling. It permits the observer to view a simulated but realistic sky as it would appear in the present, past, or future.

Planetary nebula An extremely hot star surrounded by a large shell of rarefied gas that is slowly expanding. A late stage in a star's evolution.

Planetoid A minor planet or an asteroid.

Polarized light The partial or complete alignment of the light waves so that they vibrate in one plane.

Positron A subatomic particle with the same mass as an electron, but with positive charge.

Precession The slow movement of the earth's rotation axis that causes the north celestial pole to sweep a circle of 23½° radius around the north ecliptic pole during a period of about 26,000 years.

Prime focus The point in the telescope tube where the image is formed by the objective (primary lens or mirror).

Prime meridian The meridian (great circle) that passes through the old Royal Observatory at Greenwich, England and is used as the standard reference for measuring longitude on the earth.

"Primeval atom" Lemaitre's single, superdense sphere of matter, which exploded, expanded, and formed the present matter in the universe.

Prominence Luminous gas clouds of many different shapes and sizes visible on the sun's limb projecting upward from the chromosphere.

Proper motion The rate at which a star's direction in the sky changes. Usually expressed in seconds of arc per century.

Proton One of the two fundamental units that make up the nucleus of the atom. A subatomic particle of positive charge. It is the nucleus of the ordinary hydrogen atom.

Protostar A cloud of dust and gas that is condensing at an accelerating rate, decreasing in size, and increasing in temperature and is in the early stages of becoming a star.

Pulsar A rapidly pulsating source of electromagnetic waves (usually radio), produced by the rapid rotation of a neutron star. Pulsars are believed to be formed in supernova explosions.

Pulsating variable A variable star that periodically changes its brightness because it pulsates, for example, a Cepheid.

Quadrature The configuration of a planet when the angle between the planet and the sun as viewed from the earth is 90°. This occurs only for planets whose orbits are larger than the earth's.

Quarter moon The phase of the moon in which only one-half of its illuminated hemisphere is visible from the earth. This occurs when the moon has traveled one-quarter or three-quarters of its orbit.

Quasars The abbreviated name for quasi-stellar radio sources. They are extragalactic objects emitting light and radio energy, are stellar in appearance, highly luminous, and show large red shifts.

Quiet sun The sun when the number of centers of activity are at a minimum.

Radar telescope A radio telescope that transmits a radio signal toward a celestial body. The body reflects part of the energy, which the telescope picks up, amplifies, and records.

Radial velocity The component of relative velocity of a body that is measured along the observer's line of sight.

Radiant A point in the sky from which meteors in a meteor shower appear to radiate.

Radiation The method of transferring energy through a vacuum (space). Also, the energy (electromagnetic or corpuscular) that is transmitted.

Radio telescope A telescope which makes observations in radio wavelengths. A large paraboloidal antenna which collects radio energy emitted by a celestial source, a receiver which amplifies the signal, and a recorder which records the information.

Radioactive decay The process by which a radioactive element decomposes into lighter elements and emits gamma rays and other subatomic particles. This radioactivity provides the means of determining the age of a body and contributes to the heating of the earth's interior.

Radius vector The imaginary line that joins a planet to the sun and moves as the planet revolves around the sun.

Rays (lunar) A system of long, bright streaks that appear to radiate from some of the lunar craters.

Recurrent nova A star that has been observed to erupt on several occasions. These outbursts occur on an average of once every 30 years.

Red giant A large, cool, very luminous star. It is located above and to the right of the main-sequence line of the H-R diagram.

Reddening (interstellar) The reddening of starlight as it passes through interstellar dust because the dust scatters the blue light more effectively than the red.

Red shift The shift in the spectral lines toward the red end of the spectrum of remote galaxies, which is attributed to the expansion of the universe.

Reflecting telescope A telescope that uses a mirror for its objective and the principle of reflection of light for its operation.

Reflection nebula A cloud of interstellar dust that is luminous because it reflects the light from a nearby star.

Refracting telescope A telescope that uses a lens for its objective and the principle of the refraction of light for its operation.

Resolving power The telescope's ability to resolve (separate) two objects that appear as a single source of light to the unaided eye.

Retrograde motion The apparent westward motion of a planet with respect to the stars as a background.

Revolution The motion of a body around a point in space or another body. The earth revolves around the sun. A star in a binary star system revolves around its barycenter and around the other star.

Right ascension The angular distance measured from the hour circle that passes through the vernal equinox, eastward along the celestial equator to the hour circle that passes through the celestial body.

Rille A crevasse (cleft or channel) on the lunar surface.

Roche's limit The minimum distance at which a satellite cannot survive the gravitational force of a planet and therefore disintegrates. Saturn's rings are within Roche's limit.

Rotation The motion of a body about its axis.

Saros A cycle of about 18 years in which similar eclipses recur.

Satellite A body that revolves around a larger body. The moon is the earth's satellite.

Scattering Reflection of light off small particles, the phenomenon responsible for the blue color of the sky and for twilight.

Schmidt telescope A reflecting telescope that utilizes a spherical mirror and a correcting plate to compensate for the aberrations of the mirror. This system permits a large field of view to be photographed.

Schwarzchild radius The distance from a mass at which the escape velocity equals the velocity of light. This distance defines the size of a black hole.

"Seeing" A measure of the stability of the atmosphere, establishes the quality of the appearance of celestial bodies.

Seismic waves Vibrations produced by earthquakes or subterranean explosions that travel through the earth's interior.

Semimajor axis One-half of the major axis of an ellipse. It also represents a planet's mean distance from the sun.

Separation The angular distance between two stars of a visual binary system.

Seyfert galaxy A spiral galaxy whose bright nucleus emits strong emission lines and infrared radiation, as well as some radio radiation. This type of galaxy was discovered by Carl Seyfert.

Shell star A star surrounded by a shell (sphere) of gas.

Shower (cosmic rays) Secondary high-energy cosmic particles produced by the collision of the primary cosmic ray particles with the molecules in the earth's atmosphere.

Shower (meteor) Meteors that appear to radiate from a common point (radiant) in the sky. The shower occurs when the earth passes through a stream or a swarm of meteoric material ejected by a comet and lying in the comet's path.

Sidereal day The interval of time between two successive transits of the observer's meridian by the vernal equinox or any given star.

Sidereal month The interval of time for the moon to complete one revolution around the earth, with a star or the vernal equinox as the reference.

Sidereal period The interval of time for one body to complete one revolution around another body, with a star or the vernal equinox as a reference.

Sidereal time Star time, defined as the local hour angle of the vernal equinox.

Sidereal year The interval of time for the earth to complete one revolution around the sun, with a star or the vernal equinox as the reference.

Small circle A closed curve formed on the surface of a sphere by the intersection of the sphere with a plane that does not pass through the sphere's center.

Solar activity A phenomenon such as a sunspot, plage, flare, or prominence that is visible on or above the solar photosphere.

Solar apex The point in the sky toward which the sun is moving. In this direction the stars appear to be moving toward the observer.

Solar constant The amount of solar radiation that a unit area normal to the sun's rays receives at a distance of 1 a.u. in a unit of time. Its mean value is 1.95 calories/ $cm^2 \cdot min$ (1.36 kw/m^2).

Solar day The interval of time for the earth to complete one rotation with respect to the sun. The period for the sun to make two successive transits of the observer's meridian.

Solar parallax The angle that subtends the earth's equatorial radius at a distance of 1 a.u.

Solar system The sun's family. All the celestial bodies that are held together gravitationally by the sun (planets, satellites, minor planets, comets, etc.) and revolve around the sun.

Solar time The time based on the sun. It starts when the sun makes a lower transit of the observer's meridian. It is the local hour angle of the sun plus 12 hours.

Solar wind A tenuous gas of charged particles—ions and electrons—ejected by the sun that moves through space at high speeds.

Space velocity The star's velocity with respect to the sun, expressed in mi/sec or km/sec.

Spectral class The classification of a star according to the characteristics of its spectrum.

Spectrograph The instrument for photographing the spectrum of a body.

Spectroheliogram A solar photograph taken in monochromatic light—usually hydrogen or ionized calcium.

Spectroscopic binary A binary star system that is revealed only by the shifting of its spectral lines.

Spectroscopic parallax (spectroscopic distance) A procedure for estimating the distance of a remote star by establishing its absolute magnitude from its spectral class and comparing that with its apparent magnitude.

Spectrum A band of colors like that in a rainbow, proceeding from red to violet, produced when light is dispersed by refraction or diffraction.

Speed The distance traveled by a moving object divided by the time required to travel that distance, measured in units of km/sec or mi/hr, for example.

Spherical aberration A major defect in a spherical lens or mirror. The light rays that strike the peripheral areas of the spherical surface have shorter focal lengths than those that strike near the optical axis.

Spicules Bright, short-lived, threadlike jets of material extending into the chromosphere over the entire solar limb.

Spiral galaxy A flat, rotating galaxy with spiral arms that emerge from either a bright round or a bar nucleus.

Sporadic meteors Meteors that are not associated with a

shower. They appear at random times and places in the sky.

Spring tides The highest tides during each lunar month, occurring when the sun, moon, and earth are in line, that is, when the moon is new or full.

Standard time The time kept within a 15°-wide longitude zone based on the local mean solar time of the zone's central meridian.

Star cluster A group of stars held together by mutual gravitation that are believed to have a common origin and velocity.

Steady-state universe A cosmological theory according to which the shape and density of the universe have always been the same, that is, the universe had no beginning and will have no end. To compensate for expansion of the universe without a decrease in mean density, the theory suggests there is continuous creation of matter to form new galaxies.

Stefan's law A formula which states that the rate at which energy is emitted from a unit area of a black body is proportional to the fourth power of its absolute temperature.

Stratosphere The layer of the earth's atmosphere that lies between the troposphere and the mesosphere. The ozone layer lies near the upper boundary of the stratosphere.

Subdwarf A star whose luminosity is less than that of a main-sequence star of the same spectral class.

Subgiant A star whose luminosity is less than that of a normal giant and greater than that of a main-sequence star of the same spectral class.

Summer solstice The point on the ecliptic where the sun reaches its greatest distance (23½°) above the celestial equator. It marks the longest day of the year.

Superior conjunction The planetary configuration of an inferior planet (Mercury and Venus) when the sun is between the planet and the earth.

Supernova A star that temporarily increases its luminosity millions of times when it erupts and ejects a large amount of its material into space.

Synchrotron radiation The radiation emitted by charged particles spiraling in a magnetic field at almost the speed of light.

Synodic month The moon's period of revolution with respect to the sun, which is the period of its cycle of phases.

Synodic period The interval of time between two similar successive planetary configurations.

Syzygy A lunar configuration in which the sun, earth, and moon are in line. This occurs when the moon is new or full.

T Tauri stars Extremely young variable stars with erratic variations in brightness that are associated with interstellar material.

Tangential (transverse) velocity A star's velocity with respect to the sun at right angles to the line of sight expressed in kilometers per second.

Tektites Round glassy stones found on the earth that are believed to be cooled molten material ejected by meteoritic impacts.

Telluric lines Spectral lines or bands produced by the absorption of light from a celestial body in the earth's atmosphere.

Temperature A measure of the degree of energy or motion of atoms or molecules.

Terminator The line of demarcation between the illuminated and dark portions of the moon or a planet.

Thermal equilibrium The condition under which the rate of energy generation equals the rate of energy loss.

Thermocouple A device that consists of a junction of two dissimilar metals, used to measure the intensity of the radiation absorbed by the junction.

Thermonuclear reactions The fusion of atomic nuclei under the influence of extremely high temperatures and pressures with the release of energy.

Tides The deformation of a body produced by the differential gravitational force exerted on it by another body.

Total lunar eclipse The complete obscuration of the moon's disk by the umbra of the earth's shadow cone.

Total solar eclipse The complete obscuration of the sun's disk by the moon.

Totality The interval of time during a total eclipse when the light of one body is completely obscured by another body.

Transit The passage of a small body across the disk of a larger body, e.g., Mercury transiting the sun. The passage of a body across a meridian, e.g., the upper transit of the observer's meridian by the sun.

Triangulation The method by which the distance to a remote (inaccessible) point can be determined by the solution of the elements of a triangle. The accessible elements are measured and the inaccessible ones are calculated by trigonometry.

Trojans Two groups of asteroids that revolve around the sun in Jupiter's orbit—one is leading while the other is following the planet by an angle of 60° at the sun.

Tropic of Cancer The parallel of latitude, 23½° north, that marks the northernmost position above the equator reached by the sun—June 21.

Tropic of Capricorn The parallel of latitude, 23½° south, that marks the southernmost position below the equator reached by the sun—December 21.

Tropical year The interval of time for the earth to complete one revolution around the sun with the vernal equinox as a reference. Numerically, it is equal to the ordinary year of approximately 365¼ days.

Troposphere The lowest layer of the earth's atmosphere, next to the earth's crust, where most weather phenomena occur.

Twilight The phenomenon of partial light visible after sunset and before sunrise, which results from the scattering of sunlight from particles in the earth's upper atmosphere.

Ultraviolet radiation The electromagnetic radiation whose wavelengths are shorter than those of visible violet and range from about 4000 to 100 Å.

Umbra The portion of the shadow cone of a body where the sun light is completely obscured. The central portion of a sunspot which appears the darkest.

Universal time The local mean solar time on the prime meridian at the Royal Observatory, Greenwich, England.

Universe The total space occupied by matter and radiation.

Upper transit The instant of time when a body crosses the visible portion of the observer's meridian.

Van Allen radiation belt Doughnut-shaped regions that lie in the plane of the earth's magnetic equator, where high-energy charged particles are trapped in the earth's outer magnetic field and move from one magnetic hemisphere to the other.

Variable star A pulsating star that increases and decreases its size rhythmically and exhibits changes in luminosity.

Vector A quantity that has magnitude (amount) and direction.

Velocity Speed in a given direction, e.g., miles per hour to the east or kilometers per second to the southwest.

Velocity of escape The minimum velocity at which a body must move to overcome the gravitational attraction of another body, enter a parabolic orbit, and escape into space.

Vernal equinox The intersection of the celestial equator and the ecliptic at the point where the sun crosses from south to north.

Vertical circle A great circle on the celestial sphere that passes through the observer's zenith and intersects the horizon at 90°.

Visual binary A binary star system in which the two stars are visible telescopically as two separate bodies.

Visual magnitude The magnitude of a star based on its brightness as seen with the eye.

Volume A measure of the space occupied by a body.

Walled plain (lunar) The very large lunar craters that appear as depressions with hardly any outside walls.

Wandering of the poles The actual shifting of the earth itself while its axis of rotation remains fixed with respect to the stars, that is, the drifting of the north and south poles of the earth in relation to the earth's surface.

Waning moon The moon between the full and new phase, when its illuminated surface as seen from the earth is decreasing.

Wave One method of describing how electromagnetic radiation is propagated.

Wavelength The distance between two corresponding successive points in a wave motion, e.g., the distance between two successive crests or troughs.

Waxing moon The moon between the new and full phase, when its illuminated surface as seen from the earth is increasing.

Weight The gravitational force between the earth and the body.

White dwarf A star that has exhausted either all or most of its fuel, collapsed, and has become extremely dense, hot, and faint. It is believed to be near its final stage of evolution.

Widmanstätten figures A distinctive crystalline structure seen in a ground, polished, and etched surface of a meteorite.

Wien's law The relationship between the temperature of a black body and the wavelength at which maximum radiation is emitted.

Winter solstice The point on the ecliptic at which the sun reaches its greatest angular distance (23½°) below the celestial equator. It marks the shortest day of the year.

Wolf-Rayet stars Very hot O-type stars that eject gas at high velocities.

World calendar A proposed calendar by which the year is divided in four identical quarters so that a given date of the month always falls on the same day of the week.

X-rays The radiation of short wavelengths located between the ultraviolet and the gamma rays on the electromagnetic spectrum.

X-ray stars Stars that emit radiation in the x-ray frequencies.

Year A unit of time defined by the revolution of the earth. It is the interval of time for the earth to complete one revolution around the sun.

Zeeman effect The broadening or splitting of spectral lines into several lines of slightly different wavelengths when the source is in a magnetic field.

Zenith The point on the celestial sphere that is directly above the observer.

Zodiac An 18°-wide belt that completely encircles the sky and centers on the ecliptic. It was divided by the ancients into 12 equal parts, each containing one constellation.

Zodiacal light A faint glow of light visible along the ecliptic nearest the sun on the western horizon after sunset and on the eastern horizon before sunrise.

Zone of avoidance An irregular region along the center of the Milky Way (galactic equator) where the interstellar dust is so dense that few, if any, exterior galaxies are visible.

Zone time (standard time) The solar time within a 15°-wide zone as determined by the central meridian of the zone. Zone time is the local mean time of the central meridian. Zone time is kept at sea, where the zone boundaries are meridians, and standard time is kept over land areas, where the boundaries are irregular.

INDEX